Wireless Multi-Access Environments and Quality of Service Provisioning:

Solutions and Application

Gabriel-Miro Muntean
Dublin City University, Ireland

Ramona Trestian
Dublin City University, Ireland

Information Science
REFERENCE

Managing Director:	Lindsay Johnston
Senior Editorial Director:	Heather Probst
Book Production Manager:	Sean Woznicki
Development Manager:	Joel Gamon
Development Editor:	Myla Harty
Acquisitions Editor:	Erika Gallagher
Typesetters:	Jennifer Romanchak
Print Coordinator:	Jamie Snavely
Cover Design:	Nick Newcomer, Greg Snader

Published in the United States of America by
Information Science Reference (an imprint of IGI Global)
701 E. Chocolate Avenue
Hershey PA 17033
Tel: 717-533-8845
Fax: 717-533-8661
E-mail: cust@igi-global.com
Web site: http://www.igi-global.com

Library of Congress Cataloging-in-Publication Data

Wireless multi-access environments and quality of service provisioning: solutions and application / Gabriel-Miro Muntean and Ramona Trestian, editors.
p. cm.
Includes bibliographical references and index.
ISBN 978-1-4666-0017-1 (hardcover) -- ISBN 978-1-4666-0018-8 (ebook) -- ISBN 978-1-4666-0019-5 (print & perpetual access) 1. Wireless communication systems. I. Muntean, Gabriel-Miro, 1972- II. Trestian, Ramona, 1983-
TK5103.2.W573513 2012
384.5--dc23

2011042041

British Cataloguing in Publication Data
A Cataloguing in Publication record for this book is available from the British Library.

List of Reviewers

Anwer Al-Dulaimi, *Brunel University, UK*
Christos Bouras *University of Patras, Greece*
Pethuru Raj Chelliah, *Robert Bosch Engineering and Business Solutions, India*
Bogdan Ciubotaru, *Dublin City University, Ireland*
Nuno Coutinho, *Instituto de Telecomunicações, University of Aveiro, Portugal*
Stephen Hurley, *Cardiff University, UK*
Balakrishnan K, *International Institute of Information Technology, Bangalore, India*
Mathias Kretschmer, *Fraunhofer FOKUS, Germany & Brunel University, UK*
Gabriel-Miro Muntean, *Dublin City University, Ireland*
Cristina Hava Muntean, *National College of Ireland, Ireland*
Anthony Plummer Jr., *Johns Hopkins University Applied Physics Laboratory, USA*
Rajeshkumar Ramraj, *Edith Cowan University, Australia*
Susana Sargento, *Instituto de Telecomunicações, University of Aveiro, Portugal*
Volkan Sevindik, *Airvana Inc., USA*
K. K. Shukla, *Banaras Hindu University, India*
Enrique Stevens-Navarro, *Universidad Autónoma de San Luis Potosí, México*
Ramona Trestian, *Dublin City University, Ireland*
Hrishikesh Venkataraman, *Dublin City University, Ireland*
Changqiao Xu,, *Beijing University of Posts and Telecommunications, China*

Table of Contents

Detailed Table of Contents

This chapter presents an architecture for context-aware Virtual Networks (VNs) that provides user-driven multi-access communication. The architecture is dedicated and appropriate for the flexibility provided by Wireless Mesh Networks. According to this architecture, VNs with different context characteristics such as Quality of Service (QoS), mobility, and security are built to support communications with different characteristics that best fit the users' needs. The architecture is modeled both through a probabilistic and an optimization approach to provide quantitative insights into its performance. The probabilistic model quantifies the overhead on the architecture in terms of networking delays induced due to the VN management (searching, creation, and management). The optimization model provides insights into the competition of the VNs for the limited bandwidth resources. Indicative results of the models show the feasibility of the architecture, the upper bounds in terms of number of supported VNs to achieve good quality communications, and the relative placement of the flows in different, competing VNs.

Wireless Mesh Networks (WMNs) have matured in recent years and the visibility of WMN deployments has attracted commercial operators to investigate this technology for applicability in their networks. Having their roots in the Mobile Adhoc Network (MANET) world and rather cheap off-the-shelf single-radio WLAN routers, WMN routing protocols were not designed for applicability in carrier-grade back-haul networks. For example, protocols such as OLSR or B.A.T.M.A.N. can not address the QoS-requirements of a modern operator back-haul network with its increasing demand for triple-play content. Although numerous solutions have been proposed to introduce QoS-awareness at the protocol or the technology level, traditional WMNs fail to meet commercial operator requirements in terms of reliability, traffic engineering and QoS guarantees. This chapter proposes a novel approach combining an IEEE 802.21-based control plane and an MPLS-based data plane. To provide support for ubiquitous high-bandwidth multi-media services, it seamlessly integrates unidirectional broadcast technologies such as DVB into the heterogeneous multi-radio WiBACK architecture.

The exponential growth in multimedia traffic, predominantly on UDP transport, poses a threat to the TCP's best effort throughput. This problem is more acute in last mile broadband wireless access networks. Most scheduling algorithms discuss improving the combined TCP and UDP throughput or improving the TCP throughput without studying the effects of inelastic traffic such as UDP. This chapter furthers the necessity for TCP throughput protection and proposes a novel dynamically adapting Weighted Fair Queue (WFQ) based scheduling mechanism that provides a higher degree of TCP protection. This is accomplished by differentiating between TCP and UDP flows, buffer provisioning for each flow, and prioritizing TCP ACK packets. The simulation results show that the proposed mechanism yields a relative improvement of up to 29% of TCP goodput and 7.5% of aggregate MAC throughput over the mechanism without the proposed improvements.

Current backhaul networks typically comprise a wired middle mile and a wireless last mile part. The wireless part is almost exclusively based on tree topologies. However, a lot could be gained by deploying mesh-based backhauls. Meshes allow better network capacity exploitation due to load balancing and offer inherent resilience to link degradations or failures. Yet meshes come with increased complexity in terms of radio configuration, routing or mobility management. This chapter proposes architecture and mechanisms for carrier-grade mesh-based wireless backhauls. One special focus is that it supports heterogeneous backhauls, which encompass multiple different wireless technologies. The proposition has been successfully deployed in a test network.

Wireless mobile ad hoc networks, also called MANETs, hold the key to the future generation of wireless communication, promising adaptive communication any time and anywhere. Ad hoc networks constitute

Long Term Evolution (LTE) constitutes the latest step before the 4th generation (4G) of radio technologies designed to increase the capacity and speed of mobile communications. To support Multimedia Broadcast/Multicast Services (MBMS), LTE offers the functionality to transmit MBMS over Single Frequency Network (MBSFN), where a time-synchronized common waveform is transmitted from multiple cells for a given duration. In MBSFN transmissions, the achieved Spectral Efficiency (SE) is mainly determined by the Modulation and Coding Scheme (MCS) selected. This study proposes and evaluates four approaches for the selection of the MCS that will be utilized for the transmission of the MBSFN data. The evaluation of the approaches is performed for different users' distribution and from a SE perspective. Based on the SE measurement, the approach that either maximizes or achieves a target SE for the corresponding users' distribution is determined.

This chapter presents a history-based statistical channel access mechanism for enabling traffic prioritization in wireless sensor networks. Prioritized access is realized such that low priority non-real-time sensors can access channel bandwidth that is unused by high priority real-time traffic. The key idea is for the low priority sensor nodes to first observe and statistically model the channel usage pattern by the high priority traffic, then make advantageous probabilistic transmissions so that the non-priority traffic throughput is maximized while protecting the high-priority traffic from disruptions. The chapter details a practical whitespace measurement scheme and presents a channel history based prioritization protocol. The access mechanism is implemented in a TelosB mote-based sensor testbed in which the non-priority motes continually measure the RSSI to infer the channel usage pattern and probabilistically access the channel while different types of traffic are sent by high-priority TelosB motes.

Cognitive radios are proposed as secondary users of spectrum to provision for the growth in mobile users and services. However, the dynamic changes in the wireless environment and spectrum availability are blocking the success of online communications for future cognitive mesh networks. As a solution, Cognitive Radio over Fibre (CRoF) subnet is developed through combining selected Base Stations (BSs) with the radio over fibre network. These CRoF-BSs attract the neighborhood cognitive BSs to send data through fibre whenever they are unable to formulate their own wireless links. This leads to the network splitting into many zones of services in which cognitive BSs are controlled by CRoF-BS zones. Therefore, a new paradigm for local resource sharing emerges through these architectural network modifications. In this chapter, the multi-zone structure is analyzed in order to formulate the rules of adaptation between the CRoF zones and the traditional cognitive networks.

Foreword

From a wireless networking point of view, we are in the midst of a revolution. Traditional approaches and devices, which were considered as luxuries only 15 or 20 years ago, have given way to a proliferation of wireless access technologies and an explosion in the number and variety of user devices. Mobile and wireless access is now a global necessity and underpins much of the world's societal and economic activities. People everywhere have shown a tremendous desire for running a wide range of applications on their wireless devices, from voice communications and email/text messaging to interactive video and the full Internet experience.

However, it would be premature to claim that we have reached the ultimate goal of "any application, anytime, anywhere" where the end-user gets the same quality of experience as they get in wired networks. Despite the recent advances in wireless network protocols and services, there is no clear technological path to achieving this goal. What does seem certain is that the wireless environment will be heterogeneous, with a mixture of WLAN/Bluetooth/3G/LTE/WiMAX/other technologies accessible to a range of multi-access-enabled user devices. This leads inevitably to another feature of the current and future wireless environment: it can be extremely dynamic, with very rapid changes in network characteristics and user traffic behavior. This requires adaptive approaches to network access, transmission and reception, resource allocation, and many other elements of wireless networking.

There has of course been a surge in the number of researchers working in this field, and there is now an extensive literature on wireless quality of service, wireless multimedia applications, wireless mobility management, and related areas. Gabriel-Miro Muntean and Ramona Trestian have made a significant contribution to the field by focusing attention for their book on Quality of Service (QoS) provisioning in heterogeneous wireless environments. They have identified two key aspects of this problem, namely architectural approaches and QoS provisioning solutions, and have carefully chosen and edited an excellent selection of up-to-date chapters addressing these aspects. The architectural chapters cover coexistence of different wireless technologies, mobility and handover management, network selection and virtualization, and other leading-edge topics; while the QoS provisioning chapters provide the latest results on carrier-grade QoS, multimedia broadcasting, traffic prioritization, packet scheduling, and much more.

Muntean and Trestian have selected an impressive array of experts from around the world to provide the insights contained in these chapters. I believe there is much to savor here for both students and researchers, who will get valuable details on current research and possible future approaches, as well as for those closer to the industry, who will get sight of the varied and inventive ways in which problems familiar to them are being addressed. I am sure that *Wireless Multi-Access Environments and Quality of Service Provisioning: Solutions and Application* will serve as an important bridge between academic research and industrial application, and that it will be an authoritative reference work in the field of wireless quality of service for many years.

Liam Murphy
University College Dublin, Ireland
March 2011

Liam Murphy *received a B.E. in Electrical Engineering from University College Dublin in 1985, and a M.Sc. and Ph.D. in Electrical Engineering and Computer Sciences from the University of California, Berkeley, in 1988 and 1992, respectively. He is currently a Full Professor in Computer Science at University College Dublin, where he is Director of the Performance Engineering Laboratory. Prof. Murphy has published over 150 refereed journal and conference papers on various topics, including multimedia transmissions, dynamic and adaptive resource allocation algorithms in computer/communication networks, and software development. His current research projects involve mobile and wireless systems and computer network convergence issues. Prof. Murphy is a Member of the IEEE (Communications, Broadcasting, and Computer societies) and a Fellow of the Irish Computer Society.*

Preface

During past years many people have thought much about what a next generation network environment would look like and have proposed diverse ideas. Significant steps have already been made towards merging the many and often very different visions of such an environment into a more coherent picture in the first place, and then towards proposing the mechanisms, protocols, and technologies which would make possible its real-life rolling and eventually adoption. It was like putting together a puzzle with a secret picture whose overall aspect no one knows, but everybody can imagine. A puzzle at whose completion many people with potential different visions of the overall image are working together, revealing different areas of that picture at various moments in time. A puzzle whose secret picture keeps changing, evolving during the work of putting the puzzle together, often influenced by the work of the people in charge of finishing the puzzle themselves. A puzzle whose secret image is not only a 2D representation required for the communication between different entities, but a 3D structure based on the classic layered network architecture enhanced with novel contributions. In this context almost everybody agrees that the effort required to put this puzzle together is large and complex and there should be an expectation of many more years of hard and highly innovative work to be put towards the puzzle completion in order to fully develop the next generation network architecture and its underlying mechanisms.

In this concentrated effort imaginative thinking plays an important role, and identifying various scenarios and trying to find solutions for them have helped fuel the progress. Imagine a user leaving their house for work in the morning, journey in which they have to drive then take a public means of transport and finally walk to reach the city centre's workplace. During commuting they may want to be directed via certain place (e.g. school or petrol station) following the best route, while also getting access to the latest news, listening to music and/or being informed about the most pressing work-related aspects of the day. At work regardless of the building or office the user is located in, there should be continuous communication and access to relevant applications in order to make work efficient. On the journey back home, avoiding high traffic while getting past certain landmarks (e.g. school, shopping centre), while also connecting with family and friends, performing other pending tasks and accessing infotainment are highly relevant. At home family routine may require availing from additional communication-related applications from various devices be them for talking to distant relatives or planning the day ahead or solving some other pressing problems.

In short in order to enable and support such a mobile dynamic lifestyle which is common to most of us, at least three major aspects need to be taken care of: devices, communication infrastructure, and services. Most users have access to complex computational *devices* be they desktops, laptops or simple mobile devices. With the advances in technology, computing devices such as smartphones, iPhones, and netbooks, have become more affordable, easy to use, and powerful, while enabling mobility to users while also making sure there is network connectivity. *Network connectivity* is possible via wired and wireless infrastructures. There are many Internet Service Providers which provide excellent wired connectivity to residential and business users alike. For local network access and service support wireless technologies such as WLAN (wireless local area network) and WPAN (wireless personal area network), can be employed. Additionally full wireless connectivity is supported via different radio access technologies, offered by different operators. These operators can provide connectivity by making use of different technologies such as: GSM (Global System for Mobile Communications), EDGE (Enhanced Data Rates for GSM Evolution), UMTS (Universal Mobile Telecommunications System), HSPA (High Speed Packet Access), LTE (Long Term Evolution), WiMAX (Worldwide Interoperability for Microwave Access), etc. One operator can cover different regions with different technologies.

Most of the effort put in enabling networked communications and providing the user with complex devices is done in order to support an increasingly extensive set of *service* types. Starting with the classic email and Web browsing services with basic best-effort transport requirements, continuing with online shopping and online banking which have additional requirements which include security and finishing with multimedia streaming, voice-over-IP (VoIP), mobile TV, and high definition television (HDTV) applications which put significant pressure on all components of the communication chain in terms of data amount and continuity-related demands is just an ever evolving set of services, which are increasingly hungry in terms of resources for content production, storage, delivery, processing and display. This continuous evolution is mainly triggered and fuelled by the users who are more and more interested in mobility, rich media content, novel services (e.g. social networking), etc.

In this context, there is a continuous race between the service, device and network researchers in order to propose solutions to the latest problems as response to users desire to access more complex content, in a higher quantity and in more difficult conditions. No single solution will solve all the problems the way no single device, no single service and no single network technology will be able to cope with the amount, type and requirements of data exchange. Multiple devices, multiple services and multiple co-existing access technologies, many of them in the radio space, will be fundamental for the next generation network communication in order to handle the volume and diversity of user application demands at high the quality levels expected by the users. Therefore the overall aim is to keep the user "always best connected" anywhere and anytime in this multi-technology multi-application multi-terminal multi-user environment.

This book will cover important aspects of the emerging technologies in the heterogeneous next generation network environment with focus on wireless communications and their quality. It presents network selection mechanisms, handover solutions and adaptive techniques. It identifies the research issues and challenges and presents a survey of the proposed solutions in the literature.

The manuscript is structured in two sections. The first section, on *Architectures,* describes the current wireless multi-access environment which leads towards the next generation of wireless networks. It includes important chapters written by researchers from prestigious laboratories from Australia, China, Mexico, Portugal, Greece, Germany, United Kingdom, and India which present the current state of the art in wireless communications including mobility challenges, management, security, ensuring always best experience, multimedia content distribution, etc. The second section entitled *QoS Provisioning Solutions* presents various solutions proposed by world known researchers in different areas of wireless communications. It includes research results in the area of ad-hoc networking, wireless sensor networking, cognitive and fibre networking, wireless mesh networking, etc.

The prospective audience of this book is mainly undergraduate students, postgraduate students, and researchers who are interested in learning more about the latest developments in the area of mobile and wireless communications. It also targets industry professionals who are working or are interested in this area, providing them with a reference of the latest efforts which bring the research further by addressing some of the shortcomings of the existing solutions.

Wireless Multi-Access Environments and Quality of Service Provisioning: Solutions and Application is published by IGI Global (formerly Idea Group Inc.), publisher of the "Information Science Reference" (formerly Idea Group Reference), "Medical Information Science Reference," "Business Science Reference," and "Engineering Science Reference" imprints. For additional information regarding the publisher, please visit www.igi-global.com.

The editors wish you a pleasant reading.

Gabriel-Miro Muntean
Dublin City University, Ireland

Ramona Trestian
Dublin City University, Ireland March 2011

Acknowledgment

Wireless Multi-Access Environments and Quality of Service Provisioning: Solutions and Application would not exist without the effort of many people whose names may not appear on the cover of the book. However, their hard work, cooperation, friendship, and understanding are very important to its preparation and production.

The editors would like to sincerely thank IGI Global for the publication of this book. We thank Prof. Liam Murphy for all his help in writing the foreword. As the success of this book depends on the high quality chapters submitted, the editors wish to express their deep appreciation to the authors of each chapter for their significant contribution. We would also like to thank the reviewers for the generous commitment of time and effort that they have put into the reviewing process and for providing feedback that has enhanced the overall quality of the book. We thank the Editorial Advisory Board for their implication.

Last but not least, we would like to thank our families for being the "center of gravity" and for their continuous support along the way. In particular Gabriel-Miro wishes to dedicate this book to his dear parents, Dora and Ivo, to his lovely wife, Cristina, and to his smart son, Daniel-Sasha. Ramona would like to dedicate this book to her wonderful and loving parents, Maria and Vasile, and to her dear brother, Ionut.

Gabriel-Miro Muntean
Dublin City University, Ireland

Ramona Trestian
Dublin City University, Ireland

Section 1
Architectures

Chapter 1
Multi-Access Environments in Next Generation Networks

Rajeshkumar Ramraj
Edith Cowan University, Australia

Iftekhar Ahmad
Edith Cowan University, Australia

Daryoush Habibi
Edith Cowan University, Australia

ABSTRACT

Wireless heterogeneous environment is becoming increasingly popular because of its ability to support multi access technologies, which keeps the mobile users always connected. The fourth generation (4G) of wireless environment consists of dissimilar technologies like Worldwide Interoperability for Microwave Access (WiMAX) and Long Term Evolution-Advanced (LTE-A). Coexistence of these standards presents technological challenges. Other challenges towards heterogenous 4G environment include coexistence of third generation (3G) and 4G and coexistence of Wireless Local Area Network (WLAN) and 3G/4G technologies. This chapter presents the technical overview of the 4G (WiMAX and LTE-A) standards and identifies the challenges and research issues (i.e., mobility management, network selection, handover, and Quality of Service [QoS]) toward a multi-access environment that consists of 3G, 4G, and WLAN technologies. The chapter also discusses future directions and some research solutions.

INTRODUCTION

The 3G network is the third generation of mobile networking and telecommunications, which features a wider range of services, and advanced network capacity over the previous second generation (2G) network. 3G fulfils the International Telecommunication Union (ITU) specification for mobile telecommunications and uses a software enhancement called High Speed Downlink Packet Access (HSDPA) through which the network can send data (audio, graphics and text) significantly faster than the 2G network. The ITU specifications of 3G include Universal Mobile Telecommunications System (UMTS), Digital Enhanced Cordless Telephone System (DECT), and Code Division Multiple Access 2000 (CDMA2000). The 3G technologies aim to deliver improved

DOI: 10.4018/978-1-4666-0017-1.ch001

throughput in addition to voice service. This throughput, however, is not sufficient to meet the increasing demand for wireless broadband (Figure 1). A new standard called WiMAX, developed in 2004. The IEEE 802.16e, also known as mobile WiMAX, is an enhancement over the fixed WiMAX standard and is the meeting point of mobile and fixed broadband networks through wide-area radio access technology and flexible network architecture.

The IEEE 802.16e interface utilizes Orthogonal Frequency Division Multiple Access (OFDMA) in the downlink (DL) and in the uplink (UL) to improve bandwidth scalability, efficiency and multipath performance. The IEEE working group has been working towards the development of a new standard (IEEE 802.16m) to meet the requirements of ITU—Radio Communication Sector / International Mobile Telecommunication (ITU-R/IMT) for the 4th generation wireless networks. The IEEE 802.16m promises more efficient use of the radio spectrum and delivers improved data rates.

Mobile network operators continue with technology developments to improve user experience and service opportunities. One such progressed technology is the 3rd Generation Partnership Project (3GPP) Long Term Evolution (LTE). LTE-A is an enhancement over LTE, and LTE-A is included in 4G standard in mobile network technology tree, aiming to ensure 3GPP's competitive edge over the other mobile technologies. It is designed to offer multi-megabit bandwidth (i.e., 1 Gbps DL and 500 Mbps UL) with more efficient use of the radio spectrum, reduced latency and improved mobility. LTE-A uses OFDMA for DL communication and Single Carrier-Frequency Division Multiple Access (SC-FDMA) for UL communication. LTE-A also supports multiple-input multiple-output (MIMO) antenna design. The next generation of wireless communication systems will need to support multiple services ubiquitously in heterogeneous network (Berezdivin, Breinig, & Topp, 2002; Gang, Mizuno, & Havinga, 2002).

Background of wireless network has been changing progressively from homogeneous to heterogeneous and the outlook of the future gen-

Figure 1. Broadband access worldwide

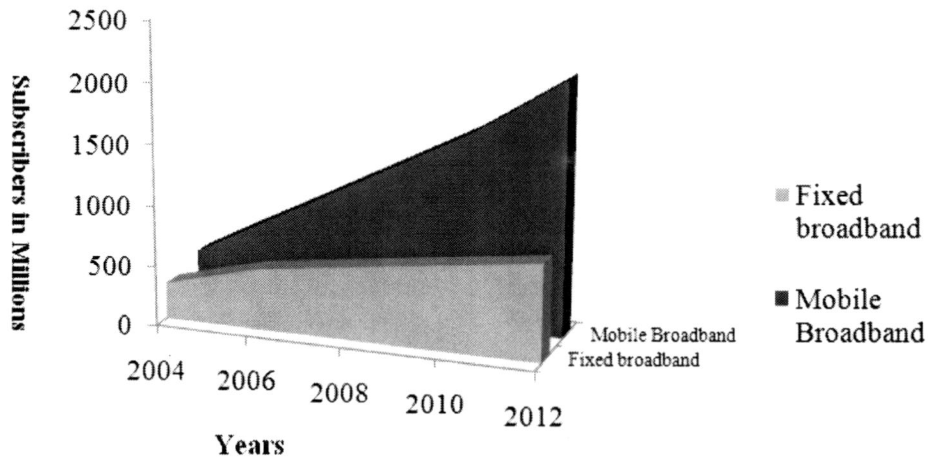

eration networks are expected to be a grouping of diverse but complimentary access technologies, like General Packet Radio Service (GPRS), Wideband Code Division Multiple Access / High Speed Packet Access (WCDMA/HSPA), Wireless Fidelity (WiFi) WiMAX, LTE and LTE-A. We will cover some key aspects of multi-access heterogeneous networks, network selection mechanism, handover solutions and their adaptive techniques.

3G NETWORKS AND ITS EVOLUTION

3G mobile wireless networks represent the next successive approximation of today's digital mobile systems. First generation (1G) systems were the old analogue-based brick sized phones sported by pinstriped executives and showbiz personalities way back in the early 80s. The 2G systems, which introduced digital technology on top of traditional telecoms circuit switching, were introduced in most industrialised nations in the early 1990. 3G networks represent the natural evolution over the previous standards developed. The 3G networks represent an international standard for Wide Area Network (WAN) in cellular communication that replaced 2G networks. Use of radio spectrum in 3G increases the rate of data transmission for advanced multimedia services and larger network capacity in addition to basic voice service.

Increasingly cellular phones and handheld mobile devices have incorporated additional services to telephony. Mobile devices these days offer wide range of services like high-speed World Wide Web access, emailing, messaging, video phone and multimedia access. Streaming of movies on their cellular mobiles phones, downloading and playing music are the major services subscribers need these days. 3G network aims to offer these services which again are not free from various challenges. 3G implementations were slower than initially anticipated, mainly because of the cost of upgrading equipment and licensing fees for additional spectrum. The earlier, 2G networks did not typically use the same frequencies as 3G (except in the United States), and licensing fees, particularly in Europe, were extremely expensive. Only Japan and South Korea were able to implement this technology quickly, largely because of the high level of government support for new infrastructure advances. The first pre-commercial 3G network branded as Freedom of Mobile Multimedia Access (FOMA) was launched in May 2001, by NTT DoCoMo in Japan. Following the first pre-commercial launch, NTT DoCoMo again made history with the first commercial launch of 3G in Japan on October 1, 2001. In Japan, by the end of 2006 the majority of customers were on 3G and upgrades to the next stage, 3.5G (with 3Mbs data rates), were underway. Implementation in the rest of the world is coming along, but at a slightly slower pace. In December 2007, 190 3G networks were operating in 40 countries, with 200 million subscribers, but that is still only around 7% of the 3 billion mobile phone subscriptions worldwide.

The three major technologies of 3G are:

UMTS

UMTS is one of the 3G technologies and has 3 interacting domains: Core Network (CN), UMTS Terrestrial Radio Access Network (UTRAN) and User Equipment (UE) which provide switching, routing and transit for traffic. Core network is divided in circuit switched and packet switched domains. UTRAN consists of a collection of NodeB's and Radio Network Controller (RNC). UMTS offers teleservices and bearer services which provide transfer service between access points. The offered data rate are 144 Kbps in rural outdoor, 383 Kbps in urban outdoor and 2048 Kbps in low range outdoor. UMTS has an air interface based on WCDMA and an Internet Protocol (IP) network based on GPRS. It also covers Time Division – Code Division Multiple Access (TD-CDMA) and Time Division – Synchronous Code Division Multiple Access (TD-

SCDMA). In UMTS, the voice and data transfer are controlled by transport layer nodes which have a base transceiver station, base station controller, GPRS support node and media gateway. The call control function is performed by call control layer nodes which have call state control function, media gateway and home subscriber server. The transport layer node can be built on an Asynchronous Transfer Mode (ATM) switch which consists of an optional interface to call control layer nodes (Holma & Toskala, 2000). UMTS brings advanced capabilities in some services like Push-to-Talk over Cellular (PoC), Voice over Internet Protocol (VoIP), multimedia in mobiles, Multimedia Broadcast Multicast Service (MBMS) and connectivity for business purposes like access to internet service using laptops. The main shortcoming of UMTS is it requires new BS for frequency allocation. UMTS borrows the concept of Global System for Mobile (GSM) and Enhanced Data Rate for GSM Evolution (EDGE) as it looks similar to those technologies. Physical constrains which limit access to highest data rates and lack of regulation and standardization are also identified as some of the limitations in UMTS.

CDMA2000

This is a part of 3G family and popularly known as IMT Multi-Carrier (IMT-MC) mobile technology. The major function of this is to send voice, data and signal between cell phones and base stations using CDMA channel access. This includes a set of standards: CDMA2000 1X, CDMA2000 EV-DO Rev. 0, CDMA2000 EV-DO Rev. A, and CDMA2000 EV-DO Rev. B. CDMA2000 1X delivers 3G service up to 140 kbps at a high rate of 1.25 MHz/carrier, which is a small amount of spectrum. CDMA2000 1X EV-DO provides data delivery at 2.4 mbps and CDMA2000 1X EV-DV delivers data at a speed of 3Mbps to 5Mbps. The latest standard approved by ITU is CDMA2000 3X which uses 5 MHz spectrum to deliver data at a speed around 2 to 4 mbps (Holma & Toskala, 2000). The chip rate for CDMA2000 is 1x: 1.2288, 3x: 3.6864 Mcps and the maximum user data rate is 1x: 144 kbps now, 307 kbps in the future 1xEV-DO: max 384 kbps – 2.4 Mbps, 1xEV-DV: 4.8 Mbps. Frame length is 5ms, 10ms and 20ms with power control rate of 800Hz Radio Access Network (RAN) in CDMA2000 has multiple base stations connected to radio network controller. The radio network controller manages several radio link protocol layers with mobile nodes and performs pre link bandwidth management function. When the mobile node moves from one radio network controller to other, the on-going radio link protocol establishes a new connection with the visiting radio network (Buddhikot et al., 2003) (Holma & Toskala, 2000). The major benefit of CDMA2000 is to reduce the technical risks by phase enhancement of networks with medium level data services and to develop higher data services in order to avoid reservations on return of investments which recently obscure the UMTS 3G visions. Some of the drawbacks of CDMA2000 are, the towers interface with each other so they are installed in a shorter heights and it reduces the performance in hilly regions. CDMA2000 are restricted on roaming and when the coverage area contract under load the performance goes poor.

HSDPA

HSDPA is an enhanced 3G version in HSPA family (Tapia, Liu, karimli, & Feuerstein, 2009). Current HSDPA deployments support DL speeds of 1.8, 3.6, 7.2 and 14.0 Mbps. It is considered to be a higher version of 3G as it allows data transfer at higher rate, which is based on W-CDMA with 8 to 10 Mbps for DL and up to 20 Mbps for MIMO systems. The implementation of HSDPA consists of Adaptive Modulation and Coding (AMC), MIMO, Hybrid Automatic Request (HARQ), fast cell search and advanced receiver design.

In the market place, there are other terms like 2.5G, 3.5G (moving forward) and 4G (the step beyond). The 4G standard is growing and the existing 4G systems are simply prototypes of technologies which may or may not become part of an eventual 4G standard. The deployment of 4G, like 3G and 2G, will involve the complete replacement of existing handsets and networks, mobile experts are not predicting deployment until the 2012-2015 time frames. By that time, even with enhancements, the 3G networks will probably be fairly congested. The GSM family of technologies like UMTS-HSPA and LTE is expected to account for more than 84 percent of 3G subscribers by 2014 (Figure 2), according to Informa Telecoms & Media. It has a drawback that high speed is available only for downlink and uplink will be at the baseline WCDMA speed. This can be fixed with the combo of HSDPA + High Speed Uplink Packet Access (HSUPA).

EVOLUTION OF 4G NETWORKS

3G systems like CDMA2000 and W-CDMA provide higher channel capacity, broadband data rates up to 2 Mbps, high speed multimedia transmission and global roaming across a cellular network (Ray & Mista, 2006). The increase of mobile users and global boom in communication services have placed enormous pressure on 3G systems, which provided the motivation behind introducing 4G systems.

4G networks consist of technologies such as WiMAX and LTE-A.

WiMAX

WiMAX was initiated by the WiMAX forum, formed in June 2001 to encourage conventionality and interoperability of the IEEE 802.16 technology standards. WiMAX is a telecommunications protocol that provides fixed and fully mobile internet access (Ahmad & Habibi, 2008b). It is IP based which provides better QoS than to 802.11/WiFi. It is also a digital communication system with the standard IEEE 802.16, which is specially designed for metropolitan area networks. It can provide access to broadband up to 30 miles for fixed stations and 10 miles for mobile stations where WiFi can give access only up to 300 feet. The primary advantage of the WiMAX standard is to enable the adoption of advanced radio features in a uniform fashion and reduce costs. WiMAX forum describes the IEEE 802.16 as the standard

Figure 2. Global mobile subscribers

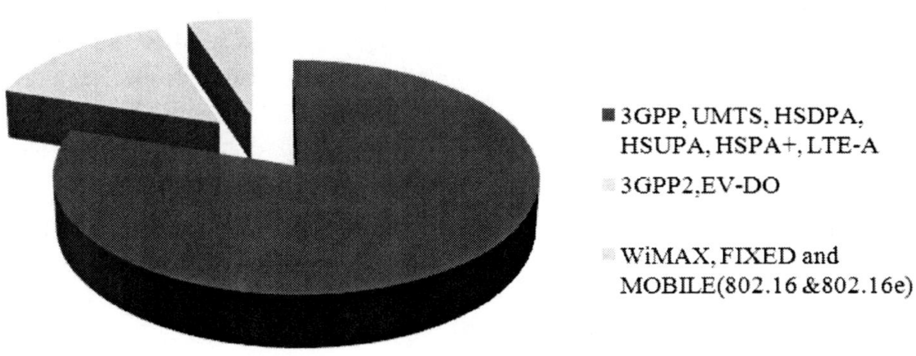

for technology enabling wireless broadband access to deliver the last mile as an alternative for cable and Digital Subscriber Line (DSL).

The end-to-end architecture of WiMAX makes the use of IETF and IEEE standards and protocols. The IP architecture of WiMAX gives flexibility in choosing the network and decides the functionality they want their network to support. It adds mobility and MIMO functionality to 802.16 standards. 802.16 standardizes air interface and adds related functions associated with wireless. The original 802.16 standard described a 10 – 66 GHz point-to-multipoint system using a traditional single carrier physical standard with QAM. The IEEE 802.16 was later developed to deliver at 2 – 11 GHz for Non Line of Sight (NLOS). The physical standard was later expanded to include OFDMA. 802.16 Media Access Control (MAC) also summarizes the air interface of Ethernet, ATM and IP technologies. The IEEE 802.16 MAC is also responsible for secure data transfer as it uses encryption standards like Advanced Encryption Standard (AES) and Data Encryption Standard (DES). WiMAX covers a wide range of spectrum in two distinct categories: unlicensed and licensed. In unlicensed spectrum WiMAX covers 2.4GHz and 5.8GHz. Since the range is unlicensed the barrier to entry is very low. In licensed spectrum WiMAX covers 700MHz, 2.3GHz, 2.5GHz and 3.5GHz.

Some of the key features of the IEEE 802.16e mobile WiMAX include:

- Voice and data session at vehicular speeds of up to 100 Miles/hr
- DL up to 63 Mbps and UL up to 28 Mbps per unit on a 10 MHz sector
- End-to-end IP based QoS
- OFMDA scalability
- Low latency and optimized handoff scheme
- Multicast and Broadcast Service (MBS)
- AMC
- Beam forming, space-time coding and spatial multiplexing.

Figure 3 shows the evolution of WiMAX before and after Orthogonal Frequency Division Multiplexing (OFDM) and MIMO technologies. WiMAX technology has a superior development from pre OFDM-MIMO to the recent OFDM-MIMO technologies. The IEEE 802.16m has advanced air interface with data rates of 100 Mbps for mobile and 1 Gbps for fixed. It is also referred as mobile WiMAX Release 2 or Wireless MAN-Advanced. In WiFi technology, the IEEE 802.11n standard is referred as high speed LAN and gives high throughput enhancements to 802.11 standards. In this the physical layer data rates are up to 600 Mbps, which is a higher improvement than the other IEEE 802.11 standards. In cellular technology, LTE – A is approved by ITU-R as a 4G technology recently, which can support peak data rate up to 1 Gbps. It is also targeting faster switching between power states and to improve the mobile performance in cell edge area.

WiMAX has some clear technological merits, which are discussed in the following section.

OFDMA

ODFMA has been around for some time and it is used extensively in WiMAX applications. It operates in the hostile multipath radio environment to deliver high data rate and allows well-organized sharing of limited resources like spectrum and transmits power between users. OFDMA is used in mobility mode of the IEEE 802.16 and is the candidate access method for the IEEE 802.22, Wireless Regional Area Network (WRAN). OFDM multiplexes the data on a multiple carrier and transmits them in parallel. The orthogonal subcarriers used in OFDM overlaps the frequency domain. Since OFDM uses overlapping subcarriers, the spectral efficiency goes high compared to conventional Frequency Division Multiplexing (FDM). OFDMA has some recent advances on channel estimation, coding and modulation, QoS and resource allocation and systems and implementation.

Figure 3. Roads lead to WiMAX evolution

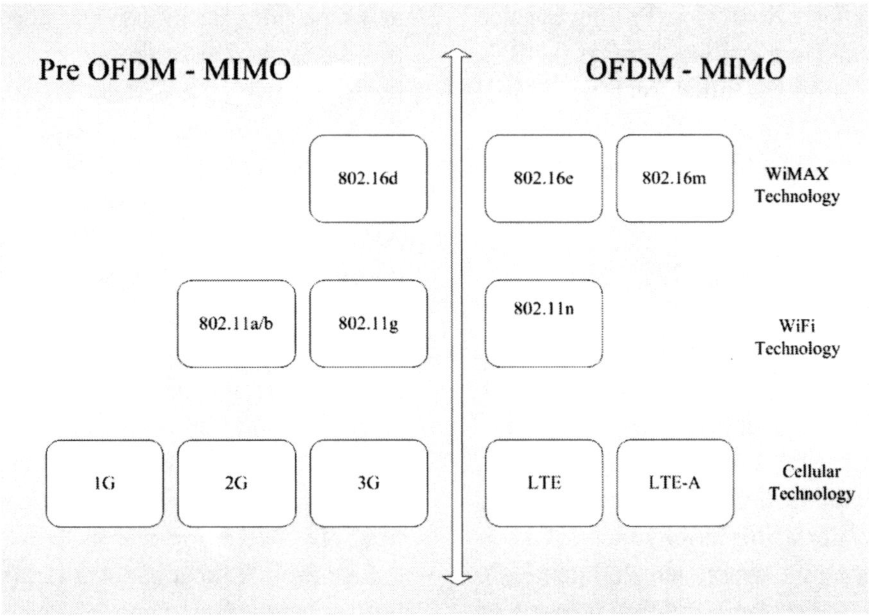

Serving channel condition without complex equalization is one of the advantages of OFDM. Linear Minimum Mean Square Error (LMMSE) (Zhou & Lam, 2009) channel estimation in OFDM proved that the system computational complexity can be reduced significantly. Partial channel estimation for OFDM using Least Square (LS) method (YiWang, Li, Zhang, & Liu1, 2009) is obtained using symmetric extension and also reduces the loss of power leakage in channel. A cross layer approach (Shao, Schiphorst, & Slump, 2009) can reduce power consumption up to 70% in Analog-to-Digital Convertor (ADC) OFDMA systems.

AMC

AMC increases the Bit Error Rate (BER) in radio channels and enables WiMAX to optimize throughput on proliferation conditions. WiMAX support AMC to regulate Signal Modulation Scheme (SMC) depending on its SNR. When the SNR goes low, the system switches to low modulation scheme and maintains the connection and the connection quality. The key element of AMC is to improve the modulation scheme in fading circumstances. This provides flexibility to modulation coding scheme to channel conditions. It becomes more effective when combined with scheduling channels. When it is combined with time domain scheduling, it gives a chance to take advantage of short term variations in UE's fading envelop. This increases the throughput of the system. Cross layer AMC designed for time block MIMO system combines with physical layer with an auto request increasing the spectral efficiency.

MAC

The IEEE 802.16 MAC provides interface between the higher transport layer and the physical layer. IEEE 802.16-2004 and IEEE 802.16e-2005 MAC includes a convergence sub-layer to interface with a variety of higher layer protocols like ATM, TDM, Ethernet and IP. Some of the key features that MAC layer incorporates with wide range of applications is; Private Key Management (PKM) for MAC security, broadcast and multicast, manageability primitives, high speed handover and mobility. Service classes like Unsolicited Grant Service (UGS), Real-Time Polling Service (rtPS), Non-Real-Time Polling Service (nrtPS), Best Effort (BE) and Extended Real-Time Variable Rate (ERT-VR) combined with OFDMA to make 802.16 suitable for high speed applications.

CHALLENGES IN WIMAX

Although WiMAX is a promising technology for fixed Wireless Metropolitan Area Network (WMAN), it still has some limitations for high speed wireless broadband services. High BER caused by multipath fading is one of the key causes for low throughput at high vehicular speed in WiMAX. Many error correction techniques have been proposed to overcome the fading issue so that the throughput can be improved. There are mainly two groups of error correcting techniques: Automatic Repeat Request (ARQ) and Forward Error Correction (FEC). ARQ is a reactive mechanism of error control and it is not suitable when the BER is high (Ahmad & Habibi, 2008a). FEC is a proactive way to control the BER. FEC code size can be adjusted based on the communication environment. If the bit error rate can be estimated at different communication environment this information can be used to adjust FEC code size for the next cycle of transmission such that the fading at the next stage can be controlled (Ahmad, Habibi, & Rahman, 2008).

Handoff occurs when the connection leaves the coverage area and loses the network connectivity. This is one of the other challenges in WiMAX. To reduce the handover latency, cell scanning can be performed prior before initiating the handover process. This increases the network optimization in real time traffic. Mobile access to the internet services requires access to packet-switched wireless networks that are generally divided into an access part and a core part. The access part contains the link-layer devices while the core part contains network-layer devices mainly acting at the network layer. A Mobile Node (MN) wishing to access the internet services needs to attach the link layer device in the access network, which is also known as link-layer Point of Attachment (PoA). Nevertheless, the communicating MN may change the link-layer PoA during an active session. This process is called handover or handoff. Alternatively, handoff may be policy driven when the change in the PoA is governed by higher data rates, better services, lower costs, and so on. Following a handoff process, a new PoA is selected to overcome the handover issue. (Krichene & Boudriga, 2008; Persaud, 2006).

WiMAX uses OFDMA on both up and DL combined with adaptive power control, modulation, channelization and system architecture to reduce mobile device power requirements. Multiple subcarriers used on UL result in high Peak to Average Power Ratio (PAPR) in WiMAX. High PAPR requires expensive and inefficient power amplifiers with high requirements on linearity, which increases the cost of the terminal and drains the battery faster. LTE uses OFDMA on the DL and SC-FDMA on the UL, is optimized for DL bandwidth and lower power consumption in portable devices. This factor of LTE overcomes WiMAX where the PAPR is reduced. In the following section, we discuss about LTE – A technology and requirements of ITU-R.

LTE-A Technology

LTE-A represents a radical new step forward for the wireless industry, targeting order-of-magnitude increases in bit rates with respect to its predecessors by means of wider bandwidths and improved spectral efficiency (Clerckx, Lozano, Sesia, Rensburg, & Papadias, 2009). Table 1 summarizes the evolution of the 3GPP UMTS towards LTE.

The main aim of 3GPP was to provide a better solution to the world of wireless communication. LTE project was launched in late 2004, and the first LTE advanced was released in 2009. LTE-A is standardized by 3GPP in their release 10. LTE-

Table 1. Evolution of the UMTS specifications (Rumney, 2008)

Release	Functional Freeze	UMTS features of release
Release 99	March 2000	Basic 3.84 Mcps W-CDMA (Frequency Division Duplex [FDD] and Time Division Duplex [TDD])
Release 4	March 2001	1.28 Mcps TDD (TD-SCDMA)
Release 5	June 2002	HSDPA
Release 6	March 2005	HSUPA Enhanced Dedicated Channel (E-DCH)
Release 7	December 2007	HSPA+ (64 QAM DL, MIMO, 16 QAM UL) LTE and SAE feasibility study
Release 8	December 2008	LTE work item—OFDMA/SC-FDMA air interface SAE work item—new IP core network Future HSPA improvements
Release 9	May 2009	Architecture aspects of EHNB - eNB enhancements

A is compatible with the first release of LTE equipment and share the same frequency and technological merits of LTE.

The standardization process of 3GPP was developed out of GSM standards. LTE was built on new standard and not from the previous UMTS standards. The term LTE is actually a project name of 3GPP. This can be noted from its new numbering and new documents. The service architecture evolution (LTE–SAE) reduces the number of nodes and supports flexible network configurations to provide high level of facility accessibility.

Figure 4 shows the architecture of LTE-SAE. LTE uses OFDMA for DL and SC-FDMA for UL. In DL from base station to terminal, OFDMA meets all the requirements of LTE for spectrum flexibility and enables cost-efficient solutions for wide carriage with high peak rate. SC-FDMA is to balance for a drawback of OFDMA, which has a very high PAPR. LTE-A uses advanced antenna solutions that are introduced in Evolved High Speed Packet Access (eHSPA) (Pospishny et al., 2010).

SC-FDMA

LTE standard for UL transmission is based on SC-FDMA. It is a novel method introduced for deployment of radio transmission in future cellular systems. The major issue in extending GSM TDMA and W-CDMA to broadband systems is the increase in system complexity with multipath signal reception. This was overcome by SC-FDMA as its robustness against multipath signal propagation, which makes the system suitable for broadband system. Additionally, low PAPR compared to OFDMA for UL in user terminals brings more benefit for choosing the system. SC-FDMA transceiver has similar structure as typical OFDM except the addition of new Discrete Fourier Transform (DFT) block before subcarrier mapping. A special case of distributed SC-FDMA is called interleaved SC-FDMA, where the occupied subcarriers are equally spaced over the entire bandwidth. Interleaved SC-FDMA shows the best PAPR with no pulse shaping filters. When compared to OFDM PAPR of Interleaved SC-FDMA with Quadrature Phase Shift Keying (QPSK) is about 10db lower, while localized SC-FDMA is about 3db lower. Consequently, interleaved SC-FDMA is a preferred modulation technique for lower PAPR.

MIMO

The combination MIMO-OFDM enables OFDM to support more antennas and provides large bandwidth. MIMO-OFDM is under intensive investigation by researchers which is a promising transmission technique. The OFDM-MIMO

Figure 4. A simple LTE-SAE architecture

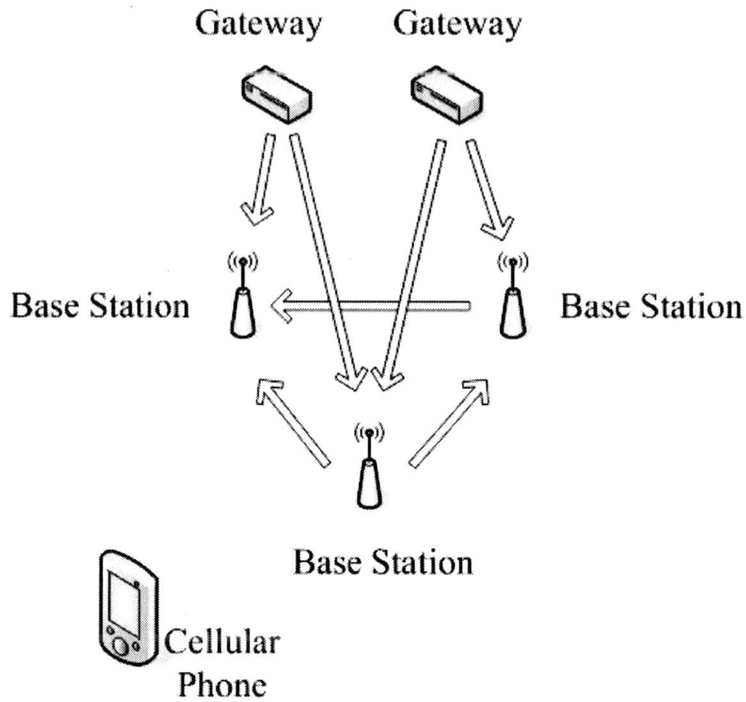

transceiver consists of N transmitter antennas and M receiver antennas. The OFDM signal is obtained using Inverse Fast Fourier Transformation (IFFT) detected by Fast Fourier Transformation (FFT). MIMO with OFDMA modulation solves the problem created by multipath channel. The IEEE 802.16e incorporates with MIMO and IEEE 802.11n (2009) recommends MIMO-OFDM. High-speed packet access plus (HSPA+) and LTE use MIMO technology to support cellular environment and also they recommend Multi-User MIMO (MU-MIMO).

In addition to the above mentioned techniques, LTE-A has the mechanism to provide the following technological advantage:

- LTE-A supports 100 MHz bandwidth and provides 1Gbps DL and 500 Mbps UL.
- LTE-A also supports Co-ordinated Multi-Point (CoMP) transmission and reception which improves the cell edge throughput.
- LTE-A supports 8x8 MIMO in DL and 4x4 MIMO in UL.
- LTE-A has relaying capacity to improve the throughput.

To get qualified for an IMT-Advanced or 4G, LTE-A is required to meet the requirements established by the ITU.

LTE-A Challenges and Targets Requirements

International Mobile Telecommunications-Advanced (IMT-Advanced) systems are mobile systems that include the new capabilities of IMT that go beyond those of IMT-2000. Such systems provide access to a wide range of telecommunication services including advanced mobile services, supported by mobile and fixed networks, which are increasingly packet-based (ITU-RReportM.2134, 2008). The main objective of ITU-R is to provide full range of IMT-Advanced requirements. The document from the report of ITU-R M.2134 presents consistent definition, specification and evaluation of the candidate Radio Interface Technology (RIT) or Set of Radio Interface Technologies (SRIT).

Some of the targeted key features of IMT-Advanced are:

- Cell spectral efficiency of 2.6Mbps.
- Peak spectral efficiency of 15Mbps for UL and 6.75Mbps for DL.
- Bandwidth of 60 MHz
- Cell edge user spectral efficiency of 0.075Mbps for UL and 0.12Mbps for UL.
- Latency of 10ms.
- Mobility up to 275 Miles/hr.
- Handover and
- VoIP capacity for high and low mobility environments of around 30 to 50 active users/sectors/MHz

The next section describes how the different technologies from different generations integrate with each other and some of the novel solutions are discussed.

MULTI-ACCESS HETEROGENEOUS NETWORKS

The wide proliferation of wireless network systems and the use of advanced wireless technologies enable the employment of a heterogeneous network. Communication networks and mobile devices integrate an increasing number of access technologies. At the same time, new business roles emerge which leads to new cooperative schemes between access providers providing different types of access connectivity (Sachs, Prytz, & Gebert, 2009). A multi access network allows connecting several terminals at same transmission point to share the same spectrum capacity. In our daily lives, we come across a lot of practical examples for multi access networks like wireless networks, ring networks, hub networks and so on. Heterogeneous wireless network enables the network to access, connect and switch between different technologies. Heterogeneous wireless networks consist of different access technologies interfacing with a common core network. The access technology consists of different standards, access systems with dissimilar scenario. The access systems such as mobile, satellite, WLAN and wired networks will be pooled to afford an optimal service at anytime from anywhere. In the following section, we discuss some of the integration between environments like WLAN to 3G/4G, 3G to 4G and 4G to 4G.

WLAN – 3G/4G

WiFi belongs to the class of WLAN technology under 802.11 standards. It is widely used for connecting broadband in small home and office networks. WiMAX, on the other hand, is a 4G technology developed to deliver high data rate with scalable bandwidth and guaranteed QoS. Integration of WiMAX and WiFi is important because an integrated WiMAX/WiFi architecture will enable to deliver low cost high speed broadband services. Integrating WiMAX with WiFi will have some significant challenges including cost for deploying WiMAX, adapting WiMAX to the spectrum and coverage, developing a gateway that supports both WiMAX and WiFi so that both technologies can work together. In WiMAX, MAC layer performs the synchronization of time blocks between different protocols and ensures spectrum allocation for different bandwidth in a time multiplexed and non concurrent basis. This feature eliminates the conflict and maintains the inherent link performance. The synchronization involves the coordination of different system clocks. With this synchronization, the next stage is to establish the bandwidth allocation mechanism which link WiMAX and WiFi Systems.

In integrating WiMAX and WiFi networks, the challenging part is to design proficient links and optimize QoS in MAC layer protocols between the component architecture. The IEEE 802.16 deals with a variety of applications via scheduling services like UGS, rtPS, nrtPS and BE. The IEEE 802.11e also achieves their QoS parameters using access mechanisms. A radio gateway (Figure 5) serves as a subscriber station for IEEE 802.16 and QoS access point gateway for IEEE 802.11e to address a real time data traffic which has a minimum latency delay, little jitter and minimum packet loss. This radio gateway has a subscriber station, mapping module and QoS access point. The mapping module records the incoming traffic and decides the priority flow of data from both the wireless technologies (Gakhar, Gravey, & Leroy, 2005). However, bandwidth assignment, scheduling and admission control was not considered in the above mentioned model. These challenges can be overcome through an adoption of a new QoS architecture (Figure 6) for the radio gateway. This architecture allows real-time services to be used in a dynamic environment, where users roam across 802.11 access points. This can be achieved by using virtual MAC addresses and auxiliary channels, implying novel functionalities inside the 802.16 network. Better QoS can be ensured by temporary allowing traffic to flow in a reserved channel. Also it manages the admission control and network resources on both wireless technologies (Neves, Sargento, & Aguiar, 2006).

The radio gateway as shown in Figure 5 has dual transceiver to support both 802.11 and 802.16 standard. Distributed Coordination Function (DCF) and Point Coordination Function (PCF) are the two mechanisms used in this model and DCF is a distributed access control mechanism to randomize the protocol employing Carrier Sense Multiple Access with Collision Avoidance (CSMA/CA). PCF is a centralized MAC algorithm which provides contention free service to provide support for delay sensitive applications (Rabbani, Kamruzzaman, Gondal, & Ahmad, 2011).

Figure 5. Radio gateway for WiMAX and WiFi

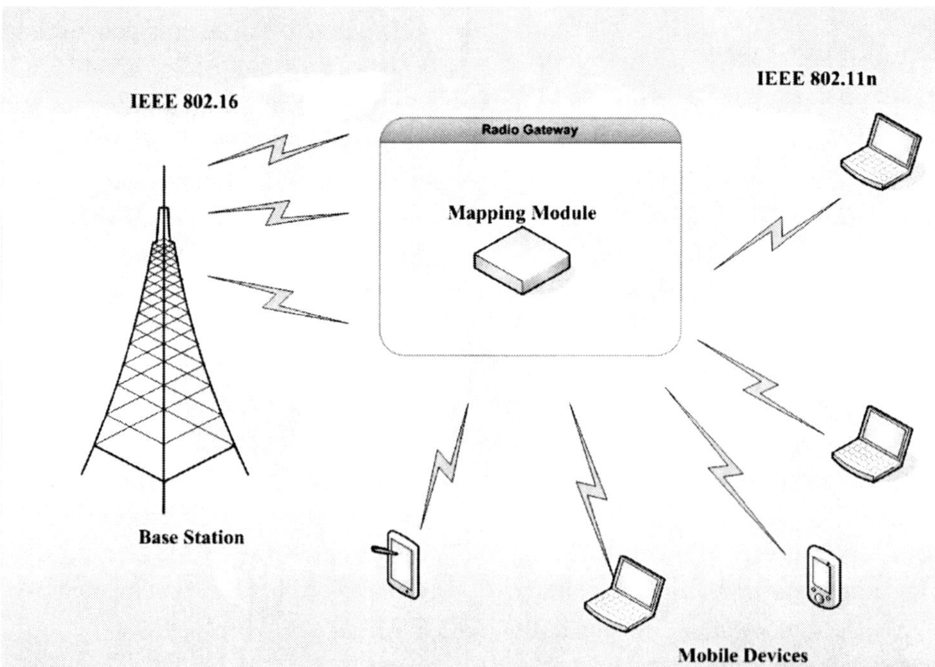

Figure 6. QoS control architecture

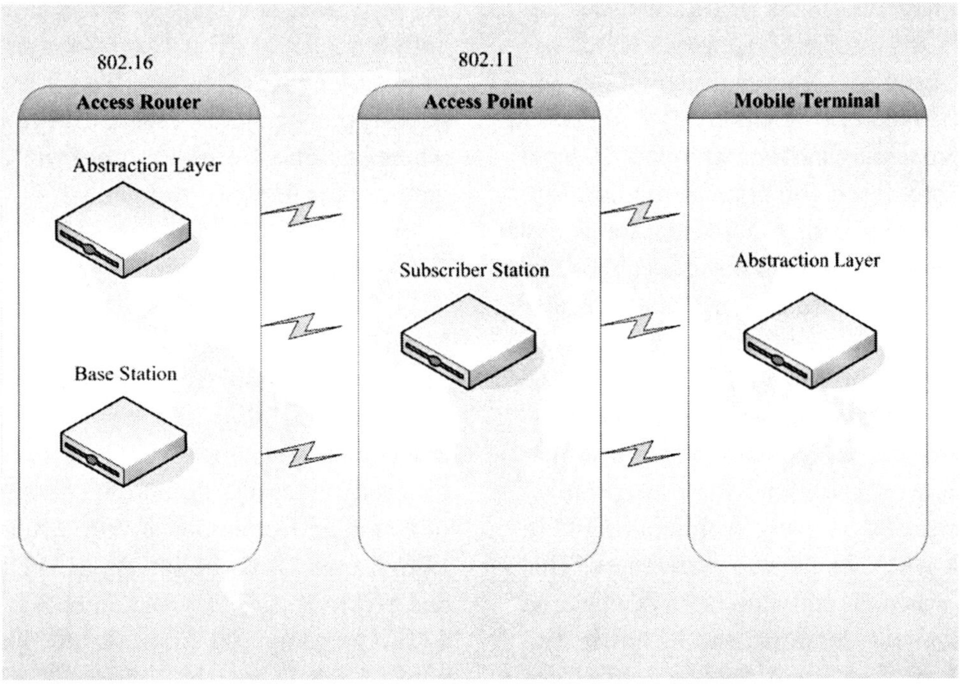

Two scheduling algorithms for DL traffic and UL traffic are used in the contention free period as "DL transmission period" and "UL transmission period" to calculate the DL and UL traffic. Finally, in a converged WiMAX/WiFi network architecture, PCF is used in the WiFi AP to allocate time slots among WiFi nodes.

In 3G mobile networks, GPRS, UMTS, and CDMA2000 support a low bandwidth over a wide geographical area and WLAN based on the IEEE 802.11 standards are able to provide comparatively high bandwidth up to 54 Mbps with a single access point over a few thousands of meters (Qingyang & Jamalipour, 2005). As a result of this integration, WLAN and cellular systems can efficiently achieve the required capacity and QoS level. It is crucial that a dynamic network section mechanism and fast handover mechanism has to be developed in heterogeneous systems to decide an appropriate radio access technology for a specific service. In fast handover mechanism, it is assumed that WiMAX BS and WiFi AP are competent of verifying and identify the neighbours network for a periodically transfer network information on all nodes connected. This will reduce the network time detection in handover process for better services to mobile users in heterogenous networks.

3G – 4G

In this multi-provider, multimedia and multi-technology environment, the management of networks and services becomes a challenging issue. The integration of 3G to 4G involves several challenges. In a multi user terminal, the system may include a Global Positioning System (GPS) for tracking locations, a Wireless LAN for the internet and a CDMA for telephony. All these systems are used in a cellular mobile with different QoS level. The challenges in integrating these systems can be grouped in to 3 types: system, service and mobile station. By using a multimode user terminal, the cellular can adapt to different networks. This multimode user terminal uses a software radio approach which has an antenna, band pass filter, noise amplifier and an ADC. The received signals are digitized after the analog processing and reprogrammed by band pass signal processor. The optimal solution is to use multiple analog parts for different frequency bands. These software radio devices can also be used in wireless system discovery for the cellular device to scan the available network for service subscription. Network selection in heterogeneous networks is more complex because it access different technologies which does not have similar QoS. This network selection issue is an optimization problem because every mobile terminal in the network will have its own preferences. This is achieved when the software radio device scan the list of available networks and its traffic data rate. Based on the current transfer traffic data rate recorded in the software radio device the appropriate network in selected.

The integration of3G with 4G networks (Figure 7) can be performed using a gateway through which the existing 3G network devices connect with the 4G networks. The gateway works as an interface between 3G and 4G networks and performs the necessary conversions at different layers so that an end point maintains a seamless connection any where anytime. Researchers are working on this gateway for further development.

4G – 4G

Multi access in 4G heterogeneous networks needs new level of mobility support as compared to traditional mobility. The way of entertainment and communication of people has changed significantly in the past decade. 4G networks consist of two main technologies: WiMAX and LTE-A. LTE-A uses OFDMA in DL, SC-FDMA in UL and WiMAX uses OFDMA in both UL and DL. LTE-A support 100 MHz bandwidth and provide 1 Gbps DL and 500 Mbps UL. LTE-A uses SC-FDMA in UL and reduces PAPR to increase the throughput. WiMAX supports up to 40MHz bandwidth and provide 128 Mbps in DL and 56 Mbps in UL. LTE-A networks consist of different

Figure 7. Integrating 3G with 4G networks(Suk Yu & Kai Hau, 2003)

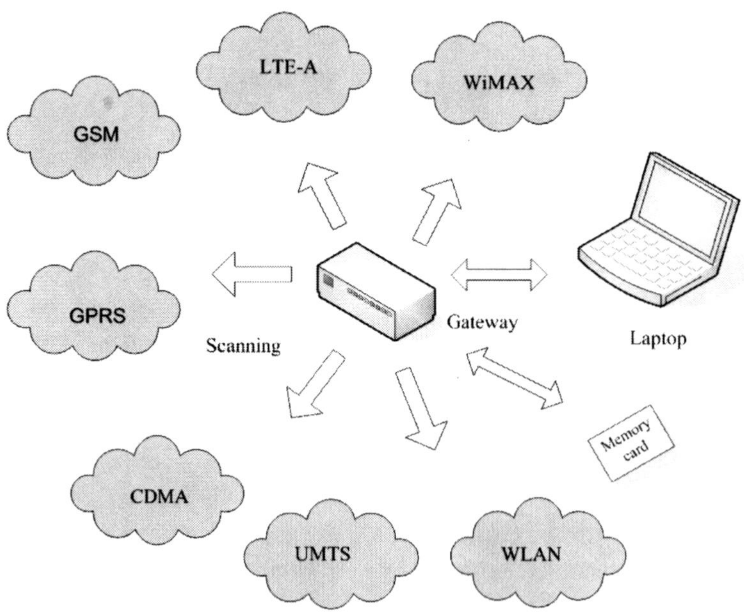

cell structures which range from macro, micro, pico and femto cells, potentially sharing the same spectrum. Both the technologies use OFDMA as a radio technology to achieve high data rates. Designing a heterogeneous network that will support both WiMAX and LTE-A is a challenging task. A common gateway can be designed where both the LTE-A and WiMAX devices can connect to support LTE mobility management entity and WiMAX access service network gateway. This could be a possible solution for blending two network technologies. The gateway (Keshava Murthy, 2008) includes physical interface (where the network elements are connected), IP security, QoS and supports control plane protocol. The user plane transport protocol is used in a common IP layer which associates functions in both WiMAX and LTE-A. Uplink in 4G networks is another challenging issue.

As LTE-A and WiMAX use OFDMA radio access, the inter-system interference must be reduced to have an interface between 4G systems. LTE-A uses SC-FDMA in UL and this will not have an impact on inter-system interference on WiMAX up to 10dB. But WiMAX will have more impact on LTE-A when the inter system interference ratio goes above 20dB. The best option (Zhang, Guo, Wang, & Zhang, 2010) to improve channel interference ratio is to improve frequency spacing between these two systems and adopting better filter in receiver and transmitter. In 4G heterogeneous networks Inter-Cell Interference Coordination (ICIC) plays a vital role. Researchers are still working on this to improve adjacent channel interference ratio. At this stage, some of the partitioning schemes such as soft reuse, factional reuse, hard frequency reuse and full frequency reuse are implemented in the Radio Resources Management (RRM) in both static and dynamic networks. However, some of the simulation results (Pauli, Naranjo, & Seidel, 2010) proved that the dynamic configuration perform well in active information exchange between 4G heterogeneous networks. This is achieved by allowing interface between eNodeB's which is referred as X2 interfacing. Further research is expected to satisfy the hot demands of heterogeneous networks.

CONCLUSION

Heterogeneous networks are promising thoughts to develop system performance to improve and to satisfy traffic demands and network performance overall the coverage area. As wireless mobile broadband service providers embrace multiple wireless technologies, it is essential to have a multi access heterogeneous environment which will provide an infrastructure to keep the mobile user "always best connected." There are many standards currently available in the market (e.g., 3G, 3.75G, and 3.9G) and many new standards are evolving (e.g., 4G). It is extremely challenging to have a single multi access heterogeneous network that will support all of the existing and evolving standards. In this chapter, we highlighted some of the challenges that need to be addressed before a true multi access heterogeneous network can be deployed. We discussed some of the potential solutions and identified the research direction to work with in the near future.

REFERENCES

Ahmad, I., & Habibi, D. (2008a). *A proactive forward error control scheme for mobile WiMAX communication.* Paper presented at the 11th IEEE International Conference on Communication Systems, Singapore, 2008.

Ahmad, I., & Habibi, D. (2008b). Quality of service in WiMAX. In Zang, Y. (Ed.), *WiMAX network planning and optimization* (pp. 16–31). Taylor & Francis Group.

Ahmad, I., Habibi, D., & Rahman, Z. (2008). *An improved FEC scheme for mobile wireless communication at vehicular speeds.* Paper presented at the Telecommunication Networks and Applications Conference, 2008. ATNAC 2008. Australasian.

Berezdivin, R., Breinig, R., & Topp, R. (2002). Next-generation wireless communications concepts and technologies. *Communications Magazine, IEEE, 40*(3), 108–116. doi:10.1109/35.989768

Buddhikot, M. M., Chandranmenon, G., Seungjae, H., Yui-Wah, L., Miller, S., & Salgarelli, L. (2003). Design and implementation of a WLAN/CDMA2000 interworking architecture. *Communications Magazine, IEEE, 41*(11), 90–100. doi:10.1109/MCOM.2003.1244928

Clerckx, B., Lozano, A., Sesia, S., Rensburg, C. V., & Papadias, C. B. (2009). 3GPP LTE and LTE advanced. *EURASIP Journal on Wireless Communications and Networking.*

Gakhar, K., Gravey, A., & Leroy, A. (2005, 7-7 Oct. 2005). *IROISE: A new QoS architecture for IEEE 802.16 and IEEE 802.11e interworking.* Paper presented at the (BroadNets) 2nd International Conference on Broadband Networks.

Gang, W., Mizuno, M., & Havinga, P. J. M. (2002). MIRAI architecture for heterogeneous network. *Communications Magazine, IEEE, 40*(2), 126–134. doi:10.1109/35.983919

Holma, H., & Toskala, A. (2000). *WDCDMA for UMTS – Radio access for third generation mobile communications.* John Wiley & Sons, Ltd.

ITU-RReportM. 2134. (2008). *Report ITU-R M.2134: Requirements related to technical performance for IMT-advanced radio interface(s).*

Keshava Murthy, K. S. (2008). *NextGen wireless access gateway analysis of combining WiMAX and LTE gateway functions.* Paper presented at the (IMSAA 2008), 2nd International Conference on Internet Multimedia Services Architecture and Applications.

Krichene, N., & Boudriga, N. (2008). Handoff management in WiMAX. In Zang, Y. (Ed.), *WIMAX network planning and optimization.* CRC Press.

Neves, P., Sargento, S., & Aguiar, R. L. (2006, 26-29 June 2006). *Support of real-time services over integrated 802.16 metropolitan and local area networks.* Paper presented at the (ISCC 2006), Proceedings. 11th IEEE Symposium on Computers and Communications.

Pauli, V., Naranjo, J. D., & Seidel, E. (2010). *Heterogeneous LTE networks and inter-cell interference coordination.* Munich, Germany: Nomor Research GmbH.

Persaud, R. (2006). *Core network mobility: Active MPLS.* Germany: Shaker Verlag.

Pospishny, I., Vasyuk, V., Romanchyk, S., Dovzhenko, O., Dony, O., & Shvaichenko, V. (2010, 23-27 Feb. 2010). *3GPP long term evolution (LTE).* Paper presented at the Modern Problems of Radio Engineering, International Conference on Telecommunications and Computer Science (TCSET).

Qingyang, S., & Jamalipour, A. (2005). *A network selection mechanism for next generation networks.* Paper presented at the IEEE International Conference on Communications (ICC).

Rabbani, M. G., Kamruzzaman, J., Gondal, I., & Ahmad, I. (2011). *A new resource distribution model for improved QoS in an integrated WiMAX/WiFi architecture.* IEEE International Conference on Communications.

Ray, S. K., & Mista, I. S. (2006). Fourth generation (4G) networks: Roadmap-migration to the future. *IETE Technical Review*, *23*(4), 253–265.

Rumney, M. (2008). *LTE and the evolution to 4G wireless*Agilent Technologies.

Sachs, J., Prytz, M., & Gebert, J. (2009). Multi-access management in heterogeneous networks. *Wireless Personal Communications*, *48*(1), 7–32. doi:10.1007/s11277-007-9431-8

Shao, X., Schiphorst, R., & Slump, C. H. (2009). An opportunistic error correction layer for OFDM systems. *EURASIP Journal on Wireless Communications and Networking*, 2009.

Suk Yu, H., & Kai Hau, Y. (2003). Challenges in the migration to 4G mobile systems. *Communications Magazine, IEEE*, *41*(12), 54–59. doi:10.1109/MCOM.2003.1252799

Tapia, P., Liu, J., Karimli, Y., & Feuerstein, M. J. (2009). *HSPA performance and evolution*. John Wiley & Sons, Ltd.doi:10.1002/9780470742068

Yi, Wang, Li, L., Zhang, P., & Liu1, Z. (2009). DFT-based channel estimation with symmetric extension for OFDMA Systems. *EURASIP Journal on Wireless Communications and Networking*, 2009.

Zhang, Y., Guo, H., Wang, X., & Zhang, X. (2010). *Co-existence analysis of LTE and WiMAX in uplink*. Paper presented at the 6th International Conference on Wireless Communications Networking and Mobile Computing (WiCOM).

Zhou, W., & Lam, W. H. (2009). A fast LMMSE channel estimation method for OFDM systems. *EURASIP Journal on Wireless Communications and Networking*, *2009*, 13. doi:10.1155/2009/752895

KEY TERMS AND DEFINITIONS

3G: International Mobile Telecommunications-2000 (IMT – 2000) popularly known as 3G or third generation. This offers mobile communication services with satisfying the specifications of International Telecommunication Union.

4G: 4G is a successor of 2G and 3G mobile technology which provides high data rate of 100 Mbps at downlink for high speed mobile communications and 1Gbps at downlink for low mobility communications.

Heterogeneous Networks: This connects different access technologies, standards, access systems with dissimilar scenario interfacing with a common core network.

LTE: It is otherwise called as 3GPP Long Term Evolution, a mobile network technology tree which follows GSM/EDGE and UMTS/HSPA technologies.

LTE-Advanced: It is a preliminary mobile communication standard, formally accepted as a 4G technology by International Telecommunication Union in early 2011.

Multi-Access: A multi-access network which allows different terminals to communicate at the same transmission point to share the same frequency at any time during the operation.

Next Generation Networks: It is a packet based network which offer communication service that every network can use multiple broadband service and handle multiple traffics such as voice, data and multimedia.

WiMAX: Worldwide Interoperability for Microwave Access is a communication protocol which provides fixed and mobile internet access.

Chapter 2
Mobility Challenges and Management in the Future Wireless Heterogeneous Networks

Jianfeng Guan
Beijing University of Posts and Telecommunications, China

Changqiao Xu
Beijing University of Posts and Telecommunications, China

Hongke Zhang
Beijing University of Posts and Telecommunications, China & Beijing Jiaotong University, China

Huachun Zhou
Beijing Jiaotong University, China

ABSTRACT

With the development of the various network technologies, the mobility management as a growing concern issue has been one of the most expected features in the future wireless heterogeneous networks environments. Both academia and the industry look forward to the deployment of mobility support, but none of the proposed solutions have been adopted in the large scale due to the various impediments. This chapter investigates the mobility challenges and management in the wireless heterogeneous environment, gives a brief description of the current mobility management solutions, and provides a comparison of the different solutions including their advantages and disadvantages.

The organization of this chapter is as follows: It begins by analyzing the mobility challenges under the multi-accessing wireless heterogeneous environments. Next the chapter gives a brief introduction of the mobility management. Afterwards it introduces the state-of-art of the mobility management solutions and describes the respective approaches and compares their advantages and disadvantages. The last section concludes this chapter.

DOI: 10.4018/978-1-4666-0017-1.ch002

INTRODUCTION

Mobility management as an important issue in current and future wireless heterogeneous environments has been got growing concern and many mobility management schemes have been proposed in the past few years (Zhu, 2010). The mobility can be classified into terminal mobility, session mobility, personal mobility and service mobility based on the different mobile entities (Strang, 2005). Terminal mobility happens when a terminal changes its attachment to the network, which may cause the change of the IP configuration. Here the terminal can be a traditional cellular device or a portable device such as notebook. Personal mobility means that a user can be globally reachable by unique personal identifier and originate or receive a service by accessing to any authorized terminals. While the session mobility is similar to the personal mobility which uses the different devices with diverse characteristics while maintaining state or transferring the session from one host to another. As for the service mobility, it enables a user to obtain subscribed and personalized services consistently even if connected to a foreign network service provider anytime and anywhere.

In this chapter we mainly concern on the terminal mobility which allows a device to move between subnets while being reachable to other host to maintain the ongoing communication. The objective of this chapter is to investigate the mobility management challenges and the relevant solutions under future wireless heterogeneous environments, and to analyze their development trends and future directions.

BACKGROUND

Mobility management born with the development of the wireless mobile communication systems, and migrates into the Internet due to the big success deployment of the IP-based services. The emerging the mobility management is due to the drawback of the traditional networks which are designed for the static hosts connected through wired networks that cannot support the mobile hosts. The mobility management solutions are generally classified into the link layer mobility support, network layer mobility support, transport layer mobility support and the application layer mobility support. IEEE 802 workgroup mainly focuses on the link layer mobility support and has published several standard specifications in the past few years include the 802.11f (Kerry, 2003), 802.11r (Kerry, 2008), 802.16e (Marks, 2006), 802.20 (Klerer, 2009) and 802.21 (Gupta, 2009). As for the network mobility support, Ioannidis et al. proposed the Columbia protocol (Ioannidis, 1991) to provide mobility on a campus, which uses the Mobile Support Station (MSS) as the default access router for all mobile nodes in that cell. The Mobile Node (MN) uses an IP address derived from a special IP prefix as its identifier. Each MSS keeps a tracking list of MNs that are currently in its cell. MSS detects the attachment and detachment events to notify the related MSS to update the records. At the same time, Teraoka et al. (1991) proposed the Virtual Internet Protocol (VIP) solution, which modify the IP header to carry the virtual IP address as the identifier and a regular IP address as the locator. After that, Bhagwat and Perkins (1993) proposed Loose Source Routing (LSR) protocol which introduces a designated router for each MN to assign an IP address as the identifier, and introduces the Mobile Access Station (MAS) entity in each network to performance the location management. Afterwards, Bhagwat et al. (1996) surveyed the mobility support solutions and compared several selected candidate proposals. Based on the proposed solutions, IETF published the Mobile IPv4 (MIPv4) specification (Perkins, 1996). After that, IETF successively standardized the MIPv4 (Perkins, 2002), Mobile IPv6 (MIPv6) (Johnson, 2004), Proxy Mobile IPv6 (PMIPv6) (Gundavelli, 2008), Dual-Stack Mobile IPv4 (DSMIPv4) (Tsirtsis,

2009) and Dual-Stack Mobile IPv6 (DSMIPv6) (Soliman, 2009). Besides, several enhanced solutions was proposed including the Fast MIPv6 (FMIPv6) (Koodli, 2005), Hierarchical Mobile IPv6 (HMIPv6) (Soliman, 2005). Especially, some mobility management solutions such as PMIPv6 and DSMIPv6 have been adopted by the 3GPP. At the same time, many transport mobility management solutions were proposed, and Atiquzzaman et al. (2005) surveyed and classified these solutions, and compared them based on a number of evaluation criteria. The application mobility management solutions are mainly based on the Session Initiation Protocol (SIP) (Rosenberg, 2002). Schulzrinne et al. (2000) describe how the SIP can help provide terminal, personal, session and service mobility to application in 2000. After that, several improved solutions were proposed such as Scalable Application-Layer Mobility Protocol (SAMP) (Pack, 2006) and Mobility Management Using SIP Extensions (MMUSE) (Salsano, 2008).

Le et al. (2006) reviewed the TCP/IP protocol stack and analyzed the problems associated with it in the mobile environment, and then investigated the mobility support techniques and existing solutions for providing mobility support on the Internet based on the protocol layers, and compared the different solutions belonging to different categories, including their advantages and disadvantages. Besides, Siddiqui et al. (2006) presented a detailed discussion of various components of mobility management and the recent handoff techniques for mobility across heterogeneous wireless access networks, and discussed about various features of multimodal mobile terminals that will enable the deployment of efficient mobility solutions, and presented several handoff implementation issues and challenges in future radio access systems. Recently, Zhang et al. (2009) reviewed the features and requirements related to the Internet mobility support, and investigated several typical implemented methods and corresponding solutions, and compared the Internet mobility and cellular network mobility. More recently, Zhu et al. (2010) surveyed the mobility solutions in the last two decades, and aimed to initiate an open discussion about the general direction for future mobility support. The development of the related mobility management solutions are shown in Figure1.

Figure 1. The development of the mobility management

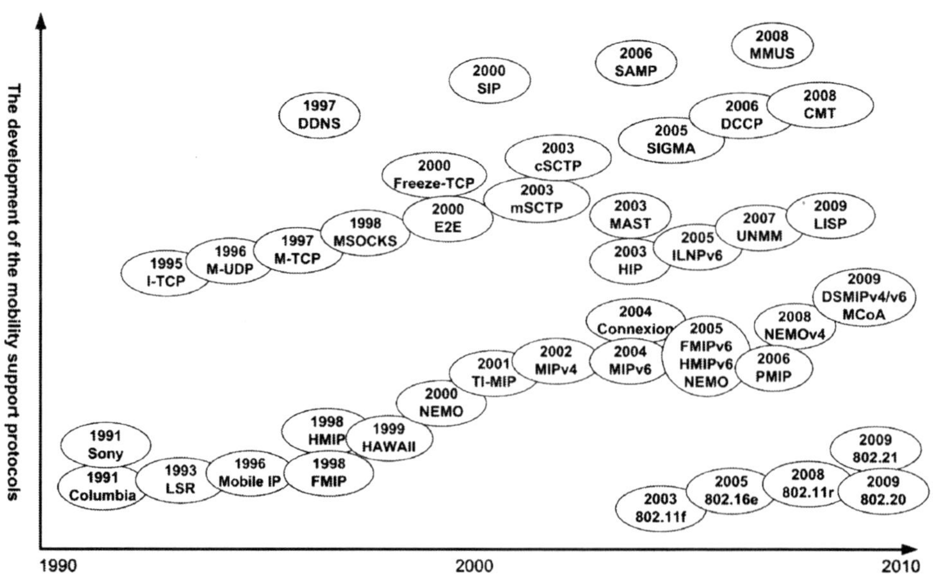

Since the mobility management can be implemented at different layers, a natural question to answer is in which layer should be the mobility management placed. Eddy (2004) attempted to answer the question of which layer should handle mobility management, and the work evaluated the characteristics of network, transport and session layers for supporting seamless handoffs, location management, and change in infrastructure, and got that in order to overcome various limitations that a network layer scheme, transport layer mobility is the most promising scheme. While in this chapter, we will discuss this problem based on the investigation of the different layer mobility management solutions, and discuss the challenges and the development trends.

CHALLENGES

The challenges of the mobility are divided into the following aspects.

Identifier and Location Split

The key issue in the mobility management is to maintain the relationship between the identifier and the locator of the MN. However, the design principle of traditional TCP/IP protocol integrates the identifier information and location information into the IP address which simplifies the implementation and the deployment but introduces the mobility support issue. The drawback the TCP/IP arises with the development of the various wireless technologies. As a result, how to separate the identifier information and location information becomes a great challenge to the future networks. Recently several mobility management approaches are proposed to separate them and set up the mapping relationship (Zhang, 2010). However, the most of them are still under study in term of the legacy Internet protocols, and many problems such as transport modification, management of the ID/locator mapping need the further research.

Seamless Mobility

The emerging multiple access technologies bring the seamless mobility challenge to achieve global mobility. Emmelmann et al. (2007) analyzed the challenges to provide seamless mobility in the standardization bodies. In the future wireless heterogeneous, the terminal will equipped with multiple network interfaces and can attach to multiple heterogeneous access networks at the same time. How to provide the seamless mobility become a challenge, which includes the network discovery, network selection and adaptive handover. The seamless mobility management should consider the multiple access technologies, multi-user, multi-application, multi-terminal, multi-user environments and has the following characters.

1. **Ubiquity:** The mobility support should not be restricted to any specific underlay network, device, or protocol, and it should be available anywhere.
2. **Continuity:** The session or communication should not be broken by the movement of the MN.
3. **Multi-homing:** The mobility management should consider the multi-homing support including the multiple network interfaces and multiple IP addresses or prefixes to provide the effective transmission and to balance the overhead among multiple access networks.
4. **Interoperation:** The mobility management solution should consider the interoperation between different networks, different technologies and different protocols.

Deployment

As for the industry, the mobility management should be deployable. However, few mobility management solutions have been deployed in commercial networks due to following reasons. Firstly, most of the mobility management solutions have the scalability problem, and they cannot support millions of mobile nodes simultaneously.

Secondly, the complexity of the most mobility support solutions arising from the routing optimization, security and compatibility prevents the commercial deployment. Thirdly, almost all the mobility support protocols assume that the mobile node has the network connectivity anywhere anytime. However, it is not always the case in the reality. So, the mobility management solutions should minimize the changes to the applications, and avoid adding the third-party, and be compatible with the existing infrastructure (Le, 2006).

Power Consumption

In the future wireless heterogeneous networks, energy and computation resources of the mobile devices usually are restricted. The physical and data link layer, designs of higher layer protocols and architecture could have implicit impact on resource consumption, especially for wireless devices. So the mobile management should be lightweight in terms of the signaling and battery cost to reduce the energy consumption.

Security

The mobility management should consider the security mechanism to prevent the mobile nodes from the various attacks. The security issues not only inherit from the traditional static networks but also arise from the mobility features in the future heterogeneous wireless networks. Authentication becomes necessary for both initial conversation setup and following movement. Address authentication might be required as well in some special scenarios. Confidentiality and integrity face more challenges as well. Roaming across the boundary where different access policies are applied would produce more sophisticated regulation requirement.

All challenges provided above are typical ones and the list is not complete. In spite of that, to solve these challenges is not an easy task. Many solutions were proposed in the past few years. However, they cannot satisfy the requirements of academia and industry due to various shortcomings.

OVERVIEW OF THE MOBILITY MANAGEMENT

Mobility management contains two components, the location management and handover (handoff) management (Akyildiz, 1999).

Location Management

Location management is used to track the location of the MN, which consists of the location registration (or location update) where mobile node periodically notifies the system to update relevant location database with its update location information, and the call delivery, where the system determines the current location of the mobile node based on information available at the system databases. Kyamakya et al. (2005) classified and compared the location management methods in cellular networks.

Handover Management

Handover management aiming to minimize the disruption time, enables MN to maintain the ongoing session or communication as the MN changes its access point, which is a core technology in the future wireless heterogeneous environments. Handover management generally consists of the initiation, new connection generation, and the data flow control. The initiation is used by mobile terminal or network to identify the need for handover, while the new connection generation is to find the new resources for handover connection and perform the additional routing operations, while the date flow control maintains the packet delivery from the old connection to the new one.

The handover management can be classified into the following kinds based on the difference metrics.

Homogeneous vs. Heterogeneous Handover

Handover that happens between the accessing points of the same network technology is called homogeneous handover (horizontal handover), while handover that occurs between the accessing points belong to different network technologies is called the heterogeneous handover (vertical handover). Compared with the homogeneous handover, the heterogeneous networks involvement makes the implementation of heterogeneous handover more challenging (Siddiqui, 2006).

Hard vs. Soft Handover

The handover that MN can be associated with only one access point at a time is the called the hard handover, or the Break-Before-Make (BBM), for example the IEEE 802.11 handover. While the handover that allows creating connectivity to the new access network or router before leaving the old access network or router is called the soft handover or Make-Before-Break (MBB). For example, if a mobile node equips with multiple network interfaces, it can simultaneously connect to multiple access points in different networks during the handover.

Micro and Macro Handover

Based on the movement range, the mobility management can be classified into the micro mobility, also called intra-cell and macro mobility, also called inter-cell (Akyildiz, 2004). The micro mobility refers to the mobility that occurs within a small scale network, and the MN moves between subnets of one domain that belong to the same ISP. As for the macro mobility, MN moves between two network domains which include different access networks. The macro mobility is suitable for the different environments and platforms to provide the global mobility support.

Host-Based Handover vs. Network-Based Handover

As for the handover management in layer 2, the handover management is generally divided into three kinds (Akyildiz, 1999), Network-Controlled Handover (NCHO), Mobile-Assisted Handover (MAHO), and the Mobile-Controlled Handover (MCHO) in wireless networks. Under the NCHO, the network makes the decision for handover. While under the MCHO, the mobile node must take its own signal strength measurements and make the handover decision on its own. Under MAHO, the decision to handover is made by the mobile node in cooperation with the network.

As for the mobility management in network layer, it can be classified into the host-based such as the MIPv6 and the network-based such as the PMIPv6. The host-based mobility management requires MN to perform the mobility management, while the network-based mobility handles mobility management on the network side without mobile node's participation.

MOBILITY MANAGEMENT SOLUTIONS

The typical mobility management can be classified based on the layer in which they operate (Banerjee, 2003), link layer, network layer, transport layer and application layer. In this section, we limit the scope to the mobility management approaches that are relative mature to be adopted by industry, and describe the relevant representative approaches in each layer based on the previous work.

Link Layer Solutions

Link layer approaches are the mobility supported from the link layer (Chiussi, 2002), in which the MN changes its point of the attachment using L2 mechanism which remains the same IP address.

IEEE 802.11f

IEEE 802.11f (IAPP, Inter-Access Point Protocol) (Kerry, 2003) is implemented on a Distribution System (DS) by extending the IEEE 802.11 to provide the necessary capabilities to achieve multi-vendor APs interoperability within the DS, to support the MN moving among the different APs.

The specification describes a set of functions and a protocol to allow APs to carry IAPP packets between APs using TCP/IP or UDP/IP, and to use the Remote Authentication Dial-In User Service (RADIUS) to obtain the information of the other APs. Besides, it also provides a proactive caching mechanism to support the fast roaming. Based on these features, Huang et al. (2006) proposed a fast and seamless handoff solution for IEEE 802.11 wireless LAN with IAPP. However, the introduction of the IAPP increases the reassociation delay between the MN and new AP (Mishra, 2003). As a result, there has been no significant deployment of 802.11f, and IEEE 802 executive committee approved its withdrawal on February 3, 2006.

IEEE 802.11r

IEEE 802.11r (Kerry, 2008) specifies mechanisms for a fast Basic Service Set (BSS) transition which addresses roaming capabilities of the real-time applications by minimizing BSS transition time while still providing the 802.11i security and 802.11e QoS (Tabassam, 2009). 802.11r permits seamless connectivity with a fast and secure handover from one AP to another within the same mobility domain, and it provides an opportunity to derive encryption keys prior to a reassociation to reduce the connectivity loss during the handover. Besides, it also provides features for resource reservation and central mobility management.

Figure 2 shows the IEEE 802.11r handover procedure. When a MN attaches to the first AP (Old-AP), the old-AP which is called the R0 Key Holder (R0KH) will cache its Pairwise Master Key (PMK) to derive session keys for others APs and it also hold the PMK-R0. When the MN handovers to another AP, the R0KH will derive a PMK-R1 and forward it to the new-AP as the new session key. After that, the new-AP is called the R1KH. In the IEEE 802.11r specification, the initial AP acts as the authenticator, communicating with the AAA server. After that, each AP interacts with the initial AP rather than directly with the AAA server.

The recently study (Clancy, 2008) compared that 802.11r with the other fast handover solutions, and concluded that the IEEE 802.11r is well suited to WLAN environments that are not centrally management or require extremely-low-latency handover performance, and its handover delay is about 50ms, which can satisfy the requirement of the VoIP service.

However, due to the delay of publication, some wireless device manufacturers have proposed their own solutions which can also reduce the handover delay. As a result, the deployment of the 802.11r is not significant. Besides, due to the scalability of the key management, IEEE 802.11r is not suitable for the large scale networks.

IEEE 802.16e

IEEE 802.16e standard (Marks, 2006) adds the mobility support based on the IEEE 802.16-2004, and the terminals can move while receiving services and preserve the ongoing sessions during handover with the help of the link layer handover mechanism. During the handover, each Base Station (BS) in the IEEE 802.16e only changes its link layer connection with the new attachment without changing its IP. However, this handover is

Figure 2. The 802.11r handover procedure

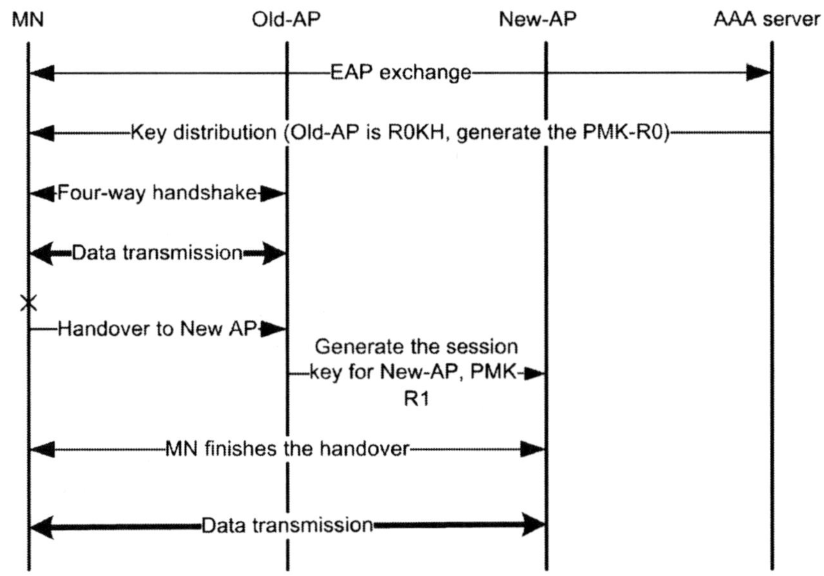

a hard handover. The terminal breaks its connection with the current base station at the beginning of the handover process and cannot resume the communication with the new base station until the handover completes. To solve this problem, Han et al. (2007) proposed a cross-layering design by introducing the FMIPv6 (Koodli, 2005) to improve its handover performance. After that, Lee et al. (2010) also proposed the enhanced scheme based on the network-assisted fast handover.

Figure 3. The basic IEEE 802.16e handover procedure

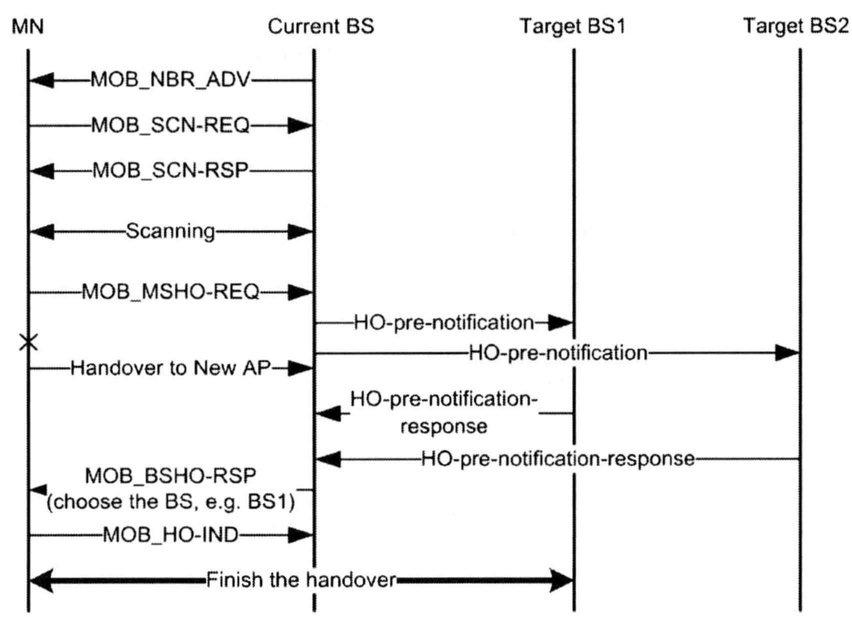

Compared with the link layer handover of the IEEE 802.11, the IEEE 802.16e is more elaborate due to its incorporation of adjustments and flexibility in choosing a target base station. The basic handover procedure is shown in Figure 3.

In the IEEE 802.16e specification, it assumes that the BS can obtain the identifying information of the neighboring BSs through the backbone network, and select the candidate BS based on the metrics such as the signal strength. Each BS advertises the periodic neighbor advertisement message (MOB_NBR-ADV), and the MN acquires the current network identification and the neighboring BSs information from this message for the future handover. Besides, the MN can also seek and monitor the suitability of neighboring BS by exchanging MOB_SCN-REQ and MOB_SCN-RSP messages with the current BS. If the MN wants to handover to anther BS, it will choose the target BS set and initiate the handover by sending a MOB_MSHO-REQ message to the current BS. The current BS exchanges the HO-pre-notification and HO-pre-notification-response messages with the target BSs and replies to the MN with an MOB_BSHO-RSP message which contains the selected BS that satisfies the QoS constraints. Otherwise, the current BS will send the MOB_BSHO-REQ message to trigger the handover. After that, the MN will initiate the handover procedure by sending an MOB_HO-IND to the current BS. Finally, the MN will finish the handover and acquires the target BS information. If MN moves into a different subnet, it has to reconfigure a new IP address and reestablish its IP connection based on the new IP address.

In addition to the default handover mechanism in the IEEE 802.16e, some improved mechanisms were proposed in the past few years. IEEE 802.16e provides a number of attractive features, but it still has some open issues in term of the network layer handover. Recently, Ray et al. (2010) investigated the potential handover-related research issues in the existing and future WiMax mobility framework, and surveyed these issues in the MAC, Network and Cross-layer scenarios.

IEEE 802.20

IEEE 802.20 standard (Klerer, 2009) is developed to meet the unique requirements for supporting high-speed data services while at the same time supporting full user mobility to provide IP-based broadband wireless access in a mobile environment. Compared with the IEEE 802.11 (Wi-Fi) and IEEE 802.16 (WiMax), the IEEE 802.20 (MobileFi) may change the direction of wireless networking, and sits on existing cellular towers, promising the same coverage area as a mobile phone system with the speed of a Wi-Fi connection. IEEE 802.20 is for truly mobile and high-speed data of 20 Mb/s, and it will increase the coverage or mobility compared with Wi-Fi and WiMax, and Bolton et al. (2007) investigated the differences among IEEE 802.20, IEEE 802.16e and 3G cellular networks, and got that the IEEE 802.20 specification is try to bridge many of the technical differences between IEEE 802.16e and 3G networks and combines them in a manner that is both powerful and accessible for the end user to provide the global mobility and roaming support of 3G with the high-bandwidth and low-latency characteristics of IEEE 802.16e. More specifically, IEEE 802.16e handles low mobility such as walking with Personal Digital Assistant (PDA) or laptop in the 2 to 6 GHz licensed bands, whereas IEEE 802.20 handles high-speed mobility up 250 km/h in bands below 3.5 GHz.

IEEE 802.21

IEEE 802.21 (Gupta, 2009) standardizes the Media-Independent Handover (MIH) framework to facilitate the integration of the heterogeneous networks by providing the upper layer uniform information about the L2 triggers to perform the handover. The media types include the 3GPP, 3GPP2, IEEE 802 wired and wireless, which refers to the method/mode of accessing a telecommunication system. IEEE 802.21 supports the handovers for both mobile and stationary users. For mobile users, handovers can occur when wireless link conditions change due to the users' movement. For the stationary user, handovers become imminent when the surrounding network environment changes, making one network more attractive than another.

According to the IEEE 802.21 specification, it consists of three main elements: (1) the framework that enables service continuity while a MN moves between heterogeneous networks, and it bases on the specific mobility management protocol and provides MIH reference models for different link-layer technologies; (2) the MIH Function (MIHF) which is an entity consisting of a set of handover-enabling functions; (3) the Service Access Points (SAPs) which includes the media-independent and media-specific interfaces. While as for the MIHF services, the IEEE 802.21 defines three services involving the Media Independent Event Service (MIES), Media Independent Command Service (MICS), and Media Independent Information Service (MIIS) to facilitate the handover across heterogeneous networks. The MIES detects change in logical link dynamically such as link status and link quality, which includes the link events originated from lower layer and the MIH events initiated by MIHF. The MICS facilitates controlling of a link state, while the MIIS gathers the information for handover. The MIHF uses the triggers from link layer to support handover initiation and handover preparation. As for the SAPs, the MIH_SAP is media independent interface, while the MIH_LINK_SAP and MIH_NET_SAP are media dependent interfaces.

Figure 4 shows the general IEEE 802.21 MIH service model, in which the MIH users are the functional entities that employ MIH services for example a mobility management application. The MIH users adopt the MIH_SAP to acquire three MIHF services (MIES, MIIS and MICS), while these MIH services can be either local or remote. Besides, IEEE 802.21 defines a management service which includes the MIH capability discovery, MIH registration, and MIH event subscription to discovery other MIHF entities. MIH_LINK_SAP enables the MIHF to obtain and transmit the media-specific information to the MIH users through the MIH_SAP. The MIH_NET_SAP exchanges MIH information and messages with the remote MIHF, while the LLC_SAP is not specified in IEEE 802.21 specification.

Figure 4. IEEE 802.21 MIH services

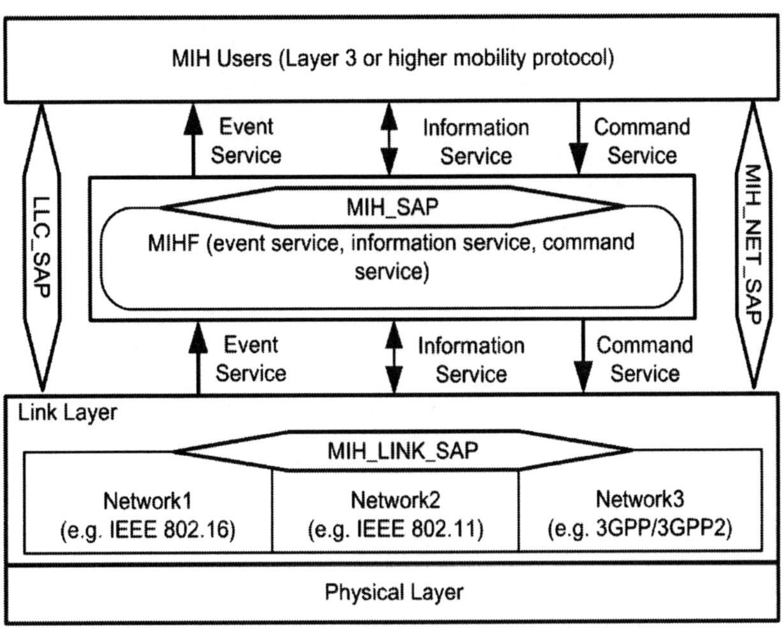

IEEE 802.21 enables seamless handover between heterogeneous access technologies, but it still has some challenges in the future deployment (Taniuchi, 2009). Firstly, it is difficult to realize the unification of all the media-specific technologies under the one abstract interface due to that the large number of technology-specific standards has to be extended to conform to the MIH_LINK_SAP. Secondly, it lacks of a conformance statement detailing the mandatory set of primitives and primitive sequences, which will cause the interoperation problem in the future.

Summary

Beside the typical IEEE 802 specifications, the other standard organizations such 3GPP/3GPP2 also published the relevant solutions. These link layer solutions can provide the better link-layer handover, however, they still needs to combine with the upper layer mobility management solution to optimize the handover performance. In a word, the link-layer mobility management solution is fundamental in the future wireless heterogeneous environments, but it constrains within a limited domain and cannot preserve higher-layer connections. It should combine with other layer mobility management solution to provide the ubiquitous mobility support.

Network Layer Solutions

Mobility management at the network layer such as MIPv6 (Johnson, 2004) and PMIPv6 (Gundavelli, 2008) have been got growing concern recently, especially in IETF and 3GPP. In this part, we mainly concern on the Mobile IP and its enhanced approaches, and describe its development and state-of-art in the standard organizations. Besides, considering the IPv4 and IPv6 migration, we will describe the IPv4-only mobility approaches, IPv6-only mobility approaches, and dual stack mobility support approaches.

Host-Based Mechanism

The host-based mechanism usually implements the mobility support function in the mobile devices. IETF has published several standards such as MIPv6 and Network Mobility (NEMO) (Devarapalli, 2005). MIPv6 as a representative standard for IPv6 mobility support can solve many problems, but it has still revealed some problems involving the handover latency, power consumption, high packet loss and signaling overhead. Furthermore, it has not got a large scale deployment. Therefore, various enhanced solutions were proposed. In this part, we will describe and compare the MIPv6 and its variants.

MIPv4/v6

MIPv4 (Perkins, 2002) specifies the IP mobility solution that allow transparent routing of IP packets to mobile nodes in the Internet. MIPv4 introduces two new entities, the Home Agent (HA) and the Foreign Agent (FA), to implement the location and handover management. MIPv4 assigns a permanent IP address called Home Address (HoA) to identify the identifier information, and a temporary IP address called Care-of Address (CoA) to identify its locator, and set up a mapping between the HoA and the CoA to maintain the on-going communications.

Figure 5 shows the basic MIPv4 operation flow. When MN is on its home network, it acts like other fixed nodes of that network and uses the traditional routing to communicate with the Correspondent Node (CN). Once it moves into a foreign network and obtains a CoA from the FA through Dynamic Host Configuration Protocol (DHCP) or other IP configuration schemes, it will initiate the binding update procedure to set up a bi-directional tunnel between HA and FA or HA and MN. After that, the HA will intercept the packets to the MN and forward them through the bi-directional tunnel to FA or MN. Finally, the MN will get the packets from the CN.

MIPv4 can provide the basic mobility support, but it introduces the triangular routing, considerable handover delay, and additional signaling overhead. As a results, some enhanced schemes were proposed such as the low latency handover (Malki, 2007), Fast handover (Koodli, 2007) and regional registration (Fogelstroem, 2007) to improve its performance.

Similar to MIPv4, MIPv6 provides a general IPv6 mobility support framework and uses the CoA and HoA to carry the location information and identifier information, respectively. Figure 6 shows the basic MIPv6 operation flow. Assume that CN transmits the packet to the MN, and the MN attaches to the HA at first, and then it leaves the home network and attaches to the Access Router (AR). When the MN locates in the home network, it receives the packets according the standard IP routing. Once it leaves the home network and enters into a foreign network, it configures the CoA at first by the stateless or stateful address configuration mechanisms, and performs the movement detection during the address configuration procedure. After that, the MN sends the Binding Update (BU) message to register the CoA with the HA. The HA, after receiving the BU message, updates the binding cache and modifies the tunnel. And then, the HA replies with a Binding Acknowledge (BA) message. The MN updates the tunnel after receiving the BA message. After that, a bidirectional tunnel is set up between HA and MN, and then CN can transmit the packets through the tunnel to the MN. In this case, the packets delivery path is triangular. If the CN supports the MIPv6, the MN can set up the binding cache in the CN through the return routability procedure to realize the optimal routing. The return routability procedure consists of four messages, called the Home Test Init (HoTI), Care-of Test Init (CoTI), Home Test (HoT), and Care-of Test (CoT). After that, the packets can delivery directly to the MN with the type 2 routing header which contains the HoA of the MN.

Compared with the MIPv4, MIPv6 does not deploy the routers as FA, and supports the optimal routing by introducing the CN function. How-

Figure 5. MIPv4 operation flow

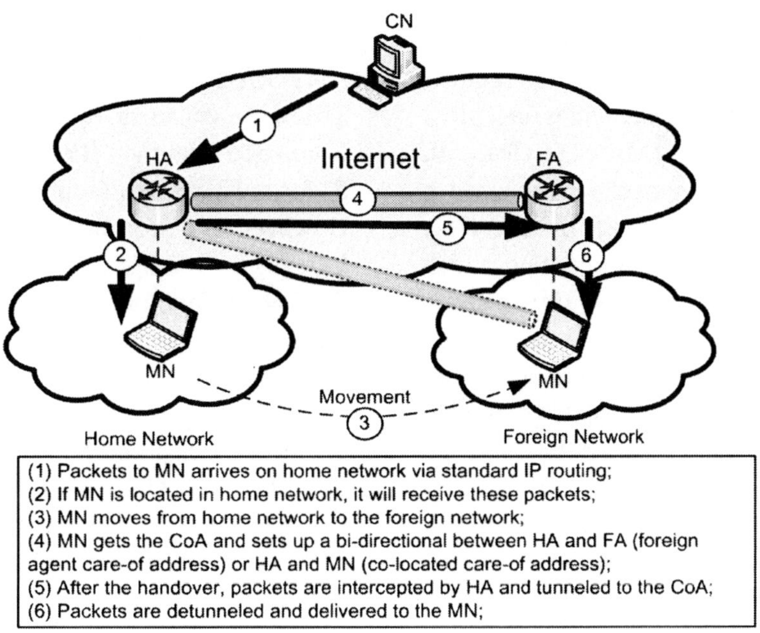

Figure 6. The MIPv6 operation flow

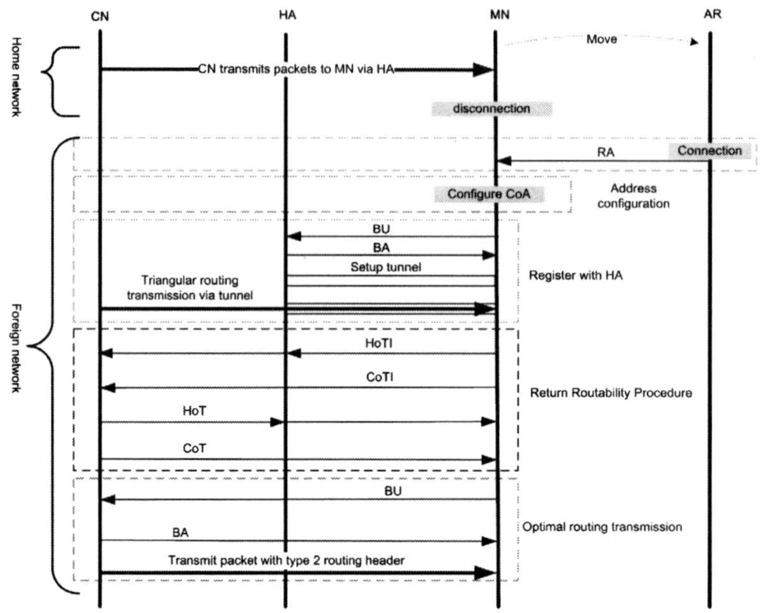

ever, similar to MIPv4, it still has the drawback in term of the handover delay and signaling cost. So, some enhanced schemes were proposed in the past few years, such as the FMIPv6, HMIPv6.

Fast Handover Enhanced Schemes

The fast handover of MIPv4 and MIPv6 are designed to reduce the built-in delay of the L2 and L3 with the help of the access network infrastructure.

The fast handover enhance schemes for MIPv4 are low latency handover and MIPv4 fast handoff. The MIPv4 low latency handover was proposed to support the delay-sensitive or real-time services by minimizing the period of time during which an MN is unable to send or receive IPv4 packets due to the delay in the Mobile IPv4 registration process. It requires the MN to communicate with the new FA (nFA) while still connected to the old FA (oFA), and supports the data delivery to the MN at the nFA even before completing the formal registration process. The MIPv4 low latency handover presented three methods involving the PRE-REGISTRATION, POST-REGISTRATION and the combined registration. The PRE-REGISTRATION method allows the MN to perform the L3 handover before it completes the L2 handover. While the POST-REGISTRATION method sets up a bidirectional tunnel between oFA and nFA through the L2 triggers, and allows the MN to continue using its oFA when attached to the nFA to reduce the packet loss. As for the combined method, it performs the PRE-REGISTRATION and POST-REGISTRATION in parallel to provide a backup mechanism for PRE-REGISTRATION. Compared with the MIPv4, the low latency handover solution can reduce the handover delay. However, its performance mainly depends on the L2 triggers, which is difficult to define for different link layer technologies (Casals, 2003). Therefore, the MIPv4 fast handover was proposed which does not require L2 triggers for protocol operation, but performance will be improved by using the appropriate triggers when they are available. The MIPv4 fast handover is based on the FMIPv6, and it provides the neighboring network information for MN to determine whether it is moving to a new subnet even before the handover. However, this solution still needs to extend the backbone network to provide the context transfer.

As for IPv6, the FMIPv6 enables the MN to perform the address configuration procedure based on the link layer information when MN is still attached to the Previous Access Router (PAR), and it uses the access routers to assist the mobility management.

Figure 7 shows signaling interaction flow in FMIPv6. At first, the MN uses the specified link layer method to get the available access points information in advance. And then, the MN sends the Router Solicitation for Proxy Advertisement (RtSolPr) message to the PAR for resolving neighbor subnet routing information. The PAR, after receiving the RtSolPr message, replies with a Proxy Router Advertisement (PrRtAdv) message. The MN configures the New Care-of Address (NCoA) based on the information included in the PrRtAdv message. And then it sends the Fast Binding Update (FBU) message including the NCoA to its PAR. The PAR sends the Handover Initiate (HI) message to the New Access Router (NAR) and begins to buffer the packets once it receives the FBU message. The NAR, after receiving the HI message, performs the Duplicate Address Detection (DAD) procedure on behalf of the MN to check the validity of NCoA. After that, the NAR replies with the Handover Acknowledge (HAck) message to the PAR, and sets up a tunnel between the Previous Care-of Address (PCoA) and the NCoA. The PAR, after receiving the HAck message, sends the Fast Binding Acknowledgement (FBack) message to the MN and the NAR. The PAR begins to tunnel the packets to the PCoA during the handover. The tunnel between PCoA and NCoA will be active until the MN completes register with the HA. The NAR buffers the packets until it receives the Fast Neighbor Advertisement (FNA) message. Once the MN attaches to the NAR, it sends the FNA message at first and begins to gets the buffered packets until it finishes the binding with the HA.

The fast handover extension of MIPv4 and MIPv6 usually incorporates with the MIPv4/MIPv6 to perform the mobility management, and they need to extend the network infrastructure to provide the related information for handover.

Figure 7. Predictive and reactive fast handover

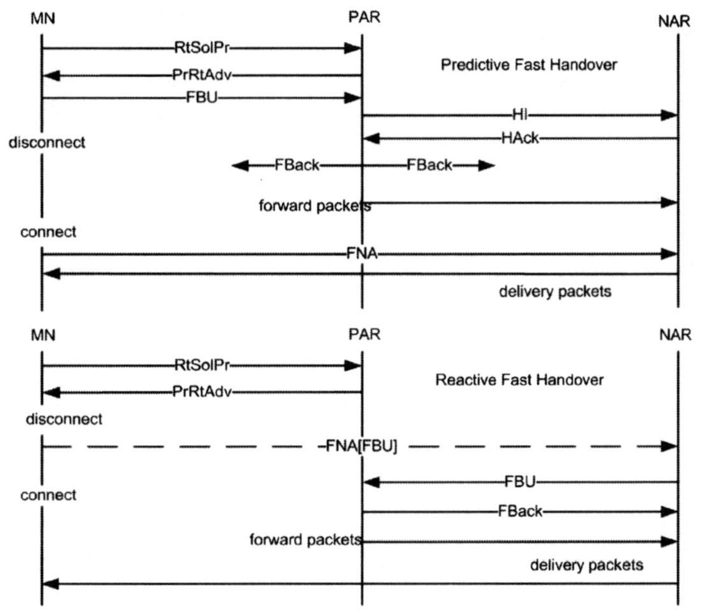

Hierarchical/Regional Enhanced Schemes

Hierarchical or regional handover is the micro mobility management solution to reduce the signaling cost introduced by the frequent movement in a small scale domain.

The MIPv4 regional registration (Fogelstroem, 2007) introduces a new network entity called the Gateway Foreign Agent (GFA) to minimize the number of signaling messages to the home network, and to reduce the signaling delay when a mobile node moves from one foreign agent to another within the same visited domain. When the MN performs the registration, it has to register with the GFA at firstly, and then the GFA will register to the HA on behalf of the MN with the GFA's address. Since the CoA registered at the HA is the GFA address, it will not change when the MN changes the FA under the same GFA. The MIPv4 regional registration will both decrease the load on the home network and speed up the process of handover within the visited domain. However, it is difficult to define the coverage of the GFA in the practice for that it needs to consider many factors such as the velocity of the MN.

Based on the MIPv6, the HMIPv6 hides the micro mobility from the HA and CN by introducing the Mobility Anchor Point (MAP) entity to reduce the signaling exchange and provide the location privacy.

Figure 8 shows the operation flow of HMIPv6. Assume that a HMIPv6-aware MN moves among the foreign networks. When the MN enters into a foreign network, it should check the MAP option after receiving the Router Advertisement (RA) messages. If the RA message does not contain the MAP options, it performs the MIPv6 handover, or else the MN has to register with at least one MAP. The MN sends the Local BU (LBU) message to the selected MAP to bind its current Local Care-of Address (LCoA) with the Regional Care-of Address (RCoA) which is an address obtained by the MN from the visited network. During the movement, the MN has to detect whether it is in the same domain. When the MN moves within the same MAP domain (intra-MAP handover), it only updates its binding with its MAP. Once the MN moves from one MAP domain to another MAP domain (inter-MAP handover), it has to send the BU message to register its new RCoA with the HA after registering the new LCoA with the new MAP. The HMIPv6 maintains two tunnels, one is MN-MAP tunnel and the other is the MN-HA tunnel. To provide the smooth inter-MAP handover, the MN may send a BU message to the previous MAP to forward the packets between previous MAP and the new MAP. The HMIPv6 improves the handover performance under the micro-mobility. However, similar to the MIPv4 regional registration, it is still in face of the MAP size selection problem.

NEMOv4 / NEMOv6

Compared with the Mobile IP which only provides the mobility support for mobile nodes, NEMO can enable mobile networks to attach to different points in the Internet and support network mobility. Here, the mobile network may be a network segment or subnet that can move and attach to arbitrary points in the routing infrastructure. NEMO extends the MIPv6 and allows session continuity for each node in the mobile network during the movement through the specific gateway called Mobile Router (MR), and it allows every node in the mobile network to be reachable while moving around. Similar to the MIPv6, NEMO maintains a bi-directional tunnel between HA and MR, and forwards the packets from or to the mobile network. The MR does not distribute the mobile network routes to the infrastructure at its attachment point.

After that, NEMOv4 (Leung, 2008) was proposed to extend the MIPv4 to support mobile networks in IPv4. The nodes in the mobile network can be fixed nodes or mobile nodes. The typical application scenario for mobile network

Figure 8. HMIPv6 operation flow

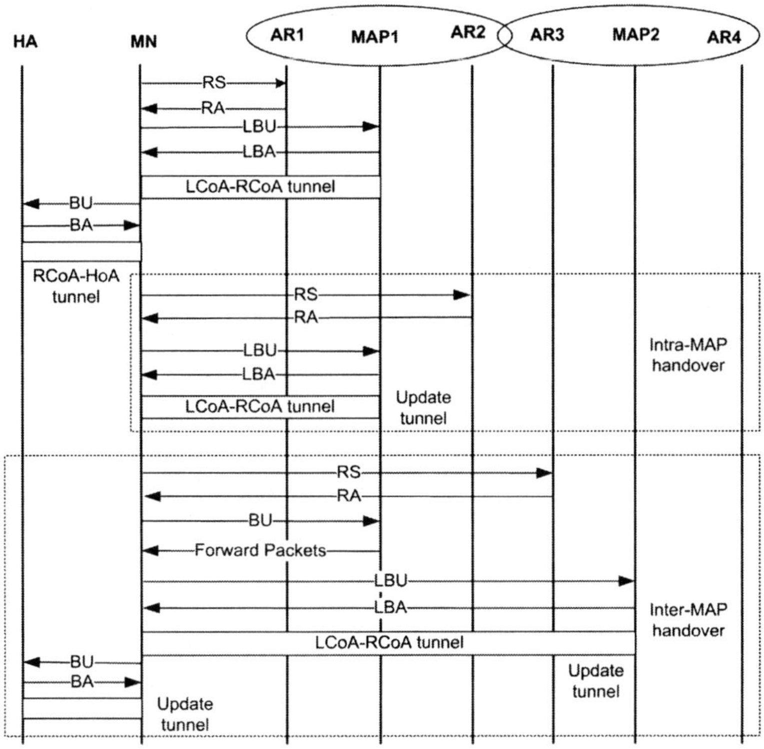

is the wireless-enabled PDA and a portable photographic camera, linked together by Bluetooth wireless link-layer technology and it is also called the Personal Area Network (PAN), in which the PDA is the mobile router.

The network mobility solution extends the MN to the mobile network which benefits in term of the cost, management and so on. However, similar to MIPv4/MIPv6, the NEMOv4/NEMOv6 still have the problems in terms of the handover delay, processing overhead and triangular routing. Therefore, several improved solutions were proposed in the past few years (Lee, 2008; Seo, 2007; Li, 2010).

MCoA

In the future wireless heterogeneous environments, mobile node or mobile router may have several network interfaces such as 802.11, 802.16, cellular radio, and attaches to multiple access networks to obtain multiple CoAs at the same time. While MIP and NEMO specifications only choose one as the primary CoA to register with its HA and CN, which cannot support the mobile node or mobile router to get the Internet access through multiple interfaces simultaneously. So, the Multiple Care-of Addresses (MCoA) (Wakikawa, 2009) was proposed to extend the MIPv6 and NEMO to use multiple CoAs, and introduces the Binding Identification (BID) to mark the binding cache entries in HA and CN.

With the help of the MCoA, MN can use various types of network interface to obtain durable and wide area network connectivity at the same time. MCoA is a multi-homing mobility solution which can provide the better performance than other mobility support solutions in term of the seamless handover. However, it introduces the additional power consumption and cost due to the use of multiple network interfaces, which is still an open issue.

DSMIPv4/v6

The repaid consumption of IPv4 address drives the deployment of the IPv6 services, and the IPv4 and IPv6 will co-exist for a long time during the migration. However, the mobility management solutions such as MIPv4 and MIPv6 only support IPv4 and IPv6, respectively. To support the mobility in the IPv4 and IPv6 co-existence scenarios, DSMIPv4 and DSMIPv6 were proposed. DSMIPv4 provides the IPv6 extensions to the MIPv4, and allows a dual stack node to use IPv4 and IPv6 HoA as well as to move between IPv4 and dual stack network infrastructures, and it separates the MIPv4 signaling from the IP version of the traffic it tunnels. However, DSMIPv4 cannot support the route optimization due to the limitation of the MIPv4. Compared with the DSMIPv4, DSMIPv6 is proposed to provide the mobility support for dual stack hosts and routers, to transport both IPv4 and IPv6 packets over the bi-direction tunnel. In the DSMIPv6, mobile node can roam over both IPv4 and IPv6 networks, and it can realize the route optimization with the help the CN. Currently, DSMIPv6 has been adopted by 3GPP to provide the dual stack support in 3GPP TS 24.303 v9.2.0.

Network-Based Mechanism

PMIPv6

The PMIPv6 uses the network to manage the mobility on behalf of the mobile nodes, and introduces the Local Mobility Anchor (LMA) and the Mobile Access Gateway (MAG) to perform the mobility signaling which makes the mobile node independent of the mobility signaling.

LMA supports the HA function of MIPv6 and extends the binding cache entry data structure which adds the new PMIPv6 flag, MN identifier, interface identifier, link local address for each interface of MN, IPv6 home prefix, bi-directional tunnel interface identifier, the access technology and 64 bits timestamp etc. The LMA supports the Per-MN-Prefix model and it assigns the unique prefix for every MN in a PMIPv6 domain. When a MN has multiple network interfaces, the LMA assigns the different prefix for every interface. MAG typically acts as a default router on the access link for mobile node and it has three functions. Firstly, the MAG detects the mobile node's movements on the access link including the attachment event and the detachment event, and it initiates the PMIPv6 signaling interworking procedure with the LMA based on these events. Secondly, the MAG advertises the mobile node's Home Network Prefix (HNP) information to emulate the home link. Thirdly, MAG sets up the data path for the MN to transmit the packets.

Figure 9 shows the signaling flow of PMIPv6. Assume that a MN attaches to the P-MAG at first, and then attaches the N-MAG. The P-MAG, after detecting the attachment event, authorizes the MN for network-based mobility management service. To do so, the P-MAG acquires the MN-ID and the policy profile which is typical configured in an AAA server. After that, P-MAG sends the Proxy BU (PBU) message to the LMA of MN. Once receiving the PBU message, the LMA checks the PBU message and allocates the home network prefix for the MN. And then, the LMA creates the binding cache entry and sets up the tunnel to the P-MAG. After that, the LMA replies with a Proxy BA (PBA) message to the P-MAG. After receiving the PBA message, the P-MAG sets up the tunnel and routing state. To emulate the mobile node's home link, the P-MAG sends the RA message with the prefix information option(s) for the MN. MN may send the Router Solicitation (RS) message at any time after the attachment and there is no strict ordering relation with the other messages. The MN configures the IP address after receiving RA messages. When the MN leaves the P-MAG, the P-MAG detects the detachment event and releases the resource. Besides, P-MAG notifies the LMA to delete the binding cache. Once the MN attaches to the N-MAG, the N-MAG will perform the same signaling flow. As long as MN moves in the same PMIPv6 domain, it can use the same address configuration.

PMIPv6 has been adopted as the IP mobility protocol for mobility between 3GPP and non-3GPP accesses and as an option for intra-3GPP access mobility in 3GPP TS402 v10.

PMIPv6 for IPv4

The PMIPv6 can only provide the mobility support for IPv6 nodes. To provide the IPv4 support for PMIPv6, PMIPv6 for IPv4 (Wakikawa, 2010) is proposed to support the MN in a Proxy Mobile IPv6 domain to operate in an IPv4-only, IPv6-only, or dual stack model. The networks between the MAG and LMA may be an IPv4 or IPv6 network. It supports the IPv4 HoA mobility in which the MN uses the IPv4 address in the PMIPv6 domain without the IPv6 address, and the IPv4 transport network in which PMIPv6 signaling message is exchanged over an IPv4 transport.

Summary

The network-layer solutions have been studied for many years, and most of them have been implemented in the test-bed. The current researches show that the host-based mechanisms are difficult to deploy due to the modification of MN which may increase the complexity, introduce the battery problem, and waste the air resource (Le, 2006). Although some improved solutions were proposed, they still cannot solve these problems. While the network-based mechanism handles the mobility management on behalf of the MN, and does not require the MN to participate in the mobility-related signaling, this makes them easy to deploy the mobility management services. Due to these attractive features, the network-based mobility management solutions have been got growing concerned, and some standard organizations have adopted them as the potential industrial mobility management solutions.

Figure 9. The signaling flow of PMIPv6

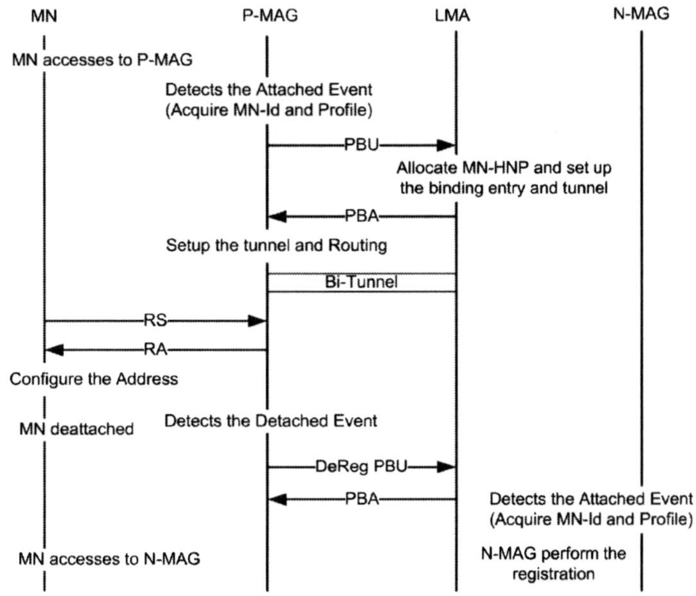

Transport Layer Solutions

The basic idea of the transport layer mobility is to maintain the end-to-end connectivity at the transport layer including the mobility support for TCP, UDP, SCTP (Stewart, 2007) and DCCP (Kohler, 2006) by using indirection, migration, tunneling, multi-homing techniques without additional infrastructure support at the network layer. The detailed survey about this domain refers to (Le, 2006; Shah, 2008).

TCP-Based Mobility Support

I-TCP

Indirect TCP (I-TCP) (Bakre, 1995) supports the mobility by introducing a gateway between the communication path of the CN and MN. I-TCP maintains a TCP connection between CN and gateway and an I-TCP connection between the gateway and MN. The TCP part remains unchanged during the lifetime of the communication and remains unaware of the mobility of MN. In the I-TCP part, when MN moves from one subnet to another, a new connection between MN and the gateway is established to replace the old one. However, I-TCP does not support IP diversity and soft handoff. Besides, the location management is not included in this scheme.

Mobile TCP

Mobile TCP (M-TCP) (Haas, 1997) is an enhanced version of I-TCP, which is implemented at MN. The gateway maintains a regular TCP connection with the CN and redirects all packets coming from CN to MN. This redirection is transparent to both the MN and CN. The enhancement of M-TCP over I-TCP is in requiring less complexity in the wireless part of the connection. Similar to I-TCP, M-TCP does not support IP diversity or location management but ensures transparency to applications.

MSOCKS

Maltz et al. (1998) proposed the MSOCKS to support the connection migration and multiple IP addresses for multiple interfaces by introducing the TCP splicing which is used to split a TCP connection at a proxy by dividing the host-to-host communication into host-proxy and proxy-host communications. In MSOCKS, when a MN leaves the previous subnet, it obtains a new IP address from the new subnet using DHCP, and establishes a new connection with the proxy using its second interface. The communication between proxy and CN remains unchanged. The data flow between MN and CN thus continues, with the CN being unaware of the mobility. Location management is done through the proxy that is always aware of the location of the MN. However, the mobility is limited due to coverage of the proxy. Besides, it requires the MN to equip multiple interfaces which increases the deployment cost and battery consumption.

TCP Migrate

Snoeren et al. (2000) proposed an end-to-end architecture (called TCP Migrate or E2E) for Internet host mobility using dynamic update to the DNS to track the MN. The TCP Migrate consists of three components including the addressing, mobile host location and connection migration. In TCP Migrate, when a MN initiates a connection with a CN, the end nodes (MN and CN) exchange a token to identify the particular connection to negotiate a change in endpoint IP address without the need for a third party. However, it depends on the DNS for location management.

Freeze-TCP

Goff et al. proposed the Freeze-TCP (Goff, 2000) which lets the MN freeze or stop an existing TCP connection during the handover by advertising a zero window size to the CN, and unfreezes the connection after handover. However, this scheme reduces the packet loss at the cost of introducing a higher handover delay. Besides, it does not provide the handover management and location management.

The above solutions take mobility as an end-to-end service without the requirement on the network layer infrastructures. However, they are not aimed at reducing the high latency and packet loss resulting from handovers, and the handover latency for them is in the scale of seconds.

UDP-Based Mobility Support

Mobile UDP (M-UDP) (Brown, 1996) is an implementation of UDP with mobility support similar to I-TCP and M-TCP for cellular networks. M-UDP is a three levels hierarchy which consists of the cell, MSS and Supervisor Host (SH). M-UDP chooses a gateway from the SHs to split the connections between MN and CN to ensure one unbroken gateway to CN connection. In the handover, MN uses any free bandwidth to retransmit lost packets to minimize the packet loss. The experimental results show that the M-UDP is suitable for the mobile environment. However, M-UDP is only designed for the cellular networks, and it does not include IP diversity or location management.

SCTP-Based Mobility Support

The SCTP is a reliable transport protocol operating on top of a connectionless packet network, and it supports the multi-homing which allows the mobile node using multiple source addresses and destination addresses for a single association between two SCTP endpoints. Multiple paths are distinguished at each endpoint by their destination addresses. Among all available paths one is selected as the primary path, whereas the rest are considered as backup or alternate paths. The multi-homing feature of SCTP can provide the mobility support without support of network routers or special agents. Currently, SCTP mobility is the core of the transport layer mobility solutions (Budzisz, 2008).

mSCTP

Mobile SCTP (mSCTP) (Koh, 2004) was proposed to provide the soft handover with the help of the ADDIP extension which can enable an SCTP endpoint to add a new IP address or delete an unnecessary IP address, and also to change the primary IP address. The ADDIP extension can notify the remote node by send an SCTP Address Configuration Change (ASCONF) chunk about the address change events which include the ADD, DELETE and CHANGE. Figure 10 shows the mSCTP handover procedure. Assume that a MN communicates with a CN. At first, the MN attaches to the AR1 and obtains MN_IP1 as its primary address, and set up association with the CN. When the MN attaches to the AR2 and obtains a new IP address MN_IP2, it adds the MN_IP2 to the SCTP association, changes the primary IP address, and deletes the MN_IP1. After that, the MN and CN use the MN_IP2 and CN_IP to communicate.

In mSCTP data packets are sent to the old IP address (MN_IP1) before the MN sets the new IP address to the primary IP address of the association. Once the old IP is not reachable, it will be sent to the new IP address. The mSCTP can maintain application transparency and provide the mobility support. However, it does not support location management, and the retransmissions during the handover will result in increased delays and degrade performance of services.

cSCTP

Compared with the mSCTP, cellular SCTP (cSCTP) (Aydin, 2003) is used to implement soft handoffs and provides better performance. During handover, the CN sends packets to the primary address of the old network and the primary address of the new network in parallel, and transfers the same data via two different paths, to reduce packet loss. Besides, to provide the location management, cSCTP uses the SIP to locate the location of the MN.

Figure 10. The mSCTP handover procedure

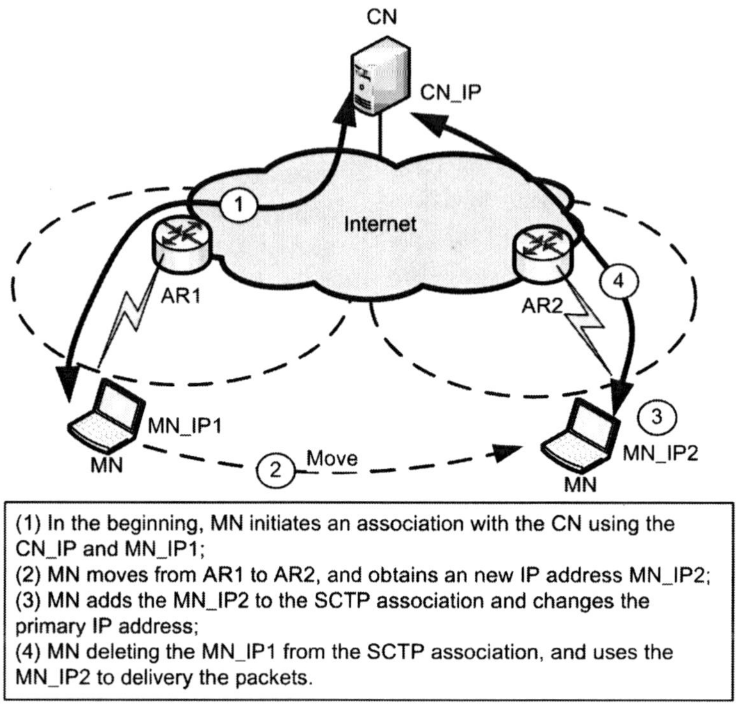

Although cSCTP can use two primary addresses in parallel to duplicate the packet transmissions to provide soft handover, it introduces the additional delivery cost, which wastes the bandwidth resource.

SIGMA

Considering that the future MN will be equipped with multiple interfaces, Fu et al. (2005) proposed the Seamless IP diversity based Generalized Mobility Architecture (SIGMA) which utilizes IP diversity to provide the seamless handover. SIGMA takes the advantage of overlay networks and changes its primary address after obtaining a new address from the new subnet. When the link quality of the old subnet goes below a certain threshold, the MN changes its primary address to the new one. Once the MN leaves the old subnet, it deletes the old address and communicates with the new address. Besides, to support the location management, SIGMA introduces the Location Manager (LM) which is typical a DNS server.

The SIGMA can realize the low latency and low packet loss mobility service, but it requires the MN equipped with multiple interfaces, which may increase the battery consumption.

CMT

Although Concurrent Multipath Transfer (CMT) which allows binding of one association to multiple IP addresses at each end of the association is used for the load balance over wired links, it is attractive for the mobility management. The temporary availability of more than one link to the destination can provide the seamless handover. Budzisz et al. (2008) after investigating several SCTP-based mobility solutions concluded that the CMT is the future research direction of the SCTP-based mobility management.

DCCP-Based Mobility Support

The Datagram Congestion Control Protocol (DCCP) was firstly introduced by Kohler et al. (Kohler, 2006) in 2001, and aimed to fulfill the gap between TCP and UDP for multimedia application requirements. The DCCP is a message-oriented transport layer protocol, which combines the connection-oriented and congestion-controlled concepts of the TCP and the unreliable data transmission feature of the UDP. DCCP is intended for applications such as streaming media that can benefit from control over the tradeoffs between delay and reliable in-order delivery. Sales et al. (2008) compared the performance of TCP, UDP and DCCP, and the results showed that DCCP and TCP can share the network bandwidth without affecting each other, while UDP can aggressively degrade TCP and DCCP flows due to the lack of the flow control mechanism.

In term of the mobility management, DCCP can provide the mobility support and the multi-homing support by introducing the DCCP-Move packet type and two new DCCP features, the mobility capable and the mobility ID (Le, 2006). Currently, the mobility management of DCCP is little considered for that DCCP is mainly aimed to provide the congestion control and congestion control negotiation for the multimedia applications.

Summary

Compared with the network layer solutions, the transport layer solutions have the following benefits. Firstly, they do not need the additional function entities such as HA and FA beside the DHCP and DNS. Secondly, they provide the inherent optimal delivery path. Thirdly, they can provide the seamless handover without knowing the lower network status. Finally, they can provide the location privacy. However, these solutions are dependent on other layers for location management. So, they have to combine with the other layer to provide the complete mobility support.

Application Layer Solutions

SIP-Based Approaches

The SIP (Rosenberg, 2002) was designed to transmit the application layer multimedia signaling. However, it demonstrates potential capabilities for mobility support. SIP can not only realize the session mobility which transfers the media of an ongoing communication session from one device to another (Shacham, 2009), but also provide the terminal mobility, personal mobility and service mobility. Figure 11 shows the basic SIP mobility operation flow.

The mainly components in SIP are the user agent, redirect server and the proxy server. The redirect server and the proxy server can provide the location management function. The basic idea of SIP mobility can be summarized as follows. The SIP server in the home network has MN's current location information, and it can redirect the message such as INVITE to the MN after receiving the message from the CN. After that, CN and MN can perform the SIP signaling procedure to establish a session. Once the MN attaches to a new network and obtains a new IP address, it will send a RE-INVITE message with the updated session description to redirect the packet to its new location. The CN, after receiving the RE-INVITE message will send the packets to the MN directly or through the tunnel based on the different transport layer protocols. Finally, the MN sends the REGISTER message to the home SIP server to update its location.

The SIP-based approaches do not depend on components such as HA and it can be deployed by any third party application service providers without much dependence on the underlying network elements. However, these solutions usually introduce a large handover delay for the signaling procedure and overhead of the IP encapsulation (Wedlund, 1999). So, some improved schemes were proposed, such as SAMP (Pack, 2006) and MMUSE (Salsano, 2008) to reduce the handover delay and packet loss.

Figure 11. The SIP mobility operation flow

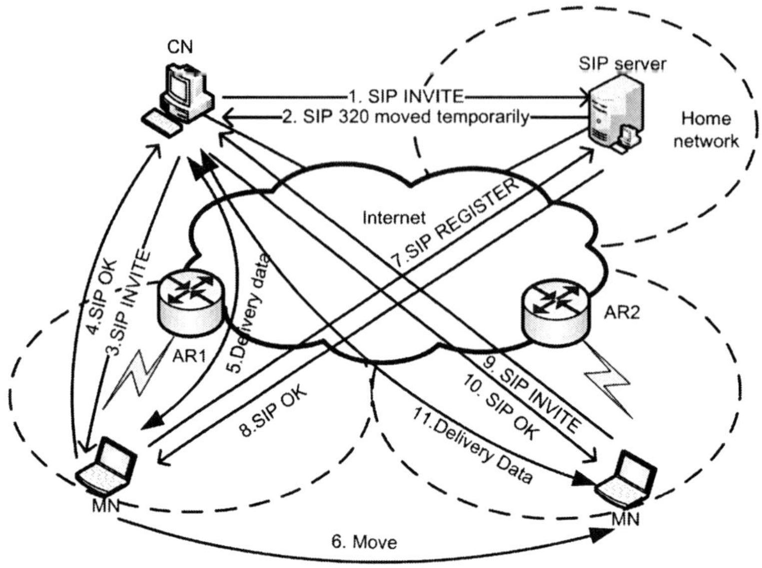

DNS-Based Approaches

The traditional DNS is only used for the fixed network, which cannot provide the mobility support. Vixie et al. (1997) proposed a method called DDNS for dynamic updating of Resource Records (RRs) or Resource Records sets (RR sets) from a specified zone by specifying the UPDATE messages. Since most applications ubiquitously resolve Fully Qualified Domain Name (FQDN) to an IP address at the beginning of communication, DDNS can be considered for location management in the mobile environment where the MN acts as a server and other nodes actively originate communication with the MN. To locate the MN as it moves to a new network, the MN dynamically registers and updates its FQDN-to-IP entry with the new IP address to DNS servers by sending DNS UPDATE messages. Then whenever the CN wants to communicate with the MN, it will query the DNS sever with the FQDN of the MN, and the DNS sever responds with the current IP address of the MN. Finally, the CN can initiate and establish communication with the MN directly. However, update delay is too long to support the real-time service. Recently, Yahya et al. (2008) based on the DNS and its ability to support dynamic updates proposed a mobility solution that locates and tracks mobile host while they change their attachment point to support mobility and real time services.

The DNS-based approaches based on the existed infrastructure, does not require the special servers. However, the DNS registration delay is too large. The future work should optimize the registration delay.

Summary

The application layer solutions are easily deployed for that they do not require modifying the lower layer protocol stacks, and they can support the heterogeneous handover. However, their handover delay is too large to support the real-time applications due to the cumulative handover delay of the lower layers. So, the future direction of the application should combine the lower layer mobility management solutions to improve their handover performance.

New Layer / Architecture Solutions

The mobility issue of the IP network is mainly due to the drawback of the traditional network architecture which is designed for the fixed topology. Several proposals about the new network architecture are proposed recently. The NewArch project (Clark, 2003) supported by Defense Advanced Research Projects Agency (DARPA) shows that the success of the current Internet is dependent on the improvement and extension. However, these small changes of the Internet only fix the specific short-term problems which will accumulate and the inconsistencies and tensions become more pronounced over time, further progress will become difficult, and eventually impossible. So, the NewArch project suggests that we should redesign the Internet. Afterward, USA national science foundation sponsored the Global Environment for Network Innovations (GENI) project to design a new Internet, which provides a security, cell phone, wireless and sensor network of pervasive computing environment, and sets up a new next generation Internet. Besides, the Future Internet Design (FIND) project was launched to solve the problems of that what the requirement in the 15 years and how to design the future Internet. In 2004 EU FP-6 invested 30 million Euros to support the new Internet research. After that, EU FP-7 proposed the Future Internet Research and Experimentation (FIRE) project to study the new Internet technologies through the large-scale experimentations, and has bounded 40 million Euros to support the proto-type system development and long-term research. At the same time, other countries also launched some similar projects such as the G-lab in German and RNRT in French. Besides, AsisFI project was launched in 2008 to promote the cooperation of new Internet research between Asia and the other countries. China also began the new network research, and supported a 973 project in 2007 which concerns on the research of universal network and pervasive service (Zhang, 2007). Based on the previous research, IRTF Routing Research Group (RRG) workgroup was set up to explore routing and addressing problems that are important to the development of the Internet. With evolvement of the Internet, the Internet routing and addressing architecture is facing challenges in scalability, mobility, multi-homing, and inter-domain traffic engineering. Thus the RRG proposes to focus its effort on designing an alternate architecture to meet these challenges, and several schemes were proposed in the past few years (Zhang, 2010).

LIN6

Location Independent Networks for IPv6 (LIN6) (Teraoka, 2003) supports multi-homing and mobility in IPv6 by introducing the 64-bit node ID called LIN6 ID in network layer and the 128-bit node ID called LIN6 generalized ID in transport layer. LIN6 generalized ID is formed by concatenating the LIN6 prefix before LIN6 ID. Mobile node can be identified by its LIN6 ID no matter where it attaches and no matter how many interfaces it has. LIN6 is similar to MIPv4/v6, but it has several advantages. Firstly, LIN6 does not have header overhead because it does not use any extension headers of IPv6 while Mobile IPv6 uses the Destination Options Header for the Home Address Option and the Routing Header for optimal routing. Secondly, LIN6 is more fault tolerant than MIPv6 in which HA cannot be replicated to the subnet other than the home link of the mobile node. LIN6 introduces the mapping agent which can be replicated anywhere in the Internet. Thirdly, LIN6 keeps end-to-end communication model, that is, LIN6 does not use any packet interceptor/forwarder such as the HA of MIPv6. So, there is no tunneling in LIN6. However, the modification of transport layer protocols and network layer protocols is required by the implementation of LIN6.

MAST

Croker (2003) proposed the Multiple Address Service for Transport (MAST) to support the mobility and multi-homing by adding a new layer between the transport layer and the network layer without introducing a new namespace. MAST uses the initial IP address for the transport layer connection or association as the identifier of the MN, and uses the other IP address as the locator of the MN. MAST uses the DNS to provide information of dynamic presence service relating to the MN through the DNS SRV (Gulbrandsen, 2000) record. MAST does not introduce any new namespace or addressing structure, and requires no change to IP modules or transport modules. Besides, it can reduce the additional processing overhead. However, it is still in face of many problems such as optimal locator selection in the deployment.

HIP

Host Identity Protocol (HIP) (Moskowitz, 2008) introduces a new host identity protocol layer between the network layer and transport layer to separate the location from the identifier. It introduces a new host identity namespace called the Host Identifier (HI) which is a public key, to set up the transport connection, and remove the IP from the transport layer to support the mobility. The HI is dynamic, mapping to one or more IP address in the HIP layer. In practice, HIP uses a host identifier tag which is the hash value of the HI. The location management of HIP is provided by the Rendezvous Server (RVS) to set up the binding between HI and IP address. With the help of the HIP layer, the change of the IP address is transparency to the upper applications. The HIP needs make a big modification for the traditional TCP/IP stack.

LISP

Locator/ID Separation Protocol (LISP) (Farinacci, 2010) is an IP-IP tunnel protocol to separate the identifier information and location information of the host by introducing the Endpoint Identifier (EID) and Routing Locator (RLOC) address spaces. EID can be a global address or an AS-domain address, while the RLOC is a global routable address of the tunnel router. Each EID has at least one related RLOC. As for the packets sent to the Ingress Tunnel Router (ITR), it will add a LISP header for all of them, while as for any arrived packets the Egress Tunnel Router (ETR) removes the LISP header.

The basic LISP architecture does not support mobility, and the mobility extension LISP Mobile Node (LISPMN) (Farinacci, 2010) in the draft-meyer-lisp-mn-03 was proposed, which enables the MN to roam among LISP and no-LISP networks while being reachable with the same address. LISPMN is a lightweight version of LISP's ITR/ETR functionality which can provide seamless mobility. Each mobile node is assigned a static EID, and a preconfigured Map Server (Fuller, 2010). When a MN roams into a network, after obtaining the RLOC, it updates its RLOC set in the Map Server, and clears the cached RLOC in the ITR of the CNs. Thus the CN can always learn the up-to-date location of the mobile node by the resolution of the mobile node's EID. However, the LISPMN does not support networks that use NAT. So, the NAT traversal for LISPMN (Klein, 2010) was proposed to make LISP mobile node behind a NAT globally reachable. Currently, the LISP is still under developing, and many issues need the further research.

UNMM

The Universal Network Mobility Management (UNMM) is based on the universal network architecture (Zhang, 2007) which is different from the traditional TCP/IP architecture. It takes the network, service and users as a whole, and divides the network into two parts, the infrastructure layer and pervasive service layer, which are shown in Figure 12.

The infrastructure layer consists of the access network and core network by introducing the Accessing Identifier (AID) and Routing Identifier (RID) namespaces, respectively. AID and RID separate the identifier information and location information by using the split-mapping mechanism to provide a universal network communication platform for data, voice, video services. The pervasive service layer identifies all the services, network resource and users with a unique identifier to support various the session, control and management. To set up the mappings between AID and RID, the universal network designs an identifier and location separation mechanism in the infrastructure layer which can support the scalable routing, security, mobility and multihoming. The universal network has already been implemented in the test-bed, which can support several typical applications.

Figure 13 shows the general communication flow, which involves the Mobile Node (MN), Fixed Node (FN), Access Switch Router (ASR), General Switch Router (GSR), Identifier Mapping Server (IDMS) and Authentication Center (AuC). ASR is the border router which connects to the access network to provide the access service. GSR locates in the core network to provide the routing service based on the RID.

Suppose that a FN communicates with a MN, and the communication procedure is show as follows.

1. Firstly, the FN uses the AID to encapsulate the packets, and sends to the ASR;
2. ASR1 after receiving the packets from the FN, initiates the authentication procedure for the FN;
3. IF authentication is successful, ASR1 will query the AID and RID mapping relation from the IDMS, or else, ASR1 will drop these packets;
4. ASR1, after acquiring the mapping relation, will de-encapsulate the packets, and encapsulates the packets with the RID, and forwards to GSR;
5. MN attaches to the access network, and sets up the RID and AID mapping relation in IDMS, and caches this relation in ASR2;

Figure 12. Two layers model of universal network

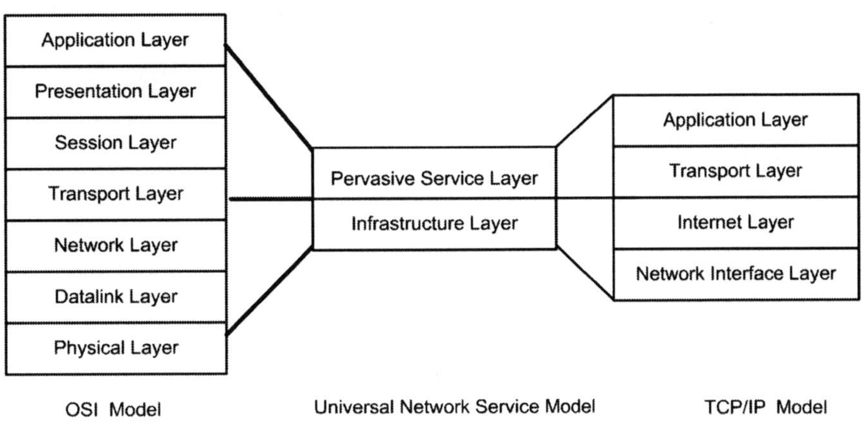

Figure 13. The UNMM general communication flow

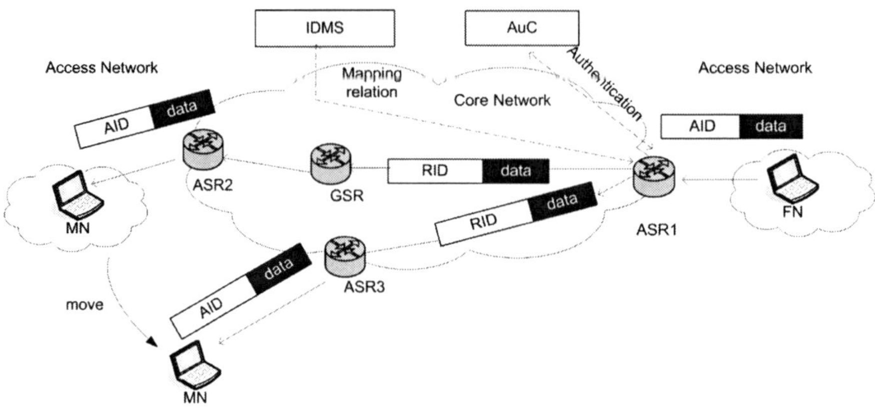

6. ASR2, after receiving the packets, de-encapsulate them and encapsulate them with the AID and forwards to the MN;
7. MN receives the packets and communicates with the FN;
8. Once the MN moves into another access network, the RID will be changed in the IDMS, while the AID is unchanged. So the upper layer session will not be disrupted.

During the communication, the header of the packets is changed in the board of the access network and core network, which can prevent the core network from the attack of the access network.

Summary

These solutions support mobility management at the beginning of the designing the network architecture, so the mobility management is inherent and easily to realize. However, they have to modify or create a new network architecture which may results in the interoperation and compatibility problems with the current network. Even so, these solutions still have many attractive features, and their future direction is to use the identifier and location split mechanism to provide the mobility management.

CONCLUSION

In this chapter, we introduce the mobility challenges and management solutions in the future wireless heterogeneous networks, and give a brief introduction of the current mobility management solutions. The above discussion investigates the development of mobility management in the past few years. From the above research, we can conclude that the mobility management is derived from the drawback of the traditional networks which are designed for the static and wired networks, and it as a key technique in the future wireless heterogeneous networks faces many challenges and its development is from the cellular network to Internet, from single layer to multiple layers, from one technology to multi-technology. From the comparison of the different kinds of mobility management solutions, we can conclude that in the short term, the network-based network layer solutions such as PMIPv6 are more attractive for that they do not modify the MN and easily deployed in the commercial networks, while in the long term, the mobility management should be set up in a new network architecture to solve the current problems. When develop a mobility management solution, the performance metrics also deserve special attention which includes the handover delay, packet loss, signaling overhead and throughput. The future research directions of mobility management include the new network architecture (identifier and locator separation), interoperation among different networks, technologies and protocols, power-effective (lightweight).

ACKNOWLEDGMENT

This work is supported in part by the National Natural Science Foundation of China (NSFC) under Grant No. 61003283, 60870015, 61001122, 60833022, and the Beijing Natural Science Foundation of China under Grant No. 4102064, and in part by the National Basic Research Program of China ("973 Program") under Grant No. 2007CB307100, and in part by the project under the Grant No. 2011ZX03002-005-03.

REFERENCES

Akyildiz, I. F., McNair, J., Ho, J. S. M., Uzunalioglu, H., & Wang, W. (1999). Mobility management in next-generation wireless systems. *Proceedings of the IEEE*, *87*(8), 1347–1384. doi:10.1109/5.775420

Akyildiz, I. F., Xie, J., & Mohanty, S. (2004). A survey of mobility management in next-generation all-IP-based wireless systems. *IEEE Wireless Communications*, *11*(4), 16–28. doi:10.1109/MWC.2004.1325888

Aydin, I., Seok, W., & Shen, C. C. (2003). Cellular SCTP: A transport-layer approach to internet mobility. In the *Proceeding of The ICCCN 2003* (285-290). Dallas, TX: IEEE Press.

Bakre, A., & Badrinath, B. R. (1995). I-TCP: Indirect TCP for mobile hosts. In the *Proceeding of IEEE International Conference on Distributed Computing Systems* (136-143). Vancouver, Canada: IEEE Press.

Banerjee, N., Wu, W., & Das, S. K. (2003). Mobility support in wireless Internet. *IEEE Wireless Communications*, *10*(5), 54–61. doi:10.1109/MWC.2003.1241101

Bhagwat, P., & Perkins, C. (1993). A mobile networking system based on internet protocol (IP). In *Proceeding of USENIX Symposium on Mobile and Location Independent Computing* (69-82). Cambridge, MA, USA.

Bhagwat, P., Perkins, C., & Tripathi, S. (1996). Network layer mobility: An architecture and survey. *IEEE Personal Communications*, *3*(3), 54–64. doi:10.1109/98.511765

Bolton, W., Xiao, Y., & Guizani, M. (2007). IEEE 802.20: Mobile broadband wireless access. *IEEE Wireless Communications*, *14*(1), 84–95. doi:10.1109/MWC.2007.314554

Brown, K., & Singh, S. (1996). M-UDP: UDP for mobile cellular networks. *Computer Communication Review*, *26*(5), 60–78. doi:10.1145/242896.242901

Budzisz, L., Ferrus, R., & Brunstrom, A. (2008). Towards transport-layer mobility: Evolution of SCTP multihoming. *Computer Communications*, *31*(5), 980–998. doi:10.1016/j.comcom.2007.12.014

Casals, O., Cerda, L. I., Willems, G., Blondia, C., & Van den Wijngaert, N. (2003). Performance evaluation of the post-registration method, a low latency handoff in MIPv4. In the *Proceeding of the ICC 2003*: Vol. 1 (522-526). Anchorage, AK, USA: IEEE Press.

Chiussi, F. M., Khotimsky, D. A., & Krishnan, S. (2002). Mobility management in third-generation all-IP networks. *IEEE Communications Magazine*, *40*(9), 124–135. doi:10.1109/MCOM.2002.1031839

Clancy, T. C. (2008). Secure handover in enterprise WLANs: CAPWAP, HOKEY, and IEEE 802.11r. *IEEE Wireless Communications*, *15*(5), 80–85. doi:10.1109/MWC.2008.4653136

Clark, D., Sollins, K., Wroclawski, J., Katabi, D., Kulik, J., & Yang, X. (2003). *New arch: Future generation internet architecture*. Technical Report. Retrieved July, 2010, from http://www.isi.edu/newarch/iDOCS/final.finalreport.pdf

Croker, D. (2003). *Multiple address service for transport (MAST): An extended proposal*. draft-crocker-mast-proposal-01. IETF draft (expired).

Devarapalli, V., Wakikawa, R., Petrescu, A., Thubert, P. (2005). *Network mobility (NEMO) basic support protocol*. IETF RFC 3963.

Eddy, W. M. (2004). At what layer does mobility belong? *IEEE Communications Magazine*, *42*(10), 155–159. doi:10.1109/MCOM.2004.1341274

Emmelmann, M., Wiethoelter, S., Koepsel, A., Kappler, C., & Wolisz, A. (2007). Moving towards seamless mobility: State of art and emerging aspects in standardization bodies. *Wireless Personal Communications*, *43*(3), 803–816. doi:10.1007/s11277-007-9255-6

Farinacci, D., Fuller, V., Lewis, D., Meyer, D. (2010). *LISP mobile node*. Draft-Meyer-LISP-MN-03, IETF draft (work in progress).

Farinacci, D., Fuller, V., Meyer, D., & Lewis, D. (2010). *Locator/ID separation protocol (LISP)*. Draft IETF-LISP-08, IETF draft (work in progress).

Fogelstroem, E., Jonsson, E., & Perkins, C. (2007). *Mobile IPv4 regional registration*. IETF RFC 4857.

Fu, S., Atiquzzaman, M., & Lee, Y. J. (2005). Architecture and performance of SIGMA: A seamless mobility architecture for data networks. In the *Proceeding of the ICC 2005*: Vol. 5 (3249- 3253). Seoul, Korea: IEEE Press.

Fuller, V. (2010). *LISP map server*. Draft-IETF-LISP-MS-05, IETF draft (work in progress).

Goff, T., Moronski, J., Phatak, D. S., & Gupta, V. (2000). Freeze-TCP: A true end-to-end TCP enhancement mechanism for mobile environments. In the *Proceeding of IEEE International Conference on Computer Communications* (pp. 1537-1545). Tel Aviv, Israel: IEEE Press.

Gulbrandsen, A., Vixie, P., & Esibov, L. (2000). *A DNS RR for specifying the location of services* (DNS SRV). IETF RFC 2782.

Gundavelli, S., et al. (2008). *Proxy mobile IPv6*. IETF RFC 5213.

Gupta, V. (2009). *IEEE 802.21-2008 IEEE standard for media independent handover services*. IEEE Std. 802.21-2008, January 2009.

Haas, Z. J., & Agrawal, P. (1997). Mobile-TCP: An asymmetric transport protocol design for mobile systems. In *Proceeding of IEEE International Conference on Communications* (pp. 1054-1058), Montreal, Canada: IEEE Press.

Han, Y. H., Jang, H., Choi, J., Park, B., & McNair, J. (2007). A cross-layering design for IPv6 fast handover support in an IEEE 802.16e wireless MAN. *IEEE Network*, *21*(6), 54–62. doi:10.1109/MNET.2007.4395111

Huang, P. J., Tseng, Y. C., & Tsai, K. C. (2006). A fast handoff mechanism for IEEE 802.11 and IAPP networks. In the *Proceeding of 63rd IEEE Vehicular Technology Conference*: Vol. 2 (pp. 966-970). Grand Hyatt Melbourne, Melbourne, Australia: IEEE Press.

Ioannidis, J., Duchamp, D., & Maguire, G. (1991). IP-based protocols for mobile internetworking. In *ACM Proceeding of the Conference on Communications Architecture & Protocols* (pp. 235-245). Zurich, Switzerland: ACM Press.

Johnson, D., Perkins, C., & Arkko, J. (2004). *Mobility support in IPv6*. IETF RFC 3775.

Kerry, S. J. et al. (2003). *IEEE trial-use recommended practice for multi-vendor access point interoperability via an inter-access point protocol across distribution systems supporting IEEE 802.11 operation*. IEEE Std. 802.11F-2003. July, 2003.

Kerry, S. J., et al. (2008). *IEEE standard for information technology—Telecommunications and information exchange between systems local and metropolitan area networks specific requirements—part 11: wireless LAN medium access control (MAC) and physical layer (PHY) Spec., amendment 2: fast basic service set (BSS) transition*. IEEE Std. 802.11r-2008. July 2008.

Klein, D., Hartmann, M., & Menth, M. (2010). *NAT traversal for LISP mobile node*. Draft-klein-lisp-mn-nat-traversal-00, IETF draft (work in progress).

Klerer, S. M., et al. (2009). *IEEE 802.20-2008 IEEE standard for local and metropolitan area networks – Part 20: Air interface for mobile broadband wireless access systems supporting vehicular mobility – Physical and media access control layer specification*. IEEE Std. 802.20-2009, 2009.

Koh, S., J., Chang, M., J., & Lee, M. (2004). mSCTP for soft handover in transport layer. *IEEE Communications Letters, 8*(3), 189-191.

Kohler, E., Handley, M., & Floyd, S. (2006). *Datagram congestion control protocol* (DCCP). IETF RFC 4340. (. *Updated by, RFC5595*, RFC5596.

Koodli, R. (2005). *Fast handovers for mobile IPv6*. IETF RFC 4068 (Obsoleted by RFC5268).

Koodli, R., & Perkins, C. (2007). *Mobile IPv4 fast handovers*. IETF RFC 4988.

Kyamakya, K., & Jobmann, K. (2005). Location management in cellular networks: Classification of the most important paradigms, realistic Simulation framework, and relative performance analysis. *IEEE Transactions on Vehicular Technology, 54*(2), 687–708. doi:10.1109/TVT.2004.842459

Le, D., Fu, X. M., & Hogrefe, D. (2006). A review of mobility support paradigms for the Internet. *IEEE Communications Surveys & Tutorials, 8*(1), 38–51. doi:10.1109/COMST.2006.323441

Lee, C. W., Sun, Y., & Chen, M. C. (2008). HiMIP-NEMO: Combining cross-layer network mobility management and resource allocation for fast QoS-handovers. In the *Proceeding of VTC Spring 2008* (282-286). Marina Bay, Singapore: IEEE Press.

Lee, L. S., & Wang, K. (2010). Design and analysis of a network-assisted fast handover scheme for IEEE 802.16e networks. *IEEE Transactions on Vehicular Technology, 59*(2), 869–883. doi:10.1109/TVT.2009.2037508

Leung, K., Dommety, G., Narayanan, V., & Petrescu. A. (2008). *Network mobility (NEMO) extensions for mobile IPv4*. IETF RFC 5177.

Li, L. S., Tsai, Y. H., & Lin, P. C. (2010). An efficient fast handoff scheme in hierarchical mobile IPv6 for NEMO. In the *Proceeding of 2010 International Symposium on Communication Control and Automation (3CA)*: Vol. 2 (pp. 90-94). Tainan, China: IEEE Press.

Malki, K. E. (2007). *Low-latency handoffs in Mobile IPv4*. IETF RFC 4881.

Maltz, D. A., & Bhagwat, P. (1998). MSOCKS: An architecture for transport layer mobility. In the *Proceeding of IEEE International Conference on Computer Communications*: Vol. 3 (1037-1045). San Francisco, CA: IEEE Press.

Marks, R. B. (2006). *IEEE standard for local and metropolitan area networks—Part 16: Air interface for fixed and mobile broadband wireless access systems*. IEEE Std. 802.16e-2005, Feb. 2006.

Mishra, A., Shin, M., & Arbaugh, W. (2003). An empirical analysis of the IEEE 802.11 MAC layer handoff process. *ACM SIGCOMM Computer Communications Review*, *33*(2), 93–102. doi:10.1145/956981.956990

Moskowitz, R., Nikander, P., Jokela, P., & Henderson, T. (2008). *Host identity protocol*. IETF RFC 5201.

Pack, S., Park, K., Kwon, T., & Choi, Y. (2006). SAMP: Scalable application-layer mobility protocol. *IEEE Communications Magazine*, *44*(6), 86–92. doi:10.1109/MCOM.2006.1668425

Perkins, C. (Ed.). (2002). *IP mobility support for IPv4*. IETF RFC 3344.

Ray, S. K., Pawlikowski, P., & Sirisena, H. (2010). Handover in mobile WiMax networks: The state of art and research issues. *IEEE Communications Surveys & Tutorials*, *12*(3), 376–399. doi:10.1109/SURV.2010.032210.00064

Rosenberg J., et al. (2002). *SIP: Session initiation protocol*. IETF RFC 3261.

Sales, L. M., Almeida, H. O., Perkusich, A., & Sales, M. (2008). *An experimental evaluation of DCCP transport protocol: A focus on the fairness and hand-off over 802.11g networks. In the Proceeding of Consumer Communications and Networking Conference 2008* (pp. 1149–1153). Las Vegas, NV: IEEE Press.

Salsano, S., Polidoro, A., Mingardi, C., Niccolini, S., & Veltri, L. (2008). SIP-based mobility management in next generation networks. *IEEE Wireless Communications*, *15*(2), 92–99. doi:10.1109/MWC.2008.4492982

Seo, J. K., Nam, S. H., & Lee, K. G. (2007). Fast route optimization for dynamic nested NEMO. In the *Proceeding of the ICPPW 2007* (pp. 49-53). Xi'an, China: IEEE Press.

Shacham, R., Schulzrinne, H., Thakolsri, S., & Kellerer, W. (2009). *Session initiation protocol (SIP) session mobility*. IETF RFC 5631.

Shah, P. A., & Yousaf, M. (2008). End-to-end mobility management solutions for TCP: An analysis. In the *Proceeding of the International Conference on Computer and Information Technology* (pp. 696-701). Venice, Italy: IEEE Press.

Siddiqui, F., & Zeadally, S. (2006). Mobility management across hybrid wireless networks: Trends and challenges. *Computer Communications*, *29*(9), 1363–1385. doi:10.1016/j.comcom.2005.09.003

Snoeren, A. C., & Balakrishnan, H. (2000). An end-to-end approach to host mobility. In the *Proceedings of the 6th Annual International Conference on Mobile Computing and Networking* (pp.155-166). Boston, MA: ACM Press.

Soliman, H. (Ed.). (2009). *Mobile IPv6 support for dual stack hosts and routers*. IETF RFC 5555.

Soliman, H., Castelluccia, C., Malki, K. E., Bellier, L. (2005). *Hierarchical mobile IPv6 mobility management* (HMIPv6). IETF RFC 4140 (Obsoleted by RFC5380).

Stewart, R. (Ed.). (2007). *Stream control transmission protocol*. IETF RFC 4960.

Strang, T. (2005). *Ubiquitous computing mobility*. Courseware of University of Innsbruck. Retrieved September, 2010, from http://www.stiinnsbruck.at/ fileadmin/ documents/ teaching_archive/ ubicomp05/09-mobility_6.pdf

Tabassam, A. A., Trsek, H., Heiss, S., & Jasperneite, J. (2009). Fast and seamless handover for secure mobile industrial applications with 802.11r. In the *Proceeding of the LCN 2009* (pp. 750-757). Zurich, Switzerland: IEEE Press.

Taniuchi, K., Ohba, Y., & Fajardo, V. (2009). IEEE 802.21: Media independent handover: Features, applicability, and realization. *IEEE Communications Magazine*, *47*(1), 112–120. doi:10.1109/MCOM.2009.4752687

Teraoka, F., Ishiyama, M., & Kunishi, M. (2003). *LIN6: A solution to multihoming and mobility in IPv6, draft-teraoka-multi6-lin6-00*. IETF draft (expired).

Teraoka, F., Yokote, Y., & Tokoro, M. (1991). A network architecture providing host migration transparency. In *ACM Proceeding of the Conference on Communications Architecture & Protocols* (pp. 209-220). Zurich, Switzerland: ACM Press.

Tsirtsis, G., Park, V., Soliman, H. (2009). *Dual-stack mobile IPv4*. IETF RFC 5454.

Vixie P. (1997). *Dynamic updates in the domain name system* (DNS UPDATE). IETF RFC 2136.

Wakikawa, R. (Ed.). (2009). *Multiple care-of addresses registration*. IETF RFC 5648.

Wakikawa, R., Gundavelli, S. (2010). *IPv4 support for proxy mobile IPv6*. IETF RFC 5844.

Wedlund, E., & Schulzrinne, H. (1999). Mobility support using SIP. In *Proceeding of 2nd ACM International Workshop on Wireless Mobile Multimedia* (76–82). Washington, DC: ACM Press.

Yahya, B., & Ben-Othman, J. (2008). Achieving host mobility using DNS dynamic updating protocol. In *IEEE Local Computer Networks* (pp. 634–638). Montreal, Canada: IEEE Press. doi:10.1109/LCN.2008.4664258

Zhang, H. K., & Su, W. (2007). Fundamental research on the architecture of new network—Universal network and pervasive services. *Acta Electronica Sinica*, *35*(4), 593–598.

Zhang, L. X., & Li, T. (2010). *Internet research task force routing research group* (RRG). IRTF RRG website. Retrieved July, 2010, from http://trac.tools.ietf.org/group/irtf/trac/wiki/RoutingResearchGroup

Zhang, P., Durresi, A., & Barolli, L. (2009). A survey of Internet mobility. In the *Proceeding of International Conference on Network-Based Information Systems* (147-154). Indianapolis, IN: IEEE Press.

Zhu, Z., Wakikawa, R., & Zhang, L. (2010). *A survey of mobility support in the Internet. draft-zhu-mobility-survey-03*. IETF draft (work in progress).

ADDITIONAL READING

Assouma, A. D., Beaubrun, R., & Pierre, S. (2006). Mobility management in heterogeneous wireless networks. *IEEE Journal on Selected Areas in Communications*, *24*(3), 638–648. doi:10.1109/JSAC.2005.862407

Bernaschi, M., Cacace, F., Iannello, G., & Vellucci, M. (2007). Mobility management for VoIP on heterogeneous networks: evaluation of adaptive schemes. *IEEE Transactions on Mobile Computing*, *6*(9), 1053–1047. doi:10.1109/TMC.2007.1003

Chiu, K.-L., Chen, Y.-S., Hwang, R.-H. (2010). Seamless session mobility scheme in heterogeneous wireless networks. *International Journal of Communication Systems*, Article first published online: 4 Nov. 2010.

Eastwood, L., Migaldi, S., Xie, Q., & Gupta, V. (2008). Mobility using IEEE 802.21 in a heterogeneous IEEE 802.16/802.11-based, IMT-advanced (4G) network. *IEEE Wireless Communications, 15*(2), 26–34. doi:10.1109/MWC.2008.4492975

George, A., Kumar, A., Cavalcanti, D., & Agrawal, D. P. (2008). Protocols for mobility management in heterogeneous multi-hop wireless networks. *Pervasive and Mobile Computing, 4*(1), 92–116. doi:10.1016/j.pmcj.2007.03.002

Gondi, V. K. (2009). Seamless secured roaming over heterogeneous wireless networks. Doctoral dissertation, University of Evry, Evry, France.

3GPP TS 23.401: General packet radio service (GPRS) enhancements for evolved universal terrestrial radio access network (E-UTRAN) access.

3GPP TS 23.402: Architecture enhancements for non-3GPP accesses.

3GPP TS 29.275: Proxy mobile IPv6 (PMIPv6) based mobility and tunneling protocols.

Guo, C., Guo, Z., Zhang, Q., & Zhu, W. (2004). A seamless and proactive end-to-end mobility solution for roaming across heterogeneous wireless networks. *IEEE Journal on Selected Areas in Communications, 22*(5), 834–848. doi:10.1109/JSAC.2004.826921

Huang, C. M., Lin, C. W., & Yang, C. C. (2010). Mobility management for video streaming on heterogeneous networks. *IEEE MultiMedia, 17*(1), 24–32. doi:10.1109/MMUL.2010.17

Hussain, S., Hamid, Z., & Khattak, N. S. (2006). Mobility management challenges and issues in 4G heterogeneous networks. In proceeding of *The First International Conference on Integrated Internet Ad Hoc and Sensor Networks* (14-19). Nice, France: ACM Press.

Kim, Y.-S., Kwon, D.-H., & Suh, Y.-J. (2008). A client-based vertical handoff approach for seamless mobility across heterogeneous wireless networks. *Wireless Communications and Mobile Computing, 10*(9), 1208–1220. doi:10.1002/wcm.673

Ma, L., Yu, F. R., & Leung, V. C. M. (2007). Performance improvements of mobile SCTP in integrated heterogeneous wireless networks. *IEEE Transactions on Wireless Communications, 6*(10), 3567–3577. doi:10.1109/TWC.2007.05403

Marques, V., Costa, X. P., Aguiar, R. L., Liebsch, M., & Duarte, A. M. O. (2005). Evaluation of a mobile IPv6-based architecture supporting user mobility QoS and AAAC in heterogeneous networks. *IEEE Journal on Selected Areas in Communications, 23*(11), 2138–2151. doi:10.1109/JSAC.2005.856825

Martinez, J. P. (Ed.). (2008). *Enable efficient and operational mobility in large heterogeneous IP networks*.

Nasser, N., Hasswa, A., & Hassanein, H. (2006). Handoffs in fourth generation heterogeneous networks. *IEEE Communications Magazine, 44*(10), 96–103. doi:10.1109/MCOM.2006.1710420

Nguyen-Vuong, Q.-T., Agoulmine, N., & Ghamri-Doudane, Y. (2007). Terminal-controlled mobility management in heterogeneous wireless networks. *IEEE Communications Magazine, 45*(4), 122–129. doi:10.1109/MCOM.2007.343621

Pack, S., Shen, X., Mark, J. W., & Pan, J. (2007). Mobility management in mobile hotspots with heterogeneous multihop wireless links. *IEEE Communications Magazine*, *45*(9), 106–112. doi:10.1109/MCOM.2007.4342864

Yee, Y. C., Choong, K. N., Low Andy, L. Y., & Tan, S. W. (2008). SIP-based proactive and adaptive mobility management framework for heterogeneous networks. *Journal of Network and Computer Applications*, *31*(4), 771–792. doi:10.1016/j.jnca.2007.09.002

KEY TERMS AND DEFINITIONS

CoA: Care-of Address which is a unicast routable address associated with a mobile node while visiting a foreign link; the subnet prefix of this IP address is a foreign subnet prefix. Among the multiple care-of addresses that a mobile node may have at any given time (e.g., with different subnet prefixes), the one registered with the mobile node's home agent for a given home address is called its "primary" care-of address.

HA: Home Agent which is a router on a mobile node's home link with which the mobile node has registered its current location (e.g. CoA). While the mobile node is away from home, the home agent intercepts packets on the home link destined to the mobile node's home address, encapsulates them, and tunnels them to the mobile node's registered care-of address.

HoA: Home Address which is a unicast routable address assigned to a mobile node, used as the permanent address of the mobile node. This address is within the mobile node's home link. Standard IP routing mechanisms will deliver packets destined for a mobile node's home address to its home link. Mobile nodes can have multiple home addresses, for instance when there are multiple home prefixes on the home link.

Identifier: A stable value that can be used to identify a mobile node. Anything could be used as an identifier as long as it is topologically and geographically independent (i.e. remains unchanged when the mobile node roams around).

LMA: Local Mobility Anchor which is the home agent for the mobile node in a Proxy Mobile IPv6 domain. It is the topological anchor point for the mobile node's home network prefix (or prefixes) and is the entity that manages the mobile node's binding state. The local mobility anchor has the functional capabilities of a home agent with the additional capabilities required for supporting Proxy Mobile IPv6 protocol.

Locator: The IP address that indicates the mobile node's current location. It could be the IP address of the mobile node itself, or the IP address of the network entity that is currently serving the mobile node.

MAG: Mobile Access Gateway which is a function on an access router that manages the mobility-related signaling for a mobile node that is attached to its access link. It is responsible for tracking the mobile node's movements to and from the access link and for signaling the mobile node's local mobility anchor.

Mobility Management: Mobility management is one of the major functions of mobile communication network that allows mobile host to work during the movement, and it generally consists of the handover management and location management. Currently, it has migrated into the Internet due to the big success deployment of the IP-based services, and extends to the link layer, network layer, session layer and application.

Chapter 3
Multiple Attributes Decision Making Algorithms for Vertical Handover in Heterogeneous Wireless Networks

Enrique Stevens-Navarro
Universidad Autónoma de San Luis Potosí, México

José D. Martínez-Morales
Universidad Autónoma de San Luis Potosí, México

Ulises Pineda-Rico
Universidad Autónoma de San Luis Potosí, México

ABSTRACT

The envisioned heterogeneous wireless systems are expected to integrate multiple access networks over a common IP platform. In such systems, since mobile users expect to achieve the Always Best Connected (ABC) experience, heterogeneous systems have to support the vertical handover of users among different access networks. On the other hand, the field of Multiple Attribute Decision Making (MADM) has proved to be a suitable tool to study the vertical handover process. This chapter presents a survey of MADM algorithms that have been proposed for vertical handover. First, the authors describe the procedures of methods such as SAW, MEW, TOPSIS, ELECTRE, and VIKOR. Then, the chapter compares them by mean of simulations and performance analysis for an heterogeneous system integrated by WLAN, UMTS, and WiMAX networks, when users conduct different applications. The chapter concludes with a summary of open issues and future research directions in the area of vertical handover.

INTRODUCTION

The envisioned heterogeneous wireless systems also called Beyond Third Generation (B3G) or Fourth Generation (4G) systems are expected to integrate multiple radio access networks over a common IP (Internet Protocol) platform (Akyildiz, 2004). Networks such as Wireless Local Area Networks (WLANs) based on IEEE 802.11/Wi-Fi standards, Universal Mobile Telecommunications System (UMTS) 3G cellular networks, CDMA 2000 3G cellular networks, Wireless Metropolitan

DOI: 10.4018/978-1-4666-0017-1.ch003

Area Networks (WMANs) based on IEEE 802.16/ WiMAX standards, etc., tend to be integrated in the coming years allowing the best connectivity to users' communication applications anywhere at anytime. This feature is the most important issue in such heterogeneous wireless environment and it is also known as the Always Best Connected (ABC) concept (Gustafsson, 2003). The ABC connectivity requirement must be performed based on several (i.e., multiple) decision factors such as Received Signal Strength (RSS), available bandwidth, service type, monetary cost, network conditions, security, and user preferences.

The current existing homogeneous wireless systems are unable to provide such ubiquitous service availability. Hence, it is desirable to interoperate between heterogeneous and complementary wireless technologies (i.e., 3G cellular, WLAN, WMAN, etc.), that is, mobile users will switch between different radio access technologies to satisfy the ABC requirement. This important mobility process is known as the vertical handover (McNair, 2004), (Nasser, 2006). The last few years, plenty of research efforts have been focussed in this important and challenging mobility process in heterogeneous wireless systems.

The field of Multiple Attribute Decision Making (MADM) (Yoon 1995) has proved to be suitable tool to study and model the vertical handover process. Several MADM methods have been proposed in the literature for vertical handover, methods such as SAW (Simple Additive Weighting) (Zhang, 2004), TOPSIS (Technique for Order Preference by Similarity to Ideal Solution) (Zhang, 2004), MEW (Multiplicative Exponent Weighting) (Stevens-Navarro, 2006), Grey Relational Analysis (GRA) (Song, 2005), ELECTRE (Elimination and Choice Translating Priority) (Bari, 2007), WMC (Weighted Markov Chain) (Ying, 2008), and VIKOR (Gallardo-Medina, 2009). Considerable amount of research to develop and modify MADM methods for vertical handover have been conducted. It is necessary to evaluate their performance under different scenarios in order to provide the best solution for a particular application.

In this chapter, we present a survey of five MADM algorithms proposed in the literature for vertical handover decision process. We compare them by mean of simulations and performance analysis for an heterogeneous wireless network, integrated by WLAN, UMTS and WiMAX networks, when the user conducts different type of applications. The rest of the chapter is organized as follows. We present some background material on vertical handover decision in heterogeneous wireless networks. Next we introduce the field of MADM techniques. It describes how a decision or selection problem consisting of several options and each described by several parameters can be formulated as a MADM problem. Here, we explain why and how the vertical handover decision or network selection in heterogeneous wireless networks can be formulated as a MADM problem. We present a survey of MADM algorithms that have been proposed for vertical handover or network selection in the last few years. The section introduces each of the MADM techniques and describes all the procedures to implement the algorithm. Some numerical results and discussion of a performance comparison among the MADM techniques introduced in previously in a specific heterogeneous scenario are presented. The MADM algorithms are implemented and evaluated by a discrete-event simulation approach using MATLAB. Next we present several open issues and technical challenges in the area of vertical handover and finally we conclude the chapter with a summary.

BACKGROUND

In the user mobility management context of cellular networks, a handover is defined as the mechanism by which an ongoing connection is transferred from one base station (BS) to another (Rappaport, 1996). The BSs are the infrastructure (i.e., antennas, towers, radio hardware) that is deployed by the cellular operator to provide service in a specific geographic area. In this case, if we consider that both BSs use the same access technology, as in current cellular systems, we can say that homogeneous wireless networks perform horizontal handovers (McNair, 2004), (Nasser, 2006). Horizontal handover mechanisms mainly use the received signal strength (RSS) measurements from the surrounding BSs to trigger and to perform the handover. The horizontal handover algorithm decides to which neighbouring BS the connection should be transferred to. It is usually the BS which provides the highest signal level to the mobile device.

On the other hand, if we consider heterogeneous wireless networks such as the B3G networks, with BSs from cellular networks using one access technology, BSs from WiMAX using other access technology, and access points (APs) from WLANs using a different one too, we can say that a vertical handover is the mechanism by which an ongoing connection is transferred from one BS to an AP and vice versa (i.e., between two different wireless access technologies) (McNair, 2004), (Nasser, 2006). The different types of handovers can be seen in Figure 1. It is worth to mention that the mechanisms to support vertical handovers required in a heterogeneous network environment have to be designed by considering the challenging tradeoff among several technical objectives such as low latency, power saving, and required bandwidth. Furthermore, the handover decision in this case becomes a more complex process than in homogeneous wireless environment because of the additional (i.e., multiple) parameters, besides the RSS, that have to be considered. A further discussion about this process will be presented in following sections.

The vertical handover process can be divided into three main steps (McNair, 2004), (Nasser, 2006), namely: system discovery, handover decision, and handover execution. During the system discovery step, the mobile devices equipped with multiple interfaces have to determine which wireless networks can be used and the communication

Figure 1. Types of handover in heterogeneous networks: horizontal (HHO) and vertical (VHO)

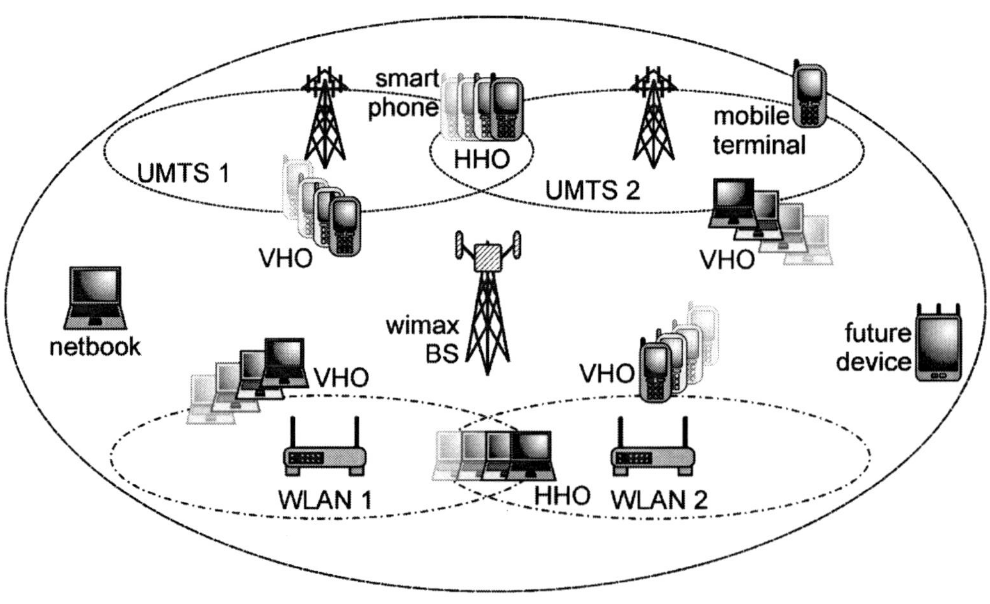

services available in each network. These wireless networks may also advertise the supported data rates for different services as well as other relevant information. During the handover decision step, the mobile device determines to which network it should connect to. The decision may depend on various parameters including the available bandwidth, packet delay, packet jitter, access cost, transmit power, current battery status of the mobile device, and even the user's preferences. Finally, during the handover execution step, the connection needs to be re-routed from the existing network in use to the new network in a seamless manner. This step also includes the authentication and authorization of the user in the new network as well as the transfer of user's context information.

There are a lot of parameters that can be considered for vertical handover decision (McNair, 2004), (Nasser, 2006). The Quality of Service (QoS) parameters of a particular connection such as bandwidth or delay bounds are usually specified by the applications. The level of security and amount of cost may be directly specified by the user itself. All this parameters are gathered by the mobile terminal during the system discovery step. Figure 2 shows several of those parameters categorized under different groups (e.g., bandwidth, delay, power, cost, security, reliability). Questions such as: "Which parameters shall the mechanism consider in the handover decision?", "How many parameters shall the mechanism consider for handover decision?", "How the mobile device shall retrieve the decision parameters from the surrounding wireless networks?", are still important open issues in this research field.

Recently, in the case of the vertical handover decision mechanisms several proposals have been presented by the research community. For the rest of the chapter, we refer to such mechanisms as vertical handover decision algorithms. An important group of such research works propose the use of Multiple Attribute Decision Making (MADM) methods and techniques (Yoon 1995). MADM methods have shown to be a suitable modeling and mathematical tool to deal with the vertical handover decision. Next, we summarize related and relevant research work in this area.

In (Zhang, 2004), the vertical handover decision is formulated as a fuzzy MADM problem. First, fuzzy logic is used to represent the imprecise information of some of the decision parameters and user preferences. Then, MADM techniques are used to rank the network alternatives for vertical handover. Thus, two important MADM methods are proposed: Simple Additive Weighting (SAW) and Technique for Order Preference by Similarity to Ideal Solution (TOPSIS). Later, in (Tawil,

Figure 2. Classification of decision parameters for vertical handover

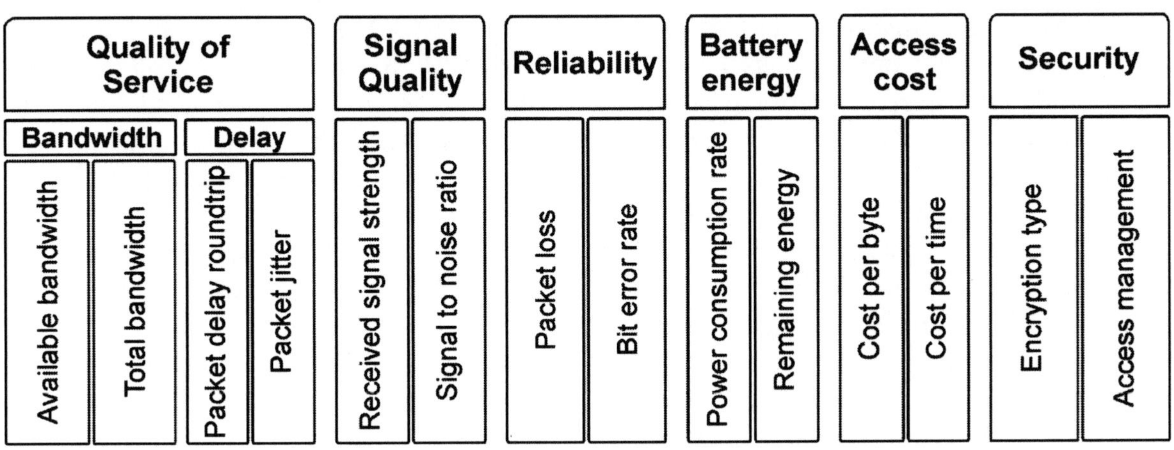

2008), SAW is used in a distributed manner for vertical handover decision to enhance mobility and reduce the processing delay.

In (Song, 2005), the network selected for vertical handover is based on the Analytic Hierarchy Process (AHP) and Grey Relational Analysis (GRA). The AHP technique decomposes the network selection problem into several sub-problems and assigns a weight value for each sub-problem. Then, the GRA method is used to rank the candidate networks and to select the one with the highest ranking which is the result of the handover decision. Furthermore, in other related works, the GRA method is also considered for network selection considering power consumption in (Joe, 2008). The AHP is modified to consider interval rages of parameters´ values rather than instant values for handover decision in (Cui, 2008), and modified to provide deterministic and probabilistic vertical handover solutions for mobile terminals with multiple communications sessions with different QoS requirements in (Ramirez, 2009).

In (Stevens-Navarro, 2006), the Multiplicative Exponent Weighting (MEW) method is introduced for vertical handover decision. In fact, in (Stevens-Navarro, 2006) and later in (Tran, 2008) and (Wang, 2009), SAW, TOPSIS, GRA and MEW methods are studied and compared by means of computer simulations for vertical handover decision. Those research works are the first with the aim to compare the performance for vertical handover of several of the proposed MADM methods. Later, in (Dhar, 2007), a ranking algorithm based on TOPSIS is presented for network selection, and in (Yang, 2009) a cooperative vertical handover algorithm using GRA which consider signalling costs and load balancing is proposed.

In (Bari, 2007), the use of MADM scheme ELECTRE for network selection is proposed, while in (Ying, 2008) the Weighted Markov Chain (WMC) technique for rank aggregation is proposed for the same problem. In (Tawil, 2008), a distributed vertical handover decision scheme based on the SAW method is proposed. Finally, in (Gallardo-Medina, 2009), we presented the use of VIKOR method for vertical handover and compared its performance among MADM methods such as SAW and TOPSIS. More recently, in (Martinez-Morales, 2010), we conducted a simulation-based performance comparison study among several decision methods proposed for vertical handover such as MADM methods as well as other rank aggregation methods.

As we can see from the previous related work revised, there are several decision techniques that have been applied to solve the handover decision problem. Most of them belong to the field of MADM methods and few others are just particular ranking techniques for decision making problems. However, for the scope of the present chapter and space limitation, we restrict our survey only to MADM methods since it is the main topic of interest. Readers interested in a broader range of handover decision methods may refer to the additional reading section and the references therein.

MULTIPLE ATTRIBUTE DECISION MAKING (MADM) METHODS

Multiple Attribute Decision Making (MADM) methods consider problems where making preference decisions over available alternatives that are characterized by multiple and usually conflicting attributes are required (Yoon, 1995). MADM is a branch of the field Multiple Criteria Decision Making (MCDM).

MADM problems are diverse in disciplines, but all share the following common characteristics: alternatives to select, multiple attributes describing the alternatives in different units of measurement, and a set of weights representing the relative importance among attributes. For notation, let M be the set of alternatives and N be the set of parameters or attributes. A MADM problem can be concisely expressed in a matrix format, where columns indicate attributes and rows list alternatives. A typical element of the matrix x_{ij} indicates the performance rating of the ith alternative with respect to the jth attribute. Thus, a MADM problem with $|M|$ alternatives and each with $|N|$ parameters is given by:

$$\begin{bmatrix} x_{11} & x_{12} & .. & x_{i|N|} & .. & x_{1|N|} \\ x_{21} & x_{22} & .. & x_{ij} & .. & x_{2|N|} \\ .. & .. & .. & .. & .. & .. \\ x_{|M|1} & x_{|M|2} & .. & x_{|M|j} & .. & x_{|M||N|} \end{bmatrix} \quad (1)$$

When the vertical handover decision or network selection problem is formulated as a MADM problem, the alternatives to select are the candidate wireless networks and the attributes are the parameters describing the network which several authors refer to them as handoff metrics.

The MADM methods use scoring techniques to rank alternatives. An index or score is calculated by taking into account the contributions from each parameter. Before the calculation of the index, normalization of the parameters is required to deal with different units (e.g., bps, sec, price, etc.)

Additionally, a set of importance weights have to be defined for the calculation of the ranking. The value of such weights should represent the different levels of importance of a parameter for the decision maker. The set of importance weights has to satisfy the constraint:

$$\sum_{j=1}^{N} w_j = 1 \quad (2)$$

In the case of the vertical handover decision, the importance weights should represent the QoS requirements of the connection as well as the user's preferences. For example, in a voice connection the importance weights of packet delay and packet jitter must have higher value than the available bandwidth of the network. It is also important to note the difference among a benefit parameter and a cost parameter. For a benefit parameter, the lager is the better, for example, the parameter available bandwidth in a data connection where 11 Mbps is better than 1 Mbps. For a cost parameter, the lower the better, for example, the parameter packet delay in a voice connection where 100 ms is better than 500 ms.

In general, we can summarize the steps required for any MADM method as follows:

STEP 1: Organize the $|M|$ alternatives and the $|N|$ parameters into a decision matrix as in (1).

STEP 2: Normalize the values of the parameters according to the procedure defined by each MADM method (e.g., benefit or cost). Such procedures are presented in the following section.

STEP 3: Assign the corresponding importance weights w_j satisfying the constraint in (2).

STEP 4: Apply the procedures defined by each MADM method to calculate the score for each alternative. Such procedures are presented in the following section.

STEP 5: Rank the $|M|$ alternatives based on the score.

STEP 6: Select the best alternative which is the one ranked in the first place.

MADM TECHNIQUES APPLIED FOR VERTICAL HANDOVER

In general, all the following MADM decision methods first calculate an index or score based on its specific procedures, then, the network selected for vertical handover is the one with the best score value or the one in the first place in the ranking.

SAW

As mentioned before, the MADM decision problem can be expressed in a $M \times N$ decision matrix, where the j–th attribute of the i–th network is represented as x_{ij}. Simple Additive Weighting (SAW) is one of the best known and most widely used scoring methods because of its simplicity (Yoon, 1995). It is also commonly called weighted sum method. It was initially proposed for vertical handover decision in (Zhang, 2004) and lately it has been used as a baseline for performance comparison with other vertical handover decision methods as shown in (Stevens-Navarro, 2006), (Martinez-Morales, 2010). If all the parameters have the same measuring units, then SAW can compute the score directly. However, if the parameters have different measuring units (i.e., as in the case of vertical handover), then the values of the parameters require to be normalized.

In SAW, the score of each candidate network i is obtained by adding the contributions from each attribute r_{ij} multiplied by the importance weight w_j. Then, the selected network A^*_{SAW} is:

$$A^*_{SAW} = \arg\max_{i \in M} \sum_{j \in N} w_j r_{ij}, \tag{3}$$

where $r_{ij} = x_{ij} / x_j^+$ for benefit parameters, and $r_{ij} = x_j^- / x_{ij}$ for cost parameters, moreover, $x_j^+ = \max_{i \in M} x_{ij}$ and $x_j^- = \min_{i \in M} x_{ij}$, and the importance weight vector must satisfy $\sum_{j=1}^{N} w_j = 1$.

MEW

Multiplicative Exponent Weighting (MEW) is another MADM scoring method and is very similar to SAW. The main difference is that, instead of addition in the method, there is multiplication (Yoon 1995). It was initially proposed for vertical handover decision in (Stevens-Navarro, 2006). In MEW, the scores of the networks are determined by the weighted product of the attributes. The score S_i of network i is determined by the weighted product of the attributes:

$$S_i = \prod_{j \in N} x_{ij}^{w_j}, \tag{4}$$

where x_{ij} denotes attribute j of candidate network i, w_j denotes the weight of attributed j, and $\sum_{j=1}^{N} w_j = 1$. Note that in (4), w_j is a positive power for benefit metrics $x_{ij}^{w_j}$, and a negative power for cost metrics $x_{ij}^{-w_j}$. Since parameter normalization is not required (i.e., it is optional), the score of a network obtained by MEW does not have an upper bound (Yoon 1995), thus it is convenient to compare the score of each network with the score of the positive ideal network A^{**}.

This network is defined as the network with the best values in each metric. For a benefit metric, the best value is the largest. For a cost metric, the best value is the lowest. The value ratio R_i between network i and the positive ideal is calculated by:

$$R_i = \frac{\prod_{j \in N} x_{ij}^{w_j}}{\prod_{j \in N} (x_{ij}^{**})^{w_j}}, \tag{5}$$

where $0 \le R_i \le 1$. The selected network A^*_{MEW} is:

$$A^*_{MEW} = \arg\max_{i \in M} R_i. \tag{6}$$

TOPSIS

In Technique for Order Preference by Similarity to Ideal Solution Algorithm (TOPSIS) with $|M|$ alternatives that are evaluated by $|N|$ decision criteria is viewed as a geometric system with $|M|$ points in the $|N|$ dimensional space (Yoon 1995). Here, the chosen candidate network is the one which have the shortest distance to the ideal solution and the longest distance to the worst case solution. The ideal solution is an "hypothetical solution" with the best values in each parameter while the negative ideal solution is the opposite. It was initially proposed for vertical handover decision in (Zhang, 2004) and lately it has been considered in performance comparison evaluations with other vertical handover decision methods as shown in (Stevens-Navarro, 2006), (Martinez-Morales, 2010). In general, to compute the final ranking-list, TOPSIS requires the following steps:

Step 1: Construct the normalized decision matrix, which allows comparison across the attributes, this matrix is given by:

$$r_{ij} = \frac{x_{ij}}{\sqrt{\sum_{i=1}^{m} x_{ij}^2}}.$$

Step 2: Construct the weighted normalized decision matrix as $v_{ij} = w_j * r_{ij}$.

Step 3: Determine ideal and negative-ideal solutions by:

$$A^+ = \{(\max_{i\in M} v_{ij} \mid j \in J), (\min_{i\in M} v_{ij} \mid j \in J')\},$$

and

$$A^- = \{(\min_{i\in M} v_{ij} \mid j \in J), (\max_{i\in M} v_{ij} \mid j \in J')\},$$

where J is the set of benefit parameters, and J' is the set of cost parameters.

Step 4: Calculate the separation measure between the networks and the positive and negative ideal networks by:

$$s_i^+ = \sqrt{\sum_{j\in N} (v_{ij} - v_j^+)^2}, \quad s_i^- = \sqrt{\sum_{j\in N} (v_{ij} - v_j^-)^2}.$$

Step 5: Calculate the relative closeness to the ideal solution.

$$c_i^* = \frac{s_i^-}{(s_i^+ + s_i^-)}.$$

A set of alternatives can now be preference ranked according to the descending order of c_i^*. Then the selected network A_{TOP}^* is:

$$A_{TOP}^* = \arg\max_{i\in M} c_i^*. \tag{7}$$

ELECTRE

Elimination and Choice Translating Priority (ELECTRE) is a MADM method which performs pair-wise comparisons among the alternatives using each of the attributes separately to establish outranking relationships between the alternatives (Yoon 1995). The philosophy behind ELECTRE has made it very popular for decision making processes requiring a deterministic and simplistic approach. It was initially proposed for vertical handover decision in (Bari, 2007). In general, for ELECTRE method, a reference attribute vector is used to adjust the raw attribute values for the alternative networks before they are compared. The value of each of the attributes in the decision matrix is compared with a corresponding reference attribute value x_j^{ref}. An absolute difference between the two values is taken to calculate a new matrix as follows.

$$r_{ij} = \left| x_{ij} - x_j^{ref} \right| \tag{8}$$

Now in this matrix all attribute values can be considered to have a monotonically decreasing utility. Since a lower value for an adjusted attribute in (2) is considered an indication of a better network in the selection process, each attribute in (2) can be normalized as follows,

$$\hat{r}_{ij} = \frac{\max_{i\in M}\{r_{ij}\} - r_{ij}}{\max_{i\in M}\{r_{ij}\} - \min_{i\in M}\{r_{ij}\}}$$

Now, is necessary take into consideration the relative importance of each of the attributes involved in the decision about network selection. For the j–th attribute is assigned a weight w_j, such that $\sum_{j=1}^{N} w_j = 1$. Using the weights, an updated matrix is calculated by,

$$\tilde{r}_{ij} = w_j \hat{r}_{ij}$$

In order to compare the network alternatives, the concept of concordance and discordance has been introduced in ELECTRE, which are measures of satisfaction and dissatisfaction of the decision maker when one alternative is compared with another. It firstly uses pair-wise comparisons of networks to obtain the concordance set *CSet*(*k*, *l*) indicating the attribute of network *k* is better than network *l* and the discordance set *DSet*(*k*, *l*) indicating the attribute of network *k* is worse than network *l*. The concordance and discordance sets are formed as follows,

$$CSet_{kl} = \{j \mid \tilde{r}_{kj} \geq \tilde{r}_{lj}\}$$
$$DSet_{kl} = \{j \mid \tilde{r}_{kj} < \tilde{r}_{lj}\}$$

Using the concordance and discordance sets, corresponding matrices are constructed. The elements of the concordance matrix C can be represented as,

$$C_{kl} = \sum_{j \in CSet_{kl}} w_j$$

The entries for the concordance matrix are not defined for the diagonal. ELECTRE defines the elements of discordance matrix as follows,

$$d_{kl} = \frac{\sum_{j \in DSet_{kl}} \left| \tilde{r}_{kj} - \tilde{r}_{lj} \right|}{\sum_{j \in N} \left| \tilde{r}_{kj} - \tilde{r}_{lj} \right|}$$

Similarly, the entries for the discordance matrix are also not defined for the diagonal. A new parameter $\tilde{C}_i$, called the net concordance index is calculated. C_i is a measure of dominance of an alternative i over other alternatives. It can be calculated as follows

$$\tilde{C}_i = \sum_{j \in N, j \neq i} C_{ij} - \sum_{j \in N, j \neq i} C_{ji}. \qquad (9)$$

Similarly, the term net discordance index D_i, is defined as a measure of relative weakness of alternative i over other alternatives, and can be calculated as

$$\tilde{D}_i = \sum_{j \in N, j \neq i} D_{ij} - \sum_{j \in N, j \neq i} D_{ji}. \qquad (10)$$

An alternative with the highest value of net concordance index $\tilde{C}$ and the lowest value of net discordance index $\tilde{D}$ would be preferred. However, if it is not the case, the alternatives are ranked based on the concordance and discordance indexes where each alternative is ranked by taking the average of these two rankings. The alternative with the highest average ranking is considered to be the best alternative. Alternatives with the same average ranking would be considered equally suited.

VIKOR

MADM method VIKOR (i.e., VlseKriterijumska Optimizacija I Kompromisno Resenje, not acronym known in English.) is a compromise ranking method (Opricovic, 2004). It works for decision making on the idea of an aggregating function representing closeness to the ideal solution. For VIKOR, a compromise solution is a feasible solution that is the closest to the ideal solution, and the compromise means that is selected by a mutual and beneficial agreement among the other solutions (Opricovic, 2004). It was initially proposed for vertical handover decision in (Gallardo-Medina, 2009). For VIKOR method the following steps are required:

Step 1: For each parameter $j=1,2,3...,N$, determine the best and the worst values given by:

$$F_j^+ = \{(\max_{i\in M} x_{ij} \mid j \in N_b), (\min_{i\in M} x_{ij} \mid j \in N_c)\},$$

and

$$F_j^- = \{(\min_{i\in M} x_{ij} \mid j \in N_b), (\max_{i\in M} x_{ij} \mid j \in N_c)\},$$

where $N_b \subset N$ is the set of benefit parameters, and $N_c \subset N$ is the set of cost parameters.

Step 2: Compute the values of S_i and R_i for $i=1,2,3...,M$ given by:

$$S_i = \sum_{j\in N} w_j \frac{(F_j^+ - x_{ij})}{(F_j^+ - F_j^-)},$$

and

$$R_i = \max_{j\in N}\left[w_j \frac{(F_j^+ - x_{ij})}{(F_j^+ - F_j^-)}\right],$$

where w_j is the importance weight of parameter j.

Step 3: Compute the values of Q_i for $i=1,2,3...,M$ given by:

$$Q_i = \gamma\left(\frac{S_i - S^+}{S^- - S^+}\right) + (1-\gamma)\left(\frac{R_i - R^+}{R^- - R^+}\right),$$

where

$$S^+ = \min_{i\in M} S_i, \quad S^- = \max_{i\in M} S_i,$$
$$R^+ = \min_{i\in M} R_i, \quad R^- = \max_{i\in M} R_i,$$

and parameter γ with $0 \le \gamma \le 1$ is the weight of the strategy. It also represents the majority of criteria. Here, we set $\gamma = 0.5$.

Step 4: Given the values for the Q, for all $i\in M$, rank the candidate networks in an increasing order. The selected network A_{VIK}^* is:

$$A_{VIK}^* = \arg\min_{i\in M} Q_i^*. \tag{11}$$

PERFORMANCE COMPARISON AMONG MADM TECHNIQUES

In this section, we present a performance comparison of the MADM decision methods for vertical handover previous introduced. Let's consider a scenario of a heterogeneous wireless environment integrated by three different types of wireless network such as WLAN, UMTS and WiMAX, and there are two networks of each type for a total of six candidate networks for vertical handover. The networks are labeled as follows: networks 1 and 2 correspond to networks UMTS1 and UMTS2 respectively, networks 3 and 4 correspond to networks WLAN1 and WLAN 2 respectively, and networks 5 and 6 correspond to networks WiMAX1 and WiMAX2 respectively. Six different decision parameters from the ones shown in Figure 2 are considered in this study and they have to be evaluated and compared in order to detect and to trigger a vertical handover. The decision parameters are: available bandwidth in Mbps, total bandwidth in Mbps, packet delay in msegs, packet jitter in msegs, packet loss in per each 10^6 packets and cost per byte. The ranges of the values of the parameters or decision criteria are shown in Table 1.

The values of the importance weights in (2) for different services or traffic classes considered in this study are the following: Case 1, all parameters have the same importance weight, this is the baseline case; Case 2, the packet delay and

Table 1. Range of values of the networks' parameters

Parameter	UMTS1	UMTS2	WLAN1	WLAN2	WiMAX1	WiMAX2
Available Bandwidth (Mbps)	0.1-2	0.1-2	1-11	1-54	1-60	1-60
Total Bandwidth (Mbps)	2	2	11	54	60	60
Packet Delay (ms)	25-50	25-50	100-150	60-150	60-100	60-100
Packet Jitter (ms)	5-10	5-10	10-20	10-20	3-10	3-10
Packet Loss (per 10^6)	20-80	20-80	20-80	20-80	20-80	20-80
Cost per byte (price)	0.6	0.8	0.1	0.05	0.5	0.4

packet jitter have 70% of importance and the rest is equally distributed among the other parameters, this case is suitable for voice connections; Case 3, the available bandwidth and total bandwidth have 70% of importance, this case is suitable for data connections; and finally Case 4, cost per byte (price) has 50% of importance, this case is suitable for cost-constrained connections.

In each vertical handover decision point (i.e., the point of the specific handover decision step which is after the system discovery step and before the handover execution step), the attribute values may be the same, increase or decrease within the range shown in Table 1. In order to varying the values of the decision criteria, a discrete event simulation is conducted where a discrete-time Markov chain is used to model the evolution on time of each decision parameter. The discrete-event simulation is implemented in MATLAB. The transition probabilities for an increment or decrement are p_c, while the probability of being in the same value is p_s. The transition probabilities for an increment or decrement are p_c=0.4, while the probability of being in the same value is p_s=0.2. For each application, we consider 20 vertical handover points in the simulation. Each vertical handover decision point is the average of 10 simulations.

Preferred Network by MADM Methods (Case 1)

Figure 3 shows the selected network for each of the MADM methods. This figure includes the

Figure 3. Selected network for vertical handover by each of the MADM methods

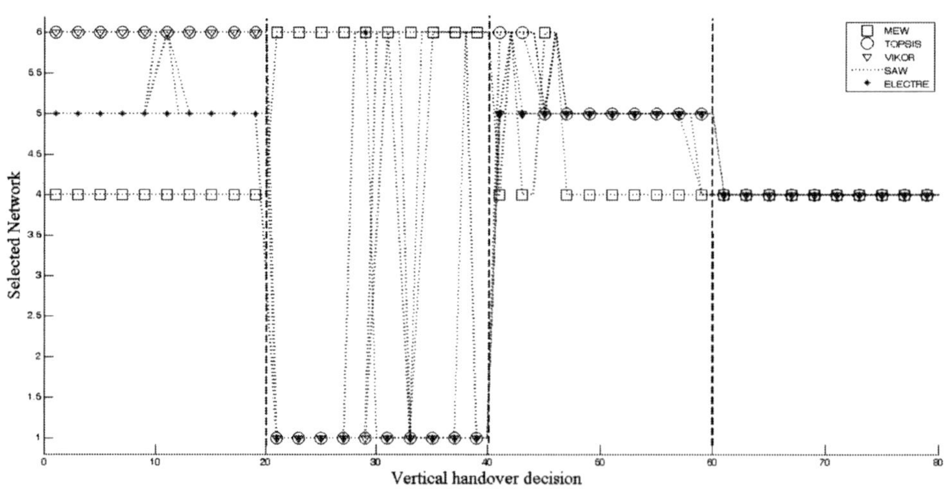

results of the four different services considered in the horizontal axis as follows: points 1 to 20 are for Case 1 (baseline), points 21-40 are for Case 2 (voice connections), points 41-60 are for Case 3(data connections) and points 61 to 80 are for Case 4 (cost-constrained connections). The numbers 1 to 6 in the vertical axis correspond to network 1 (UMTS1) to network 6 (WiMAX2) as labeled in Table 1.

At the beginning, when the importance weights are set according to Case 1 (baseline), methods SAW and ELECTRE select network WiMAX1 (network 5) in most of the vertical handover decisions, methods TOPSIS and VIKOR select network WiMAX2 (network 6), and method MEW selects WLAN2. When the importance weights are set according to Case 2 (voice connections) note that methods SAW, TOPSIS, ELECTRE and VIKOR decide to execute vertical handover to network UMTS1 (network 1). This behavior is expected since the 3G network is able to offer lower values of packet delay and jitter which are QoS parameters very important for voice connections. When the importance weights are set according to Case 3 (data connections) note that MEW decide to execute vertical handover to network WLAN2 (network 4) and methods TOPSIS, VIKOR, SAW and ELECTRE to network WiMAX1 (network 5), in both decisions the methods selected networks with higher values of available bandwidth and total bandwidth suitable for data connections. Finally, when the importance weights are set according to Case 4 (cost-constrained connections), note that the five MADM decision methods decide to execute vertical handover to the same network. The methods selected network WLAN2 since it is the one with the lowest value of cost per byte.

Results for Voice Connections (Case 2)

Figure 4 shows the packet delay in msegs achieved by the MADM decision methods. Here the figure includes the results of two different services as follows: points 1 to 20 are for Case 1 (baseline) and points 21-40 are for Case 2 (voice connections). Note that decision method VIKOR is able to achieve the lowest values of packet delay followed by SAW and TOPSIS. It is worth to mention that the five MADM decision methods in Case 2 are able to achieve lower values of packet delay compared to their performance achieved in Case 1.

Figure 4. Values of packet delay in ms achieved by each of the MADM methods. Note that vertical handover decision points 1 to 20 correspond to Case 1 (baseline) and points 21-41 correspond to Case 2 (voice connections).

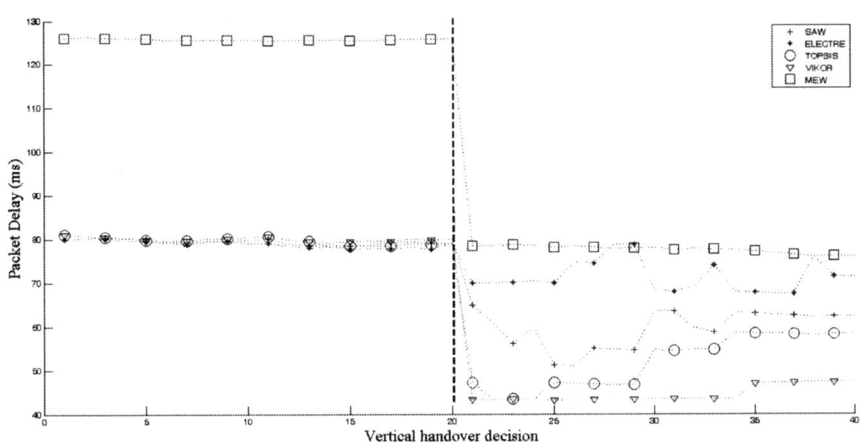

Figure 5 shows the packet jitter in ms achieved by the MADM decision methods. Here the figure includes the results of two different services as follows: points 1 to 20 are for Case 1 (baseline) and points 21-40 are for Case 2 (voice connections). Note that decision method MEW is able to achieve the lowest values of packet jitter followed very close by SAW, ELECTRE, TOPSIS and VIKOR. Note again that the five MADM decision methods in Case 2 are able to achieve lower values of packet jitter compared to their performance achieved Case 1.

Results for Data Connections (Case 3)

Figure 6 shows the available bandwidth in Mbps achieved by the MADM decision methods. Here the figure includes the results of two different services as follows: points 1 to 20 are for Case 1 (baseline) and points 21-40 are for Case 3 (data connections). Note that decision method MEW is able to achieve the highest values of available bandwidth followed very close by SAW, ELECTRE, TOPSIS and VIKOR. Note again that the five MADM decision methods in Case 3 are able to achieve higher values of available bandwidth compared to their performance achieved Case 1.

Results for Cost Constrained Connections (Case 4)

Figure 7 shows the connection cost achieved by the MADM decision methods. Here the figure includes the results of two different services as follows: points 1 to 20 are for Case 1 (baseline) and points 21-40 are for Case 4 (cost-constrained connections). Note that decision methods SAW, MEW, TOPSIS and VICKOR are able to achieve the lowest values of connection costs followed by ELECTRE. Note again that the five MADM decision methods in Case 4 are able to achieve lower values of cost per byte compared to their performance achieved Case 1.

OPEN ISSUES AND FUTURE RESEARCH DIRECTIONS

Although this chapter presents a survey of vertical handover decision algorithms based on MADM decision methods that have been proposed in the

Figure 5. Values of packet jitter in ms achieved by each of the MADM methods. Note that vertical handover decision points 1 to 20 correspond to Case 1 (baseline) and points 21-41 correspond to Case 2 (voice connections).

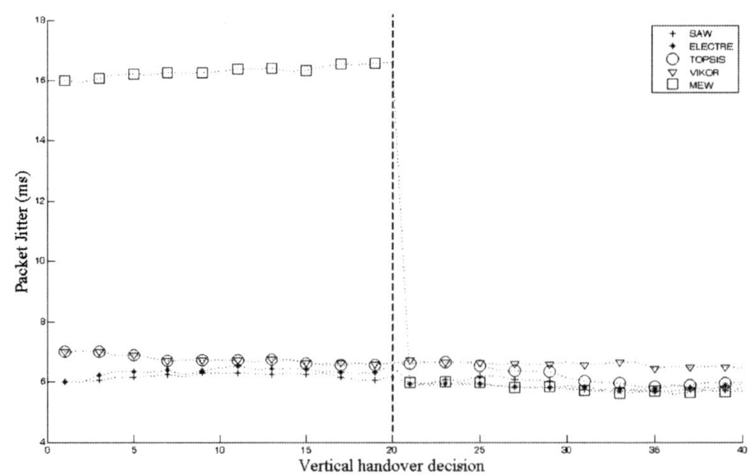

Figure 6. Values of available bandwidth in Mbps achieved by each of the MADM methods. Note that vertical handover decision points 1 to 20 correspond to Case 1 (baseline) and points 21-41 correspond to Case 3 (data connections).

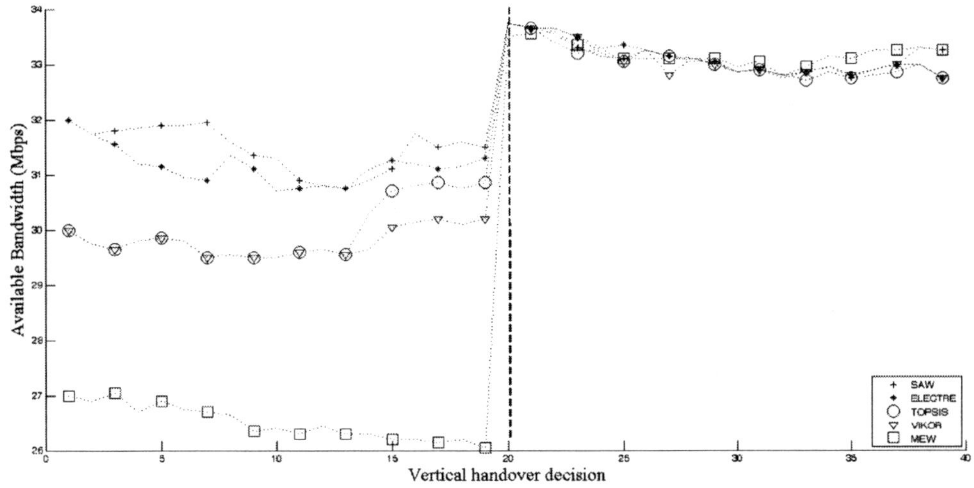

literature, there are several open issues that need to be addressed and require further investigations. We briefly summarize some of them:

- **The importance weights distribution.** As shown in the results previously presented, the performance of the MADM decision methods for vertical handover highly depends on the allocation of the importance

Figure 7. Values of cost connections achieved by each of the MADM methods. Note that vertical handover decision points 1 to 20 correspond to Case 1 (baseline) and points 21-41 correspond to Case 4 (cost-constrained connections).

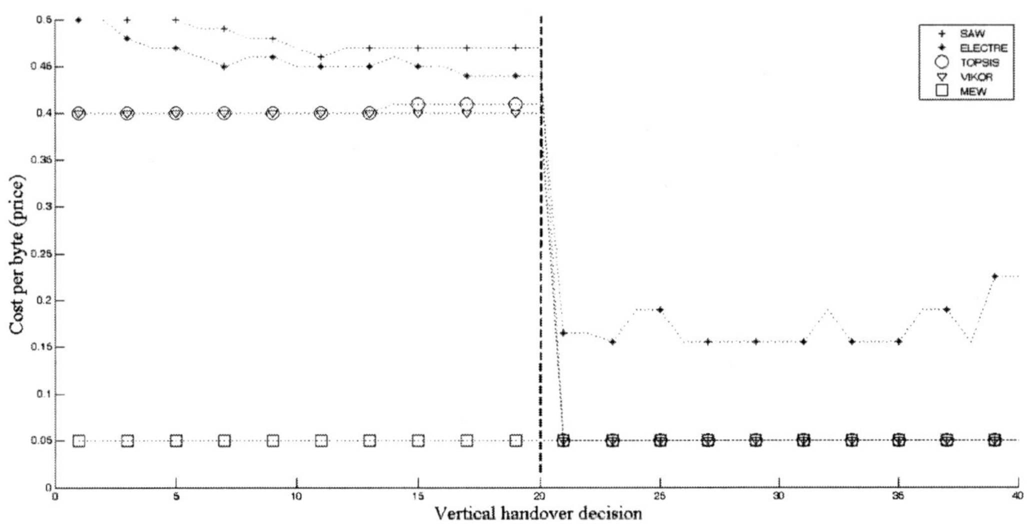

weights (e.g., Case 2, Case 3, etc.) since they indicated the type of application or service running in the mobile terminal. The exact distribution of the importance weights for any kind of current or future application (e.g., voice over IP protocol, video conference, video on demand, on-line games, on-line gambling, etc.) requires careful allocation and hence further research.

- **The impact of computational complexity.** Each of the MADM decision methods studied in the chapter defines a set of different mathematical operations that need to be executed in order to compute the ranking among the wireless networks and hence the selection of the best network for vertical handover. Arithmetic operations such that additions, subtractions, multiplications, divisions, and square root operations require different amount of computational resources (i.e., floating point operations) and time for execution on the mobile terminal thus further research is required to compare and evaluate the impact of such computational complexity in the performance of the MADM decision algorithms.
- **MADM methods for other mobility techniques.** The MADM decision methods may be used to study and analyze other mobility management techniques in heterogeneous wireless networks. Novel schemes for location management, connection admission control, resource sharing, and load balancing are required to deal with user executing horizontal and vertical handover among different wireless networks.
- **A complete framework for vertical handover.** The scope and aim of the present chapter is restricted to the decision step of the vertical handover process, however still it is required to extend and adequate the vertical handover decision methods proposed in the literature into a complete vertical handover framework able to consider and incorporate the other steps: system discovery and handover execution.
- **Implementation details and feasibility.** An important open issue in the area of the vertical handover relies in the actual implementation of the handover decision algorithms in the mobile devices. Although several of the current wireless technologies standard organizations such as Third Generation Partnership Project (3GPP), 3GPP2, IEEE 802.11 Working Group, IEEE 802.16 Working Group, WiMAX Forum are considering the inclusion and support of vertical handovers, the actual definition and implementation of the vertical handover algorithm is not clear and hence usually relies on the manufacturer of the devices. In fact, there is a specific IEEE standard (IEEE 802.21) being developed to consider vertical handovers between IEEE wireless networks and non-IEEE wireless networks.

CONCLUSION

The envisioned heterogeneous wireless systems are expected to be integrating multiple radio access networks (e.g., 3G cellular networks, WLANs, metropolitan networks, etc.) interworking over a common platform based on the IP protocol. In this networking scenario, mobile users will expect connectivity and service availability anytime and anywhere according to the ABC concept. Thus, the support of the vertical handover will be an important issue in the design and operation of such heterogeneous networks.

The vertical handover is a relevant and challenging mobility management problem in the envisioned heterogeneous wireless networks. The vertical handover decision has shown to be a more complex problem than the traditional horizontal handover decision in cellular networks. Thus, the field of Multiple Attribute Decision Making (MADM) has shown to be a suitable modeling and mathematical tool to study and analyze the vertical handover decision process.

In this chapter, we presented a survey of MADM algorithms that have been recently proposed for vertical handover decision in the literature. First, we summarized important related work in the area of vertical handover decision. After that, we described the procedures of MAMD methods such as SAW, MEW, TOPSIS, ELECTRE and VIKOR. We compared them by mean of numerical simulations and performance analysis for an heterogeneous system integrated by WLAN, UMTS and WiMAX networks, when users conduct different applications.

From the results, we found that none of the MAMD methods is the best for all cases of connections (i.e., voice, data, cost-constrained) considered in our study. However, it is shown than in general methods VIKOR, SAW and TOPSIS obtain considerable good performance with voice connections requiring lower bound of packet and jitter delay, MEW and ELCTRE achieve good performance with data connections requiring higher amounts of available bandwidth, and the five MADM methods shown good performance in achieving lower values of connection costs.

Finally, we presented a summary of open issues and future research directions for the following years in the area of vertical handover in heterogeneous wireless networks.

ACKNOWLEDGMENT

The present research work was supported by Programa de Mejoramiento del Profesorado (PROMEP) of Secretaría de Educación Pública (SEP) from the Mexican Government under grant PROMEP/103.5/09/4245 and by Fondo de Apoyo a la Investigación (FAI) under grant C10-FAI-05-10.38 from Universidad Autónoma de San Luis Potosí (UASLP) in Mexico.

REFERENCES

Akyildiz, I., Xie, J., & Mohanty, S. (2004). A survey of mobility management in next-generation all-IP-based wireless systems. *IEEE Wireless Communications*, *11*(4), 16–28. doi:10.1109/MWC.2004.1325888

Bari, F., & Leung, V. (2007). Application of ELECTRE to network selection in a heterogeneous wireless network environment. *Proceedings of IEEE Wireless Communications and Networking Conference* (WCNC'08), vol. 1 (pp. 3810-3815).

Cui, H., Yan, Q., Cai, Y., Gao, Y., & Wu, L. (2008). Heterogeneous network selection using a novel multi-attribute decision method. *Proceedings of International Conference on Communications and Networking in China* (ChinaCom'08), vol. 1, (pp. 153-157).

Dhar, J., Kiran, S., & Reddy, K. (2007). Network selection in heterogeneous wireless environment: a ranking algorithm. *Proceedings of International Conference on Wireless Communication and Sensor Networks* (WCSN′07), vol. 1 (pp. 41-44).

Gallardo, R., Pineda-Rico, U., & Stevens-Navarro, E. (2009). VIKOR Method for vertical handoff decision in beyond 3G wireless networks. *Proceedings of IEEE International Conference In Automatic Control, Computing Science and Electronic Engineering* (CCE'09), vol. 1 (pp. 1-5).

Gowrishankar, R., Raju, G. T., & Satyanarayana, P. S. (2008). Performability model of vertical handoff in wireless data networks. *Proceedings of International Conference on Wireless and Mobile Communications* (ICWMC'08), vol. 1 (pp. 309-314).

Gustafsson, E., & Jonsson, A. (2003). Always best connected. *IEEE Wireless Communications*, *10*(1), 49–55. doi:10.1109/MWC.2003.1182111

Joe, I., Kim, W. T., & Hong, S. (2008). A network selection algorithm considering power consumption in hybrid wireless networks. *IEICE Transactions on Communications. E (Norwalk, Conn.), 91-B*(1), 314–317.

Martinez-Morales, J. D., Pineda-Rico, U., & Stevens-Navarro, E. (2010). Performance comparison between MADM algorithms for vertical handoff in 4G networks. *Proceedings of IEEE International Conference In Automatic Control, Computing Science and Electronic Engineering* (CCE'10), vol. 1, (pp. 1-5).

McNair, J., & Zhu, F. (2004). Vertical handoffs in fourth-generation multinetwork environments. *IEEE Wireless Communications, 11*(3), 8–15. doi:10.1109/MWC.2004.1308935

Nasser, N., Hasswa, A., & Hassanein, H. (2006). Handoffs in fourth generation heterogeneous networks. *IEEE Communications Magazine, 44*(10), 96–103. doi:10.1109/MCOM.2006.1710420

Opricovic, S., & Tzeng, G. H. (2004). Compromise solution by MCDM methods: A comparative analysis of VIKOR and TOPSIS. *European Journal of Operational Research, 156*, 445–455. doi:10.1016/S0377-2217(03)00020-1

Ramirez, R., & Ramos, R. (2009). A vertical handoff decision algorithm which considers the uncertainty during the decision making process. *Proceedings of IFIP International Conference on Wireless and Optical Communication Networks* (WOCN'09), vol. 1 (pp. 1-6).

Rappaport, T. S. (1996). *Wireless communications: Principles and practice*. Prentice Hall.

Song, Q., & Jamalipour, A. (2005). A network selection mechanism for next generation networks. *Proceedings of IEEE International Communications Conference* (ICC'05), vol. 2, (pp. 1418-1422).

Stevens-Navarro, E. (2008). *Mobility management and admission control in heterogeneous wireless networks*. PhD dissertation, The University of British Columbia, Canada.

Stevens-Navarro, E., & Wong, V. (2006). Comparison between vertical handoff decision algorithms for heterogeneous wireless networks. *Proceedings of IEEE Vehicular Technology Conference* (VTC-SPRING'06), vol. 2, (pp. 947-951).

Tawil, R. Salazar, O., & Pujolle, G. (2008). Vertical handoff decision scheme using MADM for wireless networks. *Proceedings of IEEE Wireless Communications and Networking Conference* (WCNC'08), vol. 1, (pp. 2789-2792).

Tawil, R. Demerjian, J., Pujolle, G. & Salazar, O. (2008). Processing delay reduction during the vertical handoff decision in heterogeneous wireless systems. *Proceedings of IEEE/ASC International Conference on Computer Systems and Applications* (AICCSA'08), vol. 1, (pp. 381-385).

Tran, P., & Boukhatem, N. (2008). Comparison of MADM decision algorithms for interface selection in heterogeneous wireless networks. *Proceedings of IEEE International Conference on Software, Telecommunications and Computer Networks* (SoftCOM'08), vol. 1, (pp. 119-124).

Wang, L., & Binet, D. (2009). MADM-based network selection in heterogeneous wireless networks: A simulation study. *Proceedings of International Conference on Wireless Communications, Vehicular Technology, Information Theory and Aerospace & Electronics Systems Technology* (Wireless VITAE'09), vol. 1, (pp. 559-564).

Yang, S., Zhong, Z., & Xu, R. (2009). A cooperative vertical handover algorithm in integrated wireless and mobile network. *Proceedings of International Conference on Wireless Communication, Networking and Mobile Computing* (WiCom'09), vol. 1, (pp. 1-4).

Ying, W., Jun, Y., Yun, Z., Gen, L., & Ping, Z. (2008). Vertical handover decision in an enhanced media independent handover framework. *Proceedings of IEEE Wireless Communications and Networking Conference* (WCNC'08), vol. 1, (pp. 2693-2698).

Yoon, K., & Hwang, C. (1995). *Multiple attribute decision making: An introduction*. Sage Publications.

Zhang, W. (2004). Handover decision using fuzzy MADM in heterogeneous networks. *Proceedings of IEEE Wireless Communications and Networking Conference* (WCNC'04), vol. 2, (pp. 653-658).

ADDITIONAL READING

Balasubramaniam, S., & Indulska, J. (2004). Vertical handover supporting pervasive computing in future wireless networks. *Computer Communications, 27*(8), 708–719. doi:10.1016/j.comcom.2003.10.010

Chalmers, D., & Sloman, M. (1999). A survey of quality of service in mobile computing environments. *IEEE Communications Tutorials and Surveys*, 2nd Quarter 1999. http://www.comsoc.org/publications.

De la Oliva, A., Banchs, A., Soto, I., Melia, T., & Vidal, A. (2008). An overview of IEEE 802.21: media independent handover services. *IEEE Wireless Communications, 15*(4), 96–103. doi:10.1109/MWC.2008.4599227

Durantini, A., & Petracca, M. (2011). Performance comparison of vertical handover strategies for psdr heterogeneous networks. *IEEE Wireless Communications, 15*(3), 54–59. doi:10.1109/MWC.2008.4547523

Eng Hwee Ong & Khan, J.Y. (2010). On optimal network selection in a dynamic multi-RAT environment. *IEEE Communications Letters, 14*(3), 217–219. doi:10.1109/LCOMM.2010.03.092378

3GPP2. (2006). 3GPP2 to WLAN interworking, S.R0087-A, v1.0. http://www.3gpp2.org/public_html/specs/index.cfm.

3GPP. (2007). Requirements on 3GPP system to wireless local area network (WLAN) interworking, TS 22.243, v8.0.0. http://www.3gpp.org/specifications.

Guo, Q., Zhu, J., & Xu, X. (2005). An adaptive multi-criteria vertical handoff decision algorithm for radio heterogeneous networks. *Proceedings of IEEE International Communications Conference* (ICC'05), Vol. 4, (2769-2773).

Hou, J., & O'Brien, D. C. (2006). Vertical handover-decision-making algorithm using fuzzy logic for the integrated radio-and-OW system. *IEEE Transactions on Wireless Communications, 5*(1), 176–185. doi:10.1109/TWC.2006.1576541

Hui, S. K., & Yeung, K. H. (2003). Challenges in the migration to 4G mobile systems. *IEEE Communications Magazine, 41*(12), 54–59. doi:10.1109/MCOM.2003.1252799

Kassar, M., Kervella, B., & Pujolle, G. (2008). An overview of vertical handover decision strategies in heterogeneous wireless networks. *Computer Communications, 31*, 2607–2620. doi:10.1016/j.comcom.2008.01.044

Ormond, O., Murphy, J., & Muntean, G. (2006). Utility-based intelligent network selection in beyond 3G systems. *Proceedings of IEEE International Communications Conference* (ICC'06), Vol. 4, (1831-1836).

Pollini, G. (1996). Trends in handover design. *IEEE Communications Magazine*, *34*(3), 82–90. doi:10.1109/35.486807

Sgora, A., & Vergados, D. (2009). Handoff prioritization and decision schemes in wireless cellular networks: a survey. *IEEE Communications Tutorials and Surveys*, 11(4), http://www.comsoc.org/publications.

Song, Q., & Jamalipour, A. (2005). Network selection in an integrated wireless LAN and UMTS environment using mathematical modeling and computing techniques. *IEEE Wireless Communications*, *12*(3), 42–48. doi:10.1109/MWC.2005.1452853

Stemm, M., & Katz, R. (1998). Vertical handoffs in wireless overlay networks. *ACM Mobile Networks (MONET). Special Issue on Mobile Networking in the Internet*, *3*, 335–350.

Stevens-Navarro, E., Lin, Y., & Wong, V. (2008). An MDP-based vertical handoff decision algorithm for heterogeneous wireless networks. *IEEE Transactions on Vehicular Technology*, *57*(2), 1243–1254. doi:10.1109/TVT.2007.907072

Stevens-Navarro, E., Pineda-Rico, U., & Acosta-Elias, J. (2008). Vertical handover in beyond third generation (B3G) wireless networks. *International Journal of Future Generation Communication and Networking*, *1*(1), 51–58.

SuKyoung, L., Sriram, K., Kyungsoon, K., Yoon Hyuk, K., & Golmie, N. (2009). Vertical handoff decision algorithms for providing optimized performance in heterogeneous wireless networks. *IEEE Transactions on Vehicular Technology*, *58*(2), 865–881. doi:10.1109/TVT.2008.925301

Taniuchi, K., Ohba, Y., Fajardo, V., Das, S., Tauil, M., & Yuu-Heng, C. (2009). IEEE 802.21: media independent handover: features, applicability, and realization. *IEEE Communications Magazine*, *47*(1), 112–120. doi:10.1109/MCOM.2009.4752687

Wang, H., Katz, R., & Giese, J. (1999). Policy-enabled handoffs across heterogeneous wireless networks. Proceedings of *Second IEEE Workshop on Mobile Computing Systems and Applications (WMCSA'99)*, 1999, (pp. 51-60).

WiMAX Forum. (2008). WiMAX forum network architecture. *Stage 2: Architecture Tenets, Reference Models and Reference Points (3GPP – WiMAX Interworking)*, Rel. 1, ver. 1.2. http://www.wimaxforum.org/resources/documents/technical.

WiMAX Forum. (2008). WiMAX forum network architecture. *Stage 2: Architecture Tenets, Reference Models, and Reference Points (3GPP2 – WiMAX Interworking)*. Rel. 1, ver. 1.2. http://www.wimaxforum.org/resources/documents/technical.

Zahran, A., Liang, B., & Saleh, A. (2006). Signal threshold adaptation for vertical handoff in heterogeneous wireless networks. [MONET]. *ACM Mobile Networks and Applications*, *11*(4), 625–640. doi:10.1007/s11036-006-7326-7

Zhang, J., Stevens-Navarro, E., Wong, V., Chan, H., & Leung, V. (2008). Protocols and decision processes for vertical handovers. In Y. Zhang, L. Yang, J. Ma (Eds.), *Unlicensed Mobile Access Technology: Protocols, Architectures, Security, Standards and Applications* (123-145). Boca Raton, FL: Auercach Publications, Taylor & Francis Group.

KEY TERMS AND DEFINITIONS

Beyond Third Generation (3G) or Fourth Generation (4G) Wireless Networks: See definition of heterogeneous wireless networks.

Heterogeneous Wireless Networks: Envisioned wireless system integrated by different radio access technologies (e.g., 3G cellular, WLANs, WMANs, etc.).

Horizontal Handover or Handoff: Mobility process in charge of the transfer of a connection among cells of the same wireless system.

Multiple Attribute Decision Making (MADM) Process: Decision process consisting of selection of one alternative among several alternatives with different and usually conflicting parameters.

Vertical Handover Decision: Process of selection of network for vertical handover.

Vertical Handover Decision Algorithm: Set of rules and/or mathematical procedures in charge of performing the handover decision.

Vertical Handover or Handoff: Mobility process in charge of the transfer of a connection among cells of heterogeneous networks.

Chapter 4
Context-Aware Delivery of Multi-Party Communications:
Access and Core Network Selection

Nuno Coutinho
Instituto de Telecomunicações, University of Aveiro, Portugal

Susana Sargento
Instituto de Telecomunicações, University of Aveiro, Portugal

ABSTRACT

Heterogeneity is seen as the prevailing feature in Next Generation Networks, requiring the development of novel solutions to overcome the challenges posed by future communication environments. Thus, the major goal is to transform the diversity into an opportunity to increase user experience, which can be achieved through context-awareness. Considering this, and given the increasing demand for group-based multimedia services, this chapter focuses on providing an effective multimedia group communication in heterogeneous environments, developing an architecture driven by the available context information. It introduces a hierarchical architecture that employs the concept of abstract trees to control multiparty content distribution in a scalable way, being the trees built based on intelligent network selection schemes fed by available context information. The chapter's ultimate goal is to find the best compromise between service personalization and resources optimization, developing an architecture flexible enough to properly react to context changes.

INTRODUCTION

Future networking environments will be characterized by their diversity in features, which may be found at different levels (Figure 1), ranging from the terminals to the services: user devices have distinct processing capabilities and mobility features; the panoply of access network technologies differ on bandwidth and coverage; the co-existence of different transport technologies in the core and the variety of application requirements. Nevertheless, more than a challenge, this diversity represents an opportunity to differentiate the services provided and, as a consequence, enhance user experience. In this sense, the adoption of the increasingly popular concept of context-awareness will endow

DOI: 10.4018/978-1-4666-0017-1.ch004

future architectures with the ability to sense surrounding conditions and take advantage of that information towards better network management decisions and user service perception.

Novel architectures driven by context have been proposed, focusing on adaptable content delivery according to user characteristics and network resources. Furthermore, the recent growth of group-based multimedia applications (e.g. IPTV) poses new challenges regarding the use of context towards service personalization, which can also be an added value for service providers that might differentiate and enrich their offer. This way, multiparty content transport technologies as multicast can be enriched by employing context, enabling the development of novel applications that increase the quality perceived by users. Moreover, the same information can also be used towards better network efficiency, making a better use of available resources and adapting content distribution to the network context.

The focus of this chapter relies on the challenges to provide an effective multimedia group communication on Next Generation Networks (NGN), developing an architecture that may overcome the limitations imposed by the heterogeneity inherent to these networking environments. Our approach supports a context-driven architecture, sensitive to network resources availability and user characteristics, in order to find the best compromise between two conflicting aspects of a multiparty content delivery architecture: service personalization based on user context and network resources optimization considering group communications.

Nevertheless, context-aware networks face several challenges that must be addressed: from one side, any type of context (e.g. preferences, location, mobility, resources) can be used to select the best networks for the groups of each service; from the other side, any context change may require a complete restructuring of the network and its sessions. Thus, it is critical to wisely select the context information that triggers network adaptations, so as to avoid constant re-arrangements that can impair the quality of the service provided.

To pursue these architectural goals, we present a hierarchical architecture that employs the concept of abstract trees to control multiparty content distribution in a scalable and flexible way. The abstract trees are built based on overlay nodes distributed throughout the network, being in charge of controlling the content distribution to the multiple receivers. Using this approach, it

Figure 1. Next generation network heterogeneity

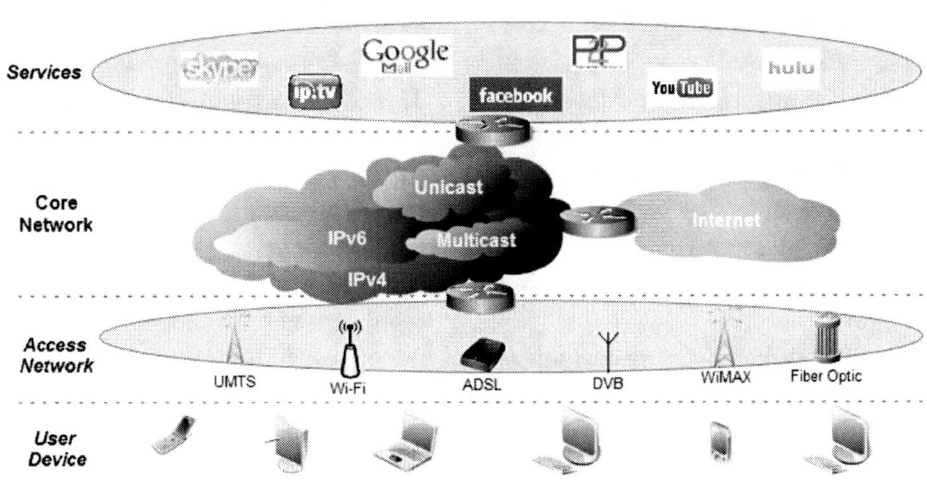

is possible to increase the flexibility of the distribution trees as well as hide network dynamics and heterogeneity of NGN. These abstract trees are built according to two main intelligent selection schemes: context-aware access and core network selection. However, group communication over heterogeneous environments increases the difficulty in deploying network selection. Considering different user interests, devices and networks characteristics, it is complex to fulfill all requirements simultaneously (different users, access networks, location).

In order to overcome this issue we introduce the grouping of interested users according to their current context, such as terminal capabilities, network conditions, location or others. This way, the same session content can be sent simultaneously to different sub-groups in distinct formats. Moreover, resilience can be improved through grouping approach, where only some sub-groups can be adapted so as to avoid affecting all interested users.

Under the scope of this work, we developed two intelligent selection schemes that take advantage of the context information sensed and capitalize the diversity of NGN. The purpose of these mechanisms is to select the networks that best fit communication requirements, and also adapt content distribution to current network, users, session and surrounding environment conditions. The first one is designed to select the best access network according to user, device, application and network context. The second one selects a core network path able to meet the requirements of the service provided. This way, our architecture is able to adapt the content distribution to user and network context, attempting to optimize the compromise between service personalization and network efficiency.

In this chapter it is also included an evaluation of the architecture presented through simulation. The results show that an architecture driven by context is feasible, being also scalable and flexible to the context changes. Beyond the network adaptations that are performed at different levels, the intelligence introduced in the network management framework is able to use the context information available towards the provision of more personalized services while handling network resources efficiently.

The remainder of the chapter is organized as follows. The Background Section presents a brief overview about the related work. Next we introduce the architecture developed, concepts associated and details the main components involved, including the overlay control strategy and an analytical analysis. We describe the selection schemes developed, both in access and core networks. The results obtained through the simulation of the architecture are also presented. Finally, we provide the conclusions about the work developed and the future work.

BACKGROUND

Context

Context-awareness is a very popular concept, originally introduced in computing environments by (Schilit & Theimer, 1994), and defined years later in a broader sense by (Abowd et al., 1999) as *"any information that can be used to characterize the situation of an entity."* Context-awareness has been highlighted as the driving force towards the dynamic adaptability of networks and services to user features and communication environment. Thus, context can be used to develop novel architectures driven by the sensed context, and flexible enough to adjust to any changing surrounding condition. Nevertheless, these sensitive architectures pose several challenges and an added complexity, since constant changing environments require continuous adaptations. Moreover, given the vague definition of context and considering that context-aware services are dependent of the information used, it is important to assess the quality of the context data used, as suggested in (Thomas Buchholz 2003).

A context-aware solution must concern three primary procedures: context acquisition, context modeling and context reasoning. The first mechanism is responsible for gathering relevant information from the several entities involved in the communication. Typically, these entities are users, network elements and surrounding environment. The collected information is usually in a raw state, and so it needs to be modeled into a suitable format so that it can be used by reasoning mechanisms. These procedures consist on inferring conclusions and decisions based on the modeled context information.

There are several proposals in the related work addressing context information management (gathering, modeling and reasoning) (Gu, Wang, Pung, & Zhang, 2004; Strang & Popien, 2004). An architecture supporting context-aware services is proposed in (Riva, Teittinen, Siikavirta, & Huovinen, 2008), which uses information about location, device and identity across multiple access networks in a seamless way. (Williams et al., 2005) introduce the Context Management Subsystem to collect and store environment knowledge through sensors, which is later used to personalize and adapt the service provided. (Mathieu et al., 2007) presents a more comprehensive architecture, employing context, not only focused on users, but also on network capabilities. It uses flexible overlay networks that attempt to meet service requirements and network adaptability based on self-management schemes.

Multicast

Context can also play an important role regarding group-based services (e.g. IPTV) and how content is delivered to users. Considering the growth of this kind of multimedia applications, multicast (Deering & Cheriton, 1990) is the most eligible transport technology to deal with group-based services, allowing a considerable saving of resources. Still, it has seen a slow commercial deployment in the Internet, justified by its original open service model and complexity (Diot, Levine, Lyles, Kassem, & Balensiefen, 2000; El Hachimi, Abouaissa, Lorenz, & Lee, 2004; Lao, Cui, Gerla, & Chen, 2007), which stalled service providers to adopt this technology. The use of multicast is clearly more attractive to network administrators than for receivers, given the bandwidth savings. For receivers, the way they receive content is irrelevant, unicast or multicast, as long as the quality of the service (QoS) is guaranteed. On the contrary, multicast is very useful for content sources, so they may offer their services to a larger number of customers in a very scalable way. However, multicast routing involves scalability issues, since each router has to maintain much more information comparing to unicast routing (groups, sources, group addresses, interfaces). Consequently, the router table lookup will be more complex and slow.

Given these issues, the research community was stimulated to develop novel strategies that could be more attractive to operators in order to adopt multicast basic principles in a very easy and scalable manner. Incremental deployment has been a common approach in order to accelerate the deployment of native multicast concepts, as proposed by (Costa, Fdida, & Duarte, 2001; R. Boivie, 2007). Application Layer Multicast (ALM) (Hosseini, Ahmed, Shirmohammadi, & Georganas, 2007) is another branch of strategies to improve the interoperability between unicast and multicast, implementing multicasting functionality at the application layer, instead of the network routers. The main idea is that by moving multicasting to the application layer, it is possible to overcome infrastructure deployment issues. Moreover, this is a more scalable solution, since routers do not need to maintain per-group state and special addresses. Novel mechanisms for congestion control and reliability can also be created in a more easy way. However, this strategy allows multiple copies of the same packet on the same link, building non-optimal distribution trees, unlike native multicast. Another drawback is that this approach is not compatible with native multicast if it is supported.

Under the ALM topic, Narada (Yang-hua, Rao, Seshan, & Hui, 2002) is one of the first and most popular. It was designed for collaborative applications with small scale multicast groups considering multiple sources. Narada starts by building a mesh control topology over participating nodes. Then, the data topology is built on top with the participants self-organized into source-rooted multicast trees. Other popular ALM proposal are NICE (Banerjee, Bhattacharjee, & Kommareddy, 2002) and Scribe (Castro, Druschel, Kermarrec, & Rowstron, 2002). NICE is a scalable application layer multicast protocol which adopts a hierarchical clustering strategy to support a large number of members. Though, it is targeted to multicast applications that require low bandwidth. Hosts are organized in a hierarchy of layers, and within each, there are different clusters of hosts achieving a scalable solution with a very low control overhead. Scribe supports a large number of groups and members, being built over a source-based multicast shared tree on top of Pastry (Rowstron & Druschel, 2001).

Unlike ALM, which only supports one group or service per overlay, overlay multicast has the advantage of supporting multiple groups and applications simultaneously in the same overlay node. This fact makes overlay multicast more suitable for environments and applications with a large number of groups. As for ALM, overlay multicast does not require changes in the infrastructure. Another advantage relies on the transparency of overlay network, hiding the dynamics of the substrate network, which is enhanced with the ability to re-route around link failures or congestions. Nevertheless, these benefits are accomplished at the cost of maintaining overlay nodes: orchestration between them.

Overcast (Jannotti, Gifford, Johnson, Kaashoek, & James W. O'Toole, 2000) is an example of overlay multicast, aiming to maximize the bandwidth from the source to all members in a scalable and reliable way. Overcast system architecture relies on a set of proxies organized in a distribution tree rooted at a central source for single source multicast. A distributed tree-building protocol is used to create this source specific tree. Then, a self-organizing algorithm is adopted to maximize the bandwidth from the root to a node. Since the main goal of Overcast is to maximize bandwidth, it potentially achieves it at the cost of delay increase, which is not suitable for real-time multicast applications.

Another approach that follows the overlay principle is Scattercast (Chawathe, 2003), which proposes routing and forwarding services on top of an application layer infrastructure. To enable this, it uses infrastructure service agents, named ScatterCast proxies, which use a specific protocol to self-organize into an application level mesh. Over these proxies of the mesh network, a routing algorithm is used to build a source-rooted distribution tree, where the routing metric used is latency.

Quality of Service Architectures for Multicast

Despite these incremental deployment solutions represent a step forward regarding the proliferation of multicast principles in real network services, they do not address mechanisms to QoS assurance. However, research under this topic has been made, as the integration of class-based QoS and IP multicast (R. Bless, 2004). A problem related with the integration of QoS and multicast is routing asymmetries (E. Cerqueira et al., 2006). However, this issue is not tackled by most IP multicast routing protocols, since these protocols build multicast trees from the receivers to the sender, while data travels in the reverse direction. The *QoS Architecture for Multi-User Mobile Multimedia* (Q3M) (Eduardo Cerqueira et al., 2007) is one of the works that addresses this issue, introducing the Multi-service Resource Allocation (MIRA) (Neto, Cerqueira, Rissato, Monteiro, & Mendes, 2007) protocol, designed to control the resources of the CoS and multicast trees in a synchronized manner taking into account the problem of asymmetric routes. A centralized structure to support QoS was developed in (Mitrano et al., 2007), enabling QoS and mobility of multicast sessions in heterogeneous environments. More recently, new algorithms and techniques were proposed, such as the QoS mobile multicast routing based on ant colony optimization (Kewen & Jing, 2008), which has a strong global optimization performance.

Context-Aware Access Selection

The heterogeneity of NGN poses, more than several integration challenges (Suk Yu & Kai Hau, 2003), a set of opportunities that can be enhanced by wisely employing context information in network management decisions. In this sense, the variety and co-existence of several access technologies in future networking environments enabled new communication paradigms as the *Always Best Connected* service envisioned by (Gustafsson & Jonsson, 2003). The idea is to offer, to each service, the access network that best fit their communication needs, considering simultaneously user preferences and network resources. Significant research under this topic was developed using different approaches to tackle a common problem: multi-constraint decisions (Iera, Molinaro, Campolo, & Amadeo, 2006; Prehofer, Nafisi, & Wei, 2003). The work developed in (Qingyang & Jamalipour, 2005) focus on a selection algorithm which constraints are related with network conditions, service application, and user preference for QoS. To deal with them, the authors initially perform an analytical hierarchy process, dividing the problem in sub-problems, and then execute a Grey Relational Analysis. (Xing & Nalini, 2005) model the access selection problem as the bin packing problem, mapping all traffic flows throughout the available accesses aiming to fulfill their requirements. Approximation algorithms are proposed to achieve a near-optimal solution and reduce computational effort.

(Chiron et al., 2008) proposes a flexible network selection decision center, which depends on mobility management policies. To scale the solution, decisions involving non critical constraints are performed at a global level, while critical parameters (signal strength, location) are managed locally. A novel architecture is proposed in (Chen & Yang, 2007), embracing mechanisms for access discovery, network selection and seamless handovers. The access selection scheme is based on a single objective optimization problem which is a simple logic comparison with user defined threshold values. More recently, (Jesus, Sargento, & Aguiar, 2008) proposed a very flexible access network selection algorithm that is able to deal with any constraint or context information. The main purpose is to provide a ranked list of a feasible distribution of the flows of each terminal. To achieve this, network characteristics and user preferences are modeled in different matrices, being the algorithm itself a simple manipulation of these matrices.

Despite all the significant work developed under the access network selection subject, we claim that these approaches are limited regarding the type of context used, given that most of them do not concern simultaneously user and network perspectives. Beyond these issues, none of the described schemes address network selection for group-based communications, which motivates our research.

ARCHITECTURE

Architecture Overview

Nowadays, context-awareness is seen as the driving force to develop novel communication paradigms aiming to increase user quality of experience (QoE). Thus, we believe that the development of a context driven architecture able to react and adapt to context changes is of crucial importance regarding future communication environments. This is the scope of our work, the development of a network architecture focused on context-aware multiparty communications, and able to overcome the challenges posed by NGN. The higher level objective of this work is to pursue the best compromise between service personalization and network resource optimization.

The architecture aims essentially to develop an effective multiparty content delivery framework that is, not only able to cope with NGN heterogeneity, but also takes advantage of context to enhance user experience. To accomplish this, we present a context-driven architecture that relies on three main frameworks for the provision of a personalized end-to-end context-aware communication, Figure 2: Multiparty Session Management, Multiparty Transport and Context Detection and Distribution.

The Multiparty Session Management is responsible for creating the session context, consisting of session's group members' identification, codec used and QoS requirements in terms of delay, loss rate and jitter. The novel feature is performed by the Group Manager element which groups or sub-groups members taking into account session context and user's preferences, device capabilities and environment context. These subgroups may use different media encodings of the same content, allowing a better match between user context and current network conditions.

The Context framework is composed by several distributed sensors, denoted as context providers, which are responsible for acquiring context information from network and terminal part, providing the information to a context database, the Context Broker (CB). This framework is transversal to the entire architecture since all other elements require some knowledge to perform their functions. The context acquired can be later retrieved from the CB which is responsible for dynamically providing context or even forecast events after performing some reasoning.

The focus of our work relies on the Context-Aware Multiparty Transport Framework, depicted on Figure 2, which comprehends a set of elements to control, in a hierarchical way, the content delivery and network resources. The central decision entity, named Network Use Management (NUM), is responsible for using the knowledge gathered towards content distribution adaptation and personalization. The second level of control is performed by an overlay network, which contains a set of nodes in the network that enforce NUM decisions on network elements, such as resource reservation and multicast tree maintenance. For each multiparty session, we define an overlay network to support an efficient and scalable network adaptation to any context changes, providing a flexible resource allocation mechanism able to offer multicast services with delivery guarantees.

The integration of the above described frameworks allows the development of a novel multiparty transport approach that takes advantage of several context sources (terminals, network and

Figure 2. Global architecture overview

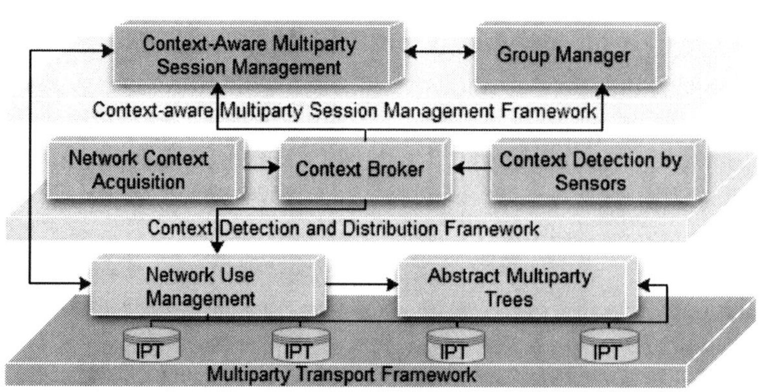

environment) and collects relevant information to be used in the optimization of the delivery process. This consciousness makes the architecture able to react to relevant events, adapting it to the new conditions. The Multiparty Transport framework and its concepts and mechanisms will be detailed below.

Multiparty Transport Framework

The Multiparty Transport framework developed aims to enable an effective multiparty content delivery in heterogeneous networking environments. Though, this effectiveness has two main perspectives regarding service provision architectures: user and operator. In view of the first, effectiveness is a good perception of the service provided (high levels of QoE), while in view of the second, it means an efficient use of the resources available. This work will pursue the optimization of the compromise between these two perspectives, comprising context-aware network selection (both access and core) and seamless content transport.

A hierarchical control model was adopted, considering scalability concerns that usually come along with context-sensitive architectures. Our strategy consists of two levels of intelligence, Figure 3: the higher level is centralized on a unique entity, NUM, and the lower level is distributed throughout an overlay network. Figure 4 depicts the context-aware procedures that are performed at NUM: grouping, access network selection and core network selection. Since it is almost impossible to meet all users' preferences simultaneously, we propose to group users according to their context. This way, the same content may be sent using different formats according to sub-groups context. Moreover, resilience can be improved through grouping approach, where only some sub-groups can be adapted, avoiding to affect all users with ongoing sessions. Given the variety of context sources from different communication actors (service, user, device and network), within each of these entities, there are several characteristics that can be used by context-aware selection schemes:

- **Application:** requirements quantified in QoS metrics (bandwidth, packet loss, delay);
- **User:** the user as a person has its own preferences and profile, which can be used to improve its QoE. Location, trajectory and

Figure 3. Hierarchical control architecture

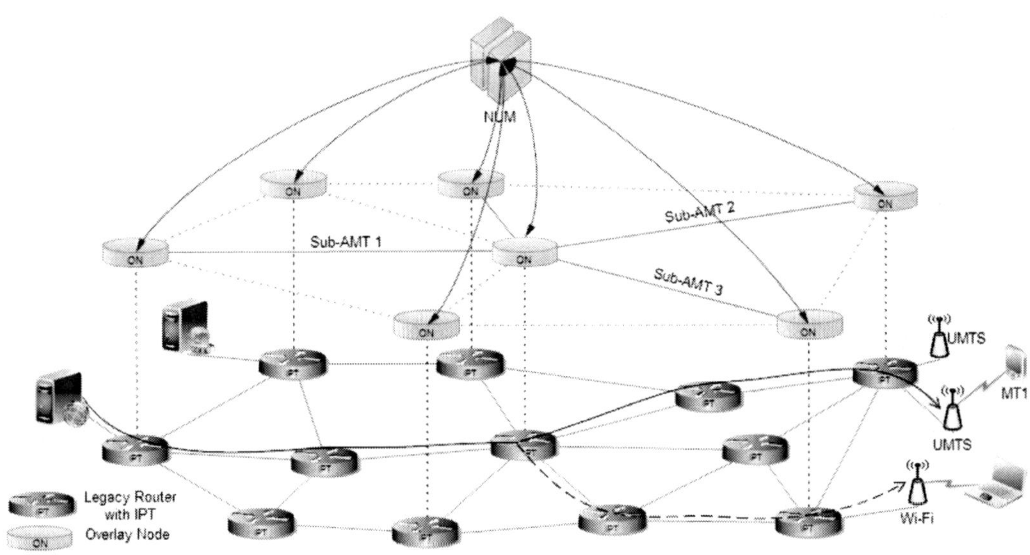

speed can also be used by proactive and predictive algorithms;

- **Device:** user's terminal related characteristics such as display resolution, available interfaces and battery status;
- **Network:** network resources, technology, performance feedback, reliability, operator policies and cost;
- **Environment:** noise, weather, type of building.

Our second level of control consists on a network of Overlay Nodes (ONs) that build the so-called Abstract Multiparty Trees (AMT). Each AMT controls the content distribution of a certain session, providing a generic and scalable transport solution for group communications. Local network segments, delimited by two ONs, define a sub-AMT (Figure 3), which can be mapped in several physical links and nodes. Each ON performs resource and QoS control in a distributed manner through the Internet Protocol Transport (IPT) components present in each network router. These abstract trees operate on top of the IP layer, allowing hiding the network dynamics, changes in the multicast session and tree. A formal definition and analytical analysis of this concept will be detailed in the following sections.

We will first introduce a formal definition of the network and the overlay structure used. We consider a network formally defined as $G=(V,E)$, where $V=\{v_1,v_2,...,v_{|V|}\}$ represents the set of nodes. E is the link set, where $e=(v_a,v_b)$ ϵE represents the link connecting v_a to v_b. Among the available network nodes (V), there is a subset of nodes $V_O \subset V$ that can perform overlay functions and being edges of AMTs. Thus, considering the substrate network G we define $O=(V_O, E_O)$ as the overlay network, where E_O is the set of logical links and $e_O=(v_{Oa},v_{Ob})$ represents the overlay connection. Moreover, an overlay connection between two ONs may be mapped into a set of physical network links $e_O=\{e_1,e_2,...,e_l\}$, being l the number of physical links in an overlay connection. We also define the set of existent multicast sessions as M, where each one has associated a content source S_{Mi} and a set of flows $F_{Mi}=\{F_1,F_2,...,F_{|F|}\}$. Each flow has a set of receivers R_{Fi} and a multicast tree G_{Fi}. The available set of access points (APs) in the network is defined as $Z=\{z_1,z_2,...,z_{|z|}\}$, having each receiver an associated access point z_j.

Figure 4. Context-aware mechanisms

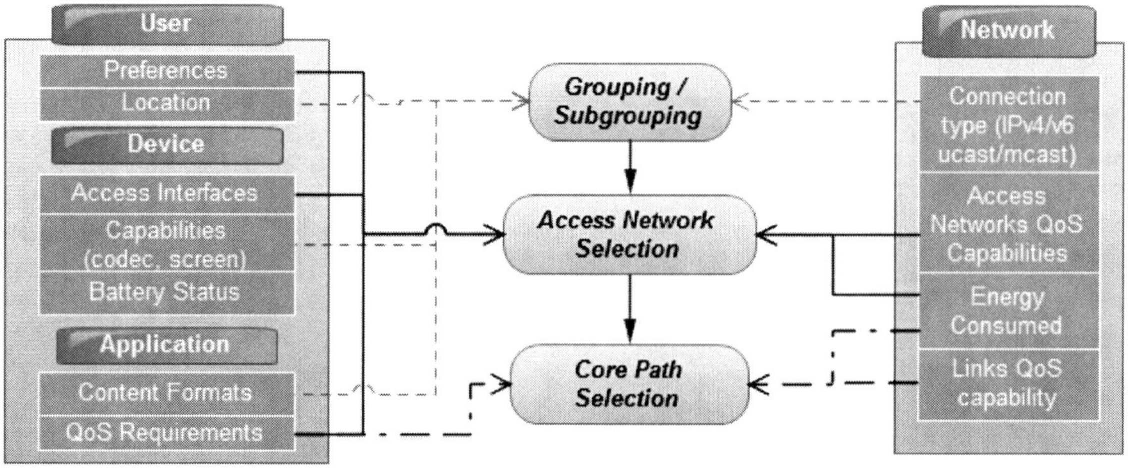

Context-Aware Grouping

Following our strategy of context adaptation and service personalization, the concept of grouping and sub-grouping was a feature added to the architecture developed. Although the same content is sent to all group members, its delivery needs to be adapted for each user based on his particular context. Considering the variety of context information available, adaptations are performed at Session and Network levels based on the context of the respective layers. At the session level, users can be arranged according to their preferences and/or device capabilities, such as media codecs or resolutions supported. In case of ongoing sessions, the grouping mechanism must also react if any relevant context change occurs (handovers, QoS degradation or device mobility), adapting user's session accordingly. This process can be performed by transferring users to a different sub-group or creating a new one more appropriate.

Our focus is on sub-grouping at the network level. At this layer, we consider two different adaptation processes that may set different sub-groups (Figure 5): access technology and network resources. The former is based on aggregating users whose share the access that best fit their communication needs. The access selection process will be described in a later section. Concerning the network resources, as at the session level users are grouped according to coding abilities, each group/coding will have different QoS requirements. Thus, at the network level, each sub-group has associated a multicast delivery tree with the proper network resources reservations that aim to meet the demands of the codecs used.

Using this approach, it is possible to provide different QoS levels to the members of a specific group that receive the same content, however in different modes. Following the same principle that was described before for the session level, network-based context changes may also trigger sub-groups re-arrangements to maintain the QoS levels required. For instance, if a user changes his access network for some reason, and the QoS levels cannot be fulfilled within the same group, different codecs may be used and the user is transferred to a more suitable sub-group.

Given the users' heterogeneity, including too many constraints in the sub-grouping process can be greedy, taking the risk of obtaining many sub-groups and do not take advantage of the network resources that could be saved using groups with more users. A fair compromise is required in order to have a fair as possible personalized group-based communication.

Network Use Management

NUM takes advantage of the information collected about user, environment, network and session to perform intelligent network selection decisions in terms of communication path, terminal interface and access technologies. The first scheme performs access network selection, aiming to achieve a compromise where all multi-homed terminals are connected to the access networks that best suit their communication needs. To accomplish this, NUM considers users' context (preferences, location, and profile), environment, session requirements and network context (QoS capabilities and capac-

Figure 5. Context-aware grouping

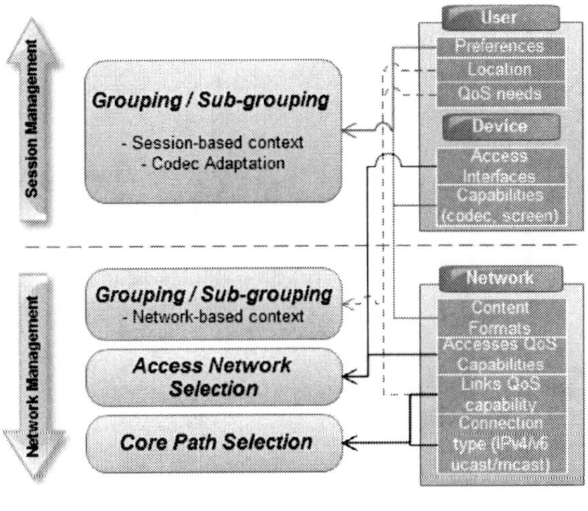

ity) in its access selection algorithm. Regarding networking context (e.g. bandwidth, delay, loss rate), a second algorithm is used to perform routing adaptations and core path selection, trying to meet the session's QoS requirements and node's multicast capability. These selection mechanisms are expected to improve the resource utilization, distributing in a more equitable way the traffic load while fulfilling simultaneously the QoS required by the session and user expectations.

The functionalities performed by NUM are detailed on Algorithm 1, which can be triggered by two main events: a new multicast session request, sent by the Session Management component, or an event that requires network rearrangement. NUM receives from the Session Management information about Multicast Session (M_i), Session Source (S_{Mi}), Session Flows (F_{Mi}) and Receivers associated to each Flow (R_{Fi}). The remaining required information is obtained from the context broker, as network knowledge (G), APs available (Z) and users' context information. In case of a new multicast session request, NUM will select the most suitable AP according to user and network context and select the several ONs to build the sub-AMTs.

In case of being an event that triggered NUM, different reaction strategies may be applied according to the cause of the event.

In the case the event is a link/node failure or QoS degradation, the recovery can be done internally to the sub-AMTs through the ONs, or the recovery needs a change on the sub-AMTs and it is required to perform a new core selection procedure to overcome the failure. On the other hand, if the event's cause is on the access network part, beyond the always necessary new access selection, it may also be required a new core selection if the selected AP is connected to a new ON (v_O). The output of NUM will always be a set of sub-AMTs O_{Fi} to control the multicast tree and the resources for a specific flow F_i.

Algorithm 1.Num Functionalities

```
input  : G,M_i,S_{M_i},F_{M_i},R_{F_i},Z,O_F = Empty
output: An AMT for each Flow: O_{F_i}, i = 1 → |F|

if New M_i Request then
    Request Context from CB;
    for i = 1 → |F| do
        for j = 1 → |R_{F_i}| do
            z_{R_j} =AccessSelection(R_j,F_i,G,Z);
            O_{F_i} =CoreSelection(O_{F_{i-1}},S_{M_i},F_i,G,z_{R_j});
        end
    end
else if Ongoing M_i then
    switch Event do
        case Core event
            O_{Fi}=CoreSelection(O_{F_i},F_i,G,z);
        case Access event
            z_{R_j} =AccessSelection(R_j,F_i,G,Z);
            if z_{R_j} is connected to a new v_O then
                O_{F_i} =CoreSelection(O_{F_i},F_i,G,z_{R_j});
            end
        endsw
    endsw
end
```

ABSTRACT MULTIPARTY TRANSPORT

The AMT principle consists in allowing end-to-end multicast content transport over network segments with different transport technologies (i.e., unicast and multicast), and also providing independence between source and listener trees (through different AMTs), and resilience support. Each AMT is composed by local network segments called sub-AMTs. All the physical nodes within the respective network segment employ the same transport technology. NUM coordinates the edges of each sub-AMT to implement proxy functionalities, mediating overlay connections and controlling the QoS of the sub-AMT independently from the others. Each sub-AMT is defined as an overlay connection (e_O) between two adjacent ONs (v_{Oa}, v_{Ob}). All core nodes composing a sub-AMT must implement the same transport technology. In the scope of a sub-AMT: (i) ingress ON, v_{Oa}, is viewed as a session source; (ii) egress ON, v_{Ob}, as leaf node; (iii) and core nodes simply perform IP forwarding operations.

Considering a multicast session flow F_i, an overlay structure $O_{Fi} \subset O$ is created, operating on top of the IP network layer to support general transport control in multicast trees (G_{Fi}). Applying this overlay paradigm at the transport layer, it is possible to provide a scalable transport service for group communications. The end-to-end AMTs will then be divided into sub-AMTs, generated between the ONs, which will split and scale the end-to-end multicast tree. This abstraction level and division of trees enables end-to-end multicast content transport over heterogeneous network segments in terms of IP multicast capability or IPv4/v6 support (each sub-AMT has associated a unique multicast address), also hiding network dynamics (the reconfigurations inside each sub-AMT can be performed locally to the sub-AMT), and providing independence between source and listeners (several sub-AMTs are supported independently). An analytical study regarding this topic will be given later.

AMT Analytic Analysis

Scalability, as previously stated, must be a major concern in the development of service-oriented architectures. In this sense, it becomes mandatory to extend the assessment of the mechanisms developed for large networks. Moreover, since we are dealing with context-sensitive frameworks, it is also important to evaluate the ability to perform constant adaptations and reconfigurations in a large scale. The Context-aware Multiparty Transport framework presented aims to efficiently respond to the dynamics imposed by the constant adaptations to context changes. The AMT strategy endows the architecture with a new level of scalability, avoiding when possible global network reconfigurations and multicast tree re-arrangements.

For a better understanding of the flexibility added by the AMT concept, we consider an AMT associated to a flow F_i formally defined as $O_{Fi}=\{V_O,E_O\}$. Assuming that an event triggers a network reconfiguration, the outcome of this process is an updated version of the previous AMT, $O'_F=\{V'_O,E'_O\}$. From a theoretical perspective, the advantages of employing abstract transport and of reducing the impact on the network depends on the amount of links and configured resources that can be reused by the new overlay structure O'_F. Thus, the efficiency of the AMT concept depends on various factors: the ratio between overlay capable nodes and normal nodes, $\lambda = V_O / V$; and the average number of physical links ($\bar{l}$) that maps an overlay connection ($e_O=\{e_1,e_2,\ldots,e_l\}$).

Considering this, and defining the previous AMT as O_{Fi} and the new one as O'_{Fi}, the number of reused overlay nodes is given by $r = |V_O \cap V'_O|$. We may then assume that $r \propto \lambda$, meaning that, increasing the percentage of ONs in the network will decrease the number of configurations required, since part of the overlay is reused. Thus, it is possible to diminish the amount of control overhead and only perform the resource reservation for the new branch of the AMT. However, this fact does not mean that a network full of overlays would be the best solution. Since the control architecture is hierarchical, there is a trade-off between the gains obtained by reusing previously allocated overlay connections and the amount of control that NUM has to perform in order to manage the overlay network.

Moreover, it is implied that $\lambda \propto \frac{1}{l}$, which reduces the physical range of each overlay connection (sub-AMT). As ONs are endowed with the ability to perform some autonomous functions within the scope of their associated sub-AMTs (e_O) (e.g. re-routing and resource allocation) increasing their number will cause a loss of the self-configuration ability of the architecture.

Numerical Results

Our work also comprehended a study on Matlab (Mathworks, 2010) to analyze the behavior of the concepts presented. Since PIM-SSM (S. Bhattacharyya, 2003) can be considered as a particular case (unique source) of PIM-SM, the most deployed multicast protocol, we decided to compare the AMT transport strategy with PIM-SSM. Despite it does not address QoS issues, it tends to have a "selfish" behavior concerning the multicast tree construction. To build its multicast distribution tree, PIM-SSM uses the unicast routing tables to forward *Join* messages until it finds the respective source tree. Thus, considering receivers of the same group sparsely located in the network, it is likely that each one has an associated tree whose set of nodes and links is disjoint from another receiver's source tree. In this case, a tree rearrangement triggered by any kind of event would require a complete setup of the path between the receiver and the source, being unaware of tree branches that could be reused. Using the AMT strategy with the control associated, we attempt to aggregate the content distribution as much as possible, while providing QoS guarantees.

To assess the validity of previous mentioned inferences, an implementation of the behavior of both approaches was implemented on Matlab in order to observe the gains that we hope to achieve regarding the distribution tree construction and maintenance. The focus of our analysis will be on the difference in the reconfigurations that each of the approaches require when dealing with terminal joins or leave, allowing to understand the ability of each strategy to deal in a scalable way with constant changing environments.

Since we are focused on assessing this behavior for large networks, we adopted the Waxman (Waxman, 1988) random network-level topology generator. Following this model, the nodes of the network are uniformly distributed in the plane and edges are added according to probabilities that depend on the distances between the nodes. The probability of connecting nodes *u* and *v* is given by:

$$P_{(u,v)} = \alpha e^{\left(\frac{-d}{\beta L}\right)}$$

where $\alpha>0 \wedge \beta<1$, d is the distance from u to v, and L is the maximum distance between any two nodes. The α parameter allows to control edge probability between nodes, and the β parameter allows to regulate edge length. In our study, we defined α =0.9 and β =0.1 in order to obtain a graph with many routing possibilities and get paths with the highest number of hops to emulate real core networks.

Considering the implementation of PIM-SSM, we adopted the Dijkstra algorithm (Dijkstra, 1959) to compute the path from the content source to the listeners, since this protocol uses the unicast routing table of the routers to build its distribution tree. The AMT approach, as explained in the previous section, builds its distribution tree based on the overlay nodes available in the network, which in our implementation, are randomly mapped in certain nodes according to a given density. Ac-

Figure 6. Comparison between SSM and AMT approaches regarding changed links

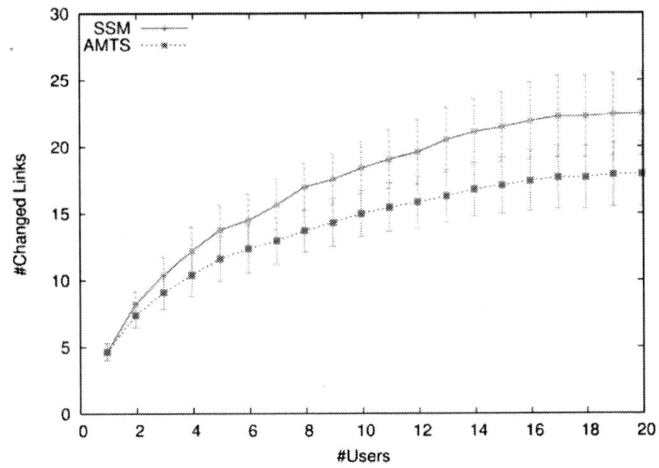

cording to their distribution, we also use Dijkstra to connect the ONs that built the abstract tree.

The methodology followed in the Matlab implementation consists in performing a set of instructions for each node that is consecutively added to the tree, keeping track of the changes observed regarding the number of links modified (added and eliminated). This way, it will be possible to compare the behavior of each approach in a large network environment and with many possible listeners.

The results depicted on Figure 6 show the advantages of the AMT strategy over PIM-SSM, which is even more notorious for a higher number of listeners (users). This fact highlights the scalability of the abstract trees for higher number of users. Although the difference between the numbers of changed links is not very high, this is related with the relative small number of hops that each path from the source to the listener has. The outcomes were determined repeating 20 times the experience for each user, obtaining the confidence intervals depicted on Figure 6 with a

Figure 7. Difference of network reconfigurations regarding different topology dimensions

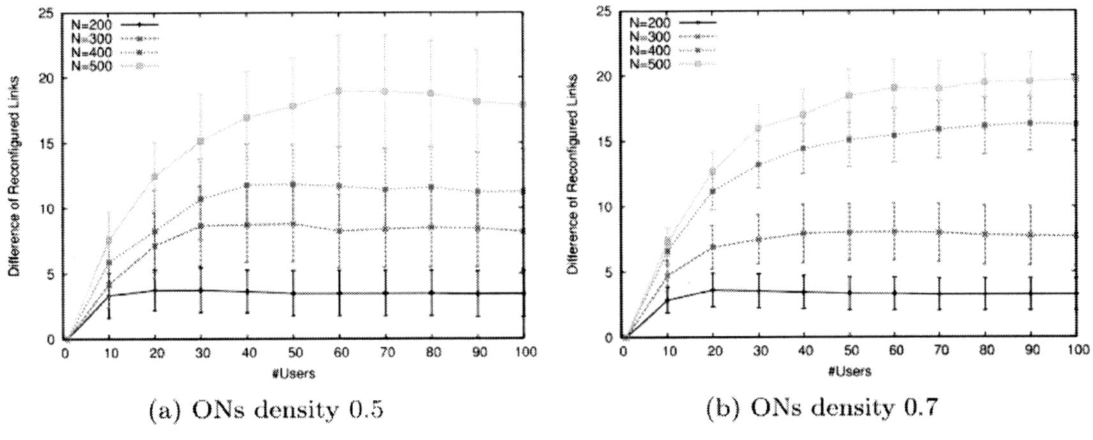

(a) ONs density 0.5

(b) ONs density 0.7

confidence degree of 95%. The superposition of the confidence intervals of each curve occurs due to the randomness of each network topology created and the random distribution of ONs throughout the network nodes.

Figure 7 shows the outcomes of our study regarding the impact of the network topology dimension and the density of ONs present in the network. As expected, as the network topology is larger, the gains obtained by the AMT approach are higher as the paths are longer and the routing possibilities increase. Comparing Figures 7 (a) and (b), one may observe that the results obtained when increasing the ONs density are slightly better. However, to the exception of the curve correspondent to N=400, the raise of ONs does not justify the gains obtained, since for a higher number of ONs it is required more orchestration between them and it will certainly increase control overhead. Moreover, one may observe that, for each curve correspondent to a network topology dimension, after a certain number of users, the gains remain constant. This fact is justified by the saturation of the network topology, since every leaf node has already associated a terminal.

This study allowed us to prove that the AMT strategy is scalable and flexible enough to deal with the dynamics of future networks. Moreover, it also showed that is not worth the excessively population of the network with ONs, since the gains obtained in terms of avoided reconfiguration not always pay the added control overhead.

OVERLAY NODE CONTROL FUNCTIONS

ONs are in charge of setting up sub-AMTs to which they are individually associated, controlling their resources and QoS through the IPT elements within the scope of the sub-AMT. They represent the second level of network intelligence which, despite being controlled and managed by NUM, present autonomous mechanisms to manage their sub-AMTs.

The functionalities performed by an ON and IPTs within its domain are detailed on Algorithm 2. Each ON receives from NUM the information about the sub-AMT to which it is associated, as well as session requirements (bandwidth, delay). Regarding ON functionalities, two different procedures may occur depending on the trigger event.

In case of a new *Sub-AMT Setup Request*, the ON starts by assigning a multicast address to the tree, since each sub-AMT has associated a unique multicast address to endow the global AMT with more flexibility. Then, for each IPT within the physical path of the sub-AMT (until reaching the other edge ON), the ingress ON will trigger network resource allocation according to the session requirements (bandwidth, delay, etc.) and then performs the router *Join* process to the multicast tree. Among the IPT enforcement tasks, it is also present the update of the multicast routing information table (MRIB), in order to employ, throughout the physical links and nodes, the route in which the sub-AMT is mapped. Finally, and regarding the different multicast addresses assigned to each sub-AMT, proxy settings need to be configured in the ONs. Thus, ONs change the destination IP address of packets that belong to the flow associated with the sub-AMT configured, updating them with the multicast address of the sub-AMT.

In case of a *Sub-AMT Rearrangement*, ONs have autonomous features that make them able to rearrange their controlled sub-AMTs without notifying NUM. To achieve this, in case of link/node failure or QoS degradation, the ingress ON of a certain sub-AMT checks in the Routing Information Base (RIB) for routing alternatives with the same characteristics to the egress ON. If a new route is available, the ON has the capability to update the MRIB and enforce the new path, rearranging the sub-AMT without NUM's knowledge. Otherwise, NUM has to be notified and select a new egress ON for the sub-AMT.

Algorithm 2. ON and IPT Functionalities

input : $G, M_i, S_{M_i}, F_{M_i}, R_{F_i}, Z, O_F = Empty$
output: An AMT for each Flow: O_{F_i}, $i = 1 \rightarrow |F|$

switch *Trigger Event* **do**
    case *Sub-AMT Setup Request*
        Allocate IP Multicast Tree;
        repeat For each IPT
            Resource Allocation according to session CoS;
            Join IP multicast tree;
            Update MRIB;
        until *Next ON*;
        Configure ON Proxy Settings;
    endsw
    case *Sub-AMT Rearrangement*
        Check RIB for route alternatives until next ON;
        if *Successful* **then** Local Rerouting;
        else Notify NUM for global rearrangement;
    endsw
endsw

SELECTION

The selection scheme aims to improve the efficiency of the multiparty content distribution based on the context information available. The adopted strategy considers different schemes for Access and Core network selection, Figure 8, since the involved constraints for both parts are different, and a unique algorithm involving all would be infeasible and most of all not scalable. Thus, two schemes with different inputs and outputs were developed, starting with the access network selection to obtain the most suitable Access Point (AP) for the user. This procedure attempts to achieve the best compromise between user preferences/requirements and network resources, giving a special attention to the group/sub-grouping concept. Based on the AP chosen, which will be the first ON of the path, the core network path selection is performed to determine the best feasible path between session content source and the selected AP, concerning network QoS metrics and available bandwidth. Both schemes will be detailed in the following subsections.

Figure 8. Overall context-aware selection scheme

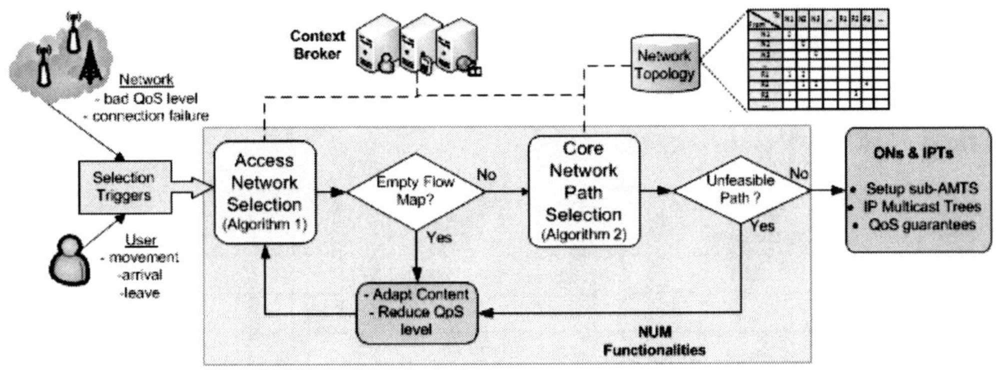

Access Network Selection

The access network selection scheme adopted is briefly described by its pseudo-code in Algorithm 2, being based on a proposed multi-constrained unicast access network selection algorithm (Jesus, et al., 2008). It was designed to determine a ranked list of candidate APs, with sufficient resources and to which the terminal is able to connect. Since the architecture proposed in this work aims at a context-aware multiparty content delivery, it is required an integration with QoS mechanisms and multicast constraints. Thus, it would be on the best interest of operators to aggregate in the same AP users that share the same content source, CoS or other features (e.g. device capabilities, location, codecs).

This considers in the access selection process the constraints depicted in Table 1: User Preference, Cost, Sub-grouping, CoS and Bandwidth Allocation. Nevertheless, a wide variety of parameters could also be used, such as signal strength, Bit Error Rate or location. Table 1 values are comprehended between *0* and *100*, representing the higher the better. The User Preference value of each AP is empirically assigned by the user, possibly taking into account its background experience. The Cost parameter is directly associated with the access price. Bandwidth Allocation is the result of a simple function that returns the value *100* when any traffic flows through the AP and *0* if there are no resources available. Considering the Sub-grouping constraint, a higher value means that the AP is serving a low number of different groups and it may support a session of a different group. The CoS parameter has a similar function, it intends to aggregate traffic with the same CoS, being assigned higher values for APs that already have flows with the same CoS. Table 2 holds, for each user profile, the weights that each one assign to the constraint involved in the access selection. These values are comprehended between 0 and *1.5*, allowing constraint differentiation according to each user.

The access selection algorithm, Algorithm 3, starts by obtaining from the CB a list of APs (AP_{CANDS}) to which user is within range. From the same repository, it also gets the session QoS demands ($Flow_{QoS}$) mapped in bandwidth, delay and loss rate requirements. This information is

Table 1. AP properties and respective values

Access Technology	User Preferences (static)	Cost (static)	Sub-grouping (dynamic)	CoS (dynamic)	Bandwidth Allocation (dynamic)
WiMAX 1	100	30	90	90	50
UMTS	70	50	60	50	100
Wi-Fi	80	80	0	0	90
WiMAX 2	100	30	30	70	10

Table 2. Weight distribution according to user profiles

User Profile	User Preferences	Cost	Sub-grouping	CoS	Bandwidth Allocation
Business man	0.5	1.0	1.5	1.0	1.0
Gamer	1.5	0.5	1.5	1.5	1.0
Groupie	0.5	1.5	0.5	0.5	0.5

Algorithm 3. Access Network Selection Algorithm

```
AP_CANDS = {AP_1, ..., AP_N}, List of candidate APs;
Flow_QoS ⇐ Get session QoS requirements from CB;
M ⇒ number of APs and W ⇒ number of constraints;
FM = 0 ⇒ Matrix Flow Map contains APs and their rank;

for i = 1 to N do
   Get QoS context of AP_i from CB;
   if AP_i resources do not meet Flow_QoS then
      Remove AP_i from AP_CANDS;
   end if
end for
if AP_CANDS == Empty then
   Abort and re-select with different content format or CoS;
end if
Require: Matrix APN_MxW
for i = 1 to M do
   Fill respective AP_i properties in APN;
end for
Require: Creation of Matrix UP_WxW according to user profile stored in CB;
FM = APN_MxW × UP_WxW;
FM = Σ_{i=1}^{W} FM;
Sort FM with highest ranked AP on top;
return FM;
```

then crossed with the QoS context of each AP candidate, executing a simple admission control mechanism. Then, the matrix *APN* is built based on the context of each AP that passes the admission mechanism, which final form will be similar to Table 1. After that, it is also created the matrix *UP* with the user context relative to each of the constraints involved. Combining both matrices, and after some simple matrix operations, it is obtained the Flow Map (*FM*) with the information about the most suitable APs regarding user and communication context.

Core Selection

Once selected the most suitable AP regarding communication needs and user context, we may then select the best path between content source and the chosen AP through the ONs. In our architecture this path is mapped in an AMT formed by a set of sub-AMTs bounded by ONs. We assume that the ONs are already defined in the network, being the sub-AMTs defined by the ONs that belong to the path selected. If we consider a network where all core nodes are capable of performing ON functionalities, we may then select them and construct sub-AMTs based on that, which would require a new intelligent selection scheme that was not addressed in this work. Though, it is our intention to assess the impact of ONs' number and positioning in the network performance, so that a better understanding of how their intelligent selection could evolve.

The core path selection scheme is presented in Algorithm 4. Available links in the network (*L*) and their characteristics in terms of bandwidth available (*BW*), routing cost (*Cost*) and mean delay (*Delay*) observed are retrieved from NUM's database, being essential that this information be updated before performing a selection. Another crucial input is a list (*dst*) of Egress ONs to which the content source is intended, including not only the AP previously selected but also all other APs that share the content source. Once again, considering session QoS demands ($Flow_{QoS}$) retrieved from CB, an admission control is executed based on links' QoS context. To each admitted link is computed a cost (*Weight*) combining link's bandwidth available, delay and routing cost. Normalization factors (NF_C, NF_B, NF_D) are associated to each of these parameters so that links' weights may be related with each other. Dijkstra's algorithm is then employed to compute the constrained shortest path to the respective APs. Based on the resultant path, and considering ONs within it, sub-AMTS are defined between two ONs. However, if there was a path previously selected for the same content source, some sub-AMTs already defined are very likely to be re-used, which saves network and resource configurations.

In both selection schemes, the network resources available may not be sufficient to accommodate a new flow with its QoS required. In order to solve this denial of service, a possible solution could be the re-establishment of a new session with a different content format (codec) requiring less resources, Figure 8. At the network level, as long as the user accept higher delays and losses, the session content could be distributed with a lower level of QoS supported by network resources.

EVALUATION

The architecture described was implemented in the Network Simulator 2 (NS, 2009). In order to assess its performance and robustness, it was created a flexible network topology regarding the number of ingress, egress and core nodes, ONs, APs, MTs, data sources and sessions. This way, we easily evaluate the response of the scheme implemented by varying the input parameters and generating different scenarios. The links in the core network were configured with a random delay between 1 and 2ms and a bandwidth ranging between 5 and 7 Mbps. The NS-2 has some limitations when dealing, in the same scenario, with multicast and wireless environments. To overcome this issue, it was considered the use of wired links between the APs and the MTs, with a dynamic error model to emulate wireless connections behavior (delay and loss). Two different traffic generators were randomly used: exponential traffic and constant bit rate. All flows have a packet size of 1000 bytes and an average rate of 100 Kbps. Regarding the QoS mechanism implemented, 6 CoS were considered: signaling, routing, Expedited Forwarding, BE, Assured Forwarding (AF) 1 and AF2. The CoS of each flow is randomly chosen, but it is guaranteed that more than an half of the traffic is Best Effort to simulate a real network.

Influence of ONs

From the results obtained in Figure 9, one may observe that packet delay decreases with the number of overlay nodes introduced in the core network.

Algorithm 4. Core Network Path Selection Algorithm

```
L = {L_1, ..., L_N}, List of available links;
(L_i={Cost,BW,Delay,Weight})
Normalization Factors: NF_C, NF_B , NF_D
dst ⇒ List of Egress ONs to which traffic is intended;
Flow_QoS ⇐ Get session QoS requirements from CB;

for i = 1 to N do
   if (L_i.BW<Flow_QoS.BW)&&(L_i.Delay>Flow_QoS.Delay) then
      Remove L_i from L;
   else
      L_i.Weight = L_i.Cost ×NF_C + L_i.BW ×NF_B + L_i.Delay ×NF_D ;
   end if
end for
if L == Empty then
   Abort and re-select with different content format or CoS;
end if
Path = Dijkstra(Session Source, dst)
Require: Sub-AMTS Definition ⇒ Divide Path according to ONs;
Search for already defined common Sub-AMTs ;
return List of new Sub-AMTs;
```

Since the number of ONs is proportional to the number of sub-AMTs, as more sub-AMTs exist in the core, less reservations and/or releases are needed to deliver content to the mobile users. These results are more notorious for a higher number of sessions, since it is likely that some sessions would use flows that already have paths reserved in the core network, only requiring an extension of the multicast tree. This way, the network reacts quickly to changes and in a scalable way, leading to lower delays. It should be stressed that, after a certain point, increasing the number of ONs does not correspond to substantial improvements and may not justify collateral costs such as overhead.

Another important result observed, although not depicted, is related with the integration support of both multicast and unicast nodes. The behavior of the control overhead is similar in the different scenarios that vary with the number of unicast nodes. When all nodes do not support multicast, the overhead of control is slightly higher than the other cases, but this difference might be neglected. Another important fact verified was that delay increases with the number of sessions independently of the number of unicast core nodes. Hence, one may conclude that the presence of unicast nodes in the core network does not affect the performance of the network, maintaining similar overhead, delay and loss.

Network Selection

This evaluation attempts to demonstrate the performance response of different access network selection methods. The "Basic" method represents the traditional AP selection based on the signal strength. The second one, "Without CoS," treats all traffic as equal using a load balancing mechanism, user preferences and cost. The last method, "With CoS," introduces the Sub-grouping and CoS in order to better manage traffic with different requirements. It also provides an admission mechanism to prevent high priority traffic to be impaired. Considering the results obtained for the different access strategies, Figure 10 shows that the method "With CoS" is obviously the better choice when dealing with EF class traffic. Moreover, the delay value maintains nearly constant due to the

Figure 9. Traffic delay time

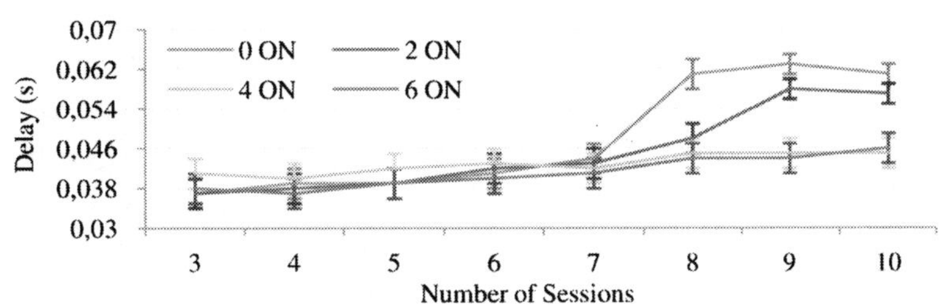

Figure 10. Delay for different access selection methods

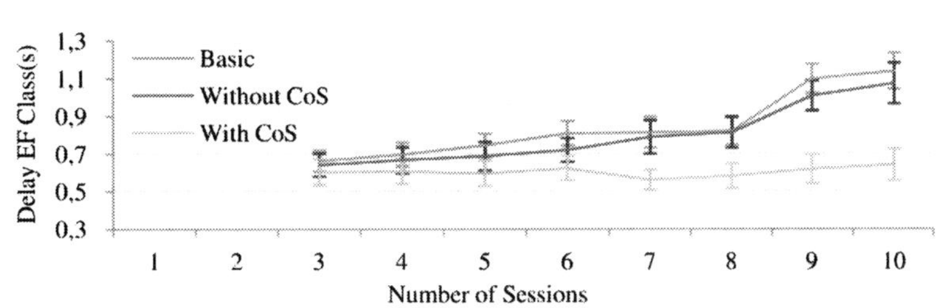

above mentioned admission mechanism. Comparing the "Basic" approach with the "Without Cos" method, we may observe better delays for each one of the sessions' number simulated. These results show that, intelligently involving more information in the access selection process, it is possible to achieve better performances while maintaining user satisfaction.

Context Dynamics

As scalability and resilience are main concerns of the architecture proposed, it is important to assess the behavior of the components implemented regarding these issues. To perform this evaluation, we considered several scenarios with different number of ONs where a mobile terminal performs a handover. Thus, the network must be reconfigured to deliver content through the new AP. The number of reconfiguration messages exchanged in this process and the links changed are presented in Figure 11.

Observing the results obtained, one may conclude that the introduction of ONs in the core network increases architecture scalability, since both metrics decrease as the number of ONs increase. We may say that core ONs facilitate the reconfiguration process of the QoS-aware multicast tree, increasing the scalability of the architecture proposed and the AMTs concept. Since 5 ONs represent a core full of ONs, the results show that there are no benefits in this scenario, as the reconfiguration messages and changed links will tend to be the same and the overhead continues to increase. A balanced number of ONs and simple nodes in the core network is the most scalable solution.

FUTURE RESEARCH DIRECTIONS

Regarding future developments for the architecture presented in this chapter, we aim to evolve it towards a decentralized system. This step will raise new challenges in what concerns the distribution of the intelligence and knowledge throughout the network elements. Following autonomic and cognitive principles, we aim to develop a cognitive cycle within the main elements of our architecture in order to incorporate intelligent decision mechanisms performing network optimization at different levels. Our investigation will focus on closing the management loop, mainly performing informed management decisions to optimize a pre-defined goal. The capabilities developed must be distributed throughout some network elements in order to address the complexity issues introduced by context-awareness in heterogeneous environments.

Following this strategy, current overlay nodes will have to be endowed with more functionalities and intelligence to compute local decisions and/or wider ones comprehending more information. Among the mechanisms that they will perform are the network resource provisioning for the multiparty content delivery trees, QoS-aware routing and

Figure 11. Impact of ONs number on network re-arrangements

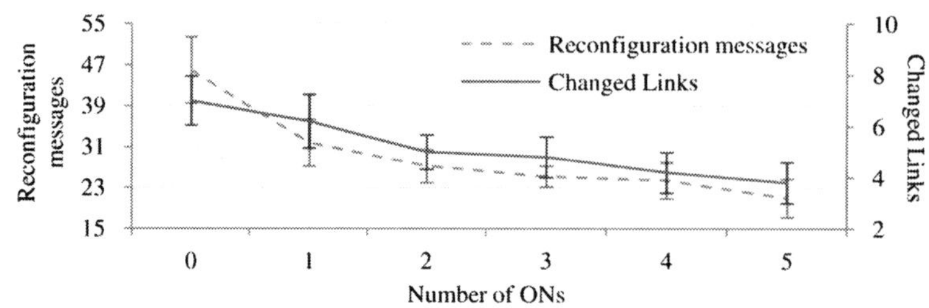

mobility mechanisms to support content sources and receivers movement. As each overlay node will only have a partial knowledge of the entire network state to compute optimization decisions, we aim to develop a distributed network protocol to exchange knowledge between nodes and feed the optimization procedures. Regarding this, some strategies have been analyzed to address the issues posed by a distributed architecture. Bloom Filters appear as good approach, considering their space and time advantages (Bloom, 1970). Following this strategy, nodes can exchange Bloom Filters to obtain available context information. When a node needs certain information, it will first check locally if the data exists. If not, a query is encoded into a Bloom filter using the same hash functions and compared to the Bloom filters of each node. If a match is verified at any layer, the query will be sent to the respective node that contains the information. This process is repeated until the query finds an exact match. However, notice that the exactly match may be false positive. To control this probability, a trade-off between Bloom filters size and false positive rates must be made, concerning always the optimal network cost. We expect to achieve, using Bloom filters, a very scalable and low resource consumption distributed knowledge protocol, able to exchange context information with low overhead.

CONCLUSION

This chapter presented a future network architecture driven by context, focused on group-based communications, and able to cope with the challenges raised by NGN. A study of the background in the scope of this topic was detailed covering the representative subjects: multicast, QoS, context-awareness and network selection. This survey allowed understanding the efforts made so far regarding the research topics of this work, and how they can be adapted or employed in our architecture.

A detailed description of the envisioned network architecture was made, introducing the functionalities, concepts and our hierarchical control strategy. The optimization achieved with the integration of the different functional framework, enables the content adaptation of multiparty communications to group members, network and environmental context. The concept of AMTs was introduced, highlighting the advantages of this control strategy, concerning the challenges of NGN. Two different network selection schemes were also presented, focused on choosing the best networks according to network, user and surrounding context.

The results obtained through the simulation allow to conclude that these mechanisms are able to not only decrease network overall delay, but also provide a more scalable solution considering necessary network re-arrangements and avoiding the necessity of global re-configurations. In general, one may conclude that the introduction of ONs in the core network benefit the overall network performance and also network reconfigurations. The results also show that the network selection schemes adopted increase overall network performance, meeting the requirements of users and efficiently managing the network resources.

The work developed allow us to conclude that context driven architectures are more suitable to face NGN challenges, providing higher levels of adaptation and personalization, enabling the development of novel applications that may be an added value as for operators as for consumers.

ACKNOWLEDGMENT

The work described in this chapter is partially supported by the Research Projects: FCT PTDC/EEA-TEL/74471/2006 "MuMoMgt: Mobility and Multicast Management in Heterogeneous Access Networks," and Internal IT Project "GEN-CAN: Generic Transport in Context-Aware Networks."

REFERENCES

Abowd, G. D., Dey, A. K., Brown, P. J., Davies, N., Smith, M., & Steggles, P. (1999). *Towards a better understanding of context and context-awareness*. Paper presented at the 1st International Symposium on Handheld and Ubiquitous Computing.

Banerjee, S., Bhattacharjee, B., & Kommareddy, C. (2002). *Scalable application layer multicast*. Paper presented at the 2002 Conference on Applications, Technologies, Architectures, and Protocols for Computer Communications.

Bhattacharyya, S. (Ed.). (2003). *RFC 3569: An overview of source-specific multicast*. SSM.

Bless, R., & Wehrle, R. (2004). *RFC 3754: IP multicast in differentiated services (DS) networks*.

Bloom, B. H. (1970). Space/time trade-offs in hash coding with allowable errors. *Communications of the ACM, 13*(7), 422–426. doi:10.1145/362686.362692

Boivie, R., Feldman, N., Imai, Y., Livens, W., & Ooms, D. (2007). *RFC 5058: Explicit multicast (Xcast) concepts and options*.

Castro, M., Druschel, P., Kermarrec, A. M., & Rowstron, A. I. T. (2002). Scribe: A large-scale and decentralized application-level multicast infrastructure. *IEEE Journal on Selected Areas in Communications, 20*(8), 1489–1499. doi:10.1109/JSAC.2002.803069

Cerqueira, E., Veloso, L., & Neto, A. Curado, l., et al. (2007). *Q3M: QoS architecture for multi-user mobile multimedia sessions in 4G systems*. Paper presented at the 10th IFIP/IEEE International Conference on Management of Multimedia and Mobile Networks and Services: Real-Time Mobile Multimedia Services.

Cerqueira, E., Veloso, L., Neto, A., Curado, M., Monteiro, E., & Mendes, P. (2006). *MMC04-3: A unifying architecture for publish-subscribe services in the next generation IP networks*. Paper presented at the Global Telecommunications Conference, 2006. GLOBECOM '06. IEEE.

Chawathe, Y. (2003). Scattercast: An adaptable broadcast distribution framework. *Multimedia Systems, 9*(1), 104–118. doi:10.1007/s00530-002-0082-z

Chen, Y., & Yang, Y. (2007). A new 4G architecture providing multimode terminals always best connected services. *IEEE Wireless Communications, 14*(2), 36–41. doi:10.1109/MWC.2007.358962

Chiron, P., Njedjou, E., Seite, P., Gosse, K., Melin, E., & Roux, P. (2008). Architectures for IP-based network-assisted mobility management across heterogeneous networks. *IEEE Wireless Communications, 15*(2), 18–25. doi:10.1109/MWC.2008.4492974

Costa, L. H. M. K., Fdida, S., & Duarte, O. (2001). *Hop by hop multicast routing protocol*. Paper presented at the 2001 Conference on Applications, Technologies, Architectures, and Protocols for Computer Communications.

Deering, S. E., & Cheriton, D. R. (1990). Multicast routing in datagram internetworks and extended LANs. *ACM Transactions on Computer Systems, 8*(2), 85–110. doi:10.1145/78952.78953

Dijkstra, E. W. (1959). A note on two problems in connexion with graphs. *Numerische Mathematik, 1*(1), 269–271. doi:10.1007/BF01386390

Diot, C., Levine, B. N., Lyles, B., Kassem, H., & Balensiefen, D. (2000). Deployment issues for the IP multicast service and architecture. *IEEE Network, 14*(1), 78–88. doi:10.1109/65.819174

El Hachimi, M., Abouaissa, A., Lorenz, P., & Lee, M. O. (2004). Scalable multicast provisioning in DiffServ with MPLS labeling. *Telecommunication Systems*, *27*(2), 253–272. doi:10.1023/B:TELS.0000041011.21556.da

Gu, T., Wang, X., Pung, H., & Zhang, D. (2004). *An ontology-based context model in intelligent environments.*

Gustafsson, E., & Jonsson, A. (2003). Always best connected. *IEEE Wireless Communications*, *10*(1), 49–55. doi:10.1109/MWC.2003.1182111

Hosseini, M., Ahmed, D. T., Shirmohammadi, S., & Georganas, N. D. (2007). A survey of application-layer multicast protocols. *IEEE Communications Surveys & Tutorials*, *9*(3), 58–74. doi:10.1109/COMST.2007.4317616

Iera, A., Molinaro, A., Campolo, C., & Amadeo, M. (2006). *An access network selection algorithm dynamically adapted to user needs and preferences.* Paper presented at the Personal, Indoor and Mobile Radio Communications, 2006 IEEE 17th International Symposium.

Jannotti, J., Gifford, D. K., Johnson, K. L., Kaashoek, M. F., & James, W. O'Toole, J. (2000). *Overcast: Reliable multicasting with on overlay network.* Paper presented at the 4th Conference on Symposium on Operating System Design & Implementation.

Jesus, V., Sargento, S., & Aguiar, R. L. (2008). *Any-constraint personalized network selection.* Paper presented at the Personal, Indoor and Mobile Radio Communications, 2008. PIMRC 2008. IEEE 19th International Symposium.

Kewen, L., & Jing, T. (2008). *A QoS mobile multicast routing algorithm based ant colony algorithm.* Paper presented at the Computational Intelligence and Industrial Application, 2008. PACIIA '08. Pacific-Asia Workshop.

Lao, L., Cui, J. H., Gerla, M., & Chen, S. (2007). A scalable overlay multicast architecture for large-scale applications. *IEEE Transactions on Parallel and Distributed Systems*, *18*(4), 449–459. doi:10.1109/TPDS.2007.1008

Mathieu, B., Meng, S., Galis, A., Cheng, L., Jean, K., Ocampo, R., et al. (2007). *Self-management of context-aware overlay ambient networks.* Paper presented at the Integrated Network Management, 2007. IM '07. 10th IFIP/IEEE International Symposium.

Mitrano, F., Guainella, E., Tamea, G., Castrucci, M., Di Giorgio, A., Benali, O., et al. (2007). *QoS management of multicast and broadcast services in next generation networks.* Paper presented at the Mobile and Wireless Communications Summit, 2007. 16th IST.

Neto, A., Cerqueira, E., Rissato, A., Monteiro, E., & Mendes, P. (2007). *A resource reservation protocol supporting QoS-aware multicast trees for next generation networks.* Paper presented at the Computers and Communications, 2007. ISCC 2007. 12th IEEE Symposium.

Prehofer, C., Nafisi, N., & Wei, Q. (2003). *A framework for context-aware handover decisions.* Paper presented at the Personal, Indoor and Mobile Radio Communications, PIMRC 2003.

Qingyang, S., & Jamalipour, A. (2005). Network selection in an integrated wireless LAN and UMTS environment using mathematical modeling and computing techniques. *IEEE Wireless Communications*, *12*(3), 42–48. doi:10.1109/MWC.2005.1452853

Riva, O., Teittinen, V. M., Siikavirta, S., & Huovinen, L. (2008). *A next generation operator environment to turn context-aware services into a commercial reality.* Paper presented at the Mobile Data Management, 2008. MDM '08. 9th International Conference.

Rowstron, A., & Druschel, P. (2001). Pastry: Scalable, decentralized object location, and routing for large-scale peer-to-peer systems. *Lecture Notes in Computer Science*, 2218.

Schilit, B. N., & Theimer, M. M. (1994). Disseminating active map information to mobile hosts. *IEEE Network*, *8*(5), 22–32. doi:10.1109/65.313011

Strang, T., & Popien, C. (2004). *A context modeling survey.* Paper presented at the Workshop on Advanced Context Modelling, Reasoning and Management, UbiComp 2004 - The Sixth International Conference on Ubiquitous Computing.

Suk Yu, H., & Kai Hau, Y. (2003). Challenges in the migration to 4G mobile systems. *IEEE Communications Magazine*, *41*(12), 54–59. doi:10.1109/MCOM.2003.1252799

Thomas Buchholz, M. S. (2003). Quality of context: what it is and why we need it. In *Proceedings of the 10th Workshop of the OpenView University Association: OVUA'03.*

Waxman, B. M. (1988). Routing of multipoint connections. *IEEE Journal on Selected Areas in Communications*, *6*(9), 1617–1622. doi:10.1109/49.12889

Williams, M. H., Roussaki, I., Strimpakou, M., Yang, Y., MacKinnon, L., Dewar, R., et al. (2005). *Context-awareness and personalisation in the daidalos pervasive environment.* Paper presented at the International Conference on Pervasive Services, ICPS '05.

Xing, B., & Nalini, V. (2005). *Multi-constraint dynamic access selection in always best connected networks.* Paper presented at The Second Annual International Conference on Mobile and Ubiquitous Systems: Networking and Services, MobiQuitous 2005.

Yang-Hua, C., Rao, S. G., Seshan, S., & Hui, Z. (2002). A case for end system multicast. *IEEE Journal on Selected Areas in Communications*, *20*(8), 1456–1471. doi:10.1109/JSAC.2002.803066

KEY TERMS AND DEFINITIONS

Context-Aware: Ability to perceive surrounding conditions.

Heterogeneity: Diversity of features and characteristics present at different levels of today's and future networks.

Multicast: Network transport technology for the delivery of information to a group of hosts simultaneously in a single transmission (whenever possible).

Multiparty Communications: Communications usually involving more than two actors (one-to-many, many-to-many, many-to-many).

Network Selection: Mechanisms/algorithms that take advantage of context-awareness in order to select the best networks and distribution paths according to current network conditions.

Overlay Network: Network built on top of a substrate (physical) network where nodes are virtual or logically connected, each logical link represents a path that can transverse several physical links.

Quality of Experience: Is an end-to-end subjective measure of a user's experiences regarding a service provided.

Chapter 5
Multi-Access Communications in Wireless Mesh Networks by Virtualization

Susana Sargento
Instituto de Telecomunicações, University of Aveiro, Portugal

Ricardo Matos
Instituto de Telecomunicações, University of Aveiro, Portugal

Karin Anna Hummel
University of Vienna, Austria

Andrea Hess
University of Vienna, Austria

Stavros Toumpis
Athens University of Economics and Business, Greece

Yahya Al-Hazmi
University of Passau, Germany

Muhammad Ali
University of Passau, Germany

Hermann de Meer
University of Passau, Germany

Yiannis Tselekounis
Athens University of Economics and Business, Greece

George D. Stamoulis
Athens University of Economics and Business, Greece

ABSTRACT

This chapter presents an architecture for context-aware Virtual Networks (VNs) that provide user-driven multi-access communication. The architecture is dedicated and appropriate for the flexibility provided by Wireless Mesh Networks. According to this architecture, VNs with different context characteristics such as Quality of Service (QoS), mobility, and security are built to support communications with different characteristics that best fit the users' needs. The architecture is modeled both through a probabilistic and an optimization approach to provide quantitative insights into its performance. The probabilistic model quantifies the overhead of the architecture in terms of networking delays induced due to the VN management (searching, creation, and management). The optimization model provides insights into the

DOI: 10.4018/978-1-4666-0017-1.ch005

competition of the VNs for the limited bandwidth resources. Indicative results of the models show the feasibility of the architecture, the upper bounds in terms of number of supported VNs to achieve good quality communications, and the relative placement of the flows in different, competing VNs.

INTRODUCTION

Wireless Mesh Networks (WMNs) are spreading both in city and rural areas and connect users in both home and mobile environments, in company areas, campuses, and a variety of other settings. The term WMN includes a variety of networks showing a mesh topology and connecting over wireless links. In this chapter, we concentrate on larger WMNs based on a variety of wireless technologies (multi-access) covering larger areas which can make best use of virtualization. As most operational WMNs consist of fixed mesh routers and fixed and mobile mesh clients, this is the setting we refer to in our further investigations. One typical example are community provided WMNs for providing Internet access in urban areas. The topology of the WMNs' physical infrastructure provides high flexibility and can support changing and switching routes and transport connections, which can be exploited to provide multi-access communication and support for a variety of different services and situations. Context information can be used to autonomously build logical networks on top of physical networks. By *context* we mean any information characterizing the situation of the user, including his/her services and preferences, such as *Quality of Service* (QoS) requirements, pricing preferences, mobility behaviors, and privacy and security demands. The logical networks are formed based on context information expressing user requirements, and are described by the context characteristics each of them supports. Users are grouped according to similarity of their environmental conditions and preferences. User groups can be assigned to the logical networks matching their context.

When building logical networks, network virtualization can be a very useful technique for a flexible utilization of a physical network infrastructure. For instance, wireless mesh links are prone to disturbances on the air interface, such as reflection and absorbing of the signal due to weather changes, new buildings, car traffic conditions, etc. A flexible solution which is capable to adapt and make use of different wireless link qualities changing over time is required (e.g. operating on different frequencies and power levels). Network virtualization is able to make parallel use of the same infrastructure network and even combine physical network components to a single logical component. Hence, network virtualization is a powerful tool to build multiple logical networks and, thus, to support different network organizations and purposes. In combination with network federation (cooperation of different network operators and parties), network virtualization is seen as a major step for managing larger sets of physical resources to provide better services.

Bringing WMNs, context-awareness, and network virtualization together, this chapter presents a novel architecture for WMNs that consists of a multitude of context-based *Virtual Networks* (VNs), each one providing personalized communications for a user or group of users, according to their requirements, which are part of the user context. This approach is able to provide user-driven multi-access communications, since a user can simultaneously access different VNs in the same infrastructure according to the services and context associated with each VN. In the presented approach, we model user context in terms of QoS, security, user mobility, cost, etc., and virtualize the WMN by introducing multiple VNs, each corresponding to a specific context. The mesh topology, together with the WMNs' self-organizing characteristics, allows for the introduction of virtualization following different approaches. First, the physical resources of a single WMN can be shared by different VNs, and, second, resources belonging to multiple WMNs can be aggregated to a single VN. For example, multiple paths owned by different providers can be aggregated providing increased reliability. Hereby, we further consider and include multi-homing supported by different providers and different WMNs in our approach.

This approach raises several challenges that need to be studied, such as: (*i*) the characterization of VNs according to their context; (*ii*) the discovery of the best fitting VN for a specific user, according to the context of all entities involved; (*iii*) the match of available context VNs against the users' context requirements; and (*iv*) the VN topology maintenance and resource management, given that the VNs are built on top of the physical networks.

This chapter presents the architecture along with its main mechanisms. In addition, two different models are introduced, and are used for performance evaluation. The first is a probabilistic model that quantifies the architecture's overhead in terms of networking delays induced because of the VN management. The second is an optimization model providing insights to the behavior that arises between VNs competing for the same limited bandwidth resources.

This chapter starts by providing a survey on user-driven multi-access communications, virtualization and context-aware approaches. It then presents the concepts for multi-access communications in the Background section, virtualization in the next section followed by the context information modeling and description. Next we present the architecture and specific mechanisms developed to address the presented challenges. We also describe the probabilistic and the optimization models, and the results achieved in terms of performance and requirements from the architecture. Finally, we discuss research challenges in this area, and present the final conclusions of the chapter.

BACKGROUND: CONTEXT AND VIRTUALIZATION

WMNs are accessed by heterogeneous mobile devices with a variety of different users' contexts and needs (e.g., some users will access location services and Web-based community networks on the street while others will use video-streaming applications at a friend's place). *Context* is here any information that can be used to characterize the situation of the user or any entity (following the widely accepted definition of (Salber 1999)). The mobile devices themselves provide different wireless interfaces (such as 3G, WiFi, Bluetooth, etc.) enabling multi-homing of these devices.

The envisioned different applications for mobile users require corresponding network quality supported by the infrastructure, which may actually consist of multiple wireless networks, where the user device is connected by multiple physical interfaces (multi-access technologies). In order to provide different services and service qualities by the resources of the physical network(s) available, *network virtualization* is proposed for WMNs. Network virtualization introduces an abstraction of the available networking infrastructure, so that, depending on the circumstances, a single physical network may be broken up to multiple virtual ones, multiple physical networks may be pooled together and appear as a single, virtual one, or combinations of these two scenarios may occur (Hsieh 2002) (Liao 2008). In this section, we discuss the major concepts of context-aware virtual networking along the following related topics: (*i*) context-aware networks, (*ii*) context information in WMNs, and (*iii*) virtualization.

Context-Aware Networks

Considering *context* in networking is a new way of adapting networks to dynamically changing requirements. The European projects C-Cast (Context Casting)[1] and SENSEI (Integrating the Physical with the Digital World of the Network of the Future)[2] are two major recent projects in this field demonstrating the timeliness of context-aware networking. The goal of C-Cast is to provide mobile multimedia multicasting in an end-to-end context-aware communication framework for everyday use in a sensor enriched mobile environment (Antoniou 2010). SENSEI focuses on a mixed mobile wireless and fixed wired Internet where context-aware services are introduced. A major focus of SENSEI is set on integrating *Wireless Sensor and Actuator Networks* (WSANs) to future networks (Lombriser 2009).

Networking research has already started to incorporate context information for providing efficient routing and transport. An architecture which adapts to current network conditions (determined by performance metrics and history data) for the support of real-time communication is proposed in (Staub 2008). The robustness of paths is determined by metrics for availability, bandwidth, delay, and outage, established by network monitoring and statistical evaluation. Wirelessly connected users require networks which handle mobility and handover and are – to some degree – delay tolerant, self-aware, and self-adaptable. These networks are able to react to network context changes (for example by accepting and rejecting traffic flows as presented in (Gelenbe 2007)) and can adapt themselves based on the set traffic goals. These networks can even move resources to reach an optimization goal such as providing the most energy-efficient routing decision or selecting the most reliable paths. Our approach extends these works by presenting a general method for managing wireless mesh networks based on virtualization and context awareness including an extended (and further extendable) set of context information considered.

Context Information in WMNs

Among the most relevant context characteristics for networking in WMNs are: user mobility behavior, user privacy, trust, security preferences, pricing, and the user's QoS demands depending on the application. The use of context in networking is manifold. It can be used to perform adaptive routing, flow control, resource reservation, and even network selection (for example, if a user has specific privacy and security needs, or has pricing constraints, it will select a network among those satisfying all these requirements).

Routing is one of the most thoroughly studied mechanisms that make use of context information. Many routing schemes consider the quality of communication links by integrating signal interference measures into their metrics, as proposed in (Subramanian 2006). In (Hu 2008), a rule-based middleware for context adaptive routing in WMNs is described. Here, routing metrics, like hop count, retransmission count, and link interference, are related with context data (i.e., security level, traffic priority, and mobility) by means of manually configured rules. Based on neural networks, such relations can be learned from past experiences, as described in (Lin 2008) for user mobility.

Mobility (i.e., movement and resulting mobility management) is another context characteristic of importance. Approaches to WMNs that are particularly aware of the mobility of mesh clients can be found in (Oh 2008) (Määttä 2009) (Langar 2009). In (Oh 2008), an adaptive routing algorithm is proposed, which reacts to the intensity of mobility within the WMN. It is assumed that the mobility in the network can be determined to be high if link breakages are detected. Depending on the mobility assumed, either a reactive or a proactive routing algorithm is applied. The current position of mesh clients is taken into account by channel assignment and router selection in (Määttä 2009). In particular the position information is used by channel assignment and router selection algorithms to adapt to significant location changes of clients. In (Langar 2009), the authors present two mobility-aware clustering schemes that aim at improving the radio resource utilization in WMNs. Here, the mesh network is divided into virtual clusters in order to restrict the majority of signaling messages to a local area. In the static clustering scheme, the optimal cluster placement is determined by using a linear programming based method to minimize resource utilization costs, whereas the second, less time-consuming clustering algorithm optimizes the cluster placement in a distributed manner. Additionally to the mobility-awareness aspect, these clustering schemes are related to our work by the fact that mesh clients are grouped "virtually."

Security is another important context parameter which encompasses confidentiality, integrity, access control, secure routing, intrusion detection mechanisms, secure key exchange, authentication etc. Like other networking paradigms, WMNs are also deeply impacted by security constraints and considerations. Numerous surveys on WMN security have been conducted and different security architectures have been proposed to overcome security constraints (Cho 2008)(Dong 2009)(Lukas 2009). In addition to approaches of overall secure architectures, research has also been conducted to address specific security problems within the context of WMNs like authentication and key exchange (Zhao 2009), anonymous communication, attack analysis and intrusion detection and corresponding defense mechanisms, as well as secure routing protocols. Closely related to security, end user privacy is another important parameter. Within the context of the *Service Oriented Architecture* (SOA) paradigm, there has been considerable work done on abstract concepts like privacy commitments, obligations, and user rights, and a number of privacy, rights and obligation representation policy languages have been developed for such purposes (OASIS)(Damianou 2001)(Ardagna 2008). Security and privacy are a crucial part of a user's trust in the network which can be further supported by reputation systems which may also cover more aspects including pricing and quality of the provided service.

Cost and billing mechanisms, trust metrics, and dispute resolution mechanisms can also make use of the user's context. If several physical networks are used by multiple VNs, billing and pricing are non-trivial issues. As in WMNs users are charged (against what they use) and may provide resources (e.g., by relaying messages), billing becomes even more sophisticated. In (Zhang 2007), an architecture is presented which addresses billing in a secure way.

Finally, QoS is another aspect important for WMNs and the user acceptance for using the WMN to access also network demanding services, like multimedia real-time services. The challenging characteristic is here, again, the parallel use of services with different quality requirements and additional service adaptations necessary given a specific user context (e.g., increased noise level when using video streaming). QoS can be characterized in terms of delay and jitter, but also in terms of reliability (loss ratio), etc. and extended by *Quality of Experience* (QoE) metrics which address the user perception accurately.

To conclude, it is possible to define an exhaustive list of context metrics which can become part of context representation. However, the more complex the structure of the context representation, the more intensive will be the processing and decision making. Thus, it is important to find an optimal level of context representation to achieve efficient performance. Having discussed the different metrics and parameters which are envisioned to become part of the context description and, thus, WMN meta-information, an appropriate way to represent context and the exchange of context information is described later on.

Virtualization

Virtualization is a concept in computer science which has multiple facets. Initially, it was proposed to abstract logical machines (virtual machines) running simultaneously on the same physical hardware. Today, virtualization is not limited to operating systems and virtual machines only, but it has been extended to concepts like network virtualization (Pfaff 2009)(Berl 2009), memory and storage virtualization, etc.

We focus on network virtualization, that is, the virtualization (abstraction) of physical network resources which enables a more flexible and dynamic network management. Network virtualization has already been implemented in technologies like *Virtual LANs* (VLANs), *Virtual Private Networks* (VPNs), and *Peer-to-Peer* (P2P) networks. However, these previous approaches are limited to specific targeted scenarios. In order to include the core network, a more comprehensive approach is needed, virtualizing both node and link resources. Recent research initiatives have progressed in that direction, studying the application of virtualization methods on core network elements like virtualizing routers and switches. Some of the mainstream network equipment vendors like Cisco and Juniper have already introduced such techniques into their products. Nevertheless, there are several issues that have not been fully investigated yet. The scalability and stability of large-scale virtualized networks has to be verified. In order to provide optimal performance, suitable mapping algorithms have to be found, mapping virtual resources onto physical ones.

Furthermore, virtualization is among the key enabler technologies to achieve the vision of energy efficient IT by consolidating physical resources (Berl 2010). Some of the very recent computing paradigms like cloud computing, which were originally envisioned for energy efficiency, sharing of resources and cost saving, became a reality because of virtualization. Thus, virtualization indirectly contributes to the goal of energy efficiency in information and communication technology.

Lastly, the security of virtualized networks still requires further investigation; indeed, VNs introduce both security benefits and security problems. In (Gartner 2010) it has been stated that by 2012, 60% of all the virtualized servers will have less security than the physical servers they replace. There are currently no substantial studies which can provide the answers to questions like: do the same security mechanisms provide the same level of security for both virtualized and un-virtualized counterparts for the same system? Or, does virtualization induce new attack vectors? We discuss network virtualization within the context of WMNs in details later on.

MULTI-ACCESS COMMUNICATIONS IN WMNs

To understand the multi-VN approach, in this section, an introduction of basic communication paradigms in WMNs and multi-homing and multi-path are detailed. Hereby, the new flexibility introduced by these technologies is emphasized.

Wireless Mesh Networks

WMNs are spreading both in city and rural areas to connect heterogeneous users. Their aim is to support (mobile) users seamlessly with cheap and easy to maintain connectivity. The topology of the WMNs' physical infrastructure provides high flexibility in changing and switching routes and transport connection, which can be exploited by network management.

The particular characteristics of WMNs (Akyildiz 2005), setting them apart from traditional wireless networks, bring up new challenges to communication protocols, network management, reliability assurance, and security. Scalability, for instance, has been identified as a major problem of most proposed WMN protocols, but there are other open issues, such as the support of multi-cast applications and the utilization of multi-radios and multi-channels. In addition, the characteristics of the nodes have to be considered in the routing protocols since they can no longer be assumed to be similar.

Considering the different contextual features and preferences of the users in current WMN environments, the users need to be linked to different wireless access networks with different bandwidth and robustness features, probably belonging to different operators and *Internet Service Providers* (ISPs) with different security policies. The customers use different devices with different capabilities, which run different applications with different QoS requirements. As WMNs are edge networks connecting (mobile) users, they are expected to play an important role when introducing the required context-based user-centric networks. Moreover, WMNs are adaptable, self-configuring, and self-organizing to a high degree. As a consequence, WMNs are well suited to demonstrate the benefit of context-based approaches considering heterogeneous node capabilities and user preferences.

Two major application scenarios for WMNs are last mile Internet access in city areas as well as low-cost Internet connectivity in developing regions. *Wireless Community Networks* (WCNs) (Efstathiou 2006) where users participate with their own mesh routers have evolved in many cities, like funkfeuer[3] in Vienna, freifunk[4] in Berlin, Seattle Wireless[5], and Consume in London[6]. In addition, there are projects like Wireless Ghana[7] attempting to connect public institutions and to provide Internet access throughout West Africa. Apart from deployments of mesh networks over larger geographic regions, small-scale WMN applications have been envisaged in domains such as *Voice over IP* (VoIP), public safety, and emergency, which has led to commercially available solutions (e.g., provided by WMN vendors like Strix Systems[8] and Cisco[9]). For example, in a public transport application in San Diego eleven solar-powered wireless mesh nodes are deployed for video surveillance and on-board communication in railway trains[10].

On the other hand, there exist several testbeds primarily targeted at experimental evaluations of new protocol and architecture ideas. MagNets (Botta, 2007; Karrer, 2006), for instance, is a next-generation wireless infrastructure consisting of 50 nodes and a 5-node high-speed wireless backbone deployed in the city center of Berlin. MagNets is seamlessly integrated with other network technologies, like GPRS, UMTS, and WiMAX. The Google WiFi is a 500-mesh node testbed deployed in Mountain View, which allows studying network usage of stationary access devices and mobile laptops and smartphones on a city-wide scale (Afanasyev 2010). RescueMesh (Dilmaghani 2006) is a hybrid wireless network testbed for investigating communication in disaster scenarios aimed at providing an easily reconfigurable alternative for first responders handling emergency response. Some of the community networks mentioned, such as the funkfeuer network in Vienna, are also partly experimentation networks besides being operational networks.

Finally, indoor WMN testbeds have also been installed (e.g., in office areas to investigate different low cost network scenarios coping without access points). In (Eriksson 2006), a 21-node testbed for evaluating network performance aspects of commonly used applications like a web browser is described, while MeshBed (Bayer 2008) is a 12-node testbed for investigating carrier-grade aspects of delay-sensitive applications, such as VoIP. The provision of QoS required by carrier-grade triple-play services over heterogeneous WMNs is further investigated within the EU FP7 project CARMEN (Azcorra 2009). Besides, a 13-node indoor testbed for the evaluation of multi-path routing in WMNs is presented in (Staub 2009).

Multi-Homing and Multi-Path

The ability to maintain connections with multiple points of network attachment (e.g., different Internet Service Providers in the Internet, different access points of a wireless infrastructure, etc.) is called multi-homing. Basically, multi-homing can be realized by using *vertical handovers* where handovers are carried out between associations to networks of different technologies (e.g., IP-based and cellular networks), or *horizontal handovers* utilizing different points of attachment of the same technology, like multiple IEEE 802.11 WLAN access points. Multi-homing schemes might therefore prefer APs providing higher performance for longer periods of time than with APs providing lower performance (Shakkottai 2007). Similar to multi-homing in WLAN networks, in Bluetooth node may be allowed to participate in multiple *scatternets*, by means of a frequency division multiplexing mechanism (Law 2003); this may also be thought of as a multi-homing mechanism.

In addition to the concept of mobile devices integrating multiple wireless cards, there are ideas to virtualize a single wireless card to connect to multiple WLAN access points by, making one AP believe that the wireless card is in "sleep mode" while it is actually sending traffic to another AP (Chandra 2004). An intermediate layer is introduced below IP, continuously switching the wireless card between APs, which are forced to buffer messages addressed to clients currently communicating with other APs.

While the term multi-homing is used to refer to the possibility of managing multiple network interfaces, the term multi-path describes the use of multiple paths in a network simultaneously. Utilizing multiple (node or link disjoint) paths simultaneously to route data packets between a pair of source and destination nodes, in order to increase effective bandwidth and reduce congestion and packet loss, has been studied for wired networks for many years now (Sidhu 1991). Due to the dynamic nature of wireless communications, caused by node mobility and changes in wireless propagation conditions, such mechanisms have been investigated in mobile ad hoc settings. Distributing the packet traffic to multiple paths allows overcoming the effects of network dynamics by improving fault tolerance, load balancing, and QoS. Thus, several multi-path solutions have been proposed since the beginning of mobile ad hoc network research, where an optimal number of paths for concurrent transmission might be selected by ranking possible paths according to their probability of failure (a metric adapted continuously to network topology changes), like in (Tsirigos 2001).

Existing multi-path routing protocols for mobile ad hoc networks can be grouped according to their goal (see, (Tarique 2009)) into (*i*) delay-aware, (*ii*) reliable, (*iii*) minimum overhead, (*iv*) energy efficient, and (*v*) hybrid multi-path routing protocols – protocols of the latter category attempt to combine shortest path and multi-path routing. Of particular interest for WMNs are reliable multi-path routing protocols since they are aimed at coping with the link breakage resulting from node mobility, link failure, and battery exhaustion. In (Wang 2005), the support of QoS requirements of real-time services is achieved through a redundant multi-path approach. To minimize the packet drop rate, the original data packets are distributed to fewer paths than the total number of available node-disjoint paths since additional ones are required for the transmission of redundant packets. In (Liang 2005), on the other hand, an *m*-path routing scheme for wireless sensor networks taking mobility and remaining battery capacity into account is proposed. At every hop of a path the forwarding node selects the *m* best fitting nodes in its communication range as forwarding nodes in a rule-based manner according to distance to destination node, remaining battery capacity, and degree of mobility.

VIRTUALIZATION AS AN ENABLER FOR MULTI-ACCESS SUPPORT

Virtualization has attracted considerable interest and is envisioned to be one of the key technologies in the future Internet, not only because of its ability to use the available physical resources in an efficient way, in terms of reduced energy consumption and optimal resource utilization, but also because of the possibility to create multiple VNs with different properties on top of the same physical infrastructure. However, by looking at the mapping between the virtual environment and the physical hardware, two different types of virtualization techniques are possible:

- **Aggregation of resources (*N*:1):** several physical instances are clustered to create one virtual instance.
- **Splitting of resources (1:*N*):** a single physical instance is shared by multiple virtual instances.

In the proposed multi-VN architecture, virtualization is exploited to provide multi-access communication support. How virtualization enables user-driven multi-access support is discussed here in two parts, first, virtualization in the physical infrastructure, and, second, virtualization in the user side (mobile device).

In the first part, network virtualization provides an abstraction of the available physical resources (e.g. routers and links between them). Such resources can be split to allow coexistence of several different VNs on the same structure. Network virtualization in this context enables the creation of multiple VNs to address the following aspects: (*i*) through a single physical machine, a user is able to access multiple service providers that are located on different VNs; (*ii*) physical resources are shared by multiple VNs, and thus, multi-accessed by users belong to different VNs simultaneously. Additionally, it gives a user device the ability to select (or create) the most suitable VNs which fit best to its requirements and preferences. Finding the best fitting VN that meets the user's context is based on the context policy model.

To define a VN, one has to first define what a physical network actually is. A communication network contains routers interconnected via links. Network virtualization is then a matter of building VNs on top of that structure. VNs, in this regard, refer to virtual routers logically interconnected via virtual links as shown in Figure 1. However, virtual links are realized as logical tunnels, if there are no direct physical links between the virtual routers. Indeed, these tunnels are established over physical links.

In such virtual environments, a user can have access to multiple VNs simultaneously which could provide different services from the same or different providers. Such multi-access can also

Figure 1. Multi-access and multipath transmission in virtual topologies

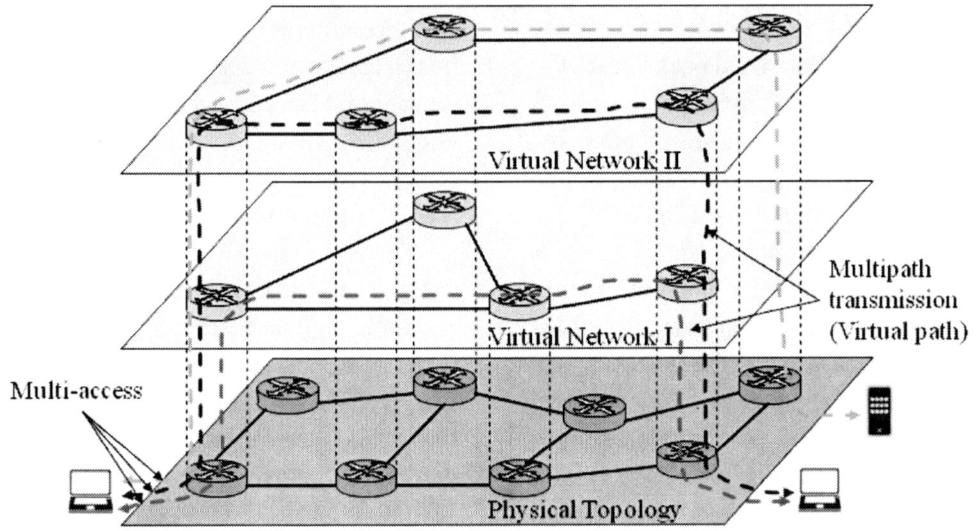

be considered as multi-homing which means that multiple paths are available for a user. These paths can be aggregated to one virtual path which is used for high-reliable data transmission. The user can divide its data stream into segments and transmit them separately through different paths offered by the VNs to which the user is connected as illustrated in Figure 1. This concurrent multi-path transmission (Zinner 2010) is expected to increase the throughput and the resilience in the networks. The *Stream Control Transmission Protocol* (SCTP) (RFC4960[11]), which is natively supporting multi-homing, can be used for concurrent multipath transfer over different end-to-end paths between multi-homing source and destination hosts to increase the application's throughput (Iyengar 2006).

When a user wants to transfer its data stream through multiple paths simultaneously, it has first to connect to multiple VNs. As a matter of fact, the user gets multiple IP addresses, one from each network. Using TCP or UDP as a transport protocol does not support the communication between users having more than one IP address in the same communication session. SCTP gives the user the ability to have multiple IP addresses and transfer the data streaming through multiple paths to the corresponding user, who is aware on the fact that the data segments received from different baths are belong to the same user as long as both use SCTP as a transport protocol and each knows the IP addresses of the other in advance.

To connect to multiple VNs simultaneously, a wireless device would have to be equipped with multiple *Network Interface Cards* (NICs). However, the usage of multiple NICs demands higher cost, large physical space and more energy consumption. Instead, we can gain from virtualization again by virtualizing NICs. Virtualization of a single NIC refers to an abstraction of the physical NIC to be appearing as multiple virtual interfaces to the user (Chandra 2004)(Coskun 2009). Thus, a user can connect to each VN via a separated virtual interface. Another NIC virtualization technique is presented in (Chereddi 2007), where each device is equipped with at least two physical NICs. One of them is switched to a fixed channel for receiving traffic, while the second interface is switched to any of the remaining channels for sending. Packets are always transmitted to the fixed channel of the receiver. Thus, a wireless node can virtually connect to multiple networks simultaneously. Virtualization of NICs refers to the aforementioned virtualization in the user side (mobile device), that is, along with network virtualization enable multi-access support.

To create VNs on demand and dynamically adapt to the changes, a management and monitoring system is required. There have been a handful of architectures developed to achieve this and additional high level goals. For instance, within the context of the EU FP7 Autonomic Internet project[12], an open source interface called *virtual Component Programming Interface* (vCPI) has been developed by university of Passau which offers mechanisms to dynamically provide context-aware VNs. The vCPI is located at every physical component and performs management (such as creating, stopping, and suspending virtual resources) and monitoring (CPU assigned to virtual router X) commands received from outside to manage or monitor that component. The creation of different context-based VNs based on the system virtualization technology Xen[13] and the vCPI has been done in UCL-Testbed in the context of Autonomic Internet project (in deliverables D1.2 and D6.3). Thus, users can gain multiple accesses to different services, providers, etc, in terms of VNs through one single machine.

Network virtualization technologies are already used by some experimental testbeds for research such as Planetlab, CoreLab, Grid5000, German-lab, GENI and others. Most of them use system virtualization techniques to create the virtual environment. Virtualization software commonly known as "hypervisor" or "*Virtual Machine Monitor* (VMM)" is run on a given hardware platform and multiplexes several virtual instances onto a single hardware instance (e.g. Xen hypervisor[13]).

Within the context of the EU FP7 4WARD project, researchers developed a virtualization architecture for the future internet in order to incorporate new networking capabilities which are not possible in the current restrictive internet architecture. The idea presented is to allow the operation of multiple non-conflicting isolated networks on the same underlying physical network infrastructure. The features of the architecture include virtual and physical nodes discovery (Nogueira 2010), on-demand VN instantiation and its management, and resource management. X-Bone (Touch 2005) is another project which focused on the development of network virtualization architecture. The architecture dynamically deploys and manages Internet overlays to reduce configuration effort and increase network component sharing. The X-Bone discovers, configures, and monitors network resources to create overlays over existing IP networks.

Beside these, there are many other recent research initiatives around the globe which focused on network virtualization and related problems like AKARI (Japan), CABO (USA), Clean Slate (USA), NouVeau (Canada), Trilogy (EU), VINI (USA) etc. We will discuss in detail the context information model and VN context representation approach.

CONTEXT INFORMATION MODEL AND DESCRIPTION

The context of a user is essential information to be integrated in novel future network approaches, to better suit the users' requirements. As a consequence, context is the driver for creating and selecting appropriate networks supported by network virtualization. In this section, we present a methodology to express VN context. We introduce *policies* that are collections of rules expressing the VN context features and capabilities. They are used as formal descriptions in the handshake process between the user's device and the virtual network.

Figure 2. Elements of the context policy

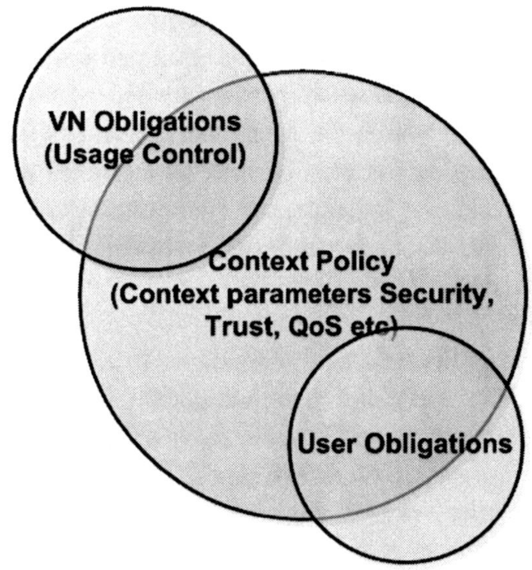

There are many existing policy languages which target different classes of problems, such as access control and privacy issues, as, for example, XACML (eXtensible Access Control Markup Language) (OASIS), IBM's EPAL (The Enterprise Privacy Authorization Language) (Ashley 2003), Ponder (Damianou 2001), and PRIME-DHP (PRIME Data Handling Policy) (Ardagna 2008). The representation of VN context is analogous to the problem of usage control or access control rules. Thus, the problem can be addressed by a policy based solution where the policy is a collection of assertions of three different forms as depicted in Figure 2:

- **VN capabilities** which are represented through a set of context parameters and describe the principle potential of the VN in terms of policy assertions.
- **VN obligations/commitments**[14] to the user which are promises made by the VN to the user about the usage of his data, pro-

vision of certain services etc. These are the actions which the VN commits to oblige.

- **User obligations** are user agreed terms of usage. These obligations are sets of assertions which the user must commit before joining (or may be later as required) any VN, for instance, the commitment to pay the cost or the commitment to share certain data.

Representation of obligations is a well researched topic and there are existing languages to represent them. Thus, our focus will remain on the representation of VN capabilities only.

Figure 3 depicts the top level policy structure consisting of the description of the overall VN capabilities, VN obligations, and user obligations. The set of context parameters represent the properties and capabilities of the VN. A basic set of important parameters should include security, trust, privacy, QoS, pricing, and the mobility (including the support for mobility management required), etc. As mentioned before, the list is exhaustive and can be extended easily as the proposed representation mechanisms give room to add new parameters and context features like geographical location, availability etc. dynamically. It is also important to emphasize that these context features and parameters may be used in both the user obligations and VN obligations description.

To illustrate the usefulness of context policies, we exemplify our concept along the VN context parameter *security*. We propose to design a more elaborated hierarchical structure where all the basic security parameters are presented within a policy as depicted in Figure 3. Each of the subcomponents of "Security" can be further elaborated and the structure evolves correspondingly.

The policy structure needs to have some basic characteristics to accommodate the foreseen and unforeseen set of requirements:

- **Expressiveness:** Ability to express wide class of context features and properties.
- **Flexibility and scalability:** Ability of the language to induct new vocabulary into it.
- **Consistent and Safe:** The structure must not be self contradicting and is free of any ambiguities which would lead policy processors into deadlock/un-decidable states.
- **Processing Efficiency:** Efficiency of the policy based systems which is dependent on the structure of the language and policy itself.
- **Usability:** The degree of ease with which the users can interact with a policy based system, can define their preferences or enter policies.

The VN capabilities will be represented by the assertions of the form

Figure 3. The top level policy structure consisting of the description of the overall VN capabilities

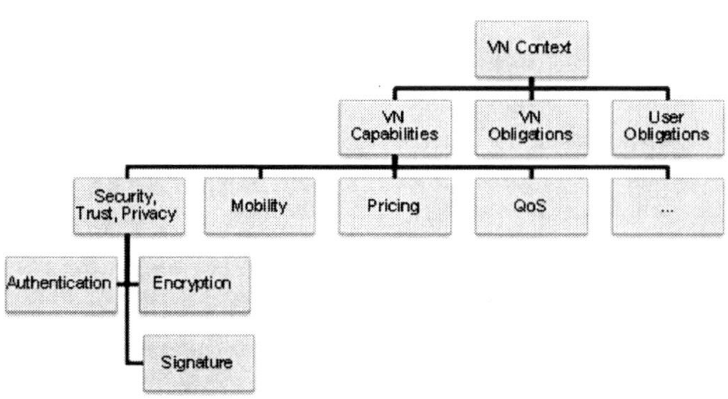

VN<*ID*> PROVIDES <*CAPABILITY (PARAM1, PARAM2…)*> IF <*C1…CN*>

As an example, we now present three basic assertions complying with the above general form followed by the description of language constructs:

```
1.        VN<1> PROVIDES Security
(        Authentication(Biometric))
2.        VN<1> PROVIDES Security
(        Encryption(AES))
3.        VN<1> PROVIDES Security
(        Authentication(X509),
Encryption (AES)
```

The two statements (1) and (2) above can be combined into one assertion (3). The same has been represented in XML encoded form in Box1.

In the formal statement above the following syntax is used:

- VN<*ID*> = "VN" construct presents the ID of the VN to which policy applies. *ID* is variable.
- PROVIDES <*CAPABILITY (PARAM1, PARAM2…)*> = "PROVIDES" constructs

Box 1. XML encoded policy (with statement 3 = encoded).

```
<context>
<VN>
        <ID>…</ID> //Other VN ID/Meta information comes here
</VN>
<Provides>
        <capability id = "1" name="url:.../Security">
        <parameters count="2">
                <parameter name="url:.../Authentication"> //Parameter1
<parameter name="url:.../scheme"/>
<value>X509</value>
</parameter>
</parameter>
                <parameter name="url:..../encryption"> //Parameter 2
<parameter name="url:.../algorithm">
<value>AES</value>
</parameter>
</parameter>
                …
        </parameters>
        <If/>
                                                  //Optional condition block
        </capability>
        < capability id="2" name="url:.../Pricing" />
…
</Provides>
<Obligations/>         //Empty block, place holder for obligations
</context>
```

presents the corresponding context capability like, Security, QoS, Mobility etc. A capability may be fully represented by appending it with optional 1…n parameters. Each of these parameters can themselves be complex with multiple sub-parameters as shown in example statements 1 and 2 above where "*Security*" is a capability with two parameters *Authentication* and *Encryption*. Each of them have one parameter themselves (e.g. *AES* for *Encryption*)

- IF <*C1…CN*> = Optional "IF" construct allows to append set of conditions. The conditions can vary from very simple to complex ones, however the condition must always resolve to Boolean state.

The collection of assertions in this form constitutes a VN context which is presented to the users on request. Ideally, we employ parametric VN context policies similarly to (Cassasa 2007), as shown in the example below. In parametric policies, the users set their preferences by overwriting the parameters within the presented VN context policy. A sample XML encoded representation of a VN context is presented below. The <value> tag can be overwritten by the user to provide its preferences to the WMN and if accepted the user then can join the VN.

When joining a VN, each mobile user tries to join the VN which matches its preferences closely. This is done by matching the preferences described by policies to the capabilities of the network considering both VN commitment and user obligation handshaking. Since it is probably not feasible to support exact VNs for all context cases, the concept foresees to manage also partial matches. If the VN policy is fully compliant to the user requirements, then this VN might be selected; in case only a partial match is found, the best match will be selected or the creation of a new VN will be enforced. The selection and creation of VNs will be described in detail.

MULTI-VIRTUAL ARCHITECTURE FOR CONTEXT-BASED WIRELESS MESH NETWORKS

In (Matos 2013), a novel architecture is proposed to contextualize wireless access environments (see Figure 4). A physical WMN, which can be provided by one or more infrastructure providers, is split into several adaptable context-based VNs, each one appropriate to certain levels of *context* of users and their services (e.g., cost, security, mobility, QoS, etc.). Therefore, the architecture is able to support, in the same infrastructure, a multiplicity of users, devices, and applications, according to their context requirements.

The set of nodes and links that are part of a VN forms a suitable topology for a specific context-based communication. Moreover, the resources assigned to those nodes and links, and the protocols running over each VN (routing, transport, security and mobility approaches, etc.) should be suitable to its context purpose.

Following this approach, the architecture aims at providing personalized networking for users, devices, or applications, which will access the network through the VN that best fits the users' and applications' requirements, mapped in context values. Through the real-time creation and reconfiguration of VNs, it will be possible to optimize the users' QoE, while dynamically meeting changing demands.

There are several specific environments that will benefit from the deployment of a WMN to serve their users according to the proposed multi-VN approach, enabling the appearance of several applications. Those environments include residential areas, schools, universities and college campus, business and industrial areas, and tourist attraction areas.

Impact of Network Virtualization

Network virtualization has been considered as a powerful tool to create, configure, and adapt

Figure 4. High-level overview and challenges

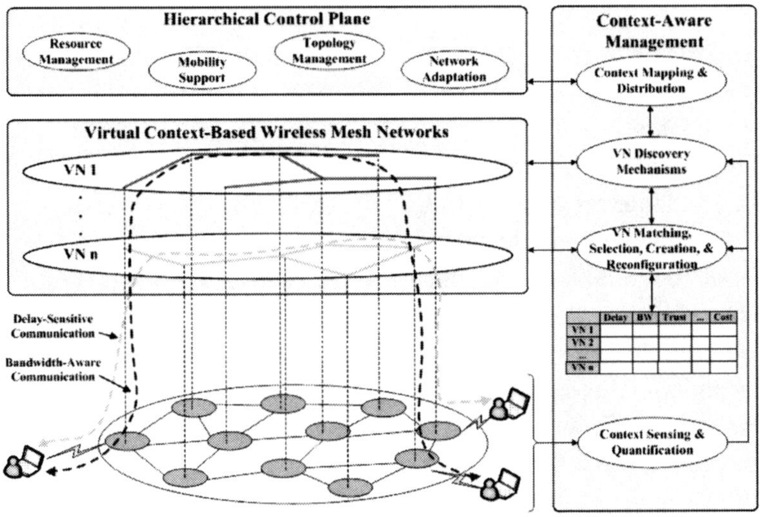

context-based VNs. In the envisioned scenario mesh clients are constantly moving from one mesh router to one of its physical neighbors, or changing their preferences and expectations. The available VNs have to be dynamically extended to be incorporated in specific physical mesh nodes.

The application of network virtualization to WMN environments brings well-known benefits, such as, less (or no) interference among different context-based communications, which, due to their specific trust or QoS requirements, are better supported by an isolated VN, without interfering with other communication requirements supported by other VNs. Nevertheless, the control overhead and latency introduced by network virtualization techniques, and the lack of resources of wireless environments are seen as potential obstacles to establish the proposed context-based virtual WMNs.

Context-Aware Management

The multi-dimensional availability of context information, and the constant changing of such information over WMN environments (e.g., preferences and location of mesh clients) or the required resources of applications, have highlighted the need for the deployment of several mechanisms to create, setup, and reconfigure the proposed context-based VNs. Such mechanisms will make the VNs usable, scalable, and adaptable to different levels of context features that characterize WMN's clients and applications.

Context Sensing and Quantification

It is of major importance to support a platform to sense, acquire, and predict context information, quantifying such information in different levels that can then be directly mapped to suitable topologies and assigned resources to specific context-based communications (and VNs).

VN/Node Discovery

When a user arrives at a WMN node, or moves inside the WMN, local and global discovery mechanisms are find an available VN/node that fits user's context requirements. Such discovery mechanisms impose the distribution of context information along the network elements.

VN/Node Matching and Selection

Semantic similarity metrics and thresholds are required to efficiently match the available context-based VNs and the set of context requirements that characterize a particular user or application (trade-off between availability and interest). During this selection procedure, the approach can adapt to changing wireless mesh link qualities since the selection is provided in real-time depending on the current status of all available nodes and links.

Even if we consider only certain context parameters characterized by a small number of levels, it is not feasible to create a distinct VN to serve a specific user, which can be very often attached to a similar semantic VN without resulting in a high QoE disturbance.

The trade-off between the performance cost of the perfect fit and feasible number of VNs, and the overall quality enhancements to users should be analyzed.

VN Topology and Resource Management

The creation of appropriate topologies for the available context-based VNs and the tailored allocation of the physical resources among them is a complex task due to the dual optimization problem, on nodes and links. Moreover, it will be required to dynamically adapt the topology and assigned resources to each VN, triggering the reconfiguration of available VNs according to different levels of context changing.

User Connectivity Establishment

When a user arrives at a WMN node where a suitable VN is not available, it triggers a discovery mechanism. Such discovery mechanism will then trigger the extension of such VN up to the user attached router, which tries to avoid the creation of the exactly fitting VN for the user (creation of a new VN is more time and resource consuming). However, if there are a large number of users with the same context patterns accessing the WMN with no exactly fitting VN, a new VN can be created).

We consider four scenarios for user connectivity establishment, which include the discovery, extension, and update of a fitting available VN for the user (see Figure 5).

Scenario 1

If router A belongs to one VN that fits the user requirements, the user can be attached to it. In this scenario, the following control processes will be performed:

- **User association** to router A, which includes the L2 association that follows the traditional IEEE 802.11 procedures, and the transference of user context.
- **Update of the chosen VN** in terms of routing tables.

Scenario 2

If router A does not belong to one VN that fits the user requirements, but one of its physical neighbors belongs to one available fitting VN, a virtual link needs to be added to this VN, in order to include router A. As an extension of scenario 1, scenario 2 also requires:

- **Discovery of one available VN in the physical neighborhood** of router A.
- **Extension of the chosen VN**, by the addition of a virtual link to include router A in the chosen VN for the user.

Scenario 3

If the discovery procedure in the physical neighborhood of router A does not provide a point of attachment of one available fitting VN for the user, a discovery procedure in the virtual neigh-

Figure 5. User connectivity establishment (Algorithm 1)

```
U: Specific Mesh Client                 A: Attached Mesh Router of U
PNA: Physical Neighborhood of A         VNA: Virtual Neighborhood of A
V: Fitting Virtual Network for U        PoA: Point of Attachment of V

Associate U with A
if PoA ⊂ A then                                          /* Scenario 1 */
|  Update V
else
   if PoA ⊂ PNA then                                     /* Scenario 2 */
      Extend V to include A
      Update V
   else
      if PoA ⊂ VNA then                                  /* Scenario 3 */
         Extend V to include A
         Update V
      else                                               /* Scenario 4 */
         P2P distributed discovery to find a PoA
         Extend V to include A
         Update V
```

borhood of router A is performed. This scenario also includes:

- **Discovery of one available VN in the virtual neighborhood** of router A.

Scenario 4

If neither router A, nor one of its neighbors belongs to one VN that fits the user requirements, a distributed discovery procedure is performed in the network to find a best fitting VN. Another control process is needed:

- **Global discovery mechanism** to find a point of attachment of one available best fitting VN for the user in the WMN (this process will be detailed below).

Global Discovery Mechanism

A centralized approach could be considered as the most obvious solution to control the architecture. In this case, if a user arrives at a WMN node where no fitting VN is available, neither in its physical neighbors, the WMN node would have to communicate with the WMN controller to trigger the selection and extension of an available fitting VN of the WMN.

Figure 6 and Figure 7 show two possible centralized approaches to find the most suitable VN for a user according to its context requirements. In Figure 6, the initial handshake operations cover the sequence of steps where the user is being authenticated. In the following, user's preferences are sent to the WMN controller. Then, the WMN controller matches the user's preferences against the context features of the available VNs, selecting the most appropriate context-based VN and returning the result to the user. Additionally, the user can ask for the full features of the selected VN, and finally sends its acceptance to join the VN. Figure 6 depicts an alternative centralized approach, where the user requests the list of VNs and their corresponding features and selects the best one by itself.

Centralized solutions have well-known drawbacks, such as their scalability limitations and

Figure 6 Centralized approach

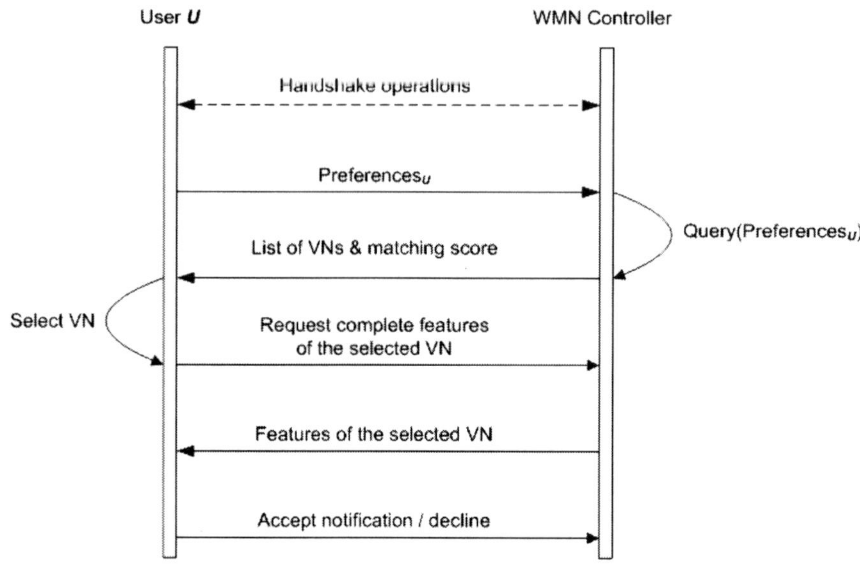

processing bottlenecks. Therefore, a distributed global discovery process is considered to find available VNs and nodes in the WMN, endowing every WMN node with intelligence to co-operate in different management processes.

The architecture adopts a DHT (*Distributed Hash Table*)-based semantic control ring to: (*i*) characterize the VNs, through the direct mapping of their levels of context features in ring nodes; (*ii*) enable a "semantic-driven" mechanism to quickly find a point of attachment of a VN; (*iii*) incorporate dynamic behavior of ring nodes to trigger distinct management processes, such as, VN creation, assignment, and reconfiguration. In this chapter, the semantic control ring enables the global discovery of a fitting VN for a user. However, in the future, this control ring will co-operate in the distribution of other management functionalities, such as, VNs' resource or topology control.

The semantic control ring is composed by a node of each available VN, which is identified by

Figure 7. Alternative centralized approach

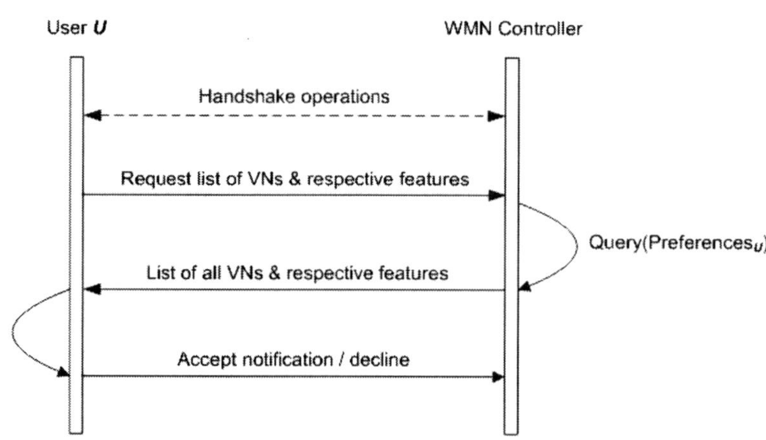

a key that is derived from the context parameters, C_p, and levels of such parameters, L_p, which characterize the VN. The ring nodes are interconnected in a consecutive order, enabling the possibility to quickly discover available fitting VNs for users. Each ring node can establish some shortcuts with other ring nodes (basics of Chord DHT (Stoica 2003)). Figure 8 presents an example of a WMN and a semantic control ring, considering $C_p=2$ (e.g., as defined in the previous section, PRICING, SECURITY_AUTHENTICATION), $L_p=4(-- \rightarrow 0, - \rightarrow 1, + \rightarrow 2, ++ \rightarrow 3$, where ++ represents a very high generic level of a context feature), *10* nodes, and *4* VNs. In this example each VN key embeds information about the levels of the VN context features, and node A is the controller of VN_1, C of VN_2, G of VN_3, and D of VN_4.

Impact on Real-World Implementations

Our approach considers the interaction and usage of various physical resources, including the support of heterogeneous technologies and the interaction between multiple network providers.

With respect to heterogeneous technologies, as long as inter-technology networks can be integrated, our multi-VN approach can be implemented, since all the control mechanisms are performed in the upper layers.

With respect to multiple network providers, the cooperation between resource providers is here a prerequisite to make full use of our multi-VN approach. Since the concepts allow to provide better quality for the users and that the resources can be efficiently operated (optimizing [e.g. for energy efficiency]), it is expected that cost models can be found which make cooperation advantageous for wireless mesh technology operators as well.

Considering virtualization itself, virtualization is already being available for wireless equipments. For example, MadWifi (MadWifi) allows to run several virtual APs on a single card, by introducing the so called Virtual APs (VAPs). With a VAP, a single WLAN card can connect to an AP, while at the same time is acting as an AP for other stations. Other combinations of operational modes are possible as well. To manipulate VAPs, MadWifi comes with a tool that is used to create and destroy VAPS with various different modes. In (Timm 2010) it is presented a virtualization framework for LTE technology.

Finally, a simplified version of the presented control mechanisms required to manage the multi-VN approach are already implemented in a small testbed in Institute of Telecommunications premises. These mechanisms address the local and global discovery of the VNs and their extension to match the location of specific users.

Figure 8. DHT-based semantic control ring

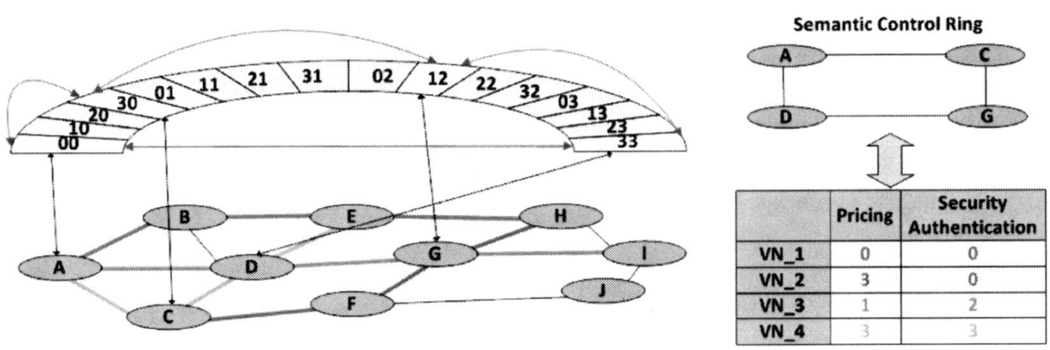

Through the following evidences of the real support of both data and control path mechanisms required to build the proposed approach, together with the definition of new business models for network providers cooperation, we see clear advantages for real-world implementations in near-term future Internet development.

ARCHITECTURE MODELING

In this section, we introduce two mathematical models that can be used for the quantitative evaluation of network topologies with multiple virtual networks. The sheer complexity of the problem, and the need to accurately capture a large number of features of such topologies, and in particular the competition and interaction of different VNs, necessitates that we follow a dual approach, whereby:

- We first introduce a probabilistic model that is based on diffusion approximations in order to derive accurate estimations of the end-to-end delay of a packet within a context-based VN, and the impact of the control approaches to the inter-VN architecture;
- We then introduce an optimization model that uses network optimization techniques to capture the evolution of the flows in VNs and the usage of the available resources, particularly as they interact and compete with each other.

Therefore, whereas the probabilistic model captures the performance impacts, on both data and control plane, the optimization model captures the performance evolution of a collection of competing VNs.

Probabilistic Model

In this subsection, an analytical model is presented and used to establish closed-form expressions for the end-to-end delay of data communications over each VN, and to study the control processes involved in user connectivity establishment and reconfiguration.

End-to-End Delay in a WMN: Data Path

The model used in (Bisnik 2006), which represents WMNs as open queuing networks, is extended to virtualized WMNs. The model adapts the diffusion approximation method that is commonly used to solve an open *G/G/1* queuing network, where all nodes in the network are single servers with infinite buffers and First-Come First-Serve service strategy (more details can be found in (Bolch 1998)). This model will be extended to define a closed-form expression for the mean end-to-end delay of a WMN data communication.

Considering a WMN as a square unit area where the total area is divided into non-overlapping square zones of area a, there are $N=1/a$ zones in the WMN, each one covered by a mesh router. Each one of the N mesh routers (numbered from $i=1... N$) has a utilization factor of ρ_i, and is equipped with two non-interfering interfaces: one for router-router communication with capacity W_T, and another for router-client communication with capacity W_{client}. The number of physical neighbors of router i, $K(i)$, will be considered constant and equal to K in this study.

The clients arrive at the WMN according to a renewal process with rate λ_T, and all routers are equally probable to be the attached router for a client. Therefore, the mean number of clients in the WMN is equal to $c= 1/\lambda_T$.

The data packet arrival rate at the WMN is a renewal process, with a mean packet inter-arrival time of $1/\lambda_e$, and *Squared Coefficient of Variance*

(SCV) of C_A^2. Each client may be the source or the destination of data packets with size *S*, and generated according to a renewal process with rate λ. Therefore, new data packets arrive at the WMN at rate $\lambda_e = c\lambda$.

Let $prob_{0i}$ be the probability that a data packet enters the WMN through the zone covered by router *i*. According to the topology, and since all routers are equally probable to be the attached router for a client, $prob_{0i}$ is defined by the area "*a*" of each zone. As soon as a data packet is generated by a client, it is forwarded to the router within its zone, which will relay it over the backbone to the zone of the target client.

Denoting *p* as the probability of locality of data traffic (probability that a data packet received by a router is destined to a client within its zone), the probability that a data packet is routed to router *i* after completing its service at router *j*, p_{ij}, is defined by:

$$p_{ij} = \begin{cases} \frac{(1-p)}{K}, j \in K(i) \\ 0, otherwise \end{cases}.$$

The visit rate of router *i*, e_i, is defined by the average number of times that a data packet visits it. Thus, and considering from symmetry $e_i = e_j$, we have:

$$e_i = p_{0i} + \sum_{j=1}^{N} p_{ij} e_j = a + \sum_{j \in K(i)} \frac{1-p}{K} e_j = \frac{a}{p}.$$

The effective data packet arrival rate at router *i*, λ_i, can be now defined by:

$$\lambda_i = \lambda_e e_i = c\lambda \frac{a}{p}.$$

Denoting *s* as the number of routers that forward a data packet before it reaches the destination, $P(s=k) = (1-p)^{k-1} p, k \geq 1$. Thus, the mean number of hops traversed by a data packet is defined by:

$$\bar{s} = E\left[s\right] = \frac{1}{p}. \quad (1)$$

A transmission made by a router *i* is successful if none of its one or two hop neighbors transmits concurrently on the same channel. The number of one- or two-hop possible interfering neighbors of router *i*, *I (i)*, is considered constant and equal to *I* in this study.

In order to have an analytical tractable random access MAC model, similar to the IEEE 802.11 DCF MAC model, (Bisnik 2006) defines expressions for the: (*i*) number of active interfering neighbors of router *i*, H_i, which is a subset of *I(i)*; (*ii*) duration of the back-off timer of router *i*, T_i (exponentially distributed with mean $1/\xi_i$); (*iii*) number of times the back-off timer of router *i* freezes during a transmission epoch, M_i, assuming that it is frozen each time that an active interfering neighbor starts transmitting (more details in (Bisnik 2006)).

With the already defined variables, the service time of router *i*, X_i, for the transmission of a data packet is a sum of the: duration of its back-off timer (T_i); time for which the back-off timer remains frozen ($M_i(S/W_T)$); data packet transmission time (S/W_T).

At this stage, and according to the diffusion approximation method:

The SCV of the inter-arrival time of data packets at router *i*, C_{Ai}^2 (where $C_{B0}^2 = C_A^2$, and C_{Bi}^2 is the SCV of X_i):

$$C_{Ai}^{\ 2} = 1 + \sum_{j=0}^{N} (C_{Bj}^{\ 2} - 1) p_{ji}^{\ 2} e_j e_i^{-1}.$$

The probability that the number of data packets at router *i* equals *k*, $\pi_i(k)$:

$$\pi_i(k) = \begin{cases} 1-\rho_i, k=0 \\ \rho_i(1-\hat{\rho}_i)\hat{\rho}_i^{\,k-1}, k>0 \end{cases}, \text{ with}$$

$$\hat{\rho}_i = \exp\left(-\frac{2(1-\rho_i)}{C_{Ai}^{\,2}\rho_i + C_{Bi}^{\,2}}\right).$$

The mean number of data packets at router i, $Z_i = \rho_i / (1 - \hat{\rho}_i)$.

According to Little's law, the mean data packet delay at router i is defined by:

$$\bar{D}_i = \frac{Z_i}{\lambda_i}. \tag{2}$$

By symmetry, the mean data packet delay at all routers is the same, and therefore, the mean end-to-end delay of a data communication, D_T, is defined by:

$$D_T = \bar{s}\bar{D}_i = \frac{\rho_i}{ca\lambda(1-\hat{\rho}_i)}. \tag{3}$$

End-to-End Delay in a WMN: Control Path

We will now model the control processes involved in user connectivity establishment, which requires the discovery, extension or update of an available fitting VN for a user. Closed-form expressions for the mean end-to-end delay of such processes are derived, through the adaptation of the proposed data path model to the control path.

User Association

In the case of user association to a particular physical router, the total service time of this router, X_{client}, can be derived from X_i, and it is dependent from the back-off timer duration distribution in the router-client interface, ξ_{client}. Since the mean number of active interfering neighbors in the router-client interface of router i, $E(H_i)$, is equal to the mean number of clients attached on it, c/N, it is easy to derive that $E(M_i)=E(H_i)=c/N$ (more details in (Bisnik 2006)). Denoting S_c as the size of the control packets, the mean delay of each control message between a client and its attached router, $E(X_{client})$, is defined by:

$$E\left[X_{client}\right] = \frac{1/\xi_{client} + S_c / W_{client}}{1-(S_c / W_{client})(c/N)}.$$

If it is considered the CSMA-CA protocol for medium access, there are *3* messages for link association (RTS, CTS, transference of context). Therefore, the delay involved in user association is defined by $T_{Association} = 3E(X_{client})$.

Mean Number of Hops of a Control Packet

Considering that the mesh clients and the virtual nodes that compose the context-based VNs are uniformly distributed over the WMN, the area and topology of the network can be used to determine the mean number of hops traversed by a control packet. Following the assumption of Kumar-Gupta model (Gupta 2000), it is considered that the mean number of hops traversed by a control packet is given by:

$$\bar{s}_c = \sqrt{\frac{N}{\log(N)}}.$$

This expression can be directly applied to the control processes that take place in the physical and control networks, since both networks are composed by one instance in every physical node. In case of the control processes over the context-based VNs, N is replaced by the number of virtual nodes that compose each VN. With $\bar{s}_c$, and from the relation achieved in (1), the probability of locality of control traffic is defined by $p_c = 1/\bar{s}_c$.

Router Service Time Model

The mean control packet delay at router i, $\bar{D}c_i$, can be translated from (2). However, $\bar{D}c_i$ will consider the specific features of the network that provide support for the specific control process: (*i*) in the context-based VNs, $\bar{D}c_i$ is totally related with the data traffic that flows over them, since the number of control messages can be neglected comparing to the volume of data traffic; (*ii*) in the physical WMN and the DHT-based control ring, $\bar{D}c_i$ only depends on the control traffic that flows over them.

VN Discovery

The VN discovery process in the physical neighborhood of the user attached router involves the transference of a message to all its physical neighbors, which answer with a response message. Therefore, the delay involved in this discovery process is defined by:

$$T_{Pneighbor} = 2\,\bar{D}c_i.$$

The VN discovery process in the virtual neighborhood of the user attached router involves the transference of a message by each virtual instance of this router to all of its virtual neighbors, which answer with a response message. This process involves a delay defined by:

$$T_{Vneighbor} = 2\,\bar{s}_c\,\bar{D}c_i.$$

In scenario 4, it is required to perform a distributed VN discovery along the DHT-based control ring. Since the architecture is able to provide a context-based VN for any set of C_P context parameters and L_P levels, in theory there are a maximum of L_P^(C_P) peers in the control ring, each one responsible to characterize each VN according to its context features. Moreover, it is assumed that: (*i*) each one of the c clients, which are uniformly distributed over the WMN, may trigger one discovery process in the control ring, that is, the starting point of the discovery process is uniformly distributed among all available nodes; (*ii*) the ring is composed by one virtual node in every physical node, and the peers are uniformly distributed among these available virtual nodes in order to enable load-balancing of the storage responsibilities; (*iii*) the ring is structured in a consecutive order of peer identifiers, based on Chord protocol (Stoica 2003). It is already known from the literature that the mean number of messages required by a lookup procedure in Chord protocol is $log\ (\#Peers) = log\ (L_P$^$(C_P))$, and so, the delay involved in this discovery process is defined by:

$$T_{Ring} = log\ (L_P\text{^}\ (C_P))\ \bar{s}_c\,\bar{D}c_i.$$

Extension of the Chosen VN

This process involves the transference of one message between the user attached router and the point of attachment of the chosen VN for the user. Therefore, the delay involved in this control process is defined by:

$$T_{Extension} = \bar{s}_c\,\bar{D}c_i.$$

In case of adding a link between the user attached router and one of its physical neighbors, $\bar{s}_c = 1$.

Update of the Chosen VN

Assuming that it is used an efficient link-state routing protocol, the number of messages required to update the routing tables of each VN, in order to integrate a new user on it, is proportional to the mean number of virtual nodes per VN. Since these messages occur in the network in a concurrent manner, the delay for the update of the chosen VN for the user is defined by:

$$T_{Update} = \overline{s}_c \overline{D}c_i .$$

Multi-VN Architecture Model for Context-Based WMNs

The analytical models for data and control communications are now adapted to the context-aware multi-VN architecture. Since the performance metric is the delay of the data and control communications, the packet data and control delays are included at each router ($\overline{D}_i$ and $\overline{D}c_i$, respectively), taking into account the requirements of data and control traffic (λ, S, p, and λ_c, S_c, respectively), and the features (W, ρ) of the networks (physical WMN, different context-based VNs, control ring for VN management) which provide support for these communications.

For simplicity reasons, we consider that each available VN is mostly specific for one of the three types of communications: Delay-sensitive, secure or high-bandwidth communications. However, these context-based VNs may be characterized by other context-features that have less importance for the VN behavior (for e.g., a high-secure VN may offer low bandwidth for communications, whereas other high-secure VN may offer a higher level). Following this approach, we describe the specific features of each network are described through its context-aware purpose: (*i*) in a delay-sensitive VN, the traffic is more located (high probability of locality of data traffic, p), and it is required to have fast communications with a small number of hops; (*ii*) a secure VN requires more capacity of utilization of virtual routers, ρ, in trust and authentication procedures; (*iii*) a high-bandwidth VN requires more bandwidth capacity, W, of the virtual links; (*iv*) the physical and control networks need a small amount of resources.

The bandwidth required by each network, W, will be a percentage of the total available bandwidth of the physical links, W_T, which will map its context-aware purpose:

$$W = \begin{cases} \alpha_1 W_T \ \ for\ high\text{-}bandwidth\ VNs \\ \alpha_2 W_T \ \ for\ secure\ VNs \\ \alpha_3 W_T \ \ for\ delay\text{-}sensitive\ VNs \\ \alpha_4 W_T \ \ for\ PHY\ and\ overlay\ networks \end{cases} \qquad \alpha_1 > \alpha_2 > \alpha_3 > \alpha_4 . \tag{4}$$

Evaluation and Discussion

We now perform the evaluation of the effectiveness of the probabilistic model. The variables of this study consider the specific features of each available network and the requirements of control and data traffic. Then, we model the proposed scenarios for user connectivity establishment based on a probabilistic approach. Next, we determine the maximum limits of the architecture to accomplish feasible delays for different data and control communication requirements. Finally, we evaluate the mean end-to-end delay of the different scenarios for user connectivity establishment.

Assumptions

Table 1 summarizes the assumptions of the study. The proposed architecture is established over a WMN composed by $N=100$ routers, containing a variable number of VNs, V, each one with a variable number of virtual nodes, n, which are uniformly distributed over the physical mesh routers. The clients arrive at the WMN at a variable rate, λ_T, and they are equally distributed among the physical routers and available VNs. Each VN is characterized by $C_p=5$ context parameters, and each parameter has $L_p=4$ levels, supporting a total number of L_p^ $(C_p) = 2048$ peers. The interfaces' capacity are $W_T=54$ Mb/s for router-router, and $W_{client}=11$ Mb/s for router-client.

The WMN follows a squared topology and each physical router has $K=8$ physical neighbors, and $I=24$ possible interfering neighbors. The control ring is composed by one virtual router in every physical router. For context-based VNs, K and I are considered equal to the total number of virtual nodes, n, that compose each VN.

Table 1 also defines the utilization factor of the routers, ρ, and the bandwidth of the links, W, of the available networks. Moreover, it provides the requirements of data traffic that flows over context-based VNs (expected arrival rate, λ, packet size, S, and probability of locality, p), and values for S_c (size of packets necessary for the proposed control processes) and λ_c (arrival rate of control traffic).

The specific requirements of data and control traffic, as well as the routers' utilization factor and link's bandwidth of the networks, are derived from their context-aware purposes: (*i*) in the delay-aware VNs, the data traffic is more located (small value for p) in order to accomplish small end-to-end communication delays; (*ii*) in secure VNs, the higher utilization factor (ρ) of their virtual routers is the parameter that differentiates these VNs from the others; (*iii*) the VNs for high-bandwidth communications require high capacity (W) of their virtual links, and the communications are characterized by packets with larger size (S) and higher arrival rate (λ) when compared with delay-sensitive or secure communications; (*iv*) the physical and control networks require less physical resources than the context-based VNs.

Note that the assigned bandwidth to each type of VN refers to the total assigned bandwidth to all VNs from this type, and it will be equally divided among the available number of VNs from this type. For simplicity reasons, the total number of VNs, V, is equally divided among the three considered types.

Modeling of the Scenarios for User Connectivity

Now, we model the four scenarios for user connectivity establishment, using the already defined closed-form expressions. We also derive the probability of occurrence of the four proposed scenarios, considering that the users, the virtual nodes that compose each VN, and the nodes of the control ring, are uniformly distributed among the physical mesh routers.

Table 1. Features of data/control traffic and available networks

	Physical WMN	Control Overlay	**Context-Based VNs**		
			Delay-Aware	**Secure**	**BW-Aware**
N	*100 (10 x 10)*				
V, n, λ_T	Number to be evaluated				
C_P	*5*				
L_P	*4*				
W_T	*54 Mb/s*				
K	*8*		N		
I	*24*				
ρ	*0.1*	*0.1*	*0.4*	*0.5*	*0.4*
W	$0.05W_T$	$0.05W_T$	$0.1W_T$	$0.3W_T$	$0.5W_T$
S	–		*64Bytes*	*96Bytes*	*128Bytes*
λ			*17.6Kb/s*	*96Kb/s*	*512Kb/s*
p			*0.4*	*0.3*	*0.3*
S_c	*48 Bytes*				
λ_c	*4 Kb/s*				

Scenario 1

In scenario 1, the chosen VN for a user is already available in the user attached router. Following the previous assumptions, this scenario has a probability to occur of:

$$p_1 = \frac{1}{N}.$$

Its end-to-end delay only includes the user association to a mesh router, and the update of the chosen VN for the user, and it is defined by:

$$T_1 = T_{Association} + T_{Update}.$$

Scenario 2

Scenario 2 is an extension of scenario 1, and then its occurrence probability includes the probability of non-occurrence of scenario 1, times the probability that one of the K physical neighbors belongs to a fitting VN for the user:

$$p_2 = \left(1 - \frac{1}{N}\right)\frac{K}{N}.$$

Its end-to-end delay is defined by:

$$T_2 = T_1 + T_{Pneighbor} + T_{Extension}.$$

Scenario 3

When there is no available VN in the physical neighborhood of the user attached router, the virtual neighborhood is checked. Since there are V context-based VNs and a mean number of n virtual nodes per VN, the mean number of virtual routers for data communications per physical router can be defined by:

$$R = \frac{Vn}{N}.$$

In scenario 3, it is checked if one of the R virtual nodes of the user attached router has a virtual neighbor that belongs to a physical router with a fitting VN for the user. Following the same approach, and including the probabilities of non-occurrence of the two previous scenarios, this scenario occurs with probability:

$$p_3 = \left(1 - \frac{1}{N}\right)\left(1 - \frac{K}{N}\right)\left(R\frac{n-1}{N}\right).$$

In this case, the end-to-end delay is defined by.

$$T_3 = T_2 + T_{Vneighbor}.$$

Scenario 4

If no fitting VN is available neither in the user attached router nor in one of its physical or virtual neighbors, a discovery process is required in the control ring to find a node with a fitting VN. This scenario is the last one to occur, and so, its probability is the remainder:

$$p_4 = \left(1 - \frac{1}{N}\right)\left(1 - \frac{K}{N}\right)\left(1 - R\frac{n-1}{N}\right).$$

The end-to-end delay of this scenario is defined by:

$$T_4 = T_3 + T_{Ring}.$$

Through this probabilistic approach, the mean end-to-end delay for user connectivity establishment is given by:

$$T_{Total} = p_1 T_1 + p_2 T_2 + p_3 T_3 + p_4 T_4. \quad (5)$$

Limits of the Multi-VN Architecture

Now, we evaluate the number of VNs that may be established over the WMN, V, as well as the number of virtual nodes that compose each VN, *n*, and mesh clients, $1/\lambda_T$.

Figure 9 shows the mean end-to-end delay of a data communication (D_T (3)) over the considered context-based VNs. The WMN with *100* physical mesh routers is able to support *24* VNs, each one composed of a maximum of *10* virtual nodes, with *200* attached clients, with delays of ~*5s* in the worst case. A smaller value of ~1s is achieved for a delay-sensitive communication; an intermediary value of ~*3s* is achieved for a secure communication, where there is the need for more processing time of the virtual routers; and a high value of ~*5s* is present in a high-bandwidth communication. Since each context-based VN meets the user requirements through the adaptation of its features to the context-aware purpose of user communication, the differences between their communication delays are expected: delay-sensitive communications present the lowest delays when compared with the others, since their packets flow over a smaller number of hops; high-bandwidth communications present the highest delays, due to the higher size and arrival rate of their packets. Furthermore, the end-to-end delay of a context-aware communication is higher when there are more VNs, which implies the reduction of assigned physical resources to each VN, and increases the processing time at each virtual node and virtual link. Moreover, the increasing number of virtual nodes per VN, and users, increases the end-to-end data communication delay, because there is more traffic per VN.

For a control communication over the physical or control networks, its delay is defined by $Dc_T=\overline{s}_c\overline{D}c_i$, which maximum value is less than *300ms*, even for the highest possible number of users (*200*).

Mean End-to-End Delay for User Connectivity

Figure 10 presents the total end-to-end delay (T_{Total} (5)) for user connectivity establishment, considering that there are *15* context-based VNs over the WMN. Figure 11 further depicts the delay of each control process involved in each scenario, when there are *15* VNs over the WMN, each one

Figure 9. Mean end-to-end delay of a data communication over a delay-aware VN (top left), secure VN (top right) and BW-aware VN (down), with a total number of 6 or 24 VNs over the WMN, a variable number of virtual nodes per VN, and a variable number of users accessing the WMN

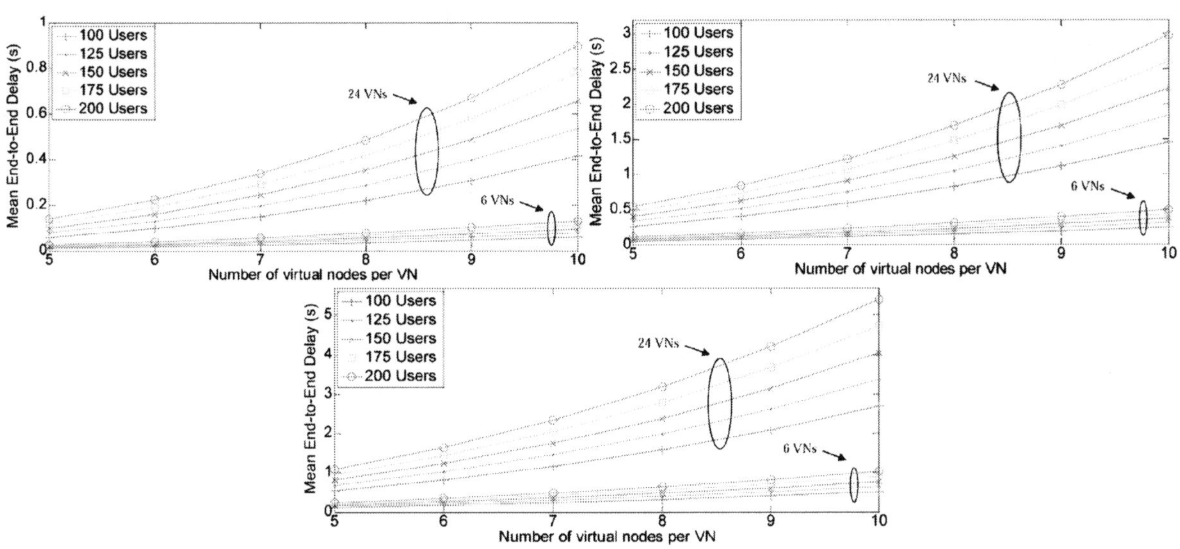

composed by *8* virtual nodes and *150* users accessing the WMN.

As it is expected, the end-to-end delay for user connectivity establishment increases from scenario 1 to 4, since each scenario is an extension of the last one. From the comparison between

Figure 10 and Figure 11, it is shown that the total end-to-end delay is clearly affected by scenario 4, since this is the most probable scenario to occur. Due to the large size of the WMN, there is often the need to start a distributed discovery process over the control ring, in order to find a point of attachment of one fitting VN for a user.

The user association time, even for the largest number of clients accessing the WMN (*200*), can be neglected comparing with the mean delays of the other control processes.

The discovery processes of a fitting VN in the physical neighborhood and in the control ring are independent from the data traffic, since they do not occur over the context-based VNs. Moreover, their delays increase with the increasing number of users in the WMN, which increases the control traffic over the physical and control networks.

The requirements of user communications, and the features of the VNs which support them, affect the control processes on these networks: discovery process of a fitting VN in the virtual neighborhood of the user attached router, and the extension and update of the chosen VN for the user. The volume of data traffic over these VNs clearly influences the end-to-end delay for user connectivity establishment: this value is lower for a secure VN than for a bandwidth-aware VN, and even lower for a delay-aware VN (as previously concluded).

Optimization Model

VNs, whether coming from the same physical network or from different physical networks that have been lumped together, will be competing for available bandwidth with each other. In this subsection, we develop a network optimization formulation that captures this competition and allows its quantitative analysis. The basic idea is to construct a model for a set of virtual networks, each with its own traffic requirements, and each with its own access to bandwidth, and also model the fact that the cost of using bandwidth in a given channel by a given VN will depend on the usage of bandwidth in that channel by competing VNs.

Figure 10. Mean end-to-end delay for user connectivity establishment (T_{Total}) over a delay-aware VN (top left), secure VN (top right) and BW-aware VN (down), with a total number of 15 VNs over the WMN, a variable number of virtual nodes per VN, and a variable number of users accessing the WMN.

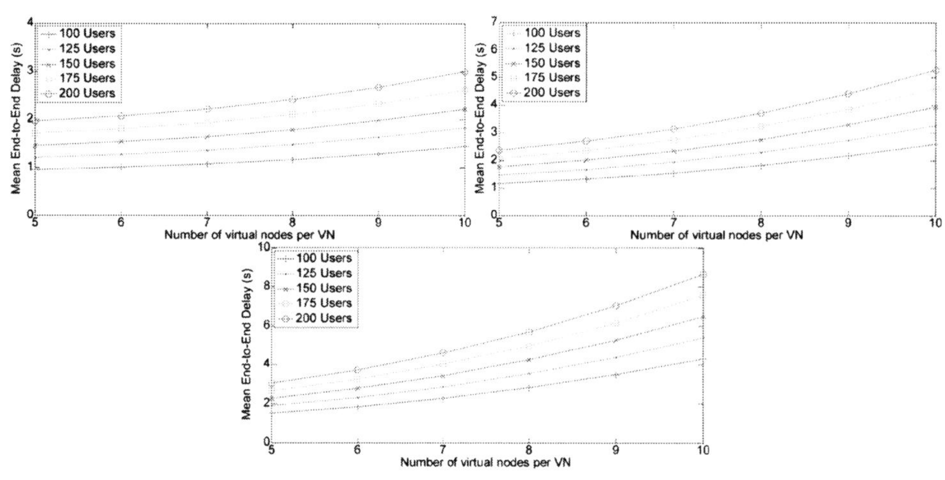

Network Elements and Quantities

We assume that there is a set $\mathcal{C}$ of C channels available for communication. We use the term channel in the sense it has in cellular contexts (i.e., as a frequency band, or a slot, that exists in a given, confined region of space and can be used for wireless communication in that region). It is possible that groups of channels are partly or completely collocated in space. This makes sense when, for example, there are two or more cellular providers operating in the same region.

We assume that the channels belong to one or more entities that are willing to lend them for the construction of one or more virtual networks. However, for each channel a cost is incurred by its owner, which is modeled in terms of a cost function $f_c(\cdot)$ that accepts non-negative arguments, and that is assumed to be non-negative and increasing. As will be seen later, the owner of a channel charges its users according to this cost. The argument of the cost function is the reservation level r_c. We define the C-dimensional reservation vector **r** to be the column vector

$$r = \begin{bmatrix} r_1 & r_2 & \dots & r_C \end{bmatrix}^T.$$

The relation of r_c to the traffic occupying channel c will be discussed soon.

We also assume a set $\mathcal{N}$ of a total of N nodes that are spread throughout a geographic region, and are interested in forming one or more virtual networks. Each node lies within the regions covered by one or more channels, and it can use these channels for its communication with the other nodes.

Communication is over a set $\mathcal{L}$ of directional links. Each link is between two nodes, one acting as a transmitter, and one acting as a receiver. There is a total of L links, and it is possible for a node pair to be connected by more than one link. The reason is that each link may be operating in different channels, and therefore may be associated with different costs. The transmitter and the receiver of a link l are denoted by $t\ (l)$ and $r\ (l)$ respectively. The sets of links that have node n as transmitter and receiver are denoted by $o\ (n)$ ('outgoing') and $i\ (n)$ ('incoming') respectively.

Let x_l be the flow of traffic through link l. As links are directional, we assume that $x_l \geq 0$. We define the L-dimensional flow vector as the column vector

Figure 11. Mean delay of each process involved in the scenarios (1, 2, 3, 4) for user connectivity establishment over a delay-aware VN (top left), secure VN (top right) and BW-aware VN (down), for a specific case: 15 VNs over the WMN, 8 virtual nodes per VN, and 150 users accessing the WMN

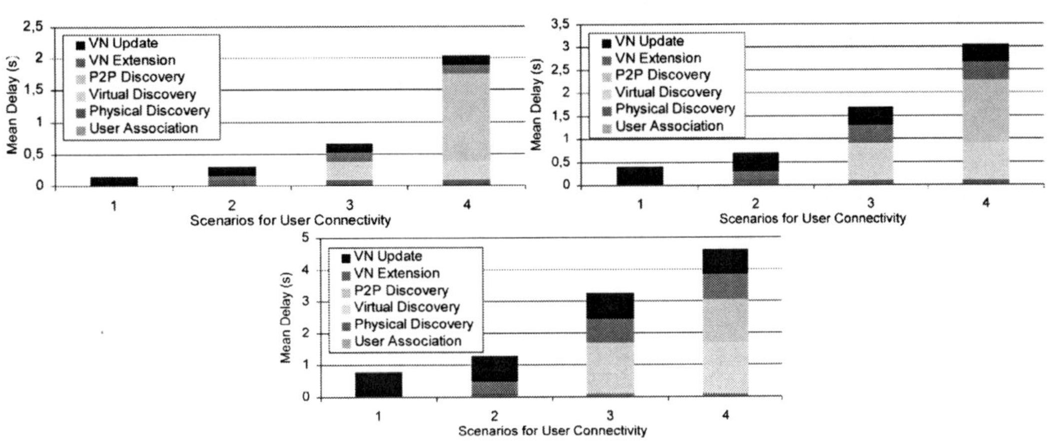

$$x = \begin{bmatrix} x_1 & x_2 & \dots & x_L \end{bmatrix}^T.$$

The reservation levels of the channels are related to the link flows through the reservation matrix

$$R = \begin{bmatrix} r_1 \\ r_2 \\ \vdots \\ r_C \end{bmatrix} = \begin{bmatrix} R_{11} & R_{12} & \dots & R_{1L} \\ R_{21} & R_{22} & \dots & R_{2L} \\ \dots & \dots & \dots & \dots \\ R_{C1} & R_{C2} & \dots & R_{CL} \end{bmatrix}$$

whose components R_{cl} are all non-negative; each such component corresponds to the reservation of channel c imposed by a unit of flow of link l. Note that we have also defined the row vectors

$$\mathbf{r}_c \equiv \begin{bmatrix} R_{c1} & R_{c2} & \dots & R_{cL} \end{bmatrix}, \qquad c = 1, 2, \dots C.$$

In particular, the following equation holds:

$$\mathbf{r} = R\mathbf{x} \Leftrightarrow \mathbf{r} = \begin{bmatrix} \mathbf{r}_1 \\ \mathbf{r}_2 \\ \vdots \\ \mathbf{r}_C \end{bmatrix} \mathbf{x} \Leftrightarrow \begin{bmatrix} r_1 \\ r_2 \\ \vdots \\ r_C \end{bmatrix} = \begin{bmatrix} R_{11} & R_{12} & \dots & R_{1L} \\ R_{21} & R_{22} & \dots & R_{2L} \\ \dots & \dots & \dots & \dots \\ R_{C1} & R_{C2} & \dots & R_{CL} \end{bmatrix} \begin{bmatrix} x_1 \\ x_2 \\ \vdots \\ x_L \end{bmatrix} \Leftrightarrow r_c = \sum_{l=1}^{L} R_{cl} x_l, \qquad c = 1, 2, \dots C. \tag{6}$$

In the simplest case, there could be one link for each channel, so that $L=C$, and the reservation matrix could be a diagonal matrix of size $C \times C$. However, with the current, significantly more general formulation, we can capture the following cases:

- The reservation level of each channel depends on the flows through multiple links (because they are all collocated with the channel);
- Using a link creates reservation levels to multiple channels. This is the case if, for example, we would like to use a link that straddles a region between two channel regions, and therefore interferes with both channels;
- Combinations of the above two cases.

Finally, observe that the elements R_{cl} are non-negative but otherwise arbitrary. Therefore, two links carrying the same amount of information can create different reservation levels for the same channel. This makes sense if, for example, the two links have different channel gains, so that, for the same data rate, they end up using the channel for a different percentage of time or frequency.

In this setting, we assume the existence of a set V of V virtual networks. The networks are virtual, in the sense that they do not have their own communication resources, but are willing to rent them from the owner(s) of the set of channels C.

Within network v, node n creates/absorbs traffic with a divergence $s_{n,v}$, implying that: if $s_{n,v}>0$ then information is inserted in the network, if $s_{n,v}<0$ then information is removed from the network, and if $s_{n,v}=0$ then the node simply relays the traffic. Divergences are not fixed, but are subject to optimization, in a manner to be discussed soon.

Also, let $x_{l,v}$ be the rate of information flow of the network v through the link l, and let the flow vector for network v be

$$\mathbf{x}_v = \begin{bmatrix} x_{1,v} & x_{2,v} & \dots & x_{L,v} \end{bmatrix}^T.$$

Obviously, we must have:

$$\mathbf{x} = \sum_{v=1}^{V} \mathbf{x}_v \Leftrightarrow x_l = \sum_{v=1}^{V} x_{l,v}, \qquad \forall l. \tag{7}$$

In words, the link rate is the sum of the link rates of all virtual networks. Furthermore,

$$s_{n,v} = \sum_{l \in o(n)} x_{l,v} - \sum_{l \in i(n)} x_{l,v}, \quad \forall n, \quad \forall v. \tag{8}$$

This stems from the fact that the difference between what comes out of each node and what comes in at that node must equal the divergence (i.e., the externally imported data traffic).

Condition (8) is the only condition we place on the traffic of a particular virtual network. Therefore, we are implicitly assuming that each such network aims at transporting traffic of a single type. In other words, packets of the same size are interchangeable, and if we allow the existence of multiple sinks of information, each packet is allowed to be extracted from the network through any of them. This makes sense in many wireless networks, notably wireless mesh networks (which is the focus of this study), but also wireless sensor networks and cellular networks. Indeed, the objective is often to transfer a packet "somewhere" through the wireless infrastructure, and subsequently to its final destination. Similarly, if there are multiple sources of information, each packet can be inserted in the network through any of them.

We associate with each node a utility function $U_{n,v}(s_{n,v})$, which expresses the satisfaction of network v when it manages to insert into or remove from the network, at that node, a divergence $s_{n,v}$. The shape of the utility function at each node and for each network is left arbitrary. For example:

- If the network v wants to send data to a node n, one possibility would be that the utility function $U_{n,v}(s_{n,v})$ is zero for positive divergences, positive for negative divergences, and decreasing (i.e., increasing in the negative direction).
- If, on the other hand, the network v wants node n to insert data in the network, one possibility would be for the utility function $U_{n,v}(s_{n,v})$ to be zero for negative divergences, positive for positive divergences, and increasing.

Having broken the data rate flowing through each link into the flows of many different virtual networks, we can do the same for the reservation levels of the channels. We therefore define the reservation level $r_{l,v}$ of channel c from the flow of the network v and the associated reservation vector

$$\mathbf{r}_v = \begin{bmatrix} r_{1,v} & r_{2,v} & \cdots & r_{C,v} \end{bmatrix}^T$$

through the equations

$$\mathbf{r}_v = R\mathbf{x}_v \Leftrightarrow \mathbf{r}_v = \begin{bmatrix} \mathbf{r}_1 \\ \mathbf{r}_2 \\ \vdots \\ \mathbf{r}_C \end{bmatrix} \mathbf{x}_v \Leftrightarrow \begin{bmatrix} r_{1,v} \\ r_{2,v} \\ \vdots \\ r_{C,v} \end{bmatrix} = \begin{bmatrix} R_{11} & R_{12} & \cdots & R_{1L} \\ R_{21} & R_{22} & \cdots & R_{2L} \\ \cdots & \cdots & \cdots & \cdots \\ R_{C1} & R_{C2} & \cdots & R_{CL} \end{bmatrix} \begin{bmatrix} x_{1,v} \\ x_{2,v} \\ \vdots \\ x_{C,v} \end{bmatrix}$$
$$\Leftrightarrow r_{c,v} = \sum_{l=1}^{L} R_{cl} x_{v,l}, \qquad c = 1,2,\ldots C, \qquad v = 1,2,\ldots V. \tag{9}$$

Observe that using (6), (7), (9), we have,

$$\mathbf{r} = \sum_{v=1}^{V} \mathbf{r}_v \Leftrightarrow r_c = \sum_{v=1}^{V} r_{c,v}, \qquad c = 1,2,\ldots C.$$

Therefore, the reservation level of each channel is the sum of the reservation levels of all VNs at that channel.

Network Optimization

We will now define an optimization problem using the setting defined so far. If network v has a reservation level $r_{c,v}$ at channel c, we assume that the owner of the channel is charging *that* network an amount equal to

$$\frac{r_{c,v}}{\sum_{v'} r_{c,v'}} f_c\left(\sum_{v' \in V} r_{c,v'}\right).$$

In other words, the owner of the channel divides the cost of the operation of the channel among the users of the channel, in proportion to the reservation levels of each network.

Given this method of charging, each virtual network would like to solve the following optimization problem:

Problem 1: optimization for virtual network v

maximize:

$$F(\mathbf{s}_v, \mathbf{r}_v) = \sum_n U_{n,v}(s_{n,v}) - \sum_c \frac{r_{c,v}}{\sum_{v'} r_{c,v'}} f_c\left(\sum_{v'} r_{c,v'}\right),$$

subject to:

$$s_{n,v} = \sum_{l \in o(n)} x_{l,v} - \sum_{l \in i(n)} x_{l,v}, \quad \forall n,$$
$$r_{c,v} = \sum_l R_{c,l} x_{l,v}, \quad \forall c.$$

In words, the virtual network would like to maximize its net benefit, namely the utility it gets out of inserting (or removing) traffic in the network, minus the cost of renting off the channels, subject to the constraints that the flow conservation equations are satisfied (i.e., traffic does not disappear at nodes) and the fact the amount of channel reserved is exactly the one needed, given the traffic flows.

Observe that, for now, we take the reservation levels, along with the flows, of all *other* virtual networks to be fixed[15].

Treating the flows of other VNs as constant, the above problem belongs to the class of single commodity flow problems, for which the body of literature is extensive (see (Bertsekas 1999) and references therein). One way to solve the above problem is to dualize the constraints, using standard duality theory. As the resulting Lagrangian and dual problem are very intuitive, we now pursue this approach to some detail.

In particular, let us assume that the network is allowed to violate the flow conservation constraint, as follows: At each node, the network is allowed to set the external flow $s_{n,v}$ coming into that node and the traffic flows on links adjacent to it without the flow conservation constraint being satisfied, provided that:

1. If any excess traffic flow remains at the node, the network must sell it off to an external location, at a price λ_n per unit flow.
2. If there is a deficit of traffic flow, the network is required to buy traffic from the same external location, at the same price λ_n.

Let $\lambda = \begin{bmatrix} \lambda_1 & \lambda_2 & \dots & \lambda_N \end{bmatrix}$ be the node price vector, of dimension N.

Similarly, the network is allowed to violate the constraint that it must reserve as much of the channels as it needs. In particular, at each channel,

1. The network is allowed to reserve a reservation level higher than the one corresponding to actual usage by links, selling off its surplus reservation level to an external broker, for a price of μ_c.
2. In addition, the network is allowed to reserve less, but must buy the reservation level it lacks from the external broker, at the same price.

Let $\mu = \begin{bmatrix} \mu_1 & \mu_2 & \dots & \mu_N \end{bmatrix}$ be the channel price vector, of dimension C. The new maximization problem has the following objective function, which we call the Lagrangian:

$$L(s_v, r_v, \lambda, \mu) = \sum_n U_{n,v}(s_{n,v}) - \sum_c \frac{r_{c,v}}{\sum_{v'} r_{c,v'}} f_c\left(\sum_{v'} r_{c,v'}\right)$$
$$+ \sum_n \lambda_n \left\{ s_{n,v} - \sum_{l \in o(n)} x_{l,v} + \sum_{l \in i(n)} x_{l,v} \right\} + \sum_c \mu_c \left\{ r_{c,v} - \sum_l R_{c,l} x_{c,l} \right\}.$$

The first two terms come from the original problem, the other two from the fact that the constraints can now be violated. Observe that the maximum of the Lagrangian is greater than or equal to the maximum of the original optimization function, as, in the second maximization, the network is given more options with respect to the first optimization while the value of the optimal point of the original problem yields the same value of the optimization function in the new problem.

After rearranging terms, the objective function becomes:

$$L\left(\mathbf{s}_v,\mathbf{r}_v,\lambda,\mu\right)=\sum_n\left\{U_{n,v}(s_{n,v})+\lambda_n s_{n,v}\right\}+\sum_l\left\{x_{l,v}\left[-\lambda_{t(l)}+\lambda_{r(l)}-\sum_c\mu_c R_{c,l}\right]\right\}$$
$$+\sum_c\left\{\mu_c r_{c,v}-\frac{r_{c,v}}{r_{c,v}+A}f_c(r_{c,v}+A)\right\}$$

where

$$A=\sum_{v',v\neq v} r_{c,v}$$

is the total reservation of channel c from all networks except network v, and is assumed, for now, constant, as we have already discussed.

Observe that each of the terms of the three sums only contains a single optimization variable. Therefore, the dualization of the constraints converted a high dimensional problem into many *single* dimensional ones. In particular, the single dimensional problems are as follows:

1. N problems in which each of the nodes brings an amount of traffic in the network and immediately sells it off to the external traffic source.
2. L problems in which each link decides on how much traffic it will transport to maximize its profits, taking into account that it will buy it as some price at the transmitter, it will sell it at some price at the destination, and must also pay for using the channels.
3. C problems, one for each channel, in which the network decides how much will be the reservation level, assuming that it plans to sell off the excess amount (or buy the amount it lacks) immediately.

The usefulness of this dualization becomes clear if we consider the following facts: consider the dual function

$$g(\lambda,\mu)=\max_{s_v,r_v} L(s_v,r_v,\lambda,\mu),$$

which, as discussed, can be calculated by solving $N+L+C$ one-dimensional problems. The dual function can be shown to have the following subgradient (Bertsekas 1999):

$$\begin{bmatrix}\mathbf{s}_v^*-\sum_{l\in o(n)}\mathbf{x}_v^*+\sum_{l\in i(n)}\mathbf{x}_v^*\\ \mathbf{r}_{c,v}^*-\sum_l R_{c,l}x_{l,v}^*\end{bmatrix}$$

where s*, x*, r* are the solutions of the one-dimensional problems for the choices **λ**, **μ** of the price values. Note that the subgradient is an $N+C$-sized vector. Also note that each term equals the violation of the corresponding constraint. Using this equation for the subgradient, we can easily find the minimum of the dual function. Duality theory specifies that if the original problem is convex, then the minimum of the dual function coincides with the maximum of the original problem (prior to the dualization), and furthermore, the optimal variables for the original problem are the s*, x*, r* when the prices are the optimal ones of the dual function minimization problem. Therefore, it is possible to solve the original problem through solving the dual one. This is a very appealing approach, since the dual one can be solved in a *distributed* fashion, by placing agents in the entities involved in the solution of the one dimensional problems: the channels, the links, and the nodes. Moreover, each of these agents solves a simple problem according to its own best interest and to individually known information. These features are highly desirable for such large environments such wireless mesh networks.

Simulations

In order to provide proof-of-concept, and to gain intuition, we present one case where the optimization framework we developed can lead to an efficient sharing of the available bandwidth among competing virtual networks. While this is intuitively expected, it will also come as the outcome of the optimization problem defined above.

We consider the setting of Figure 12, where two partly overlapping virtual networks exist. We assume that the topology is a 12 by 12 grid, and any node must be in one of the 12×12=144 locations. Each virtual network consists of exactly 49 nodes, arranged in a 7 by 7 grid. We call the lower left network VNA and the upper right network VNB. Observe that the VSs are partially overlapping, so that at certain locations there are nodes of both VNs.

Figure 12. An example topology of 2 partially overlapping VNs, VNA (lower left) and VNB (upper right). Each VN consists of a single Base Station (BS) (denoted by a square) creating data, a number of Mobile Nodes (MNs) (denoted by hollow circles) receiving, and also relaying data, and a number of relays (denoted by solid circles), that have no communication needs of their own, but are there to support the transport of data from the BS to the MNs. The letters 'a' and/or 'b' next to mobile nodes denote the virtual network to which the nodes belong. Note that some of them, placed in the central overlapping region, belong to both virtual networks. Note that, of the two BSs, that of VNB is outside the common region (at location [9,5]), and that of VNA is inside (at location [5,6]).

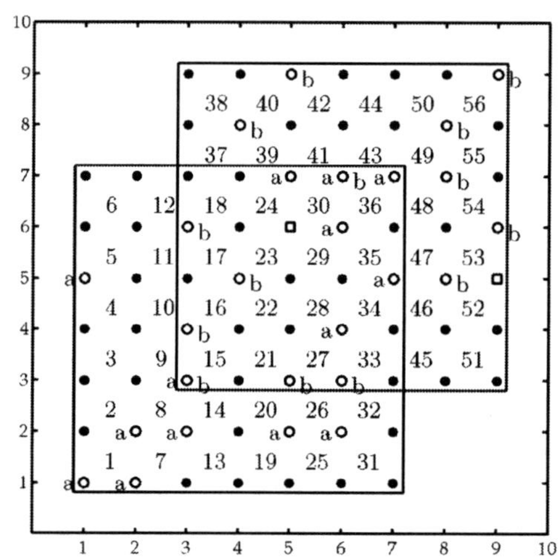

In each VN, we have the following types of nodes:

- **A Base Station (BS)** that is creating data. The two BSs are located in positions (9,5), (5,6), and are denoted by squares. Observe that one of the base stations is within the region where the virtual networks overlap, and the other one is outside this region. Each BS is associated with a utility function that is minus infinity for negative divergences, and zero for positive divergences. Therefore, each BS is not allowed to receive data, or our objective function will be minus infinity.
- **12 Mobile Nodes (MNs)** that must receive data. Locations where MNs exist are denoted by hollow circles, and the letter 'a,' if there is a VNA MN, 'b,' if there is a VNB MN, and both 'a' and 'b,' if there are MNs of both VNs there. The MNs are associated with a concave utility function that is minus infinity for positive divergences, and positive and decreasing (i.e., increasing in the negative direction) for negative directions. Therefore, MNs would like to receive data from the BS.
- **A number of relays** filling up the grid, denoted by solid circles. Relays do not have

Figure 13. The traffic flow of VNA, as determined by our iterative process. The thickness of the arrows connecting two locations is proportional to the total flow through the link connecting these locations. For clarity in the figure, we do not differentiate between the flows destined for each mobile node, instead only plotting their sum.

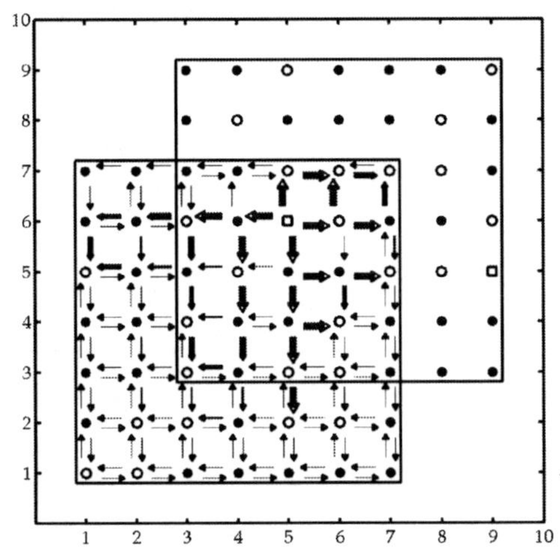

communication needs of their own, and their divergence must be zero.

To model wireless communication, we assume that two nodes can communicate directly with each other if and only if they are in adjacent positions in the grid. To model the limited bandwidth resources, we assume that there is a number C of channels (C=58 in this example) such that whenever two nodes communicate, they are using two channels that exist in either sides of their link. The channels are numbered in the figure. This is just one modeling assumption that can be generalized using the concept to reservation matrix. The cost of using a channel increases quadratically with its usage.

In this setting, it is clear that, in order to keep the cost down, the virtual networks should plan their traffic so that the overlap between flows belonging to different networks is minimal. This is indeed the case, as we now report.

In Figure 13 and Figure 14, we are plotting the flows that each VN arrives at, if they both solve the optimization problem we defined above, *iteratively*, with each one placing its best response to the other, until convergence is attained. In other words, starting from some arbitrary but feasible flow, the first VN finds the optimal traffic given the traffic of the second network, and then the second network updates its traffic based on the traffic of the first one, etc. Eventually, the networks reach a Nash Equilibrium (Nash 1950) (Fudenberg 1991), under which none of the two networks is interested in updating its traffic.

In Figure 15, we have plotted the evolution of the corresponding net benefit versus the iteration. Note that, for this example, the number of iterations needed is very modest, namely in the order of 5. Also, although in the second iteration the net benefit of VNB drops, due to competition from VNA, the net benefit increases again in subsequent iterations, and eventually VNA finds a way to distribute its traffic without causing unnecessary interference to VNB.

As already mentioned, VNB has its BS in the region that is not overlapping with VNA. VNB keeps the usage of the channels in the common area at a minimum, even completely starving one mobile node. On the other hand, VNA has its BS right in the middle of the common region, and as the utilities associated with the flow coming out of the base station are large, that BS ends up using its surrounding channels almost exclusively.

FUTURE RESEARCH DIRECTIONS AND IMPACTS

Before concluding the chapter, it is highly important to outline the near-term steps directly related to the work presented in this chapter, and the research directions in this area and how we see the technology to evolve. Therefore, we first

Figure 14. The traffic flow of VNB. Observe that VNB avoids sending significant amounts of data to its MNs within the region shared with the other VN, VNA. This is because the BS of VNA is within this shared region, and is more pressed to use the common channels, in order to send traffic to multiple of its own MNs.

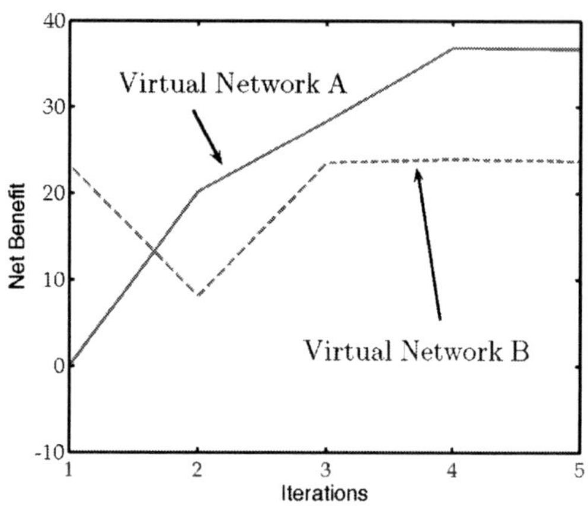

present our planned direction of future work in the upcoming months. We close this section by presenting a set of emerging trends and future research directions from a general perspective, and our vision about the evolution of the field.

We presented our initial approach about the context representation and proposed a rule based system. The VN context is represented through a policy which is a collection of rules representing different aspects of context. Policies are written within a specific policy language designed to target a specific problem (e.g. XML schema or SQL like constructs). The design of such a formal policy language which can represent most of the context features is part of our future research directions. Additionally, it includes the formal reasoning on the language constructs so the resulting policies are consistent and safe. An architecture of a rule based system being developed which interacts with the users, parse the policies and preferences, manage contexts for individual users and integrate itself with the overall architecture of the infrastructure.

Figure 15. The net benefit of the networks versus the iteration. Note that, for this example, convergence is very fast.

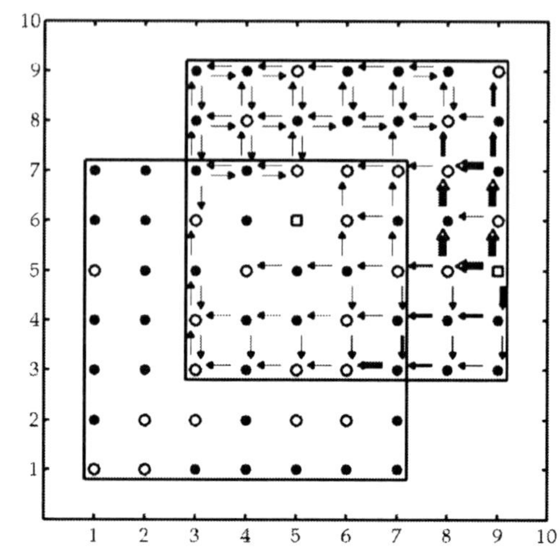

The rules based system sits between the user and rest of the infrastructure at the startup. Once the user joins a specific VN, the control is handed over to the other respective components to handle user requests.

The probabilistic model is able to show the impact of the architecture in the data and the control processes. However, it is not built to tackle dynamics of the network, such as the studying of the reconfiguration effects and the optimal reconfiguration approaches.

The optimization model, as presented before, does not currently cover the user side parameters. We plan to extend the model to capture the user-WMN parameters. Some of the obvious parameters include user billing and usage calculation, throughput control, accounting for the bandwidth offered by the user etc.

Next we turn to a broader perspective and present our perceived vision about the emerging research trend and evolution of different domains of technology which were part of this project.

- **Communication over Heterogeneous Virtual Networks:** The application of virtualization techniques to commodity network infrastructures has several advantages, chief of which is the creation of context specific virtual networks. The interest of having such diverse virtual networks is related to the needs as well as the demands of certain application classes. However, the major challenge that arises from various heterogeneous virtual networks is their interconnectivity and interoperability.
- **Context-Aware Networking:** Control and manage the changes and reconfigurations in the context-aware networks to optimize the network resources while maximizing the experience of the users. There is a clear trend to capture more and more context information by sensors integrated in mobile smart devices, such as GPS receivers, accelerometers, light sensors, surrounding sound measures, etc. Since these parameters clearly influence the quality of experience of a user, it is envisioned that more and more environmental parameters will be considered for dynamic service adaptation for mobile networked services. Additionally, supporting challenged networks by WMN access networks, such as sensor networks supporting the "*Internet of Things*" (IoT) in mobile, pervasive environments or exposed to fire and extreme weather conditions will require even more information about the current context of the (mobile) nodes connected and the networking possibilities.
- **Optimized Resource Usage:** Considering and optimizing resource mobility in multi-virtual networks considering network conditions, benefits and costs for movement and resource management adaptations. The number of active powered-on physical resources has to be determined in order to provide enough resources to the overlaying virtual network to fulfill certain requirements in a given context. Once, the number of active hardware resources is determined, the optimal placement of virtual resources has to be found that supports an optimal resource usage.
- **Security, Trust, Privacy and Identity Management in the Envisioned Architecture:** Another challenge is the security assurance in each individual virtual network and in the entire networking architecture. Using virtual instances on top of the same physical infrastructure requires security measures to enable a secure isolated usage. Otherwise, it can be a suitable environment for some attacks for example denial of service attacks. Secure communication between users in different heterogeneous domains are one of the novel inter-virtual networking architecture challenges that should be addressed and therefore new mechanisms are needed to avoid violating security objectives in the emerging virtual context-aware networking.
- **Mobility-Aware Networking:** Support users changing their communication and networking requirements depending on location, direction and movement. Here, the multi-virtual networks approach offers promising means to support changes seamlessly. The challenge here is how to provide fast, reliable interoperation based on anchor points in the multi-virtual network architecture.
- **Monitoring of Energy Efficiency:** In order to achieve an energy-efficient operation of the network, the monitoring and management of virtual and physical resources has to be extended by energy awareness. The monitoring needs to collect data concerning the utilization of virtual and physical resources and information about current power consumption and temperature of physical components of the network in

order to get an energy-based view on the system. Based on this energy-related monitoring, an autonomous and energy-efficient management of physical and virtual resources needs to be supported.

- **Higher Reliability and Performance:** Context-based path selection for in concurrent multipath transmissions as well as multi-topology routing strategies. The use of multiple topologies and connections will reduce the probability of getting disconnect and the use of concurrent paths will increase the throughput. Another important aspect is to apply new networking techniques especially network coding to see the performance improvements. Network coding allows intermediate nodes to mix input packets and forward them down stream to achieve higher throughput and side benefits of security and energy efficiency. There has already been some work done to employ network coding within WMNs to achieve higher performance. Incorporating it within multi VNs may be another new direction of work.

CONCLUSION

This chapter presented an approach to support context-aware communication, building different virtual networks for different users with different sets of requirements. To have an insight on the presented architecture, two mathematical models were presented, to evaluate the impact of the control processes on data traffic quantitatively and how network topologies deal with multiple virtual networks. The sheer complexity of the problem, and the need to accurately capture a large number of features of such topologies, and in particular the competition and interaction of different VNs, required the dual approach. The presented results showed the limits on the number of VNs to provide acceptable delays for communications, the evolution of the flows in VNs and the usage of the available resources. These results show that this architecture is able to accommodate a significant number of VNs with diverse characteristics and users with diverse requirements accessing the WMN without degrading the communication quality.

ACKNOWLEDGMENT

The work described in this paper is partially supported by the Euro-NF network of excellence. More specifically, it is integrated in two of its specific joint research projects: CAMYM—Context-Aware Mobility in Wireless Mesh Networks; MOMO—Multi-Overlays for Multi-hOming: A Wireless Mesh Network Use Case.

REFERENCES

Afanasyev, M., Chen, T., Voelker, G. M., & Snoeren, A. C. (2010). Usage patterns in an urban WiFi network. *IEEE/ACM Transactions on Networking*, *18*(5). doi:10.1109/TNET.2010.2040087

Akyildiz, I., Wang, X., & Wang, W. (2005). Wireless mesh networks: A survey. *Computer Networks*, *47*(4), 445–487. doi:10.1016/j.comnet.2004.12.001

Antoniou, J., Pinto, F., Simoes, J., & Pitsillides, A. (2010). Supporting context-aware multiparty sessions in heterogeneous mobile networks. *Journal on Mobile Networks and Applications*, *15*(6), 1–14.

Ardagna, C., Cremonini, M., Capitani di Vimercati, S. D., & Samarati, P. (2008). A privacy aware access control system. *Journal of Computer Security*, *16*(4), 369–397.

Ashley, P., Hada, S., Karjoth, G., Powers, C., & Schunter, M. (2003). *Enterprise privacy authorization language (EPAL 1.2)*. World Wide Web Consortium (W3C).

Azcorra, A., Banniza, T., Chieng, D., Fitzpatrick, J., Von-Hugo, D., & Natkaniec, M. (2009). Supporting carrier grade services over wireless mesh networks: the approach of the European FP-7 STREP CARMEN. *IEEE Communications Magazine*, (April): 14–16. doi:10.1109/MCOM.2009.4907397

Bayer, N., Hock, D., Roos, A., Siebert, M., Xu, B., Rakocevic, V., & Habermann, J. (2008). VoIP performance in "meshbed": A wireless mesh networks testbed. *Vehicular Technology Conference (VTC Spring 2008)* (pp. 2218-2222). IEEE.

Berl, A., Fischer, A., & de Meer, H. (2009). Using system virtualization to create virtualized networks. *Electronic Communications of the EASST*, *17*, 1–12.

Berl, A., Race, N., Ishmael, J., & de Meer, H. (2010). Network virtualization in energy-efficient office environments. *Computer Networks*, 54(16). Elsevier B.V.

Bertsekas, D. P. (1999). *Nonlinear optimization*. Belmont, MA: Athena Scientific.

Bisnik, N., & Abouzeid, A. (2006). Delay and throughput in random access wireless mesh networks. *IEEE International Conference on Communications* (pp. 403–408). IEEE, Istanbul.

Bolch, G., Greiner, S., de Meer, H., & Trivedi, K. S. (1998). *Queueing networks and Markov chains: Modeling and performance evaluation with computer science application* (1st ed.). Wiley Blackwell. doi:10.1002/0471200581

Botta, A., Pescape, A., Ventre, G., & Karrer, R. P. (2007). High-speed wireless backbones: Measurements from MagNets. *4th International Conference Broadband Communications, Networks and Systems. (BROADNETS 2007)* (pp. 680-689). IEEE.

Cassasa, M., & Beato, F. (2007). *On Parametric obligation policies: Enabling privacy-aware information lifecycle management in enterprises. Technical Report*. HP Labs, Advanced Research at HP.

Chandra, R., Bahl, P., & Bahl, P. (2004). MultiNet: Connecting to multiple IEEE 802.11 networks using a single wireless card. *23rd Annual Joint Conference of the IEEE Computer and Communications Societies (INFOCOM 2004)* (pp. 882- 893). IEEE.

Chereddi, C., Kyasanur, P., & Vaidya, N. H. (2007). Net-X: A multichannel multi-interface wireless mesh implementation. *SIG-MOBILE Mobile. Computer Communication Review*, *11*(3), 84–95. doi:10.1145/1317425.1317435

Cho, Y.-B., Woo, S.-H., & Lee, S.-H. (2008). *Efficient key distribution scheme for secure routing in wireless mesh network environment* (pp. 79–82). World Scientific and Engineering Academy and Society.

Coskun, H., Schieferdecker, I., & Al-Hazmi, Y. (2009). *Virtual WLAN: Going beyond virtual access points* (p. 17). ECEASST.

Damianou, N., Dulay, N., Lupu, E., & Sloman, M. (2001). *The ponder policy specification language* (pp. 18–38). Lecture Notes in Computer Science Springer-Verlag.

Dilmaghani, R. B., & Rao, R. R. (2006). *Hybrid wireless mesh network deployment: A communication testbed for disaster scenarios*. 1st International Workshop on Wireless Network Testbeds, Experimental Evaluation & Characterization (WiNTECH '06). ACM.

Dong, J., Curtmola, R., & Nita-Rotaru, C. (2009). Practical defenses against pollution attacks in intra-flow network coding for wireless mesh networks. *Proceedings of the Second ACM Conference on Wireless Network Security* (pp. 111-122). ACM.

Efstathiou, E. C., Frangoudis, P. A., & Polyzos, G. C. (2006). Stimulating participation in wireless community networks. *25th IEEE International Conference on Computer Communications* (pp. 1-13).

Eriksson, J., Agarwal, S., Bahl, P., & Padhye, J. (2006). Feasibility study of mesh networks for all-wireless offices. *4th International Conference on Mobile Systems, Applications and Services (MobiSys '06)* (pp. 69-82). New York, NY: ACM.

Fudenberg, D., & Tirole, J. (1991). *Game theory*. Cambridge, MA: MIT Press.

Gartner (2010). *Technology research & business leader insight*. Retrieved September 27, 2010, from http://www.gartner.com/it/page.jsp?id=1322414

Gelenbe, E., Sakellari, G., & D'Arienzo, M. (2007). Controlling access to preserve QoS in a self-aware network. *International Conference on Self-Adaptive and Self-Organizing Systems* (pp. 205-213).

Hsieh, H.-Y., & Sivakumar, R. (2002). pTCP: An end-to-end transport layer protocol for striped connections. In *10th IEEE International Conference on Network Protocols (ICNP 2002)*, Paris, France.

Hu, P., Portmann, M., Robinson, R., & Indulska, J. (2008). Context-aware routing in wireless mesh networks. *ACM International Conference on Context-Awareness for Self-Managing Systems* (pp. 16-23).

Iyengar, J. R., Amer, P. D., & Stewart, R. (2006). Concurrent multipath transfer using sctp multihoming over independent end-to-end paths. *IEEE/ACM Transactions on Networking, 14*(5), 951–964. doi:10.1109/TNET.2006.882843

Karrer, R., Zerfos, P., & Piratla, N. (2006). *Magnets: A next generation access network*. Poster at IEEE International Conference on Computer Communications (INFOCOM 2006).

Langar, R., Bouabdallah, N., & Boutaba, R. (2009). Mobility-aware clustering algorithms with interference constraints in wireless mesh networks. *Computer Networks, 53*(1), 25–44. doi:10.1016/j.comnet.2008.09.012

Law, C., Mehta, A. K., & Siu, K. (2003). A new Bluetooth scatternet formation protocol. *Mobile Networks and Applications, 8*(5), 485–498. doi:10.1023/A:1025133710426

Liang, Q., & Ren, Q. (2005). Energy and mobility aware geographical multipath routing for wireless sensor networks. *IEEE Wireless Communications and Networking Conference*, vol. 3, (pp. 1867-1871).

Liao, J., Wang, J., & Zhu, X. (2008). A multipath mechanism for reliable VoIP transmission over wireless networks. *Computer Networks, 52*, 2450–2460. doi:10.1016/j.comnet.2008.04.008

Lin, T., Wang, C., & Lin, P. (2008). A neural-network-based context-aware handoff algorithm for multimedia computing. *ACM Transactions on Multimedia Computing, Communications, and Applications, 4*(3), 1–23. doi:10.1145/1386109.1386110

Lombriser, C., Marin-Perianu, R., Roggen, D., Havinga, P., & Troester, G. (2009). Modeling service-oriented context processing in dynamic body area networks. *IEEE Journal on Selected Areas in Communications, 27*(1), 49–57. doi:10.1109/JSAC.2009.090106

Lukas, G., & Fackroth, C. (2009). WMNSec: Security for wireless mesh networks. *Proceedings of the 2009 International Conference on Wireless Communications and Mobile Computing Connecting the World Wirelessly* (pp. 90-95). ACM.

Määttä, J., & Bräysy, T. (2009). A novel approach to fair routing in wireless mesh networks. *EURASIP Journal on Wireless Communications and Networking*, (1): 1–13. doi:10.1155/2009/379059

MadWifi. (n. d.). *Multiband atheros driver for wireless fidelity*. Retrieved from http://madwifi-project.org/

Matos, R., Sargento, S., Hummel, K., Hess, A., Tutschku, K., & de Meer, H. (2013). Context-based wireless mesh networks: A case for network virtualization. *Special Issue on Future Internet Services and Architectures: Trends and Visions, Telecommunication Systems, 52*(3). Springer.

Nash, J. (1950). Equilibrium points in n-person games. *Proceedings of the National Academy of Sciences of the United States of America, 36*(1), 48–49. doi:10.1073/pnas.36.1.48

Nogueira, J., Melo, M., Carapinha, J., & Sargento, S. (2010). *Creating virtual networks: An approach for virtual network discovery*. Workshop on the Network of the Future, IEEE Global Telecommunications Conference (IEEE GLOBECOM 2010).

OASIS eXtensible Access Control Markup Language (XACML) TC. (n.d.). *OASIS: Advancing open standards for the global information society*. Retrieved September 23, 2010, from http://www.oasis-open.org/committees/tc_home.php?wg_abbrev=xacml#XACML20.

Oh, M. (2008). An adaptive routing algorithm for wireless mesh networks. *10th International Conference on Advanced Communication Technology* (pp. 2087-2091). IEEE.

Pfaff, B., Pettit, J., Koponen, T., Amidon, K., Casado, M., & Shenker, S. (2009). *Extending networking into the virtualization layer*. In ACM HotNets, New York, NY, October 2009.

Salber, D., Dey, A., & Abowd, G. (1999). The context toolkit: Aiding the development of context-enabled applications. In *SIGCHI Conference on Human Factors in Computing Systems* (pp. 434–441).

Shakkottai, S., Altman, E., & Kumar, A. (2007). Multihoming of users to access points in WLANs: A population game perspective. *IEEE Journal on Selected Areas in Communications, 25*(6), 1207–1215. doi:10.1109/JSAC.2007.070814

Sidhu, D., Nair, R., & Abdallah, S. (1991). Finding disjoint paths in networks. *ACM SIGCOMM Computer Communication Review, 21*(4), 43–51. doi:10.1145/115994.115998

Staub, T., & Braun, T. (2008). *ATOM: Adaptive transport over multipaths in wireless mesh networks*. 2nd ERCIM Workshop on eMobility.

Staub, T., Ott, S., & Braun, T. (2009). Experimental evaluation of multi-path routing in a wireless mesh network inside a building. *5th Workshop on Mobile Ad-Hoc Networks - WMAN 2009 - in conjunction with the 16th Conference on Communication in Distributed Systems (KiVS)* (pp. 1-12).

Stoica, I., Morris, R., Liben-Nowell, D., Karger, D. R., Kaashoek, M. F., Dabek, F., & Balakrishnan, H. (2003, February). Chord: A scalable peer-to-peer lookup protocol for internet applications. *IEEE/ACM Transactions on Networking, 11*(1), 17–32. doi:10.1109/TNET.2002.808407

Subramanian, A. P., Buddhikot, M. M., & Miller, S. (2006). Interference aware routing in multi-radio wireless mesh networks. *2nd IEEE Workshop on Wireless Mesh Networks* (pp. 55-63).

Tarique, M., Tepe, K. E., Adibi, S., & Erfani, S. (2009). Survey of multipath routing protocols for mobile ad hoc networks. *Journal of Network and Computer Applications, 32*(6), 1125–1143. doi:10.1016/j.jnca.2009.07.002

Timm-Giel, A., Zaki, Y., Zhao, L., & Goerg, C. (2010). *A novel LTE wireless virtualization framework*. Second International ICST Conference on Mobile Networks and Management (MONAMI 2010), Sept. 2010.

Touch, J. D., Shun Wang, Y., Pingali, V., Eggert, L., Zhou, R., & Finn, G. G. (2005). A global x-bone for network experiments. In *Proceeding in IEEE Tridentcom* 2005, (pp. 194-203).

Tsirigos, A., & Haas, Z. J. (2001). Multipath routing in the presence of frequent topological changes. *IEEE Communications Magazine, 39*(11), 132–138. doi:10.1109/35.965371

Wang, L., Jang, S., & Lee, T. Y. (2005). Redundant source routing for real-time services in ad hoc networks. *IEEE International Conference on Mobile Adhoc and Sensor Systems Conference* (pp. 7-87), IEEE.

Zhang, Y., & Fang, Y. (2007). A secure authentication and billing architecture for wireless mesh networks. *Wireless Networks, 13*, 663–678. doi:10.1007/s11276-006-8148-z

Zhao, X., Lv, Y., Yeap, T. H., & Hou, B. (2009). A novel authentication and key agreement scheme for wireless mesh networks. *International Conference on Networked Computing and Advanced Information Management,* (pp. 471-474).

Zinner, T., Tutschku, K., Nakao, A., & Tran-Gia, P. (2010). Using concurrent multipath transmission for transport virtualization: Analyzing path selection. In *Proceedings of the 22nd International Teletraffic Congress (ITC)*, Amsterdam, The Netherlands.

ENDNOTES

1 http://www.ict-ccast.eu/
2 http://www.sensei-project.eu/
3 http://www.funkfeuer.at/
4 http://www.freifunk.net/
5 http://www.seattlewireless.net/
6 http://www.consume.net
7 http://www.cuwin.net/projects/wghana
8 http://www.strixsystems.com
9 http://www.cisco.com
10 http://www.nctimes.com/news/local/article_903bb6df-abf4-577a-a419-6bff-cb2309f2.html
11 http://tools.ietf.org/html/rfc4960
12 http://ist-autoi.eu/autoi/
13 http://xen.org
14 **Obligations** and **Commitments** can be used interchangeably. However, we stick to the term "Obligation" throughout this chapter to avoid confusion.
15 In reality, if all networks solve this problem, we expect these to change, leading to a coupling of the optimization problem in a non-trivial, game theoretic sense. We will present an example of this.

Chapter 6
Heterogeneous Meshed Wireless Back-Haul Network Integrating Unidirectional Technologies

Mathias Kretschmer
Fraunhofer FOKUS, Germany & Brunel University, UK

Christian Niephaus
Fraunhofer FOKUS, Germany

George Ghinea
Brunel University, UK

ABSTRACT

Wireless Mesh Networks (WMNs) have matured in recent years and the visibility of WMN deployments has attracted commercial operators to investigate this technology for applicability in their networks. Having their roots in the Mobile Adhoc Network (MANET) world and rather cheap off-the-shelf single-radio WLAN routers, WMN routing protocols were not designed for applicability in carrier-grade back-haul networks. For example, protocols such as OLSR or B.A.T.M.A.N. can not address the QoS-requirements of a modern operator back-haul network with its increasing demand for triple-play content. Although numerous solutions have been proposed to introduce QoS-awareness at the protocol or the technology level, traditional WMNs fail to meet commercial operator requirements in terms of reliability, traffic engineering and QoS guarantees. This chapter proposes a novel approach combining an IEEE 802.21-based control plane and an MPLS-based data plane. To provide support for ubiquitous high-bandwidth multi-media services, it seamlessly integrates unidirectional broadcast technologies such as DVB into the heterogeneous multi-radio WiBACK architecture.

INTRODUCTION

The visibility of successful WMN deployments has raised interest regarding a commercial exploitation of this technology. Compared to traditional wireless operator back-haul networks, WMNs potentially offer easier deployment and flexible self-configuration at lower costs (Chieng, 2010). For a WMN-type architecture to be considered as an alternative for an operator back-haul network, it must meet strict requirements such as guaranteed Quality of Service (QoS), high availability and

DOI: 10.4018/978-1-4666-0017-1.ch006

predictable behavior in high load situations. QoS support for WMNs has received a lot of attention from the research community (Kone, 2007)(Abu-Ali, 2006)(Zhou, 2009), but has yet to show its applicability in operator deployments.

Compared to OLSR(Clausen, 2003) or B.A.T.M.A.N. based community mesh networks, such as FREIFUNK, which rely on low-cost off-the-shelf hardware, a carrier- grade Wireless Back-Haul (WiBACK) network may assume more reliable and higher performing multi-radio hardware potentially yielding more stable and orthogonal links. Thus, a WiBACK network, as developed by the EU FP7 CARrier grade wireless MEsh Network (CARMEN) project (Azcorra, 2009)(Banchs, 2008), can be compared to a typical operator back-haul network with two major limiting factors: the severe capacity constraints due to scarce wireless spectrum resources and temporarily varying link conditions. A modern operator back-haul network is required to support triple-play services (web, video, voice) which emphases the need to provide QoS guarantees while efficiently utilizing the resources in a capacity-constrained network. Wireless cell resource utilization can be optimized by using the most suitable technology for a given payload. For example, the delivery of live multimedia content is facilitated by only one transmitter, hence the overhead of MAC protocols can be avoided. Unidirectional technologies such as Digital Video Broadcast - Terrestrial (DVB-T) offer a very efficient distribution medium for 1-to-N traffic and usually span across multiple mesh hops.

Typically, WMNs rely on the bi-directionality of links and can therefore not utilize Unidirectional Links (UDLs). For example, Ad hoc On-Demand Distance Vector Routing Protocol (AODV) (Perkins, 2003) black-lists UDLs since they break the distance vector algorithm. What's more, most MANETs and WMN protocols cannot differentiate between, for example, a faulty IEEE 802.11 link and an inherently unidirectional link provided by a unidirectional technology. In this chapter we discuss our IEEE 802.21-based approach (Kretschmer, 2009) to seamlessly integrate Unidirectional Technologies (UDTs) into a heterogeneous WiBACKs network so that higher layer protocols and services can transparently utilize them when beneficial for a given payload or receiver distribution.

This chapter is structured as follows. In the next section we discuss work related to our WiBACK architecture focusing on the support for UDTs. In the following section we compare two approaches to seamlessly integrate UDTs into the WiBACK architecture and identify which protocols or components require adaptation. In section four we present our approach to seamlessly integrate UDTs below the Network layer and show that by properly designing the architecture, UDL support only requires a limited amount of modifications. In return, the architecture has become more flexible supporting potentially any kind of wireless technology.

BACKGROUND

The cross-layer WiBACK architecture builds on proven technologies which have been extended to support heterogeneous wireless technologies. The core of the control plane builds on an extended version of IEEE 802.21 standard, which al- lows for a hardware-independent and modular architecture design, see Figure 7. The data plane builds on Traffic Engineering (TE) principals incorporating Multi Protocol Label Switching (MPLS)-based forwarding and constraint-based path computation following the Path Computation Element (PCE) concepts. Those established technologies have been adapted to exploit the heterogeneity of wire- less technologies including unidirectional broadcast technologies such as DVB.

The typical wireless network management components such a topology discovery, radio resource management, link monitoring, path computation and mobility management can be implemented as modules using the IEEE 802.21 messaging mechanism. This differentiates our approach from typical Network layer routing protocols, which integrate similar functionality in one protocol and are often agnostic to physical hardware capabilities.

IEEE 802.21

The main goal of the IEEE 802.21 standard is to facilitate a handover between heterogeneous access networks including wired and wireless technologies as well as 802 and non-802 networks by providing link layer intelligence for the upper layers and, thus, improving the user experience of mobile devices by enabling a seamless handover wherever this is supported by the underlying network environment. This includes for example a scenario where a user gets seamlessly handed off from a IEEE 802.3 network to an 802.11 or a mobile terminal which is handed off from a 3G cell to a 802.11 network during an ongoing call. For this purpose IEEE 802.21 defines a media-independent abstraction layer which provides a uniform interface to the higher layers allowing for implementing and designing upper layer modules in a technology-independent way.

Figure 1 depicts the general 802.21 architecture. The IEEE 802.21 standard specifies a framework consisting of three main components, namely the Media In- dependent Handover Function (MIHF), the Media Independent Handover (MIH) Users and several Service Access Points (SAPs). The MIHF is the central entity in each 802.21-enabled device. It provides a common set of media-independent primitives to the upper layer, the MIH Users, via the MIH SAP and maps those to the corresponding technology-specific link layer primitives via the MIH LINK SAP. In addition to that, 802.21 defines the MIH NET SAP which provides an abstract interface of MIHF allowing the MIH Users and the MIHF itself a resilient MIH message exchange with a remote MIHF utilizing L2 or L3 transport mechanisms. It is important to note that the actual handover decision as well as the handover execution is done by the MIH User(s) and outside of the scope of the Institute of Electrical and Electronics Engineers (IEEE) 802.21 standard. The purpose of 802.21 is to enable a (seamless) handover by providing appropriate information, event notifications and commands to an MIH User. In order to do so MIHF defines three main services:

- The Media Independent Event Service (MIES) provides a mechanism for classification, filtering and reporting of events related to changes in link status, link quality or link characteristics. Events originate at the lower layers or the MIHF. A MIH User

Figure 1. IEEE 802.21 architecture

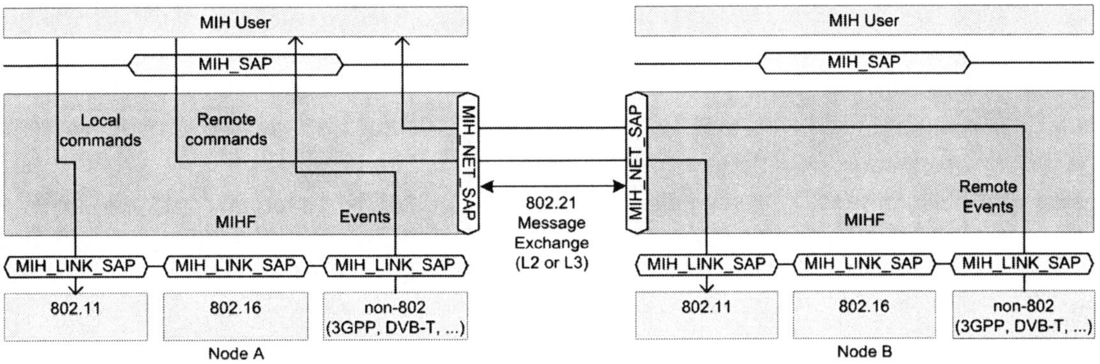

needs to explicitly subscribe to a particular event in order to receive it. For this purpose the MIHF maintains an appropriate subscription list. A MIH User can also subscribe for remote events which are generated on a different node in the network. It should be noted that events are delivered asynchronously by the MIHF and the receiving MIH Users don't need to take any action when being notified of an event.

- The Media Independent Command Service (MICS) allows the upper layer to manage and control mobility related functions in the lower layers (e.g. initiate the handover or turning off a certain radio). Command primitives are sent from an MIH User either to the lower layers or the MHIF itself and can be delivered locally or remotely. In the latter case, the MIH User sends the command to the local IMF which will transport it to the appropriate destination node. In contrast to the MIES command primitives are transmitted synchronously and are following the request-response message exchange pattern.
- The Media Independent Information Service (MIIS) enables the MIH Users to obtain information on the serving network and neighboring networks in a geographical region. This includes even information on networks of other technologies than the network to which the mobile node is currently attached (e.g. a node which is connected to a 3G network is able to obtain information on accessible 802.11 WiFi networks within the same geographical region). MIIS either supports a pull mode, which is primarily used by a mobile node to gather information from the network or push mode which gives, for example, the operator the possibility to push handover policies on the mobile devices.

Besides IEEE 802.21, there are other approaches which are providing similar functionality: WiOptiMo (Team, 2006) for example provides a small Java application allowing for a handover particularly of client/server applications by providing a special socket that switches the underlying network invisible for the actual application. In contrast to that Unlicensed Mobile Access (UMA) or Generic Access Network (GAN) (Arjona, 2007) is offering a proprietary solution for operators to allow access to their networks not only via cellular but also via unlicensed radio access technologies such as IEEE 802.11 by tunneling the connections back to the operator networks through a secure connection. Thus, UMA provides an easy solution for offloading traffic from cellular networks such as GSM or 3G to a Wifi Hot-Spot. However, as opposed those approaches 802.21 is aiming at enabling fast handovers by providing appropriate Data Link layer mechanism which makes it fairly flexible in terms of extending the foreseen use case as described in the following paragraphs.

The basic idea behind IEEE 802.21 can be described as Media Independent paradigm since it hides the technology specificities from higher layers by pro- viding a common set of primitives and, thus, reducing their complexity. The concept of media independence, however, is not limited to supporting (seamless) handovers between heterogeneous networks but rather can be easily exploited to simplify a larger variety of applications and operations in nowadays operator networks, such as bootstrapping, routing or network management since typically all of those require operations of high complexity when being applied to heterogeneous network environments.

IEEE 802.21 WiBACK Amendments

The WiBACK architecture design follows the general IEEE 802.21 architecture and thereby implements the Media Independent paradigm.

The main difference between the IEEE 802.21 standard and the WiBACK architecture is the respective target application. Whereas the scope of IEEE 802.21 is clearly limited to facilitating handovers between heterogeneous networks the WiBACK architecture is designed to provide solutions to manage heterogeneous wireless networks in a media independent fashion. For this purpose, as depicted in Figure 2, the WiBACK architecture extends the IEEE 802.21 architecture at several points: First, instead of considering only MIH Users with a single use case, namely handover, WiBACK is able to deal a variety of higher layer modules, which are required in the foreseen scenario. Second, WiBACK extends the MICS in order to allow for sending module-to-module commands locally as well as remotely (see dashed arrows in Figure 2). Third, WiBACK adds primitive specific to wireless network management to the IEEE 802.21 MIHF and therefore the name Interface Management Function (IMF) in- stead of MIHF has been chosen to reflect its responsibilities which goes beyond Media Independent handovers. However, the messaging services provided by the MIH NET SAP or the IMF NET SAP respectively as well as the majority of the primitives defined by IEEE 802.21 can also be utilized for non-handover related purposes, such as managing local and remote radio technologies in a media in- dependent manner. In the following the IEEE 802.21 standard and the outlined extensions are referred to as 802.21++.

Some of those additions have been proposed for standardization which has led to the forming of the 802.21b task group to address broadcast technology support and a recently started study group investigates management of heterogeneous wireless networks.

Following the approach of reusing the IEEE 802.21 standard where possible and extending it with mesh specific requirements the WiBACK architecture allows for designing a control plane for networks with heterogeneous technologies which can be implemented in a technology independent manner.

Heterogeneous Technologies

As mentioned above, the WiBACK architecture integrates heterogeneous technologies via MPLS on the data plane while the control plane relies on an extended set of IEEE 802.21 primitives and remote messaging. The additional primitives to extend IEEE 802.21 beyond support for Media Independent Handover have been defined by the CARMEN FP7 EU-project and in this work form the so- called Abstract Interface (AI) which includes primitives needed to add native UDT support.

Figure 2. IMF architecture based on IEEE 802.21

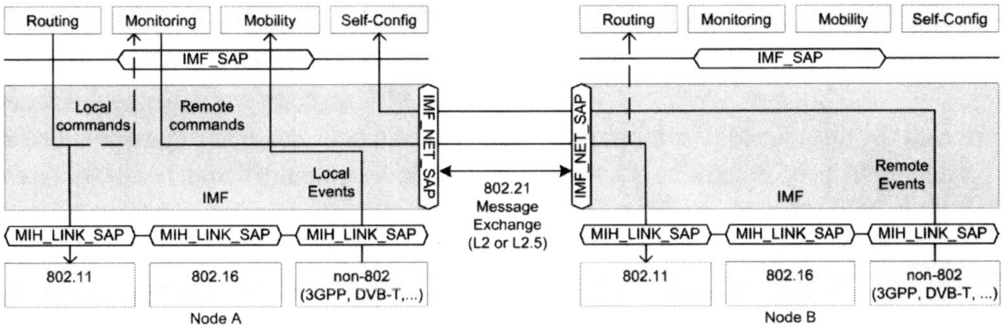

While the majority of technologies provide bidirectional communication, the minimum requirement for a technology to be integrated is the ability to address and transmit a frame from node A to node B or to receive a frame on a specific interface while being able to identify the sender address. Also, each node must be equipped with, at least, one transmit-capable and one receive-capable interface. This could be one bidirectional interface, two unidirectional interfaces or a combination thereof.

The AI abstracts the capabilities of common radio technologies in order to provide a common interface to the upper layers. The parameters covered include:

- Channel frequency
- Channel bandwidth
- Transmit power
- Modulation and Coding Scheme properties
- Media Access Control (MAC) scheduler properties
- Directionality
- Antenna properties

An important parameter within the scheduler properties is the operational mode, which describes the different modes that common wireless technologies may implement:

- Infrastructure mode, where a Base Station controls a set of Subscriber Stations
- Ad-Hoc mode, where all stations operate in the same mode without a coordinator
- Broadcast mode, where only one sender is active on a channel and an arbitrary number of receivers may be present

Together with the directionality information, the mode parameter helps to determine if a technology implements bidirectional or unidirectional communication. This can also be used to detect anomalies such as, for example, IEEE 802.11 links that seem to only communicate unidirectional due to unstable radio conditions, which is a major issue for typical WMN or MANET protocols.

Traffic Engineering

Traffic Engineering (TE) deals with network engineering and performance optimization of networks. One of its main objectives is to facilitate reliable and efficient network operations while at the same time considering optimizations of network resource utilization and traffic performance. The Internet Engineering Task Force (IETF) has issued numerous Request for Comments (RFC) on this topic, such as RFC 3272 (Awduche, 2002) which introduces the scope of Internet Traffic Engineering or RFC 2702 (Awduche, 1999) which addresses traffic engineering requirements over MPLS.

TE addresses how technologies and scientific principles can be utilized to perform measurement, modeling, characterization as well as control of network traffic in order to achieve specific overall network performance objectives, as defined by the network operator. In the MPLS context, measurement and control are seen as the most important aspects of TE.

In many of today's operator networks, TE has become an indispensable function because of the high cost of network assets and the commercial and competitive nature of the Internet (Awduche, 1999). In the WiBACK context, the main focus of TE is to optimize the rather limited wireless spectrum resources to meet the ever increasing bandwidth demand of wireless applications and devices. Another important objective of TE in the WiBACK architecture is to support the conservation of resources such as power in order to build and operate cost-efficient nodes which eventually can be driven with alternative energy sources such as solar or wind.

Constraint-Based Path Computation

Constraint-based path computation computes paths through a network topology fulfilling a set of constraints. Instead of simple hop count metrics those constraints maybe be expressed, for example, as minimum bandwidth or maximum delay. In the WiBACK context additional metrics such as the energy consumption of a link are considered in order to reduce the power consumption footprint of the network. As such, Constraint-based path computation is a core component of networks implementing TE principals such as MPLS or Generalized Multiprotocol Label Switching (GM-PLS) networks. Path computation in larger multi-domain networks can be a complex task and often requires customized computational components and cooperation between management entities of the different network domains.

The PCE architecture as described in RFC 4655(Farrel, 2006) consists of a set of building blocks which allow an implementer of such an architecture to tailor it to his specific needs. For example, the architecture allows for centralized or distributed computation, stateful and stateless implementation as well as multi PCE operation. It also specifies the communication among PCEs or between PCEs and their clients.

In the WiBACK context a centralized stateful approach is considered, which may use multiple PCEs in order to divide a WiBACK network into smaller areas.

Multi Protocol Label Switching

Unlike hierarchical prefix routing which is typically associated with Internet Protocol (IP) packet routing in the Internet, Multi Protocol Label Switching (Rosen, 2001)(Rosen, 2001) tags packets with a 32 bit wide label based on which forwarding is then performed in intermediate routers. MPLS operates at an Open Systems Interconnection (OSI) layer which is often referred to as Layer 2.5, between the data link and the network layer and MPLS labels are positioned between link layer headers and the headers of the encapsulated packet; usually a layer 3 protocol header. The name MPLS refers to its independence from both layer 2 and layer 3 protocols, which makes MPLS an interesting candidate for data forwarding in a WiBACK network that integrates heterogeneous technologies such as IEEE-based ones and those developed by 3GPP or Digital Video Broadcast (DVB), as well future emerging technologies. MPLS paths are unidirectional; hence MPLS could be used on top of UDT without any need for modifications. Three bits in the MPLS header are designated for QoS prioritization and congestion notification which is especially important in capacity-constrained network architecture such as a WiBACK.

When a packet enters its ingress router, which is called Label Edge Router (LER) in MPLS terminology, the MPLS header is prepended to the outgoing datagram. Or, in MPLS terminology, a label is pushed onto the label stack of the packet. Downstream routers simply look up the first label in the stack of the packet in a table that maps labels to outgoing interfaces and therefore next hop routers. Forwarding is then performed based solely on the table entry referenced by the label. In order to avoid scalability limitations of globally unique labels, intermediate routers usually rewrite the label of a packet to be forwarded with a different label which has only local significance for the next hop. This exchange is referred to as label switching.

When a packet reaches its egress router in the MPLS network, the label is removed. Or, in MPLS terminology, the last label is popped from the label stack, and further packet handling must be performed based on the header in- formation of the actual packet. MPLS can carry numerous traffic types, incl. IP, Asynchronous Transfer Mode (ATM), or IEEE 802.3(Ethernet) (Martini, 2006).

The paths which are given by the label based forwarding states in all routers between ingress and egress routers are called Label-Switched Paths

(LSPs). LSPs are conceptually similar to virtual connections in ATM, although unlike ATM, MPLS supports variable packet lengths. MPLS can be used to support and enforce Traffic Engineering decisions and to provide resource isolation when for- warding data along traffic-engineered computed paths, such as path computed via PCE.

In a low-power WiBACK scenario, the smaller footprint of MPLS label switching compared to IP-Routing is important, since on the data plane label switching is less complex than Network Layer forwarding. On the control plane in the centralized WiBACK architecture no complex distributed routing or management protocols are required at each general mesh node. Hence an intermediate forwarding node merely has to switch Data Link layer frames, which allows for very low power and therefore solar-friendly hardware designs.

An additional important feature of MPLS is Fast Reroute (FRR) (Pan, 2005) which can provide protection against link breakage in an LSP by a backup path that circumvents the broken links. In the WiBACK architecture, backup path are crucial to maintain reliable management connections, as well as to provide fall back connectivity for user data aggregates. Each LSP can have dedicated backup paths, as depicted in Figure 3, or all LSPs sharing network segments can also share the same backup path. Maintaining backup paths introduces management overhead, but provides faster recovery in case of node or link failures when compared to higher layer protocol recovery. Hence FRR is particularly useful for real-time traffic or WiBACK management connections. In a capacity constrained WiBACK network backup paths might block precious capacity needed for regular payload traffic and the trade-offs need to be considered carefully. Different resource allocation strategies may be used for backup capacity such as statistical overbooking, multi-path rope-ladder structures (Lessmann, 2010) or the admission of regular Best Effort traffic within the stand-by backup capacity. In the centralized WiBACK approach, the additional complexity of the backup mechanisms is almost exclusively introduced at the management node, while the general mesh nodes merely store extra LSP state and perform passive monitoring as detailed later in this section.

For IP-based networks two protocols have initially been developed by the IETF to setup and maintain MPLS LSPs, namely Constraint-based

Figure 3. MPLS one-to-one backup and MPLS facility backup

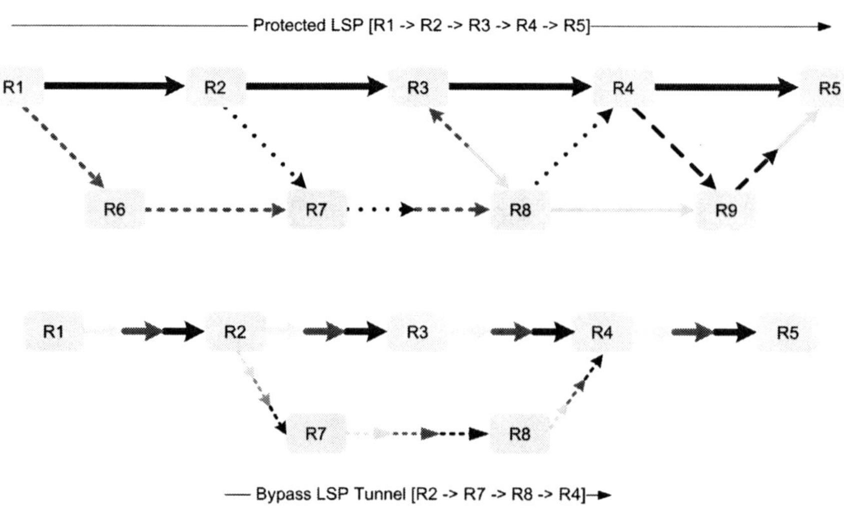

Routing Label Distribution Protocol (CR-LDP) or Resource ReSerVation Protocol - Traffic Engineering (RSVP-TE) (Awduche, 2001). In 2003 a decision was taken by the IETF to obsolete CR-LDP in favor of RSVP-TE.

In today's IP-based operator networks, the Resource ReSerVation Protocol (RSVP) (Braden, 1997) protocol suite is used to facilitate flow resource reservation and management while flow forwarding is typically performed via MPLS or GMPLS (Mannie, 2004). The RSVP-TE extension adds support for MPLS downstream- assigned label distribution via the RSVP PATH and RESV messages. Plain RSVP is a soft-state protocol and therefore relies on the periodic retransmission of messages to maintain its state and does not use Message IDs or sequence numbers nor does it provide an explicit ACK or NACK service. Hence, it performs poorly over links with higher loss probabilities; see for example (Komolafe, 2007). Reliable messaging via Message IDs was introduced with RFC 2961(Berger, 2001) which allows for Acknowledgments as well as automated retransmissions of lost messages.

RSVP messages are typically forwarded via regular IP routing. In order to support TE forwarding along pre-computed paths can be enforced using the Explicit Route Object (ERO) which describes the hop-by-hop route of RSVP messages. Hops can be specified as IPv4 or IPv6 addresses. Support for unnumbered links was added with RFC 3477 (Kompella, 2003) which allows for signaling over links without IP addresses using the Explicit Route Object (ERO).

The RSVP protocol suite addresses most functional requirements regarding path setup and management within the WiBACK architecture, except for upstream-assigned label distribution and limited support for UDT. Support for UDTs could be improved using for example the Link Layer Tunneling Mechanism for UDL (Duros, 2001). The advantages and disadvantages of such an approach are discussed in thc following section.

Unidirectional Technology Support

Multiple proprietary solutions to support inherently unidirectional technologies have been developed. They are often used to provide downstream IP-connectivity via satellite while the upstream is implemented using terrestrial connectivity. Different scenarios where UDLs have been encountered are WMNs or MANETs. Here UDLs occur when link-local neighbors of typical bi-directional technologies such as IEEE 802.11 can communicate from A to B, but not the other way around. The reasons for this may be different transmit power setting or local interferences.

The standardized Link Layer Tunneling Mechanism (LLTM) for UDL has been develop by the Unidirectional Link Routing (UDLR) working group within the IETF which aims at providing a mechanism to emulate full bidirectional connectivity between all nodes that are directly connected by a unidirectional link. The receive-only nodes use a tunneling mechanism to forward link layer datagrams back to send-only nodes via separate bidirectional IP connectivity. A typical tunneling protocol used in combination with UDLR is Generic Routing Encapsulation (GRE) (Farinacci, 2000), which is an IP protocol. Since the tunnel encapsulates the data link layer frames, UDLR is considered to be transparent to higher layer protocols. In the context of the WiBACK architecture, this would mean that in order to tunnel Data Link layer frames, Network Layer (e.g. IP) routing mechanism would have to be operational, which cannot be assumed.

The Link Layer Tunneling Mechanism uses Dynamic Tunnel Configuration Protocol (DTCP) embedded in its downstream UDL to announce to receive- only nodes at which IP address the tunnel endpoints for the return channel can be reached. Each node with a send-only interface can signal where the tunnel endpoints for a specific UDL can be found. Since the Link Layer Tunneling Mechanism transparently provides bidirectional link layer connectivity between the nodes, normal routing protocols could be used without a need for modifications incorporating unidirectional link support.

The Link Layer Tunneling Mechanism may cause problems, though, with mechanisms such as Internet Protocol, Version 6 (IPv6) Duplicate Address Detection (DAD) which sends out a Neighbor Solicitation packets to detect if other nodes already use the probed address. The standard DAD mechanism has been designed for regular link-local broadcast domains, such as an Ethernet segment. It assumes that only other nodes will respond to this solicitation packet, while in the UDLR case, it will receive its own solicitation packet and may therefore assume the address to be already allocated. Similar issues may arise during the WiBACK self-configuration phase.

However, the transparent return links provided by the Link Layer Tunneling Mechanism poses a major problem especially for WiBACK networks since the higher layer protocols are not aware of the special nature of such links which often cross numerous heterogeneous networks and wireless cells until reaching the tunnel endpoint, see Figure 4. It is therefore not possible to perform per-hop resource reservations or link monitoring. Hence including such links in a QoS- constrained path computation algorithm would requires major adaptations. Also, in order to provide load-balancing and redundancy, DTCP allows for multiple tunnel endpoint addresses to be announced, but cannot control which endpoint is eventually chosen by a receive-only node.

In the MANET context, Bidirectional Routing Abstraction (Ramasubramanian, 2008) uses a reversed distance vector algorithm to detect UDLs. Bidirectional Routing Abstraction (BRA) provides an abstraction layer so that higher layer protocols only see bidirectional links. Support for QoS-constraint path computation suffers from similar implications due the hidden heterogeneity of the return link as described for the UDLR approach.

Wireless Link Monitoring

Detailed wireless link monitoring is a difficult task due to the dynamic nature of wireless links caused by temporary fading and interferences, but also due to often very dynamic per-frame transmitter configurations. The wireless technologies currently considered in the WiBACK architecture

Figure 4. Link layer tunneling hides the underlying heterogeneity of the virtual return path V(BA)

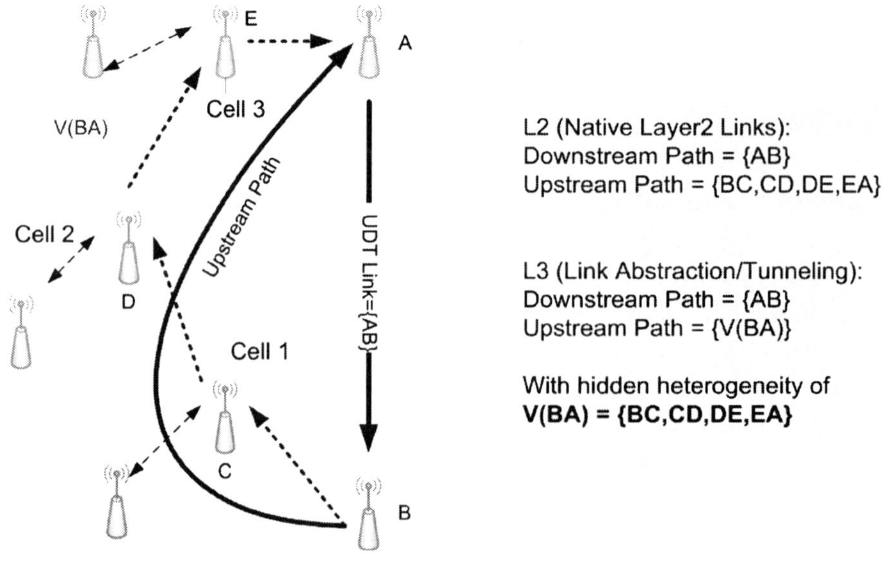

range from satellite (i.e. DVB-S) over 3GPP to IEEE 802.16 and 802.11. Therefore the nominal characteristics, as well as the parameters that can be analyzed may vary substantially.

Due to the volatile nature of radio links and the tight QoS-requirements of, for example, VoIP and multimedia services the monitoring component of the WiBACK architecture is crucial to assess the overall network state. Addition- ally, RFC 3272 (Awduche, 2002) suggests that the individual QoS assurances given to LSPs be monitored. In [23], we have presented a modular passive multi-layer monitoring architecture addressing radio, link and well as LSP monitoring. This architecture serves as the fundamental WiBACK monitoring component which provides configurable measurement, data aggregation and threshold-based triggering of so-called node-local rating agents which can create events to signal radio, link or LSP state changes to the centralized WiBACK network management function.

As shown in (Kretschmer, 2009), passive receiver-side wireless monitoring can be used to facilitate radio, link and LSP measurements. This means that WiBACK monitoring can be implemented with a very low overhead using piggy-back injection of measurement headers to aid with delay and loss measurements while minimizing the transmission of additional monitoring frames, such as RSVP HELOs. Receiver-side monitoring can also be applied to Unidirectional Technologies (UDTs), where the transmitter cannot receive any immediate feedback from its receivers.

The WiBACK monitoring component, as depicted in Figure 5 needs to sup- port the heterogeneous characteristics of the various wireless technologies including unidirectionality. Its modular design allows it to be tailored to specific wireless technology characteristics as well as payload QoS-requirements to en- sure the precise extraction of the relevant monitoring data addressing different aspects of WiBACK monitoring.

The monitoring component performs radio as well as per-link and per-LSP analysis on every frame received on any of a node's interfaces. This measured raw information is then transformed, categorized and interpreted by one or more of the monitoring levels, namely the radio level, the link level and LSP level. Information processing can be done node local or at a centralized entity which

Figure 5. The multi-layer monitoring architecture is integrated via 802.21 signaling

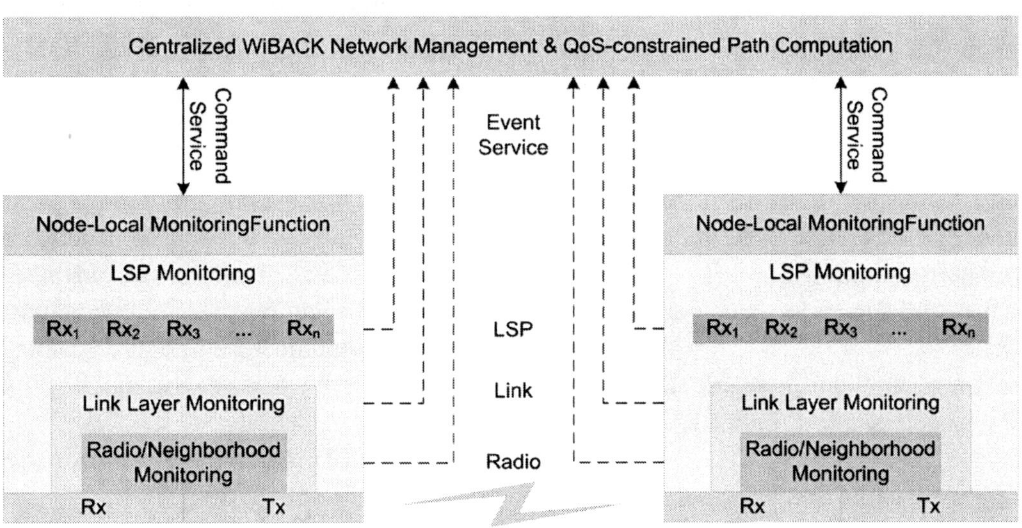

Figure 7. A WiBACK node consists of an 802.21-based control plane and an MPLS-based data plane

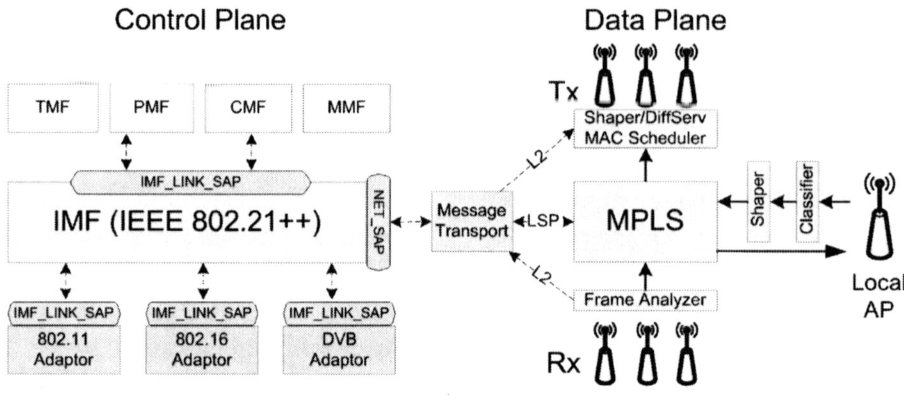

can also correlate data in order to, for example, locate sources of interference. As depicted in Figure 7, the node-local monitoring modules communicate with the centralized WiBACK network management component via 802.21++ messages. This allows the WiBACK network management component to install local event triggers or to adjust thresholds to tune the local monitoring modules to follow the overall monitoring goals, while maintaining a small footprint on the forwarding nodes.

The node-local monitoring modules can create the following set of events when the state of the monitored object changes which may trigger actions of other WiBACK management components:

- Change of Neighboring mesh node statistics
- Mobile Terminal (MT) presence detection and handover indications
- Indication of interference or intrusion
- Change of link state (i.e. up, down, or under-performing)
- Change of LSP end-to-end performance statistics
- Creation of QoS and network state aware LSP events to support (i.e. FRR)

As suggested in RFC 3272, the LSP level monitoring performs end-to-end monitoring of individual LSP statistics. In addition to the actual bandwidth utilization, it also maintains PHY status, loss, signal quality, delay and activity statistics which can indicate wireless link stability with a varying significance depending on the QoS requirements of the payload. This receiving side monitoring measures the actual end-to-end characteristics of an LSP and is therefore mandatory to verify that an LSP receives the agreed end-to-end QoS handling. The per-LSP QoS requirements are installed at each node on the path during the LSP setup procedure together with the LSP forwarding state.

WIBACK SIGNALING AND UNIDIRECTIONAL TECHNOLOGY INTEGRATION

The goal of seamlessly integrating UDTs into the WiBACK architecture is to allow higher layer services to transparently utilize UDTs when they are beneficial to address physical radio or deployment conditions, given payload characteristics or receiver distributions. The following fundamental WiBACK architecture design decisions need to be considered when integrating UDTs:

- Control plane and hardware abstraction are based on IEEE 802.21++
- Unidirectional interfaces can be identified and their capabilities described

- Topology and resource management are performed by centralized entities
- Individual flows are aggregated into QoS-aware pipes of four different traffic classes
- Pipes are unidirectional resources established between any pair of nodes
- In the multicast case, a pipe is extended into a 1-to-N tree

Since regular payload traffic is forwarded via unidirectional QoS-aware pipes across a WiBACK network, the data plane should be agnostic to UDTs as long as the chosen tunneling protocol provides the required QoS-differentiation. In the following sections, we therefore focus on the control plane and its signaling and management protocols such as monitoring, topology and resource management, connectivity among WiBACK nodes as well as pipe signaling.

We have identifies two approaches for our 802.21++-based control plane signaling in a WiBACK network. The below the network layer approach assumes no routing state to be present among the WiBACK Nodes (WNs), while the at-the-network-layer approach would depend on an operational routing protocol for signaling. In the following sections we will discuss the advantages and implications of the two approaches for the centralized WiBACK architecture and the seamless UDT integration in particular.

As already discussed in a previous section, WiBACK relies on passive receiver-side neighborhood, link and QoS-aware pipe monitoring. This monitoring subsystem supports UDTs and can be utilized in both approaches as long as its signaling requirements are addressed which include REQ and RESP messages exchanged between management nodes their WNs, as well as direct signaled WN-to-WN event indications.

In the centralized cross-layer WiBACK architecture, topology information describing all available physical links among nodes is maintained by centralized Topology Management Function (TMF) entities located at so-called management nodes and stored in topology graph data structures. A subset of these possible physical links is activated and made available to the Capacity Management Function (CMF) entity as logical links. The CMF may then place pipes onto them to facilitate the transport of QoS-differentiated user payload. The TMF and CMF functionality is identical in both approaches, but the integration and signaling implementation complexity may vary significantly.

The Below-the-Network-Layer Approach

This approach considers signaling at and above the Data Link layer and integrates UDTs where they occur as a result of MAC protocol design decisions or physical characteristics of the radio technologies. The QoS-aware pipe concept could be supported via MPLS-based forwarding on the data plane which also inherently supports the forwarding along centrally pre-computed paths. Since MPLS is considered a layer 2.5 technology, the approach proposed here would not be a pure Data Link layer approach, so the name Below the Network Layer approach was chosen. The use of MPLS on the data plane would complement the support for heterogeneous technologies, with an IEEE 802.21-based control plane providing support for local as well as remote management of heterogeneous nodes and their radio interfaces.

Due the centralized WiBACK architecture, management connectivity is mainly required between the management nodes and its associated WNs. Direct messaging between WNs may be used for MPLS FRR signaling or to optimize Mobile Terminal (MT) mobility among neighboring Access Points (APs). Connectivity between the management nodes and its WNs could be facilitated by installing MPLS LSPs among the centralized management nodes and each WN during the association procedure. Disjoint redundant backup paths could be installed using the MPLS FRR feature to provide resilience to intermediate WN or link failures. Since the topology information to compute such paths is maintained at the management nodes in the form of topology graph data structures, this information would be reused, thus avoiding potential duplication and coherence issues.

MPLS path setup is typically implemented via Network layer protocols such as RSVP-TE, which also provides the so-called Explicit Route Object (ERO) to enforce the setup along centrally computed path. A similar mechanism would be required in the case of the Below the Network Layer approach. It could be implemented via explicit source routing among the IMFs along the path to be configured, using encapsulation MIH messages describing the path to ensure that such messages would be processed at each hop along the path of the LSP to be configured. If a link provided by a UDT is present in the path, the confirmation message would have to be sent on a different path around the UDT while still traversing the downstream path in reverse order; see for example, Figure 9. Since local neighbor discovery mechanisms cannot be relied upon when signaling over UDTs, both the Data Link layer source and destination addresses would need to be explicitly specified.

The destinations for direct WN-to-WN messaging are known when the topology is configured or modified, therefore the management node could either set up LSPs among the affected WNs or install source routing information to allow for the delivery via the MIH encapsulation messages. The installation of LSPs requires the affected nodes to store additional state, while the use of MIH encapsulation messages introduces an additional overhead onto the actual messages being sent.

Similar approaches to provide Data Link layer connectivity among non- neighboring nodes exist and are widely used, such as the IEEE 802.1d Spanning Tree Protocol (STP). Or, the IEEE 802.11s standard which is based on a variant of AODV at the 802.11 MAC layer and thus provides IEEE 802.11s mesh clouds transparently to the Network layer. Both are distributed algorithms and cannot support the integration of UDT nor do they allow their state to be centrally configured.

The At-the-Network-Layer Approach

This approach considers signaling at the Network layer which already addresses Data Link layer heterogeneity in order to integrate different Data Link layer technologies. The support for UDTs could be seen as an extension of that concept. In the case of IP-based protocols the data link technologies are typically assumed to provide bidirectional means of communication, hence an integration of UDTs would require modification to proven IP-related protocols, such as interface auto-configuration, neighbor discovery, routing or flow signaling.

Alternatively, the Link Layer Tunneling Mechanism (LLTM) could be deployed to provide tunneled return channels for UDTs to provide virtually bidirectional interfaces. Since LLTM assumes a rather static network configuration regarding, for example, the return channel tunnel endpoints, an additional proto- col would be required to allow for LLTM self-configuration as well as adaptation in the case of underlying topology changes.

Following the LLTM concept, an IP-tunneled return channel would be detected as one hop by the routing protocol even though it may span across a large set of heterogeneous links and wireless cells. This would complicate QoS consideration, but this issue could be ignored, since the tunneled return links would only be used for management traffic, while user payload would be for- warded via regularly allocated data plane pipes. Under the assumption that a WiBACK node always has at least one bidirectional interface, UDTs could be ignored at the cost of potentially lower mesh connectivity, as shown for BRA. This is the default behavior of common WMN protocols such as AODV or link state protocols such as Open Shortest Path First (OSPF). In this case, nodes without a bidirectional interface could not be supported.

As a mesh network has no default routing hierarchy, a routing protocol is required in order for non-neighboring nodes to communicate with each other for management purposes. This is required independently from the centralized TMF and CMF entities. Hence, the routing protocol would have to rediscover the topology made available by the TMF and would also be required to track any modifications administered by TMF as a reaction to link or network state changes. During transition periods, the routing protocol might not be synchronized with the TMF state and signaling attempts may fail. To optimize the co- existence proven routing protocols could be adapted to accept hints from TMF or a special routing protocol for the WiBACK use case could be developed.

To support the QoS-aware pipe concept, MPLS-based forwarding on the data plane could be utilized. Alternatively, IP-in-IP tunnels or forwarding based on IPv6 flow labels could be used. The use of flow labels might, though, collide with other services already using the flow labels. IP-in-IP tunneling would require some modifications to enforce forwarding along the centrally pre-computed path instead of the relying on the routing information from the signaling routing protocol.

RSVP-TE could be used to set up LSPs and the ERO could be used to enforce centrally computed paths. This object may need to be extended so that it could hold the downstream and upstream path, which would need to be specified for UDTs where no LLTM is configured.

WiBACK Signaling Below the Network Layer

Conceptually, signaling and seamless UDT integration either at or below the Network layer would be possible, with each approach requiring a distinct set of modifications and considerations regarding the overall WiBACK architecture. Taking the aforementioned WiBACK design decisions into account, we propose an integration below-the-Network-layer. This appears to be the more natural fit, since UDTs can be seen as a Data Link layer or Physical layer phenomenon and are already described and handled by the WiBACK Abstract Interface (AI) as such.

Since the WiBACK architecture is build upon a centralized management approach, the management nodes already maintain all required topology information in the form of a topology graph data structure and the replication via a Network layer routing protocol would require extra state to be kept and causes coherence issues to be addressed. Since the TMF is the authority for configuring radio interfaces and assigning channels, etc, a Network layer protocol could only try to keep up with the decisions set forth by TMF entities at the management nodes without the knowledge or ability to influence them.

Resource allocation and book-keeping as well as path computation in the below-the-network-layer approach are straight forward, since UDTs are fully integrated and represented as regular links, identified by their source and destination Data Link layer addresses. In the Network layer approach, UDTs would require special handling with a varying complexity depending on the level of integration and interaction between the Network layer routing protocol and the centralized path computation module as well as the radio and topology management module.

Traffic Engineering is usually performed based on logically unidirectional resources and MPLS forwarding is a unidirectional task. Our WiBACK passive monitoring component has been shown to function properly relying solely on receiver-side passive monitoring which includes UDT. Therefore, the main WiBACK components that require modifications to fully support and utilize UDTs are the Topology Management Function (TMF) and the Pipe Management Function (PMF), which will be described in the following paragraphs.

Topology Management and Maintenance

The Topology Management Functions (TMFs) is built upon IEEE 802.21++ primitives and is responsible for node discovery, radio configuration and link management. This information is maintained at a centralized per-area topology graph data structure. Topology Discovery and Maintenance in a centralized WiBACK architecture is initiated and controlled by TMF instances executed by so-called management nodes. Each TMF instance aims at discovering general WNs in its vicinity and also forms areas or clusters to establish a two-level hierarchy in order to keep topology changes local. WNs are relatively simple devices which mainly forward packets along pre-computed path and perform threshold-based passive receiver-side monitoring as instructed by the TMF at the management node to provide input for the TMF's overall network maintenance.

The proposed topology discovery and maintenance mechanism is a continuous process that supports bidirectional as well as unidirectional technologies. The optimization goal of the TMF is to maintain a well balanced meshed network of multi-radio nodes with disjoint redundant connections for each mesh node towards their management node. Management nodes executing a TMF instance are either assumed to be designated manually as a result of off-line network planning or may be elected at run-time. The criterion to qualify as a TMF node may be the availability of sufficient resources to execute the TMF functionality or the presence of a back-bone connection, since in many practical use cases, a node with back-bone connectivity can be assumed to be operated more reliably. In the cases of network partitioning, other leader election mechanisms may be used to provide connectivity within the network partitions.

The discovery of new nodes is performed by forming rings around the management node. The first ring consists of mesh nodes with a direct radio association with the management node. Nodes belonging to the next outer rings connect to nodes on their next inner ring. This approach avoids unnecessary flooding by bootstrapping nodes attempting to actively detect a management node, but needs to be considered carefully in order to avoid a storm of association requests, especially after larger network outages. An issue here is that, depending on node distribution and connectivity, the number of potentially newly discovered WNs may increase exponentially with the hop distance of the ring from the management node. An exponentially increasing random back-off timer is used to distribute such association requests.

The discovery process starts at the TMF at a management node which brings up its radio interfaces. This starts a technology specific beaconing or periodic broadcast of management tables. In addition, the management node broadcasts information about itself via the IEEE 802.21 Information Service. Unassociated nodes first detect their own capabilities in terms of number and type of radio interfaces, power supply and processing capacity. They then perform passive Neighborhood Scanning on all available interfaces in order to discover potential management nodes or already associated mesh nodes. Evaluating the information collected via the neighborhood scans, they choose the most suitable node and attempt to associate with it. Since this selection is based on local knowledge only, it might be suboptimal or not fit at all with the management node's policies. Hence, once the node is associated, the management node might reconfigure its association. Once a node has been associated, monitoring of all configured interfaces switches from neighborhood scanning into link monitoring mode, while interfaces not configured by the TMF may be powered down or remain in neighborhood scanning mode to support scanning for interfering nodes, jamming attacks, etc.

Path Setup and Management

The WiBACK Pipe Management Function (PMF) is tasked to establish LSPs between any two arbitrary WiBACK nodes. Those may be regular LSPs carrying user payload or LSPs providing a management pipe between management nodes and general mesh nodes. In the latter case, Pipe Management Protocol (PMP) is used by the

TMF in order to establish a communication path between a management node and its associated mesh nodes which otherwise have no means of communication, since no distributed routing state is maintained in a WiBACK network. The PMF is therefore a crucial component of the WiBACK architecture which must ensure quick and reliable LSP setup. Hence, PMF must support the following requirements:

- Support for Heterogeneous Technologies including UDTs
- Use of IEEE 802.21 Messaging via IMF
- Support for downstream (unicast) and upstream (multicast) assigned labels
- Quick and reliable LSP setup

The Pipe Management Protocol (PMP) builds upon proven RSVP-TE concepts where possible. Since WiBACK relies on the IEEE 802.21 message format and IMF-based messaging among nodes, new IEEE 802.21-compatible primitives have been defined which contain the relevant Type-Length-Value (TLV)-encoded information defined by RSVP-TE. The RSVP PATH and RESV messages have been implemented via PipeSetup. Request and PipeSetup.Response primitives. The RSVP PathTear message has been implemented via a PipeRemove.Request primitive, which PMP explicitly confirmes with a PipeRemove.Response primi- tive. In order to support the Explicit Route Object concept, special encapsulation primitives have been introduced which allow any WiBACK module to send ex- plicitly hop-by-hop forwarded Requests, Responses or Indications along a route specified by a vector of links to be traversed.

In order to support UDLs, each element of a link vector needs to specify both ends of a link, the source address and the destination address. This is because with unidirectional technologies, local address resolution or neighbor discovery protocols cannot be relied upon. WiBACK describes its logical links via LinkIDs which consist of the source and the destination link layer address. Using the link vector forwarding mechanism and the LinkIDs as link vector element, any path in a WiBACK network can be explicitly specified and therefore LSP state can be configured. To integrate UDTs such as DVB, the upstream signaling path might be different from the downstream path while traversing the all downstream nodes, see Figure 6.

The encapsulation primitives allow a module to specify a single or multi- hop route. On the forward path, PMF uses single hop forwarding, since the Request messages need to be actively processed at each downstream node. On the return path, the Response message might take a different route around UDTs. Here multi-hop forwarding is used, since the Response message only needs to be processed by the nodes in the downstream path. In Figure 9, node F would be such a forwarding node, not part of the actual LSP being signaled, but merely acts as a forwarding hop for the response message to circumvent a UDT.

Figure 6. Path setup signaling around a UDL

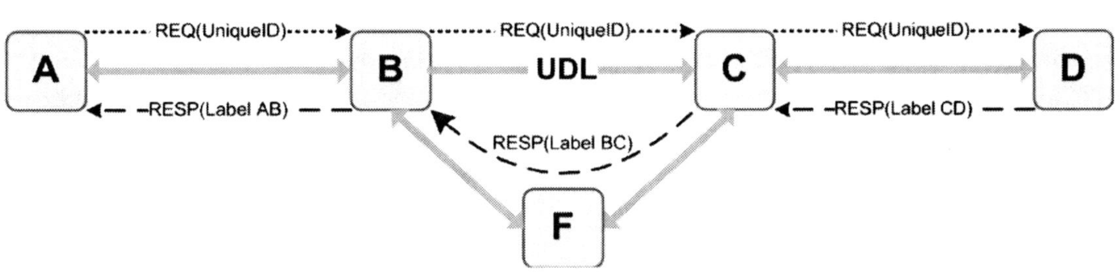

RSVP-TE implements an LSP-alive check by exchanging HELO messages among RSVP-capable nodes. Those messages may be sent as frequent as once every 1ms. This would introduce a high load on rather bandwidth limited wireless links. Additionally, this mechanism cannot be used over UDTs. In WiBACK this LSP monitoring task is provided by the passive monitoring component described in a previous section.

MPLS Fast Reroute (FRR) state is configured by the PMF. If a link failure has been detected by the monitoring component, the fail-over event is sent as a source routed IEEE 802.21 Indication message directly to the Point of Local Repair (PLR). Multiple source routes might be configured to increase the chance of a successful delivery in the case of surrounding link instabilities.

Implementation and Evaluation

The WiBACK architecture is currently being implemented using our C++ SENF[1] Framework and its Network Emulator component which provides an abstraction layer in order to support real or emulated wireless radio interfaces using the same executable built from identical source code. At runtime the interfaces are specified and configured. It is possible to operate in fully emulated mode, real hardware mode or mixed mode. Hence the Network Emulator enables us to evaluate an identical implementation of our protocols in an emulated environment or a real-world hardware test bed.

This allows on one hand a larger scale evaluation of our mesh protocols and on the other hand provides a reliable substitute for the missing real-world inter-area links/communication in our test bed. To emulate wireless links, multicast groups are used and a (frequency, channel, bandwidth) pair is mapped to a multicast group. Thus, all interfaces which are tuned to the same frequency and channel bandwidth can exchange packets among their group while being in each other's emulated communication range. Each node is executed as a separate Linux process. Real-time extensions of the Linux-kernel are used to ensure low-latency packet processing emulation. The emulation allows us to study the real-time interactions between different modules. It is also possible to connect real-world stimuli for evaluation or debugging purposes. A limitation of the real- time emulation is that it does not provide MAC layer timing resolution, hence collisions and contention situations cannot be studied accurately using emulated interfaces. Such issues could however be studied using mainly emulated scenarios where the links in question are realized via real hardware interfaces. The goal of our multi-radio WiBACK architecture is to operate links in collision-free point- to-point mode where possible. Therefore the study of contention issues it not a major focus of our work.

A WiBACK node consists of an MPLS-based data plane which also incorporates passive receiver-side monitoring and a control plane which is implemented based on our amended IEEE 802.21 specification, see Figure 7. The higher layer control plane modules combined provide the same functionality as a typical routing protocol used in nowadays operator networks and beyond.

Implementation and evaluation of WiBACK components is an iterative process where new features are first implemented and integrated and then evaluated. So far, evaluation results have been reported, among others, here (Kretschmer, 2010) and, here (Kretschmer, 2009). As first step the data plane and the IMF have been implemented. Then the monitoring component was added. Currently our efforts are focused on topology discovery as well as path computation and management. In the centralized WiBACK approach a small number of nodes act as designed management nodes which maintain the topology and link state information of the area they administer. The state information is stored in a graph which is implemented the boost graph library. The graph is maintained by the TMF while the Path Computation Function (PCF) uses this information to compute constrained path through the network and also stores its resource allocations in the graph. The boost graph library provides an API to dump graphs in.dot format. Our implementation builds on this feature to provide live snapshots of the topology and resource allocation state of a management node, see Figure 8. Here the TMF at node 1 has detected and configured four other nodes. The link between node 1 and node 3 is unidirectional. Hence, node 3 needs to communicate via node 5 to reach the TMF.

FUTURE RESEARCH DIRECTIONS

Our current architecture provides the foundation to manage heterogeneous wire- less back-haul networks. Future work will look into optimizing the algorithms of the main components such as radio planning, topology forming, path computation and mobility support. We're planning to enhance the path computation module to consider the efficient use of longer range overlay cells, with a focus on unidirectional broadcast cells to improve the support for high-bandwidth multi- cast services.

A new direction of research is energy-awareness of WiBACK nodes and the network as a whole to lower the energy consumption of WiBACK nodes in order to allow self-sustained operations using only alternative energy sources such as solar or wind. This would allow a deployment of WiBACK in rural areas of, for example Sub-Sahara Africa.

CONCLUSION

We have presented the key components of our WiBACK architecture which aims at providing carrier grade services in wireless networks utilizing heterogeneous technologies for operators by adopting and enhancing proven standards like

Figure 8. Live snapshot of the TMF topology graph with five nodes incl. an UDL

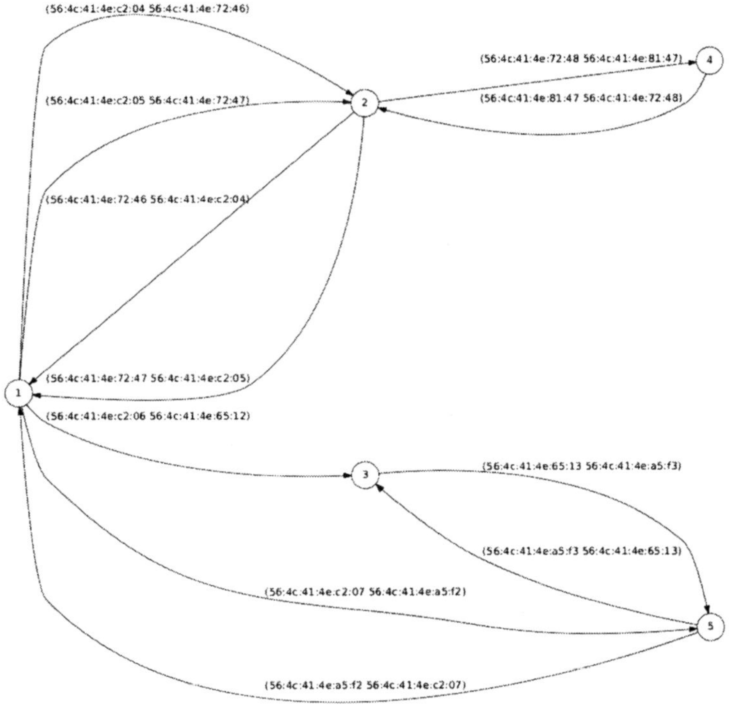

IEEE 802.21 or MPLS. Moreover, we have outlined our signaling concept and the seamless integration of UDTs such as DVB into the WiBACK architecture in order to enhance the triple-play support of the network. The Below the Network Layer signaling concept complements the support of heterogeneous technologies provided by the Abstract Interface and fits well with the centralized WiBACK design. The seamless integration of UDT required modifications to the monitoring subsystem, the topology management component and the path signaling protocol. In return, by not relying on link bi-directionality the architecture has become very flexible and can now easily support numerous radio technologies.

REFERENCES

Arjona, A., & Verkasalo, H. (2007). Unlicensed mobile access (UMA) handover and packet data performance analysis. In *Proceedings of the Second International Conference on Digital Telecommunications, ICDT 2007*. San Jose, CA: IEEE Press.

Awduche, D., Berger, L., Gan, D., Li, T., Srinivasan, V., & Swallow, G. (2001). *RSVP-TE: Extensions to RSVP for LSP tunnels*. RFC 3209 (Proposed Standard). Updated by RFCs 3936, 4420, 4874, 5151, 5420.

Awduche, D., Chiu, A., Elwalid, A., Widjaja, I., & Xiao, X. (2002). *Overview and principles of internet traffic engineering*. RFC 3272 (Informational). Updated by RFC 5462.

Awduche, D., Malcolm, J., Agogbua, J., O'Dell, M., & McManus, J. (1999). *Requirements for traffic engineering over MPLS. RFC 2702*. Informational.

Azcorra, A., Banniza, T., Chieng, D., Fitzpatrick, J., Von-Hugo, D., & Natkaniec, M. (2009). Supporting carrier grade services over wireless mesh networks: The approach of the European FP-7 STREP CARMEN. *IEEE Communications Magazine*, *47*(4), 14–16. doi:10.1109/MCOM.2009.4907397

Banchs, A., Bayer, N., Chieng, D., de la Oliva, A., Gloss, B., & Kretschmer, M. ... Zdarsky, F. (2008). Carmen: Delivering carrier grade services over wireless mesh networks. In *Proceeding of IEEE 19th International Symposium on Personal, Indoor, and Mobile Radio Communications PIMRC 2008*, (pp. 1-6). Cannes, France: IEEE Press.

Berger, L., Gan, D., Swallow, G., Pan, P., Tommasi, F., & Molendini, S. (2001). *RSVP refresh overhead reduction extensions. RFC 2961*. Proposed Standard.

Braden, R., Zhang, L., Berson, S., Herzog, S., & Jamin, S. (1997). *Resource reservation protocol (RSVP): Version 1 functional specification*. RFC 2205 (Proposed Standard). Updated by RFCs 2750, 3936, 4495.

Chieng, D., Hugo, D., & Banchs, A. (2010). A cost sensitivity analysis for carrier-grade wireless mesh networks with Tabu optimization. In *INFOCOM IEEE Conference on Computer Communications Workshops, 2010*, (pp. 1-6). San Diego, CA: IEEE Press.

Clausen, T., & Jacquet, P. (2003). *Optimized link state routing protocol (OLSR). RFC 3626*. Experimental.

Duros, E., Dabbous, W., Izumiyama, H., Fujii, N., & Zhang, Y. (2001). *A link-layer tunneling mechanism for unidirectional links. RFC 3077*. Proposed Standard.

Farinacci, D., Li, T., Hanks, S., Meyer, D., & Traina, P. (2000). *Generic routing encapsulation (GRE)*. RFC 2784 (Proposed Standard). Updated by RFC 2890.

Farrel, A., Vasseur, J.-P., & Ash, J. (2006). *A path computation element (PCE)-based architecture. RFC 4655*. Informational.

Komolafe, O., & Sventek, J. (2007). Analysis of RSVP-TE graceful restart. In *IEEE International Conference on Communications, 2007*, (pp. 2324-2329). Glasgow, Scotland. IEEE Press.

Kompella, K., & Rekhter, Y. (2003). *Signaling unnumbered links in resource reservation protocol - Traffic engineering (RSVP-TE). RFC 3477*. Proposed Standard.

Kone, V., Das, S., Zhao, B. Y., & Zheng, H. (2007). *Quorum: Quality of service routing in wireless mesh networks*. In ICST Conference on Heterogeneous Networking for Quality, Reliability, Security, and Robustness (QShine). New York, NY: ACM Press.

Kretschmer, M., & Ghinea, G. (2009). Seamless integration of unidirectional broadcast links into QoS-constrained broadband wireless mesh access networks. In *Proceedings of the 4th International Conference for Internet Technology and Secured Transactions*. London, UK: IEEE Press.

Kretschmer, M., & Ghinea, G. (2010). *An IEEE 802.21-based approach for seamless wireless mobile integration using QoS-aware paths supporting unidirectional links*. In IEEE Globecom 2010 Workshop on Seamless Wireless Mobility (SWM 2010), Miami, FL: IEEE Press.

Kretschmer, M., Niephaus, C., & Ghinea, G. (2009a). *QoS-aware flow monitoring and event creation in heterogeneous MPLS-based wireless mesh networks supporting unidirectional links*. In 9th IEEE International Conference on Communications 2009, Kuala Lumpur, Malaysia. IEEE Press.

Kretschmer, M., Niephaus, C., & Ghinea, G. (2009b). *Towards QoS provisioning in a heterogeneous carrier-grade wireless mesh access networks using unidirectional overlay cells. Lecture Notes of the Institute for Computer Sciences, Social Informatics and Telecommunications Engineering, 22*. Berlin, Germany: Springer.

Lessmann, J., Schoeller, M., & Zdarsky, F. (2010). Rope ladder routing: Position-based multipath routing for wireless mesh networks. In *IEEE International Symposium on a World of Wireless Mobile and Multimedia Networks (WoWMoM), 2010*, (pp. 1-6). Montreal, Canada: IEEE Press.

Mannie, E. (2004). *Generalized multi-protocol label switching (GMPLS) architecture. RFC 3945*. Proposed Standard.

Martini, L., Rosen, E., El-Aawar, N., & Heron, G. (2006). *Encapsulation methods for transport of ethernet over MPLS networks. RFC 4448*. Proposed Standard.

Pan, P., Swallow, G., & Atlas, A. (2005). *Fast reroute extensions to RSVP-TE for LSP tunnels. RFC 4090*. Proposed Standard.

Perkins, C., Belding-Royer, E., & Das, S. (2003). *Ad hoc on-demand distance vector (AODV) routing. RFC 3561*. Experimental.

Ramasubramanian, V. & Mosse, D. (2008). Bra: A bidirectional routing abstraction for asymmetric mobile ad-hoc networks. *IEEE/ACM TON, 16*(1), 116-129.

Rosen, E., Tappan, D., Fedorkow, G., Rekhter, Y., Farinacci, D., Li, T., & Conta, A. (2001a). *MPLS label stack encoding*. RFC 3032 (Proposed Standard). Updated by RFCs 3443, 4182, 5332, 3270, 5129, 5462, 5586.

Rosen, E., Viswanathan, A., & Callon, R. (2001b). *Multiprotocol label switching architecture. RFC 3031*. Proposed Standard.

Team, W. (2006). *A solution for seamless handover among multi-provider heterogeneous networks*.

Zhou, H., Huang, C., Cheng, Y., & Wang, G. (2009). A new multi-metric QoS routing protocol in wireless mesh networks. In *2009 International Conference on Networks Security, Wireless Communications, and Trusted Computing,* volume 1, (pp. 459–467). Wuhan, China: IEEE Press.

KEY TERMS AND DEFINITIONS

Constraint-Based Path Computation: Constraint-based path computation computes paths through a network topology fulfilling a set of constraints.

Handover: The term handover refers to the process of transferring ongoing sessions of a mobile terminal from one point-of-attachment to another. A handover may be considered *hard* or *soft* and may be considered an *intra* or *inter* technology handover.

Media Independence: The goal of media independence is to provide a uniform interface to the upper layers hiding the specificities of each network technology. To facilitate this, a common set of primitives have been specified which allow the higher layers to use the underlying network technologies in an abstract way.

Multi Protocol Label Switching: Multi-protocol Label Switching (MPLS) refers to a mechanism in telecommunication networks which forwards data packets among network nodes using so-called labels. MPLS allows for the creation of "virtual channels" between distant nodes that can carry packets of different protocols such as IP or Ethernet.

Traffic Engineering: Traffic Engineering (TE) deals with network engineering and performance optimization of networks. One of its main objectives is to facilitate reliable and efficient network operations while at the same time considering optimizations of network resource utilization and traffic performance.

Unidirectional Technology: In the context of this work, *unidirectional technology* refers to inherently unidirectional wireless technologies where the sending end can send a frame to the receiving end, but not vice versa. Common examples are broadcast technologies such as DVB.

WiBACK: A heterogeneous Wireless Back-haul network aims at extending an existing operator infrastructure to provide an increased of wireless coverage in often rural areas.

ENDNOTE

1 http://senf.berlios.de

Chapter 7
Adaptive Scheduling for TCP-Fairness in Wireless Broadband Networks

Balakrishnan K
International Institute of Information Technology-Bangalore, India

Ritesh Kumar Kalle
International Institute of Information Technology-Bangalore, India

Debabrata Das
International Institute of Information Technology-Bangalore, India

ABSTRACT

The exponential growth in multimedia traffic (Cisco Visual Networking Index, 2010), predominantly on UDP transport, poses a threat to the TCP's best effort throughput. This problem is more acute in last mile broadband wireless access networks (Bakshi, Krishna, Vaidya, & Pradhan, 1997). Most scheduling algorithms discuss improving the combined TCP and UDP throughput or improving the TCP throughput without studying the effects of inelastic traffic such as UDP. This chapter furthers the necessity for TCP throughput protection and proposes a novel dynamically adapting Weighted Fair Queue (WFQ) based scheduling mechanism that provides a higher degree of TCP protection. This is accomplished by differentiating between TCP and UDP flows, buffer provisioning for each flow, and prioritizing TCP ACK packets. The simulation results show that the proposed mechanism yields a relative improvement of up to 29% of TCP goodput and 7.5% of aggregate MAC throughput over the mechanism without the proposed improvements.

INTRODUCTION

Internet is a huge ensemble of highly distributed networks where most of the communication is carried over Transmission Control Protocol (TCP), an elastic transport protocol. TCP provides reliability along with flow and congestion control between the endpoints. User Datagram Protocol (UDP) is another transport protocol which is inelastic in nature and enjoys a small share when compared with TCP in today's Internet. UDP is mainly used for real time communications which is delay sensitive (Ex. A VoIP phone call).

DOI: 10.4018/978-1-4666-0017-1.ch007

Due to the advances made in multimedia and infotainment services over the Internet, use of UDP is seeing a meteoric rise (The Cooperative Association for Internet Data Analysis, 2010). If we assume that the total bandwidth available per base station (BS) is a constant value (e.g., say 100 Mbits) and the bursty UDP traffic increases, then the existing TCP traffic will get squeezed and the user experience for applications using TCP will get deteriorated. It is interesting to note that even few of the real time services, (Ex. Skype) use TCP for its transport (Baset & Schulzrinne, 2006). They will also incur a severe degradation in voice/video quality due to increasing UDP flows, which is bursty in nature. The negative impact of the increasing UDP traffic in Internet ranges from unfairness against competing best effort TCP traffic to the potential for congestion collapse which severely impacts the responsiveness of TCP based services (Floyd & Fall, 1999). UDP traffic also suffers with jitter by TCP's continuously changing window size. However, in this chapter we specifically focus on the challenges in protecting TCP throughput in presence of UDP flows by means of isolation of TCP and UDP traffic.

In the past decade, the problem of TCP protection has been attracting attention of researchers around the world and a number of mechanisms/architectures have been suggested on the Internet core and edge routers to guarantee a minimum performance to the TCP services. A number of router mechanisms have been suggested with an essential goal of end-to-end congestion control that includes identification and restriction of bandwidth hungry flows that are unresponsive and competing with TCP. But unfortunately, this problem is not fully solved in access network which connects the end user with the Internet backbone. The problem is common on both wired and wireless, but more acute in wireless because:

1. Wireless medium is prone to have more losses when compared with wired medium.
2. Wireless medium has a limited shared bandwidth where wired medium has the luxury to increase bandwidth with increased hardware installations.

Most solutions proposed till date, mainly focus on transport, network layer algorithms to overcome this challenge. (Balakrishnan, Padmanabhan, Seshan, & Katz, 1997) analyzed and proposed few transport layer mechanisms for improving the TCP throughput in wireless links. (Todorovic & Lopez-Benitez, 2006) discussed various kinds of TCP in Wireless networks. (Capone, Fratta, & Martignon, 2004) analyze and propose bandwidth estimation algorithm for the congestion control algorithms in TCP specifically for wireless links where random losses are very high. These papers do not consider the effect of UDP on their model. (Cottrell, et al., 2005) test various variants of TCP in real time enviroment. (Xylomenos & Polyzos, 1999) measure TCP and UDP performance separately in a Wi-Fi LAN based environment. (Xylomenos & Polyzo, 2003) simulate the performance of TCP and UDP over GSM and WLAN with different variants of TCP. Even though these papers consider TCP and UDP traffic, it does not consider the combined effect of TCP and UDP. In (Bruno, Conti, & Gregori, 2008), the authors model concurrent TCP and UDP in WLAN but with a finite load of UDP. Most of the literature models and simulates wireless traffic based on Wi-Fi which uses Exponential Backoff Algorithm. But all the new generation technologies like WiMAX, LTE, etc uses more sophisticated schedulers and design to give more support for QoS. The design of the techniques needs to be modified to exploit the advantages of the latest Wireless technologies. A few research works have also observed that there can be potential performance gains if efficient protocol solutions are designed on the MAC layer of next generation wireless broadband networks (Discussed in Section 4). TCP ACK prioritization is a technique where the ACK packets are given more priority than the data packets. But this technique is not well integrated with next generation wireless technologies like IEEE 802.16m based Mobile WiMAX, LTE- Advanced, etc. Most of the available literature discusses various solutions for improving throughput by designing new schedulers. But they have not considered about the buffer provisioning for the scheduler which is one of the main factor that can affect throughput.

In this chapter, we propose and discuss new TCP protection mechanism which isolates the TCP and UDP traffic in MAC layer, prioritizes TCP ACK and do buffer provisioning for each queue. The proposed mechanism is flexible enough to be implemented in any next generation wireless technology. We explain the current wireless trends and technologies with the commonalities between them. Even though the technologies are developed independently by standard bodies and also at different time periods, we may find a similar design paradigm in certain features and mechanisms. If these common features or mechanisms can be leveraged to achieve our objective, the same techniques can be applied in all technologies.

Next we study the nature of TCP and UDP used on Internet. TCP and UDP are transport level protocols which work on the endpoints of the network. Co-existence of TCP and UDP may affect each other because of their unique properties. This problem can be solved by isolating both TCP and UDP traffic in the network. In order to optimize TCP and to achieve objectives described in the abstract above, we need to apply cross layer optimization techniques amongst transport layer and MAC layer (i.e. by availing knowledge of the transport layer parameters in the MAC layer for a better throughput performance). We simulate the characteristics of TCP and UDP and also discuss how TCP gets affected with increasing UDP flows. We also investigate various kind of packet schedulers present in most of the networking devices. Irrespective of the technique employed, the common objective of the schedulers discussed in this chapter should achieve minimum queuing delay and high resource utilization. Further we will discuss about the buffer provisioning in schedulers. Modern schedulers have strict call admission control based on which the resources are allocated. This information available during call admission control can also be used to allocate buffer space for the flow. We discuss about provisioning buffer for individual flows in scheduler to minimize the queue overflows. We also discuss about dropping packets in buffer when there is an overflow. We propose a novel dynamically adapting WFQ based scheduling mechanism in wireless networks that provides a higher degree of protection to the best effort TCP traffic. This is accomplished by differentiating between TCP and UDP flows by introducing proposed hierarchical scheduling in all the QoS classes of WiMAX, provisioning buffer for individual flows and additionally prioritizing the ACK packets of TCP. The simulation setup and discussions about the obtained results are also presented. The simulation results give light on the advantages of our new proposed mechanism. Finally we conclude the chapter with a future scope of developments and research in this area.

TRENDS IN THE EVOLUTION OF WIRELESS TECHNOLOGIES

In the world of information and communication technology, optical backbones are energizing the core of the network and wireless networks are empowering access networks through their last mile connectivity. Wireless network has become an inevitable part of today's IT revolution because of its lower capital expenditure, flexibility of deployment and mobility support. The growth of telecommunication industry is fuelled by cellular communication and the growth in IP based broadband wireless technologies, like WiFi/WiMAX/LTE. This growing trend of networking and telecommunication has brought numerous socio-economic changes across the world. But the lack of core infrastructure has remained a hindrance to this growth in developing and under developed countries. This infrastructure gap is addressed by the advent of wireless broadband networks that has lower investment and deployment cost as compared to its wired counterpart.

Table 1 illustrates the comparison between various popular wireless technologies. This is followed by the discussion of QoS support in the technologies.

QUALITY OF SERVICE (QOS) SUPPORT

Wi-Fi

802.11e Wireless Multi Media (WMM) provides QoS by classifying the packets into four different types of flows (Voice, Video, Background, and Best Effort) based on the type and priority of the packets. Each of these queues is treated differently to satisfy the priority requirements of the queue. More over the Distributed Coordination Function (DCF) defines different time intervals to enable priority in the medium access mechanism.

WiMAX

WiMAX has three major parts in the MAC layer. The Convergence sub layer is responsible of accepting various kinds of higher layer protocols like ATM, IPv4, IPv6, etc. Security sublayer takes care of the authentication and encryption mechanisms. Common part sub layer has the scheduling mechanism and per flow management of the MAC layer. It has classified the services in five major groups: Unsolicited Grant Service (UGS), real-time polling service (rtPS), enhanced real-time polling service (ertPS), non real-time polling service (nrtPS), and best-effort (BE) in decreasing order of priority as regards to their

Table 1. Comparison between wireless technologies (LAN/MAN Committee, 2007; Cudak, 2007; 3GPP, 2009)

	Wi-Fi (802.11g)	WiMAX (802.16m)	LTE (LTE -Advanced)
Standard body	IEEE & Wi-Fi forum	IEEE & WiMAX forum	3GPP
Spectrum type	Unlicensed	Licensed	Licensed
Frequency	2.4 GHz	<6 GHz	700 - 2500 MHz
Bandwidth	20 MHz	Scalable, 5 - 20 MHz	1.4 - 20 MHz
Network type	All IP	All IP	All IP core, supports classical telephony protocols
PHY technology	MIMO, OFDM	MIMO, SOFDMA	MIMO, OFDMA, SC-OFDMA
Data rate	56 Mbps	150 Mbps downlink & 56 Mbps uplink	around 100 Mbps in downlink & 50 Mbps in uplink
Security	WEP, WPA & WPA2	EAP enabled authentication & AES-CCM based encryption	Diameter based authentication & Snow3G based encryption
Mobility	Not-supported	Supported	Supported
NLOS communication	Supported	Supported	Supported
Coverage area	100 meters	upto 10 Kilometers	upto 10 Kilometers
Protocol support	IP only	IP, ATM, MPLS	IP, ATM, MPLS and also classical telephony protocols
QoS support	Yes	Yes	Yes
Power saving	Yes	Yes	Yes

QoS requirements. In WiMAX, each MAC layer connection is unidirectional and is identified by a unique connection identifier (CID). The connection from the base station (BS) to the mobile station (MS) is referred to as the downlink (DL) and the connection from MS to BS is known as the uplink (UL). A service flow is a unidirectional flow of packets with a particular set of QoS parameters identified by a service-flow identifier (SFID). QoS guarantees are given individually per flow with the help of the above mentioned Flow Identifiers.

LTE

LTE also has three major parts in its MAC layer. PDCP (Packet Data Convergence Protocol) and RLC (Radio Link Control) form layers in the top of MAC layer. As LTE is an all packet network in the core, the packets are processed with Header suppression with security and integrity check in PDCP. Sequencing, retransmission and segmentation are taken care by RLC. MAC takes care of the scheduling of packets. The RLC passes data as logical channels to MAC layer. MAC layer passes data as transport channel to PHY layer. Within each channel there are separate subchannels for control and data packets with different priorities. There are separate channels for downlink and uplink. QoS guarantees are given per flow in the MAC. LTE uses different schemes for scheduling such as Frequency selective Scheduling, Frequency Diverse Scheduling and Proportional Fair Scheduling for assuring best QoS for the packets.

Observation

From the above description of different wireless technologies, one can notice that all the wireless technologies, irrespective of its evolution and application, use similar class of techniques for providing QoS assurances to varying degrees of reliability.

Quality of Service is a very broad term which has different metrics based on the application. In the field of networking, it is basically measured with the metrics of throughput, jitter, delay, packet loss, etc. In the field of media transportation, it is also measured based on the end user experience of viewing which is called as QoE. In our case, our objective is to differentiate and protect TCP traffic. TCP by itself takes care of packet loss. The jitter and all other network parameters for TCP boil down finally to throughput. So, in our discussion, we use throughput as the metric for describing the QoS (i.e. higher QoS is provided by achieving a high throughput).

For assuring QoS, the MAC scheduler is designed with different queues which differentiate the traffic into different classes having different QoS requirements. For assuring strict QoS, fine granularity in queuing is provided by allocating a separate queue per connection. We use this idea of traffic differentiation by different queues as the basic model for our problem and try to apply new techniques to differentiate and protect TCP data.

By the strict layering architecture followed in Internet, the MAC layer is agnostic to the information of higher layer transport protocols, but for the purpose of obtaining the QoS requirements of each queue, it needs some higher layer information too. IETF has defined Intserv (Braden, Clark, & Shenkerl, 1994) for flow based service guarantees and Diffserv (Blake, et al., 1998) for class based coarse grained QoS in Internet. The information of Intserv and Diffserv is available in "Layer 3" as IP packets. The information of whether the packets are TCP or UDP is available in transport layer. All of the information should be made available in MAC layer for designing the scheduler by applying cross layer messaging mechanism. We use WiMAX technology to illustrate the ideas for providing TCP bandwidth differentiation and protection. The same concept can be used in any of the next generation technologies by making small modification according to its specification.

TCP VS. UDP

TCP

TCP is a transport layer protocol which is used predominantly in Internet. TCP uses various algorithms for achieving end to end flow control and congestion control. This elastic property of TCP allows the Internet to work in a distributed way. TCP is also a reliable protocol which ensures all the packets from the source reaches the destination properly. However it does not guarantee on the timing and order of delivery of the packets. Flow and congestion control is achieved by varying the rate at sender with respect to the feedback of buffer available at end receiver and packet-loss/time-out at network, respectively. Reliability is achieved by resending the lost or erroneous packets based on the acknowledgement from the receiver.

UDP

UDP is a non-elastic protocol which is mainly used for sending control packets and real time packets in Internet. It is having a simple design without any considerations about network environment. Once the packets are generated at the source, they are simply injected in the network. There is no assurance for the delivery. This kind of protocol will be very useful when less delay is more important than reliability. Since the protocol is not self adjusting, the network may lose its stability, when it is used in a larger share in the network without knowing its bandwidth and load. Luckily today's internet has a very small share of UDP when compared to TCP.

Effect of UDP on TCP

In this section, we analyze the effect of UDP on TCP traffic. TCP session starts from a minimal transmission rate and increases till it reaches the maximum possible rate in steps. If the connection suffers from congestion or buffer overflow in the network or destination node, there will be packet loss. The sender will sense the packet loss or congestion and reduce its sending rate. Once again it will start increasing the rate till it hits the maximum rate, in case of continuous successful transmission. It does the above step continuously throughout its entire duration of transmission so that the transmission rate may continuously adapt to network conditions. There are various algorithms available to change the transmission window size to maximize its throughput and achieve stability. But irrespective of its algorithm, it is clear that the transmission rate is continuously changing and is difficult to predict. Moreover the sender may have idle time where it will wait for acknowledgement from receiver.

On the other hand, UDP does not see the current traffic situation (whether network is free or congested) in the network. By simply pushing the packets in the network, there is a chance that the network will get fully congested and it may even loose its own throughput. Further, it causes heavy impairment to the TCP passing through the network. Figure 1 illustrates the degree of reduction in TCP goodput because of UDP traffic. In Figure 1, When UDP traffic rate is increased from 1460 to 1606 Kbps, TCP's goodput reduces from 1627 Kbps to 601 Kbps (1026 Kbps TCP goodput reduction on just 146 Kbps increase in UDP flow!).

It is interesting to note that UDP always try to send its data without considering about network and also impacts other traffic (particularly TCP). TCP tries to be always conservative and reduces its rate if it senses congestion. So the usage of such heterogeneous protocols in the common network channel results in undesired behavior in performance. This problem is not recognized much in wireless access networks since it was not carrying much traffic when compared to core networks. However, the advent of Broadband wireless networks with a very high throughput

Figure 1. Effect of TCP on incremental UDP flows in a typical WiMAX network scenario with 5 MHz bandwidth

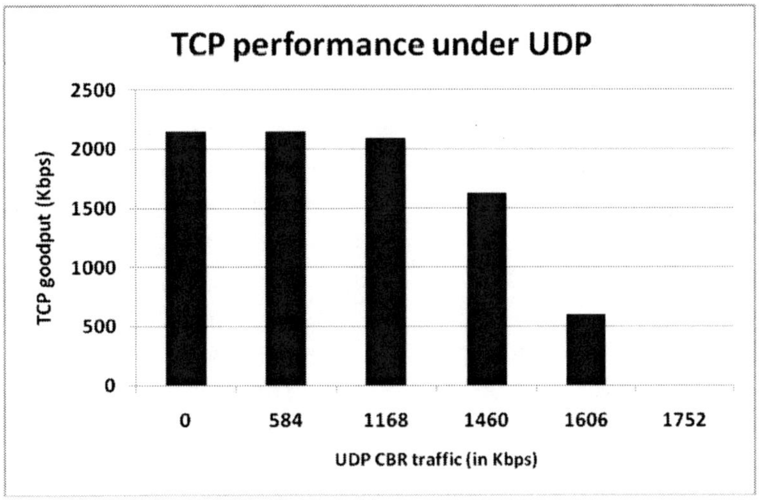

requirement, and an error prone wireless channel (when compared with wired channel), it is now important to address this problem to protect the TCP traffic.

SCHEDULERS

Scheduler forms a crucial part in the design of network devices. Scheduler allocates network resources to individual queues waiting in the network buffer for transmission/reception. The schedulers described in this chapter are expected to satisfy two kinds of requirements in a network.

- Maximum bandwidth utilization
- Differential treatment of queues to allocate bandwidth according to priority

Statistical multiplexing formed the first kind of differential allotment of time divided slots to each flow, based on the statistics of the usage by the particular flow. If we reserve bandwidth for a high priority flow, we may waste network resources when the flow is underutilizing the bandwidth for a period of time. This opposes our other requirement for a scheduler. So it is tricky to design a scheduler which satisfies the two requirements. The theoretical foundation for such a scheduler is laid out by GPS (Generalized Processor Sharing) scheduler which is explained below.

Generalized Processor Sharing

GPS assumes a fluid traffic (which can be divided infinitesimally) getting accumulated in the queue of the system with several flows. Since a fluid traffic is assumed, packets of desired variable length can be drawn from the queue. Each flow has a weight associated to it which decides the priority of the flow. A flow is considered to be active when it gets some backlogged data in its queue. The weight for N flows is taken as Φ_1, Φ_2..... Φ_n. If $S_i(\tau,t)$ is defined as the amount of traffic served for i_{th} flow between time interval τ and t, the server is defined as,

$$S_i(\tau,t) / S_j(\tau,t) >= \Phi_i / \Phi_j \text{ for } j = 1,2,3,...n$$

The guaranteed rate in session j is given by g_i

$$g_i = r\Phi_n / \sum_j \Phi_n$$

The increase in weight of a flow decreases its latency in transmission at the cost of increasing the delay in other flows.

Even though it looks like a comprehensive model, GPS does not consider the length of the packets which are practically defined as per different networking standards and are of variable sizes.

Packet-by-Packet Generalized Processor Sharing (PGPS)

PGPS is equivalent of GPS with consideration on packet sizes. WFQ (Weighted Fair Queue) scheduling is approximation of PGPS. In PGPS, a packet is considered as arrived or departed only after the last bit is received/sent. So the variable sized packets are processed according to the weight of the queue. The major advantage of PGPS is that it also considers the length of the packet unlike in Weighted Round Robin (Chaskar & Madhow, 2003). The equations applicable for GPS can also be applied for PGPS. WFQ and other packet approximations of GPS model support differentiated bandwidth allocation to each of its queues with a memory complexity of $O(N)$ where N is the number of flows (in this chapter, N is the number of service class queues) and per-packet-processing complexity usually varies from $O(1)$ to $O(N)$ (Wen, Fang, & Sun., 2009)(Parekh & Gallager, 1993) (Ramabhadran & Pasquale, 2003).

Hierarchical Schedulers

Till now, we have discussed about the evolution of schedulers based on the differentiation of flows and types of packets. As we have seen, one of the objectives of the scheduler in our context is to achieve maximum bandwidth utilization. This may be difficult to achieve without apriori knowledge of the kind of data arrival/dispatch in a queue. If we are able to group the queues with similar traffic characteristics, it will be more useful to design a more efficient algorithm in terms of TCP protection and maximum throughput utilization in a scheduler. This may be achieved by a hierarchical scheduler.

A hierarchical scheduler has multiple levels of differentiation in queues. First the queues of same category are clubbed into a first level of classification. This can be further classified into granular flows in the next levels and scheduling algorithm can be applied at each level to decide the best way of utilization of bandwidth while giving priority. We are going to use hierarchical queues for providing TCP bandwidth differentiation and protection in future sections.

In Internet, most classical services and applications were built over reliable and rate adaptable TCP transport protocol (such as FTP, WWW), newer and multimedia rich services are being increasingly deployed over UDP. For solving the problem of TCP protection in core network, a number of router mechanisms have been suggested with an essential goal of end-to-end congestion control that includes identification and restriction of bandwidth hungry flows that are unresponsive and competing with TCP. These approaches include deployment of algorithms for per-flow scheduling that individually regulate the bandwidth used by each best-effort flow with an added cost of per-flow state maintenance. Essentially, mechanisms for both bandwidth differentiation and TCP protection are important for guaranteeing Quality of Services (QoS) in TCP/IP networks (Wen, Fang, & Sun, 2009). (Yilmaz & Matta, 2001) discusses about differentiation and isolation techniques for TCP and UDP traffic for routers. (Lee, Ahn, & Campbell, 2001) discusses about QoS and bandwidth differentiation techniques in mobile Ad-Hoc network with TCP and UDP flows seperately.These sophisticated mechanisms have been mostly studied and implemented on the edge routers.

In the context of wireless access networks however, the above discussed problem was not the primary issue of attention till recent past. However, with the advent of wireless broadband access networks such as WiMAX, the problem of TCP congestion collapse has re-emerged as a significant issue for research. The IEEE 802.16m evaluation methodology document for next generation WiMAX includes the TCP layer throughput metric as a mandatory performance evaluation criterion in addition to the PHY and MAC layer throughput measurements (IEEE 802.16m, 2009).

In context of scheduling TCP over WiMAX, most of the research works try to achieve maximum MAC layer throughput, optimization for TCP only traffic. In (Yang, Venkatachalam, & Mohanty, 2007) and (Mishra, & Sivalingam, 2008), the authors propose cross layer optimization of MAC layer parameters such as congestion window adaptation and ARQ settings for overall TCP throughput enhancement. In (Kim & Yeom., 2007), the authors proposed bandwidth estimation and grant mechanism with a goal to provide max-min fairness to uplink TCP flows.

Buffer Design

Scheduler uses buffer to queue up the received packets. When the traffic arrival pattern is bursty, queue helps in storing the packets which are processed then. Most of the theoretical scheduler model assumes an indefinite memory for buffers. But in practice, the buffer size is limited and there will be a packet loss when there is a buffer overflow. There are also algorithms available like RED (Random Early Detection) for detecting overflow and dropping packets in a queue. (Ranjan, La, & Abed, 2010) discuss about TCP and UDP with RED.

Buffer design is generally out of scope of the communication standards and is fully left to the implementers. Most of the buffers in networked devices today use a common pool of memory to allocate storage space for queues. Whenever a packet enters the system, memory is allocated for the packet and it is recovered back when the packet is dispatched. This system works efficiently for a device without any service differentiation mechanism. But consider the following scenario: there are two flows A and B in a system with A assigned higher priority and B having a lower priority when compared to A. Suppose, if B accumulates a large number of packets because of its low priority and bursty packet inflow, there is a chance that it will occupy most of the space in the buffer. Under this circumstance, the high priority flow A might not get buffer to store its packet and drop the packet even though it is having a high packet processing capacity and transmission priority.

From the above example, it is clear that the buffer design for a scheduler should have differentiated packet processing capability. The design of scheduler and buffer is explained in the following section.

PROPOSED SYSTEM DESIGN

In this section, we discuss a hierarchical dynamically adapting Weighted Fair Queuing (WFQ) based scheduling mechanism in WiMAX with its buffer design that provides bandwidth differentiation to the TCP best effort traffic by introducing new hierarchical TCP and UDP queues in all the QoS scheduling classes of WiMAX.

Additionally, in case of asymmetric links (most of the times in WiMAX UL<DL bandwidth), when the ACKs of best effort TCP connections compete with data packets since they share a common buffer in the network elements, they introduce a bottleneck effect. The effect of ACK bottleneck and the resulting dynamics of transport protocols under two-way traffic over rate controlled channels were studied in (Kalampoukas, Varma, & Ramakrishnan, 1998). Furthermore, we propose a new technique to overcome the ACK bottleneck (UL or DL) to facilitate better TCP flows for the BE services in WiMAX. In our proposed solution, we schedule the TCP ACK packets for individual BE TCP service flows over separate UL connections through the nrtPS service which enjoys a higher priority than BE.

In particular, the proposed scheduler has the following three properties

1. Integrating TCP protection and bandwidth differentiation mechanism in one single architecture of WiMAX based access network (i.e., in BS and MS) from UDP flows.
2. Enhancing BE-TCP's service performance under ACK bottleneck situation in WiMAX, by using prioritized scheduling of ACK on the UL over the nrtPS service class.
3. Differentiated buffer design to reduce packet drops.

We perform extensive simulations of our proposed mechanism through implementation on the Qualnet platform. The results are discussed later on.

Scheduling Mechanism

Figure 2 show the standard WiMAX MAC architecture which is divided into three distinct components: the service-specific convergence sublayer (CS), the common part sublayer, and the security sublayer. The MAC layer interfaces with the Physical layer for the transmission and reception of data over the medium.

Figure 3 illustrates the architecture of a single service class in the scheduler (for example BE). Each service class can have any number of queues where each queue represents a single unidirectional connection with a unique CID (i.e., each flow in a service class has a CID). We will use the term '*bin*' to denote each service class for standard architecture (Figure 2).

In this chapter, we consider a WFQ scheduler that draws packets from the five service bins (UGS, rtPS, ertPS, nrtPS, BE). Additionally, it may also be noted that, the MAC architecture is common for both UL and DL flow at BS and each DL scheduling service will be queuing packets meant for all the MS(s) under that BS. However, for the sake of easier understanding, we consider the DL flows of TCP and UDP to one MS under the BS.

In Figure 4, we illustrate the proposed architectural changes with some of its design considerations to make the scheduling more efficient for TCP protection and differentiation. Since we

Figure 2. WiMAX MAC QoS architecture

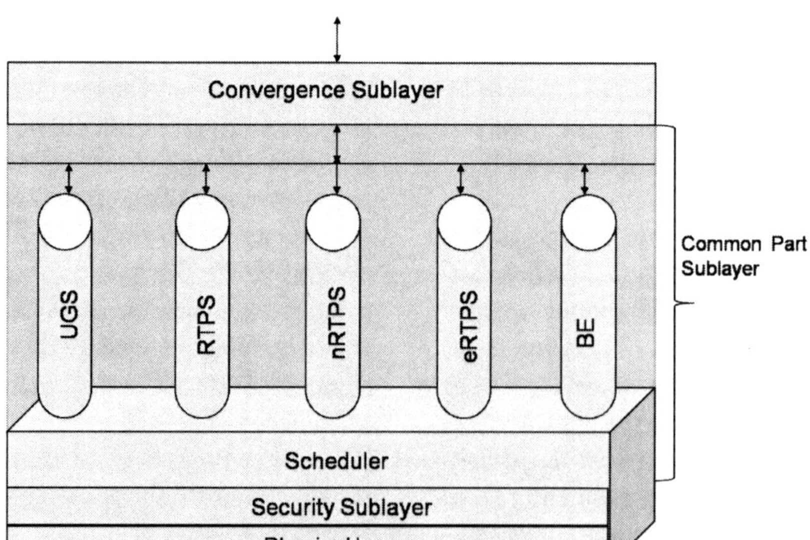

Figure 3. Architecture of a single service class in standard WiMAX scheduler at BS

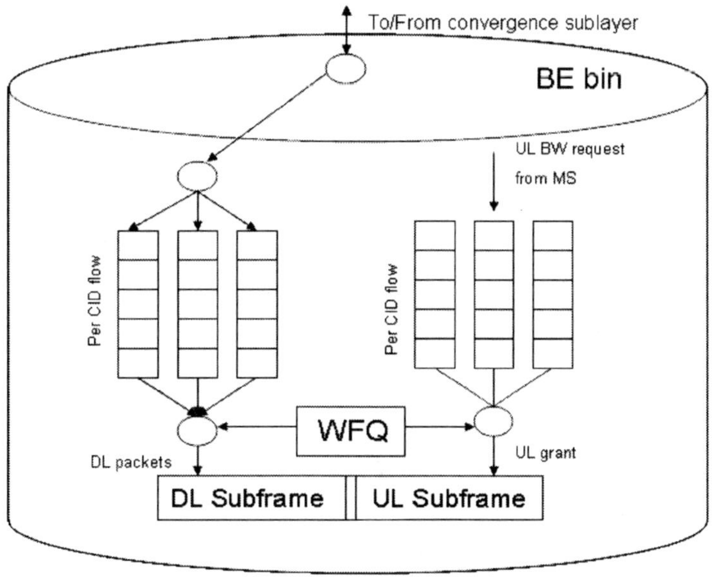

represent a hierarchical model of scheduler, we will refer to each service class in the proposed mechanism as *'macro bin,'* In our proposed mechanism (Figure 4) inside a macro-bin, we club all the flows of TCP as TCP '*micro-bin*' and all the UDP flows as UDP '*micro-bin*,' As mentioned before, the WiMAX standard (IEEE 802.16e) specifies five scheduling services with different QoS characteristics. Each micro bin (TCP or UDP service) can support multiple unidirectional queues, where each represents an active flow.

Figure 5 shows the architectural change within each service class in our proposed mechanism. When a connection is admitted, the classifier identifies whether it belongs to TCP or UDP. For example, if a connection belongs to BE TCP, then it will insert it in TCP-micro-bin of BE. The above mentioned admitted connection will create a new queue in TCP-micro-bin. By having different scheduling micro-bins of TCP and UDP, we can protect TCP from UDP, by allocating certain minimum guarantees in bandwidth usage (Figure 5).

The major advantage of this model is that it will work over the current standard without requirement of any fundamental changes in the IEEE 802.16e standard per se. TCP assures reliability of data communication by using ACK messages for the transmitted data segments. But unfortunately, the sender waits for the ACK after sending some data segments and will be idle during the intermediate period. The transmission window size may also change because of the flow and congestion control mechanism in TCP. In the mean time, UDP traffic may steal the allocated channel bandwidth used by TCP. In case of WiMAX, the BE TCP data packet's ACK needs to follow a contention based bandwidth request and allocation mechanism over the UL channel. It not only competes with data packets, but also with other MS(s) in the network connected to the same BS. This bottleneck situation created increases the delay of the TCP ACK packets being received at the sender end. One of the better mechanisms to reduce the impact of this problem is to use prioritized ACK messages. Under this

Figure 4. Proposed WiMAX QoS scheduling with TCP protection and bandwidth differentiation at BS

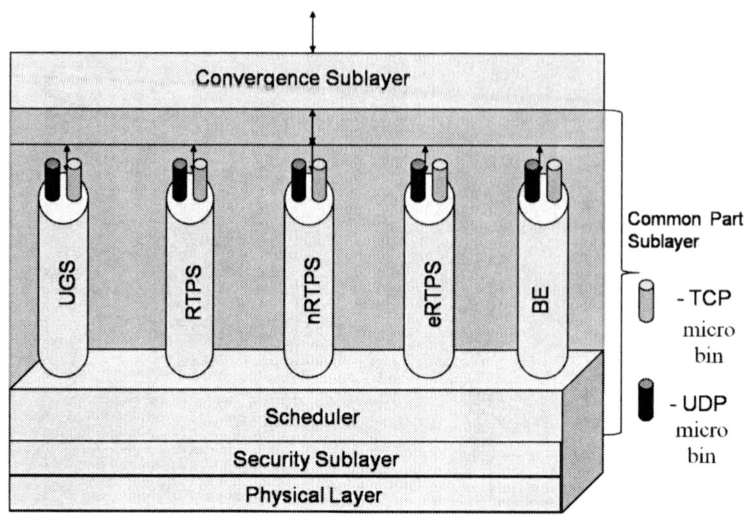

design paradigm, the ACK packets sent by the MS will be treated with higher priority than the competing data packets and will be sent back quicker to the sender end. As a result, the sender will be able to pump data with minimum delay into the channel. This mechanism can be implemented easily in WiMAX, since the standard already has five differentiated service flows with different priorities for each of them. For example, when TCP traffic uses a BE connection, the ACK packets can be made to follow an nrtPS flow, which will increase bandwidth utilization of the TCP flows and also enhance the overall throughput of the network.

Figure 5. Base Station Architecture of a single service class in proposed WiMAX scheduler (shows only DL)

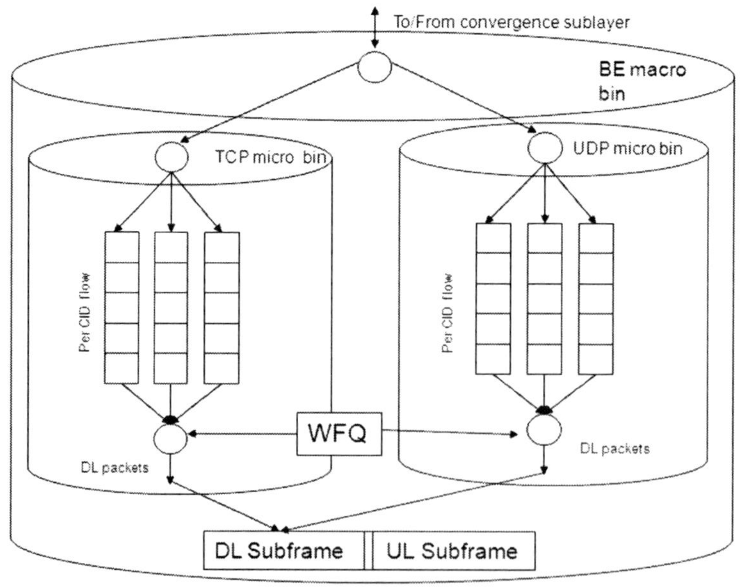

Scheduling Algorithm for TCP Protection and Bandwidth Differentiation in WiMAX

The proposed WFQ based scheduling algorithm allocates resources on a frame by frame basis in the OFDMA frame structure of WiMAX. A Time-Division Duplex (TDD) WiMAX frame consists of an uplink subframe and a downlink subframe separated by a Transmit Transition Gap (TTG). Each subframe, in turn consists of several OFDMA bursts (either UL-Burst or DL-Burst) specified by a burst profile information element (IE). The allocated resource can be defined as the number of slots and then these slots are mapped into a number of subchannels (each subchannel is a group of multiple physical subcarriers) and time duration (OFDM symbols).

In the scheduling Algorithm 1, we allot generic physical slots for every UL or DL burst. Initially, we obtain the number of physical slots available for the burst. Later, based on the admission control records for bandwidth reservation, we allot physical slots to the priority based service classes (UGS, ertPS, rtPS and nrtPS). After computing the total slots allocated for the priority classes, we classify it again into two classes called TCP and UDP. The weight of both the scheduling bins can be allocated by using various factors such as number of active queues, length of each active queue, etc.

Steps 15 and 16 of Algorithm 1 calculates the weights of TCP and UDP bins and is denoted as w_{TCP} and w_{UDP} in the algorithm. These are the differentiated weights calculated in the WFQ weight computation procedure.

In yet another scenario, these weights can be computed through a WiMAX network policy statement on the bandwidth differentiation and protection for TCP flows.

We have seen that there is a need for more buffer space if the priority of a flow is less compared to other flows. Theoretically, if serving rate < arrival rate, the queue will overflow irrespective of the queue size at some point of time. But short timed flows can overcome this limitation by finishing the purpose of the queue before the buffer overflows. Our buffer allocation and packet dropping algorithm has four different allocation schemes for different flows

Figure 6. TCP with ACK prioritization

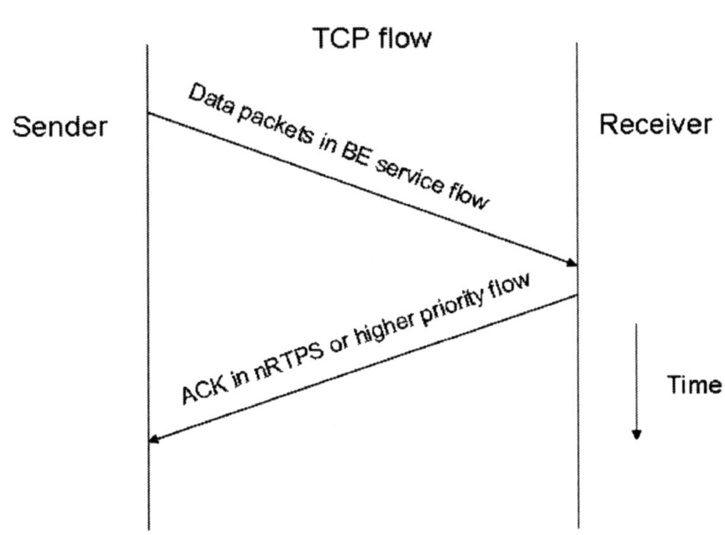

Algorithm 1. Proposed WiMAX scheduling algorithm

```
phy_slots ← ⊥ //Number of available physical slots
s_class ← ⊥ //Current scheduling class
allocated_slots ← ⊥ //Physical slots for a macro bin
alloc_slot_TCP ← ⊥ //Slots for TCP micro bin
alloc_slot_UDP ← ⊥ //Slots for UDP micro bin
remaining_slots ← ⊥ //Remaining free slots

For each DL/UL subframe
phy_slots ←Number of physical slots available in the
  subframe
while phy_slots > 0 do
   for s_class ← UGS to BE do
      if s_class = UGS or eRTPS then
         allocated_slots ← maximum physical slots ac-
         cepted at the time of admission;
      else if s_class = RTPS or nRTPS then
         allocated_slots ← average physical slots accepted
         at the time of admission;
      else
         allocated_slots ← residual physical slots avail-
         able;
      end if
      if (TCP_PROTECT = TRUE) then
         w_TCP ← Network_policy_fn(TCP), 0 ≤
         w_TCP ≤ 1;
         w_UDP ← Network_policy_fn(UDP), 0 ≤
         w_UDP ≤ 1, w_TCP + w_UDP = 1;
      else
         w_TCP ← weightToSlots(s_class, TCP)), 0 ≤
         w_TCP ≤ 1;
         w_UDP ← weightToSlots(s_class, UDP)), 0 ≤
         w_UDP ≤ 1, w_TCP + w_UDP = 1;
      end if
      phy_slots ← phy_slots − allocated_slots;
      alloc_slot_TCP ← w_TCP * allocated_slots;
      alloc_slot_UDP ← w_UDP * allocated_slots;
      Dequeue packets from corresponding micro bins ac-
      cording to alloc_slot_TCP and alloc_slot_UDP
      and put it in burst subframe
   end for
   if phy_slots > 0 then
      if weight(all service bins) = 0 then
         Allocate phy_slots to next frame;
         break;
      end if
   end if
end while
```

- UDP flows with priority and traffic assurance
- UDP flows with best effort service
- TCP flows with priority and traffic assurance
- TCP flows with best effort service

One may reason the need for different allocation and dropping schemes for UDP and TCP separately as explained below.

1. When a TCP connection is established, two flows, one for sending packets and another

for sending ACKs are created. ACK packets contain receipt for a number of packets and so more important from a network perspective. So those packets are given more importance than a normal TCP packet.

2. When a network gets congested, we need to drop packets from the queue. In case of TCP, if we deploy a random drop of packets, the throughput is unpredictable. When we drop packets in the tail of the queue, the loss of the packets will be reported to the server very late within which the server will increase its packet sending rate. Since the queue in the network device is full, it will drop further packets from the network and this reduces the overall goodput of the node in the network. If packets are dropped at the head of the queue, the loss message will go faster to the sender and it will reduce its rate which in turn reduce the congestion and increase the overall goodput delivered by the node. So in TCP, it is better to drop the packets in the front of the queue for a TCP connection.

In case of an UDP connection, at transport layer it is difficult to separate packets with more and less importance in a single flow without knowing its payload. But we have general information that UDP is used to carry real time traffic such as voice and video. Most of the media codecs have the tendency of recovering from network errors by constructing the missing packet from the neighboring frames. Research works have also shown that spreading the packet loss may vary the user perception in a media stream (Varadarajan, Ngo, & Srivastava, 2000). So in UDP, it is better to spread out the packet loss in case of a small overflow situation and drop as a chunk of contiguous packets in case of a bigger overflow situation. The decision of smaller or bigger overflow is made based on the length of the queue.

We have one more classification for prioritized and non-prioritized flows in the buffer. The reason is that prioritized flows are having some kind of bandwidth requirement information on hand before establishing the flow. So during the time of admission control itself, we can allocate buffer for the flow according to its requirements. The calculation for allocating buffer for prioritized traffic is as follows

Let us consider,

r_{max} = Maximum bandwidth or maximum rate at which the node can transmit/receive.
r_i = Assured bandwidth or rate for the i_{th} flow
m_{Max} = Maximum memory available for buffer in the system in bytes
m_i = Memory allocated for i_{th} flow in bytes

$$m_i / m_{Max} = r_i / r_{Max}$$
$$m_i = m_{Max} * r_i / r_{Max}$$

Calculation of memory allocation for prioritized flow having is very simple because of a strict requirement at the time of admission control. But if a TCP or UDP flow is created in a best effort flow, there is no admission control and the bandwidth information is not available in hand. Under the scenario, the buffer allocation is done in the following way.

Let there are '*n*' best effort flows which has '*x*' TCP flows and '*y*' UDP flows such that $n = x + y$. m_{BE} and r_{BE} be the total memory and bandwidth available in the system for allocating to best effort traffic. The allocation is done as follows.

First the memory for whole TCP connections and UDP connections flows are allocated separately

$$m_{TCP} = x * m_{BE} / n$$
$$m_{UDP} = y * m_{BE} / n$$

Since separate allocation is done for TCP and UDP, UDP will not affect TCP's resources. Now there are two ways to allocate memory for per flow. First one is simply allowing the UDP connections to contend for space allocated for all best effort UDP and the same for TCP.

In the second method,

w_i = Average waiting time of packets in i_{th} flow since its creation.

$w_{TCP} = \sum w_i$ where i = 1, 2, …, x

$w_{UDP} = \sum w_i$ where i = 1, 2, …, y

For TCP, $m_i = m_{TCP} * w_i / w_{TCP}$

For UDP, $m_i = m_{UDP} * w_i / w_{UDP}$

BANDWIDTH DIFFERENTIATION

In the standard WiMAX architecture (Figure 2), we allocate a queue for each connection. Whether it is TCP or UDP connection, the bandwidth allocation is based on the bandwidth requirement request and final allocation done at the time of admission control. After admitting the connection, the scheduler allocates the OFDMA bandwidth resource on a per connection basis. However, the MAC is unaware of the incoming traffic rate of TCP or UDP connections. In the standard mechanism, inter-service-class scheduling has been accomplished based on deadline, priority and fairness to each packet. Within each service bin (as presented in Figure 3) of given standard architecture, there may be multiple connections. The connections in a service bin are scheduled based on an intra-service-class-scheduling.

In our proposed mechanism, we introduce intra-service-class scheduling between TCP and UDP micro-bins (Figure 5) based on the weights defined for the micro bins. The weight of each micro bin is computed dynamically based on the WFQ algorithm. Because of the implementation of intra-class-scheduling, every scheduling within a service class limits and differentiates between the TCP and UDP traffic based on the weight of the micro bins.

One major tradeoff in this mechanism is that the link layer needs to know about the transport layer information too. This increases the processing and memory requirement of the base station. Since the processing and memory capacity of base station hardware are increasing due to continuous innovation in hardware, it is worth having this tradeoff for a better QoS.

In next sub-section, we present the TCP throughput protection technique.

TCP Protection

We envisage a situation in which a minimum guaranteed throughput must be allocated to the connections in the TCP micro-bin. For example, a WiMAX network policy might require a minimum BE TCP throughput assurance. This would however require computation of the scheduling parameters (in this case, weights of the micro-bins of TCP and UDP). In our proposed scheduling mechanism, we introduce a feature to compute the weights from the WiMAX network policy for TCP protection.

Table 2. System level simulation parameters

Frequency	2.4GHz
Propagation Model	Two Ray
Propagation Limit	-111dBm
Shadowing Model	Lognormal with std deviation of 8dB
Channel Bandwidth	5MHz
FFT Size	512
Symbol Period	102.9us
Frame Duration	5ms
OFDM Symbols/Frame	48
Data OFDM Symbols	44
BS/SS Maximum Transmission Power	30 dBm
UL/DL ratio	1:1

Once the weights are computed, Algorithm 1 can allocate the required number of physical slots in the OFDMA frame structure of WiMAX. This is done in step (12) and (13) of Algorithm 1. These physical slots would be reserved for the connections in the TCP micro-bin even when the number of concurrent UDP flow increase in UDP micro-bin. Within the micro-bin, we use WFQ scheduler to allocate slots for packets queued in each connection. After scheduling of micro-bins the same set of steps are repeated for scheduling of each macro-bin. After the completion of the above process, the remaining unused physical slots (if any) can be utilized for prioritized micro-bin(s) with active queues.

SIMULATION RESULTS AND DISCUSSION

In this section, we explain the simulation setup and parameters considered.

The simulations are carried out on the well accepted and high-fidelity simulation tool Qualnet. We consider a WiMAX connection with a single BS and MS as the basic setup for this study. The TCP traffic due to web-browsing applications and FTP has a lower priority and passes mainly through BE/nrtPS service class of WiMAX. For our simulations, we assume all TCP flows passing through BE service flow. We study the effect of concurrent UDP flows over a single TCP flow from an FTP source to an MS. We use standard PHY/MAC parameters throughout the simulation as given in Table 2.

Given the simulation setup for the new scheduler, we analyze the effect of TCP under unidirectional VoIP traffic. For all VoIP sessions, we use constant bit rate CBR unidirectional UDP traffic of 8 Kbps as per the ITU-T telephony standard G.711. We considered a TCP variant based on FreeBSD's TCP, TCP-LITE for all our simulations. It supports features such as Slow Start, Congestion Avoidance, Fast Retransmit and Fast Recovery. Though the results would vary when using other TCP variants such as Tahoe, Reno etc, we would like to emphasize that, in the goal is to indicate the qualitative effect of proposed MAC QoS scheduling mechanism on the end-to-end throughput protection of BE TCP flows. To simulate the effect of increasing concurrent UDP traffic load, in these simulations we have used unidirectional VoIP sessions in the five service classes of WiMAX. Though in WiMAX, ertPS and UGS service classes are preferred for VoIP sessions, it may be noted that we are using VoIP

Figure 7. Simulation setup for proposed TCP protection architecture

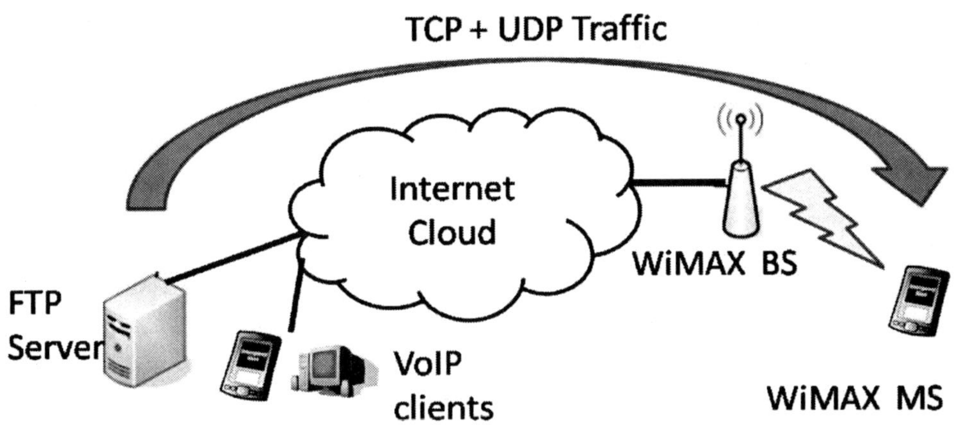

sessions in other service classes as well to reflect constant bit rate UDP traffic characteristics. The numbers of VoIP calls are increased for each simulation and the corresponding TCP goodput (end user application throughput), MAC level overall packet throughput is measured. The results are compared with a standard WiMAX architecture simulation with the reference scheduling mechanism (Figure 2). The experiment is repeated with UDP traffic passing through different kind of service classes in WiMAX. Each simulation is carried out for a duration of100 seconds. We also use a similar setup for measuring TCP protection based on WiMAX network policy based configuration. Here we analyze the behavior of TCP traffic with a fixed percentage of bandwidth allocated for the aggregated TCP flows in the TCP micro-bin.

In this section, we describe the results obtained through simulations of our proposed QoS scheduling mechanism for TCP protection and differentiation from UDP traffic for WiMAX with respect to the standard architecture. The results are categorized as follows:

A. Bandwidth differentiation aspect of the proposed scheduling mechanism (Figures 8-13).
B. TCP throughput protection aspect of the proposed scheduling mechanism (Figure 14).

In overall, we observe that, as the overall TCP goodput (application level throughput) degrades at around 201 simultaneous VoIP calls load, the proposed mechanism still shows a relative improvement of about 29% over a normal scheduling mechanism without TCP differentiation. At this point, the MAC throughput saturates and shows a relative improvement of 7.5%. We strongly believe that our proposed mechanism's unique contribution is that it can attain the dual goals of the BE TCP traffic differentiation and maximization of MAC throughput even in case of concurrent UDP flows in higher priority service classes. In case of simultaneous UDP flows sharing the BE service class, the TCP goodput degrades rapidly in normal scheduler but is remarkably stable at around 1.7 Mbps for the BE TCP flows. At 201 simultaneous VoIP calls, the relative percentage improvement shown by the proposed design is around 65.6% in this case. Also, by means of WiMAX traffic policy specification, our proposed mechanism is able to provide minimum guaranteed BE TCP goodput.

Bandwidth Differentiation Aspect of the Proposed Scheduling Mechanism

Figure 8 depicts the variation of measured application goodput (in bits per second) with increasing UDP load (in terms number of concurrent VoIP calls) in UGS/ertPS. It may be observed that, till 146 simultaneous VoIP calls, the goodput performance of the proposed mechanism closely matches that of the standard scheduling. However, beyond this point, the proposed mechanism shows a relative performance improvement. At 183 simultaneous VoIP calls load, the proposed mechanism shows an improvement of about 16% over the standard scheduling (1.74 Mbps in proposed mechanism, whereas 1.46 Mbps in standard architecture). Further, as the overall goodput falls at 201 simultaneous VoIP calls load, the proposed mechanism shows an improvement of about 29% over the standard scheduling (765 Kbps in proposed mechanism, whereas 543 Kbps in standard architecture).

Figure 9 shows the effect of increasing UDP load (in terms number of concurrent VoIP calls) in rtPS/nrtPS on the goodput. It can be seen that, in this case too the goodput performance of the enhanced maximum closely matches that of the standard scheduling till 146 VoIP calls. In the above cases, when the UDP flows belonged to the higher priority service classes (UGS, ertPS, rtPS, nrtPS), we see a relative improvement of around 16% in the goodput under congestion.

Figure 8. BE TCP goodput in presence of UDP flows in UGS/ ertPS

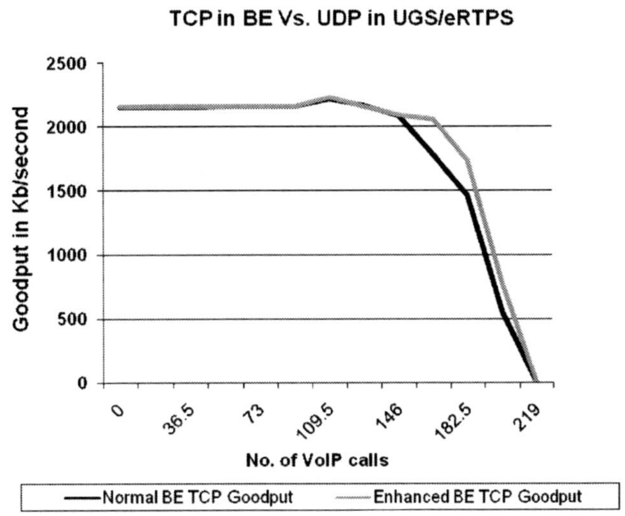

Bandwidth allocation is reserved for these prioritized flows at the admission time according to the two-step bandwidth allocation mechanism in WiMAX. Under this mechanism, the inter-scheduling-class (amongst the five service classes) resource allocation is first done and in the second step the intra-scheduling-class (amongst individual flows within a service class) allocation is done as per the WFQ algorithm. Generally BE traffic works with the residual bandwidth after other service classes are scheduled according to priority. Hence, with our proposed bandwidth differentiation and ACK prioritization mechanism, we are able to allot more residual bandwidth to the BE TCP flows. In Figure10, the goodput variation w.r.t the increase in UDP load (measured in VoIP calls) is plotted. It may be noted that, since both BE TCP flow and the UDP

Figure 9. BE TCP goodput in presence of UDP flows in rtPS/ nrtPS

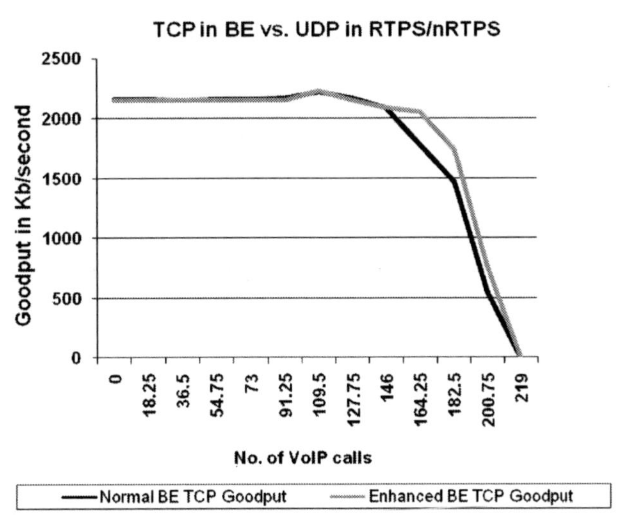

Figure 10. BE TCP goodput in presence of concurrent BE UDP flows

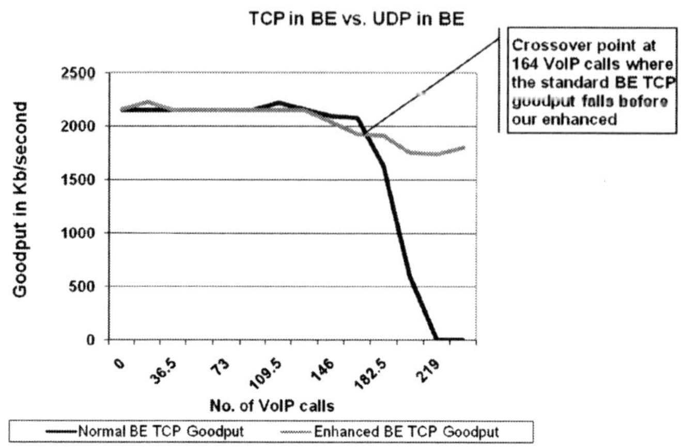

flows are competing to gain channel resources over the same service class, the goodput performance of the standard scheduling deteriorates rapidly after the threshold point of 164 VoIP calls.

However, there is a remarkable stability in the performance of the TCP goodput hovering around 1.7 Mbps. Furthermore, as the TCP goodput degrades rapidly in standard architecture, at 201 simultaneous VoIP calls the percentage improvement shown by the proposed mechanism is around 65.6%). The reason is, when both TCP and UDP traffic occupies the same service class in our proposed mechanism; it is differentiated into two categories (TCP or UDP micro-bins). After differentiation, the two aggregated micro-bins are weighted accordingly for the WFQ mechanism. Hence, based on the weights of the TCP micro-bin, a particular number of slots are always allocated to TCP. This number may be fixed or dynamically varied based on the sending rate of the source and buffer occupancy at the BS. It may also be noted that, when multiple TCP and UDP

Figure 11. Aggregate MAC throughput in presence of UDP flows in UGS/ ertPS

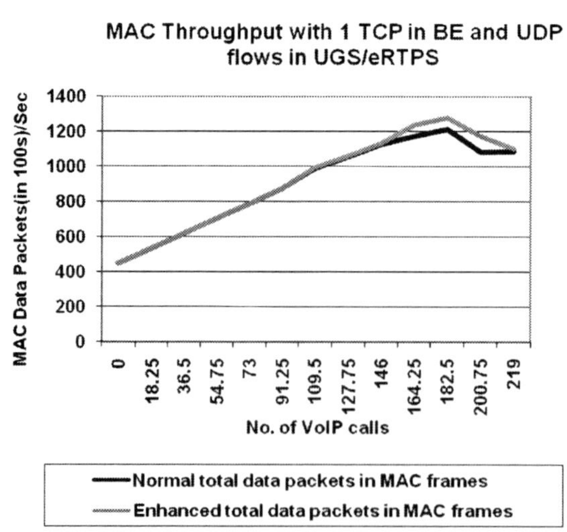

Figure 12. Aggregate MAC throughput in presence of UDP flows in rtPS/nrtPS

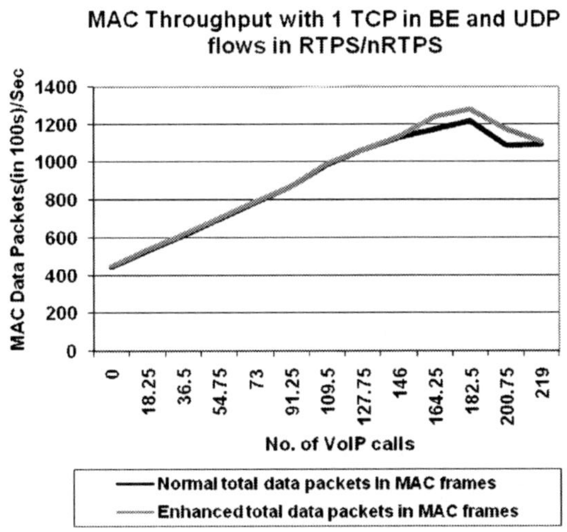

flows are competing to gain channel resources over the BE service class, a bottleneck situation gets created for the TCP ACK. In our proposed mechanism, we use TCP ACK prioritization by scheduling these packets over nrtPS service flow to relieve the TCP flow from getting compressed.

Figures 11, 12 and 13 study the performance of our proposed scheduling mechanism with respect to the MAC layer throughput achieved. In particular, Figure 11 depicts the variation of average MAC layer throughput (in packets per second) with increasing UDP load (in terms no. of concurrent VoIP calls) in UGS/ertPS service classes. It may be observed that the MAC throughput increases linearly as a function of added VoIP calls till 183 simultaneous VoIP calls are attained. Beyond this point there is a noticeable degradation (close to 10%) in the MAC throughput both in the standard

Figure 13. Aggregate MAC throughput in presence of concurrent UDP flows in BE

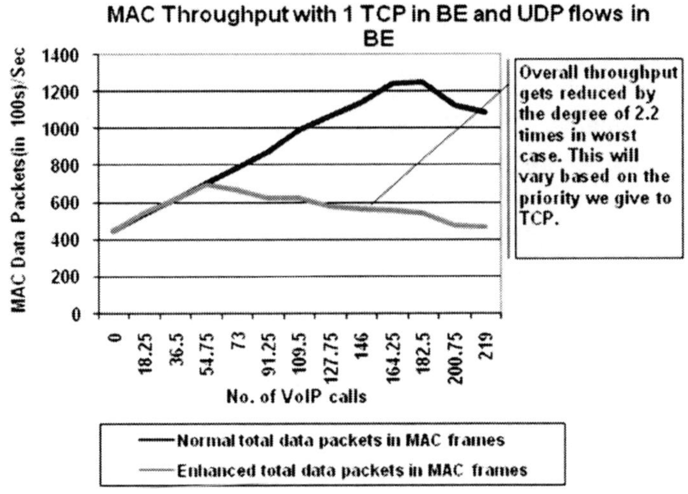

scheduling and in our proposed mechanism. This is due to the bottleneck effect induced on achievable MAC throughput due to the limited bandwidth characteristic of WiMAX channel.

However, at the peak of 164 VoIP calls, the proposed mechanism performs better than the standard ones by approximately 5% (MAC throughput for standard mechanism is 1173 packets/sec, whereas for proposed mechanism is 1239 packets/sec). Further, as the MAC throughput saturates and shows a slight degradation at 201 VoIP calls load, the relative improvement due to the proposed mechanism is around 7.5% (MAC throughput for standard mechanism is 1084 packets/sec, whereas for proposed mechanism is around 1172 packets/sec). Figure 11 shows the effect of increasing UDP load (in terms number of concurrent VoIP calls) in rtPS/nrtPS on the MAC throughput. In this case too, the MAC throughput performance is very close to that shown in Figure 10. It may be noted that our objective is to protect and differentiate TCP traffic as far as possible from disproportionate UDP load without affecting the other concurrent traffic flows. The above Figures 12 and 13 illustrate that, in case the concurrent UDP flows occupy any of the higher priority service flows (UGS, ertPS, rtPS, nrtPS), our proposed scheduling mechanism attains both its goals of differentiating the BE TCP traffic and maximizing the MAC throughput. Hence, we strongly believe, this is a special contribution of our scheduling mechanism.

In Figure 13, the aggregate MAC throughput performance with increasing number of VOIP calls over the BE traffic is depicted. In this case, however it may be observed that beyond 55 VoIP calls, the aggregate MAC throughput degrades in our proposed mechanism as compared to the standard WiMAX scheduling mechanism. It is interesting to correlate this degradation with Figure 8(c), where the BE TCP goodput was maintained at around 1.7 Mbps with marginal degradation even when the standard mechanism failed to protect TCP. Putting this fact into context, we can draw the conclusion that by protecting BE TCP goodput under contending inelastic UDP flows, we are trading off the achievable raw MAC throughput. However, the unfairness in this contention is removed with our proposed mechanism. The inherent property of TCP to not utilize the full available bandwidth is the main reason for the throughput drop in the proposed mechanism. This can be a potential research topic in future to estimate the aggregate TCP bandwidth and schedule it accordingly to increase the overall throughput.

It may also be noted that BE service class is more suitable for TCP based services and an optimal bandwidth sharing policy may be achieved to balance between the required level of TCP throughput protection and peak MAC throughput targeted.

TCP Throughput Protection Aspect of the Proposed Scheduling Mechanism

We illustrate the TCP protection aspect of the proposed mechanism by envisaging a situation where in the WiMAX network policy (for an example, if a service provider wants to give minimum 20% guaranteed protected bandwidth to TCP connection) would wish to provide a minimum guaranteed TCP throughput protection to the users. In Figure 14, policy based configuration is set for allocating minimum 20% of the total bandwidth to TCP traffic.

It may be observed that, TCP throughput degrades continuously under increasing UDP flows till it reaches the threshold of 20% of the whole bandwidth (in this simulation study, we observe that even after rapid degradation, the TCP throughput is consistently maintained above 25 – 27% of the peak throughput). In the peak load condition

Figure 14. TCP Protection through Service Provider configuration of minimum guaranteed TCP throughput

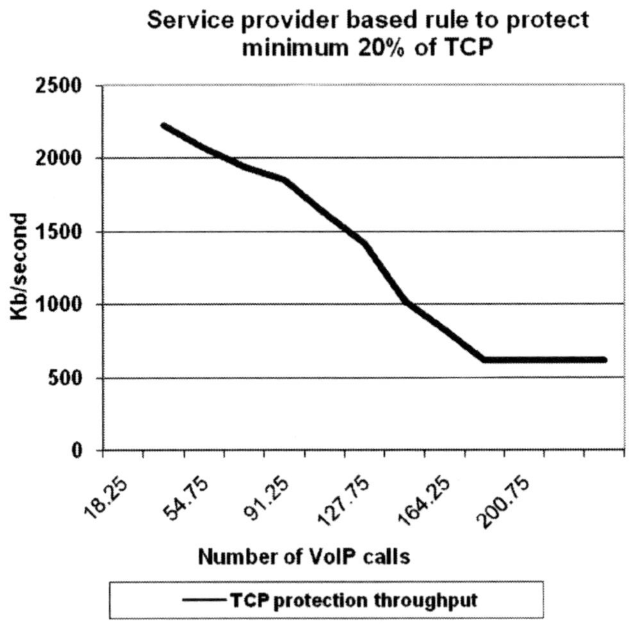

of 219 simultaneous VoIP sessions, the TCP goodput is stabilized at around 612 Kbps.

CONCLUSION

In this chapter, we proposed a dynamically adapting scheduling mechanism that provides bandwidth differentiation and throughput protection to best effort TCP traffic at multiple priority levels for WiMAX based next generation broadband wireless access network. The simulation results indicate that our mechanism is capable of providing higher bandwidth and protection to TCP in congestion conditions, even when the concurrent inelastic UDP flows occupy higher priority levels (UGS, ertPS, rtPS, and nrtPS) or equal priority levels (BE). This indicates that our mechanism is robust under disproportionate growth of inelastic traffic. Though, WFQ scheduler has a complexity of *O(N)*, we feel that at the access network this is not a major reason of concern due to the rapid improvement of computational resources at the BS. The simulations also demonstrate that the proposed mechanism is able to achieve more efficient MAC resource utilization than standard mechanism when the competing UDP flows are of higher priority levels.

Our simulation studies were designed to provide qualitative insights into the performance of the bandwidth differentiation and TCP protection mechanism. It would also be beneficial to study the energy cost of TCP throughput maximization techniques, their fairness issues and more efficient implementation of scheduling algorithms with respect to other higher or lower priority flows.

We have discussed about TCP protection mechanisms that can be implemented in MAC layer in wireless access technologies. Further comprehensive analytical modeling and optimization of resource allocation methods such as OFDMA burst computation, weight computation algorithms and subcarrier allocation are needed in the future to make the scheduling decisions more precise in assuring QoS.

REFERENCES

Bakshi, B., Krishna, P., Vaidya, N. H., & Pradhan, D. K. (1995). *Performance of TCP over wireless networks* (Technical Report 96-014). Computer Science & Engineering. College Station, TX: TAMU Press.

Balakrishnan, H., Padmanabhan, V., Seshan, S., & Katz, R. (1997). A comparison of mechanisms for improving TCP performance over wireless links. *IEEE/ACM Transactions on Networking*, *5*(6), 756–769. doi:10.1109/90.650137

Baset, S. A., & Schulzrinne, H. (2006). An analysis of the Skype peer-to-peer internet telephony protocol. In *Proceeding of IEEE INFOCOM*, Barcelona, Spain. IEEE Press.

Blake, S., Black, D., Carlson, M., Davies, E., Wang, Z., & Weiss, W. (1998). *An architecture for differentiated services*. Retrieved September 29, 2010, from http://www.ietf.org/rfc/rfc2475.txt

Braden, R., Clark, D., & Shenker, S. (1994). *Integrated services in the internet architecture: An overview*. Retrieved September 29, 2010, from http://tools.ietf.org/html/rfc1633

Bruno, R., Conti, M., & Gregori, E. (2008). Throughput analysis and measurements in IEEE 802.11 WLANS with TCP and UDP traffic flows. *IEEE Transactions on Mobile Computing*, *7*(2), 171–186. doi:10.1109/TMC.2007.70718

Capone, A., Fratta, L., & Martignon, F. (2004). Bandwidth estimation schemes for TCP over wireless networks. *IEEE Transactions on Mobile Computing*, *3*(2), 129–143. doi:10.1109/TMC.2004.5

Chaskar, H., & Madhow, U. (2003). Fair scheduling with tunable latency: A round-robin approach. *IEEE/ACM Transactions on Networking*, *11*(4), 592–601. doi:10.1109/TNET.2003.815290

Cisco Visual Networking Index. (2010). *Global mobile data traffic forecast update, 2009-2014*. Retrieved September 29, 2010, from http://www.cisco.com/ en/US/solutions/ collateral/ns341/ns525/ns537/ns705/ns827/white_paper_c11-520862.html

Cottrell, R. L., Ansari, S., Khandpur, P., Gupta, R., Hughes-Jones, R., & Chen, M. … Leers, F. (2005). *Characterization and evaluation of TCP and UDP-based transport on real networks*. Third International Workshop on Protocols for Fast Long-Distance Networks, Lyon, France: IEEE Press.

Cudak, M. (2007). *IEEE 802.16m system requirements* (IEEE 802.16m-07/002r4).

Floyd, S., & Fall, K. (1999). Promoting the use of end-to-end congestion control in the Internet. *IEEE/ACM Transactions on Networking*, *7*(4), 458–472. doi:10.1109/90.793002

3GPP. (2009). *LTE Advanced*. Retrieved September 29, 2010, from http://www.3gpp.org/LTE-Advanced.

Kalampoukas, L., Varma, A., & Ramakrishnan, K. K. (1998). Two-way TCP traffic over rate controlled channels: Effects and analysis. *IEEE/ACM Transactions on Networking*, *6*(6), 729–743. doi:10.1109/90.748085

Kim, S., & Yeom, I. (2007). TCP-aware uplink scheduling for IEEE 802.16. *IEEE Communications Letters*, *11*(2), 146–148. doi:10.1109/LCOMM.2007.061509

LAN/MAN Committee. (2007). *Part 11: Wireless LAN medium access control (MAC) and physical layer (PHY) specifications* (IEEE Std 802.11-2007).

Lee, S., Ahn, G., & Campbell, A. T. (2001). Improving UDP and TCP performance in mobile ad hoc networks with INSIGNIA. *IEEE Communications Magazine*, *39*(6), 156–165. doi:10.1109/35.925684

Mishra, M., & Sivalingam, K. M. (2008). Enhancing TCP performance in AMC based broadband wireless access networks. In *Proceedings of IEEE International Conference on Communications* (pp. 2984-2989). Beijing, China: IEEE Press.

Parekh, A., & Gallagher, R. (1993). A generalized processor sharing approach to flow control in integrated services networks: The single-node case. *IEEE/ACM Transactions on Networking*, *1*(3), 344–357. doi:10.1109/90.234856

Ramabhadran, S., & Pasquale, J. (2003). Stratified round robin: A low complexity packet scheduler with bandwidth fairness and bounded delay. In *Proceedings of the 2003 Conference on Applications, Technologies, Architectures, and Protocols for Computer Communications. SIGCOMM '03* (pp. 239-250). Karlsruhe, Germany: ACM Press.

Ranjan, P., La, R., & Abed, E. (2010). *Bifurcations of TCP and UDP traffic under RED*. Retrieved September 29, 2010, from www.glue.umd.edu/~priya/template.pdf

The Cooperative Association for Internet Data Analysis. (2010). *Analyzing UDP usage in internet traffic*. Retrieved September 29, 2010, from http://www.caida.org/research/traffic-analysis/tcpudpratio/

Todorovic, M., & Lopez-Benitez, N. (2006). Efficiency study of TCP protocols in infrastructured wireless networks. In *ICNS '06: The International Conference on Networking and Services*, (p. 103). New York, NY: IEEE Press.

Varadarajan, S., Ngo, H. Q., & Srivastava, J. (2000). An adaptive, perception-driven error spreading scheme in continuous media streaming. In *Proceedings from 20th IEEE International Conference on Distributed Computing Systems (ICDCS'00)*, Taipei, Taiwan: IEEE Press.

Wen, S., Fang, Y., & Sun, H. (2009). Differentiated bandwidth allocation with TCP protection in core routers. *IEEE Transactions on Parallel and Distributed Systems*, *20*(1), 34–47. doi:10.1109/TPDS.2008.71

Xylomenos, G., & Polyzo, G. (2003). Wireless link layer enhancements for TCP and UDP application. In *Proceedings of IPDPS '03 the International Parallel and Distributed Processing Symposium*. Nice, France: IEEE Press.

Xylomenos, G., & Polyzos, G. (1999). TCP and UDP performance over a wireless LAN. In *Proceedings from INFOCOM '99. Eighteenth Annual Joint Conference of the IEEE Computer and Communications Societies*, vol. 9 (pp. 442-446). New York, NY: IEEE Press.

Yang, X., Venkatachalam, M., & Mohanty, S. (2007). Exploiting the MAC layer flexibility of WiMAX to systematically enhance TCP performance. In *Proceeding IEEE Mobile WiMAX Symposium*. Orlando, FL: IEEE Press.

Yilmaz, S., & Matta, I. (2001). On class-based isolation of UDP: Short-lived and long-lived TCP flows. *Proceedings of the 9th International Symposium on Modeling, Analysis, and Simulation of Computer and Telecommunication Systems*. Boston, MA: ACM Press.

KEY TERMS AND DEFINITIONS

ATM: Asynchronous Transfer Mode

IP: Internet Protocol

MAC: Medium Access Control

MIMO: Multiple In Multiple Out

MPLS: Multi Protocol Label Switching

OFDM: Orthogonal Frequency Division Multiplexing

OFDMA: Orthogonal Frequency Division Multiple Access

SFDMA: Synchronous Frequency Division Multiple Access

SC-FDMA: Single Carrier Frequency Division Multiple Access

TCP: Transmission Control Protocol

UDP: User Datagram Protocol

VoIP: Voice over Internet Protocol

WFQ: Weighted Fair Queue

Chapter 8
Towards Carrier-Grade Quality in Heterogeneous Wireless Mesh Networks

Johannes Lessmann
NEC Europe Ltd., Germany

Paulo Loureiro
NEC Europe Ltd., Germany

John Fitzpatrick
University College Dublin, Ireland

Sebastian Robitzsch
University College Dublin, Ireland

Pablo Serrano
University Carlos III of Madrid, Spain

Albert Banchs
University Carlos III of Madrid, Spain

ABSTRACT

Current backhaul networks typically comprise a wired middle mile and a wireless last mile part. The wireless part is almost exclusively based on tree topologies. However, a lot could be gained by deploying mesh-based backhauls. Meshes allow better network capacity exploitation due to load balancing and offer inherent resilience to link degradations or failures. Yet meshes come with increased complexity in terms of radio configuration, routing, or mobility management. This chapter proposes architecture and mechanisms for carrier-grade mesh-based wireless backhauls. One special focus is that it supports heterogeneous backhauls, which encompass multiple different wireless technologies. The proposition has been successfully deployed in a test network.[1]

INTRODUCTION

Providing backhaul solutions for radio access networks with mobility support is very costly. Current solutions have significant limitations which will become even more pronounced with increased use of mobile data applications. New networking paradigms can result in a number of advantages when it comes to realizing such networks: mesh networks provide lower-cost solutions with greater flexibility, something that is highly desirable for operator deployments. The focus of the approach in this chapter, then, is on realizing backhaul mesh networks which can be used to deliver carrier-grade services to users, some of which are mobile.

DOI: 10.4018/978-1-4666-0017-1.ch008

Current backhaul solutions are largely divided into two categories: wired backhaul via leased lines (copper, fiber) or point-to-point, high capacity radio links. Both of these solutions have their limitations - in the first case, provisioning of the wired connection can take some time, be expensive and, in some cases, there are issues with availability of wired connectivity, especially in remote areas; in the second case, expensive radio engineers are required to install and maintain costly point-to-point backhaul links. Both of these solutions lack flexibility when it comes to deployment of new base stations in a radio access network.

Mesh networking technology represents a cost-effective and efficient alternative for realizing backhaul networks to provide mobile users with high quality services. The multi-hop wireless network architecture of mesh networks enables them to efficiently cover large areas without requiring many interconnections to a wired infrastructure. Further, mesh networks offer better capacity exploitation via statistical multiplexing effects as well as inherent resilience to link failures via higher path diversity, which ultimately results in reduced up-front cost (CAPEX) and lower network maintenance costs (OPEX) for the operator.

While mesh-based solutions can be highly beneficial for fixed backhaul scenarios, their usage is also advantageous in situations where the network infrastructure is only needed for short time periods, as is the case for large public events or emergency scenarios. In these situations, the deployment of a traditional backhaul infrastructure would require a very high investment which is usually uneconomical for such a short duration; a solution based on low-cost technologies such as Wi-Fi, deployed in mesh topologies, provides a much more cost-effective way of satisfying this short-term demand.

A critical concern for operators is to provide access to the typical service bundle via all the radio access networks offered by the operator. Specifically, access to typical voice, video and data services must be provided. This imposes some constraints on the performance of the mesh backhaul: it must support *carrier-grade* service provisioning. While carrier grade, in general, has many aspects, including equipment reliability, security, AAA, QoS, management, standard's compliance, mobility support, service integration, etc., the contribution in this chapter will focus on those aspects that are most related to reliable service delivery: specifically, on issues related to ensuring that typical carrier service offerings, such as voice, video and data, are delivered with appropriate service quality to diverse mobile devices. This is a real challenge which is not supported by current mesh network technology. Hence, we call our system approach a Carrier-grade Wireless Mesh Network (CARMEN).

Generally, there is no single optimal wireless technology for all use cases at the same time. Backhauls for mobile cellular networks are well served with point-to-point microwave radios. Communication coverage in emergency scenarios is probably better provided with WiMAX of Wi-Fi meshes. Hence, in this chapter, we propose a mesh solution that is able to combine different technologies in order to realize a heterogeneous mesh backhaul solution (see Figure 1).

In our considered scenario in Figure 1, we distinguish between CARMEN Access Points (CAPs) which are nodes that (in addition to relaying traffic within the mesh backhaul) provide wireless access to user terminals (UTs), CARMEN Mesh Points (CMPs) that are pure backhaul router nodes and CARMEN Gateways (CGWs), that connect the mesh backhaul to the operator core network or the Internet.

In summary, then, the mesh based backhaul solution has the advantage of providing lower cost, more flexible backhaul for operators wishing to provide access to mobile users. Heterogeneous technologies can be used in the mesh to adapt to the various coverage/capacity requirements. The support for carrier grade services (including broadcast services) is a key challenge that needs to be addressed in order to realize this vision.

Figure 1. CARMEN's heterogeneous wireless mesh network scenario

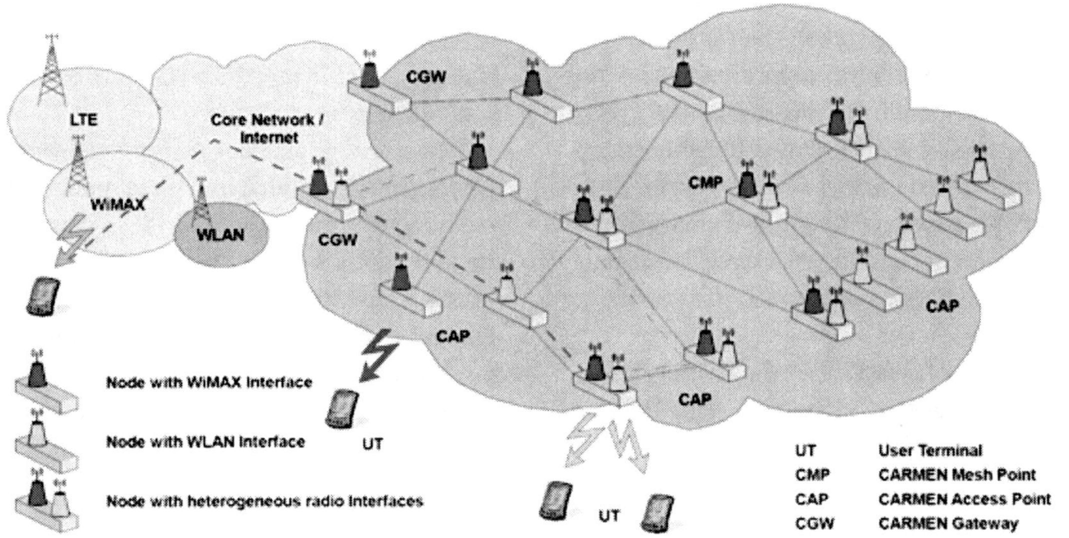

BACKGROUND

Cleary, there is a plethora of previous works in the general context of wireless mesh networks. It would not help the reader to list them all in this section, since the focus might be completely different, even though the overall subject is similar. Hence, to stay concise, we will limit the discussion to works specifically related to one of the individual key contributions of our CARMEN mesh backhaul solution.

Self Configuration

Topology Discovery and Formation

IEEE 802.1ab (Congdon, 2010), link layer discovery protocol (LLDP), is a topology discovery protocol specifically designed for IEEE 802 standards. There are three distinct characteristics of LLDP distinguishing it from all relevant studies in this area. First of all, LLDP is a one way protocol in the sense that the announcements are connectionless, passive, and unidirectional. LLDP agents do not explicitly respond to any LLD PDUs that they receive. This means that the only possible way to understand if any neighboring device received a specific advertisement is to receive another advertisement from that device in which there is information related to neighbors. However, LLD PDUs do not include any information about the neighborhood either. The main idea behind this characteristic is that LLDP's target is to provide a central network administrator, not individual devices, with the topology information.

IEEE 802.11s (Wang, 2008) also defines mesh discovery, mesh authentication, peer link management and channel selection. IEEE 802.11s supports passive or active scanning to discover neighbor mesh points. However, the focus is on building an IEEE 802 LAN (i.e. a layer-2 mesh) while our work leads to layer-3 meshes. IEEE 802.11s does not consider heterogeneous technologies.

(Li, 2006) introduces a neighborhood information based topology discovery protocol that targets limiting interference as much as possible while keeping the communication graph connected with high probability. This protocol is called the *k-Neighbor protocol.* The focus is on transmission power control and optimization. The focus in our work is on capacity-optimality and stability.

In (Hamida, 2008), the authors adopt a connectivity oriented approach to topology discovery in wireless ad hoc networks. They propose two localized topology control algorithms based on non-uniform maximum transmission power. Those algorithms are Directed Relative Neighborhood Graph and Directed Local Spanning Graph. Both algorithms are based on the graph theory and the target is to adjust transmission power such that connectivity among mesh points and bi-directionality are guaranteed.

An overview of works for topology information dissemination can also be found in (Ho, 1999).

Radio Resource Management

Generally, one can distinguish between fixed channel assignment (FCA) and dynamic channel assignment (DCA). In FCA, the set of channels are permanently allocated based on a pre-estimated traffic design (Papagiannaki, 2007),(Li, 2007). Here, we consider only DCA approaches.

DCA in cellular networks has been extensively studied and, as formulated in (Beckmann, 1999), is an NP-complete problem. Therefore, some heuristic assignment strategies and combinational optimisation tools such as Genetic Algorithms (Chakraborty, 1999), (Smith, 1998), (Banerjee, 2006). Simulation Annealing (Mathar, 1993),(Kunz, 1993), multi-colouring schemes (Sparl, 2004), (Sudeep, 2005) and neural networks (Funabiki, 1992), (Kunz, 1991), (Smith, 1997) have emerged to deliver a near optimal solution.

For DCA in IEEE 802.11 WLANs, a well-known scheme is called Least Congested Channel Search (LCCS), which (Mishra, 2005) showed to be not efficient with continued growth of WLANs. The authors formulate the channel assignment problem as a weighted vertex coloring problem and proposed two distributed algorithms that outperform the LCCS in terms of significant interference reduction. In (Riihijarvi, 2004), (Riihijarvi, 2005), the authors introduced additional techniques based on graph colouring algorithms. Other approaches are (Luo, 2004) (maximize per-user throughput), (Mishra, 2006) (maximize fairness), or (Mishra, 2005) (improve the utilization of wireless spectrum).

The existing channel assignment schemes in WMNs can be divided into three main categories — fixed, dynamic, and hybrid — depending on the frequency with which the channel assignment scheme is modified. In a fixed scheme the channel assignment is almost constant, while in a dynamic scheme it is continuously updated to improve performance. A hybrid scheme applies a fixed scheme for some interfaces and a dynamic one for others. (Marina, 2005) and (Skalli, 2007) are examples for static schemes, (So, 2004) and (Bahl, 2004) for dynamic schemes, and (Kyasanur, 2005) and (Kyasanur, 2006) for hybrid ones. Our approach is a dynamic scheme.

Radio Abstraction Layer

One of the fundamental contributions of our proposal is to introduce a radio abstraction layer in order to cope with heterogeneous wireless meshes. This layer is supposed to expose a number of technology-independent primitives to higher layers so that these can interact with the wireless interfaces in a homogeneous way.

IEEE 802.21

IEEE 802.21 (IEEE 802.21, 2010) defines a standard for handovers across heterogeneous access networks (e.g. from an 802.11 to an 802.16 network or to a 3GPP network). Part of the handover process consists in preparing the target base station to accommodate the mobile terminal – ideally before the latter's existing radio connection has been terminated. In order to enable these kinds of handovers involving multiple technologies, there must obviously be a media-independent interface allowing higher-layer functions (called "MIH Users" in IEEE 802.21) (i.e. mobility management) to obtain information from the underlying involved radios about link state measurements or to tell the target base station to allocate certain radio resources for the new terminal. For this purpose, IEEE 802.21 defines a media-independent layer called Media-Independent Handover Function (MIHF). The standard also defines a mapping between the technology-independent handover primitives and their technology-specific counterparts. Since in handovers, always multiple network nodes are involved, IEEE 802.21 also defines a protocol and transport mechanism over layer-2 or layer-3, via which primitives (i.e. handover signaling) can be transported between nodes. In Figure 2, this is called MIH_NET_SAP.

Our CARMEN radio abstraction solution is actually based on the IEEE 802.21 architecture. However, it significantly generalizes it, and adds a number of primitives to enable more comprehensive mesh functions beyond just handovers. This will be explained in more detail later.

Ambient Networks

Ambient Networks (AN) is an EU project which focuses on technology which will allow people to communicate using the most cost-effective and appropriate combination of wireless networks and services ("always best connected"). Its goal is to provide seamless communication and entertainment through wireless devices of any sort. Since it is to support all wireless technologies, one part of its solution is a so-called Generic Link Layer (GLL) which has a similar function than the radio abstraction layer defined in this chapter. However, the focus is fairly different (it is not specifically targeted at mesh networks, or carrier-grade quality, it focuses on "user-centric" vs. our node-centric in-mesh communication).

Routing and Capacity Handling

Most of the existing works that consider routing in mesh networks are targeted at distributed routing approaches, where the goal is to find paths between source and destination that are optimal with respect to a number of metrics, by deferring decisions to individual intermediate nodes or groups of nodes.

However, for carrier-grade mesh networks, localized decisions may too easily lead to suboptimal solutions on a network-wide level. Besides, we have performed studies comparing distributed with centralized routing approaches

Figure 2. IEEE 802.21 architecture

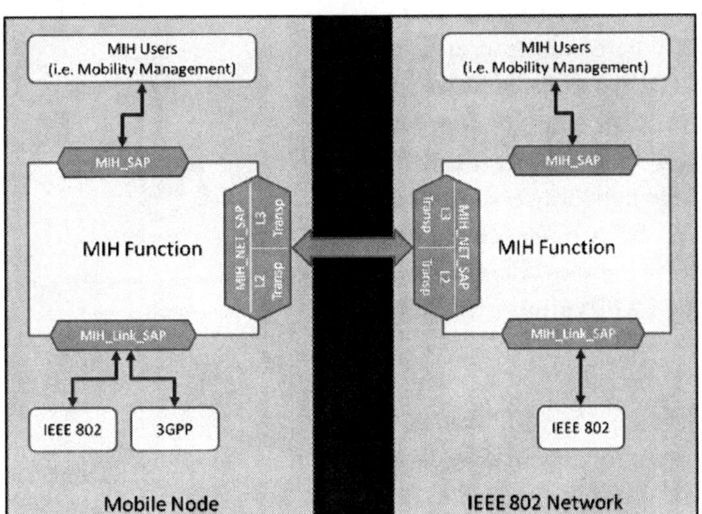

in wireless mesh networks and have come to the conclusion that convergence time after link state changes is likely to be too slow with distributed approaches. Hence, we have opted for a centralized routing solution where link state information is propagated to a centralized path computation element which then computes paths on behalf of the network nodes in a capacity-optimal way. The computation is done based on Integer-Linear Programming (i.e. leads to really optimal results for the given network situation).

Thanks to our radio abstraction layer, the central path computation element has an abstracted logical topology view of the whole network at any time. This means that its task is reduced to a pure algorithmic path search. Hence, the core contribution of routing, as far as heterogeneous wireless network support (the scope of this book) is concerned, is actually done by the abstraction layer, not by the computation element. For this reason, we skip going into long discussion of works on routing algorithms.

Mobility Management

Our mobility management solution is partly based on a well-known solution called Proxy Mobile IP (PMIP) (IETF RFC5213, 2010). PMIP is a variant of Network-based Localized Mobility Management (NetLMM) approaches, which, unlike host-based localized mobility where mobile terminals (MTs) signal a location change to the network to update routing states, provide mobility support to moving hosts without their involvement. A proxy mobility agent in the network performs the signaling and does the mobility management on behalf of the MT attached to the network. The core functional entities in the PMIPv6 infrastructure are (see Figure 3):

- Mobile Access Gateway (MAG). This entity performs the mobility related signaling on behalf of an MT that it is attached to its access link. The MAG is usually the access router for the MT (i.e. the first hop router in the LMM infrastructure). It is responsible for tracking the MT's movements in the access link. There are multiple MAGs in a Localized Mobility Domain (LMD).
- Local Mobility Anchor (LMA). This is an entity within the backbone network that maintains a collection of routes for individual MTs within the LMD. These routes point to the MAGs managing the links in which the MTs are currently located. Packets for an MT are routed to and from the MT through tunnels between the LMA and the corresponding MAG.

Work on 'MPLS Tunnel for PMIPv6' proposes Multiprotocol Label Switching (IETF RFC3031, 2001) as an alternative tunnel technology used between MAG and LMA beside IPv6, IPv4 or IPv4-UDP encapsulation headers. As MPLS has become quite popular and MPLS support is an important function for commercial edge and core routers, the integration of MPLS and Proxy IP

Figure 3. Proxy mobile IPv6 domain

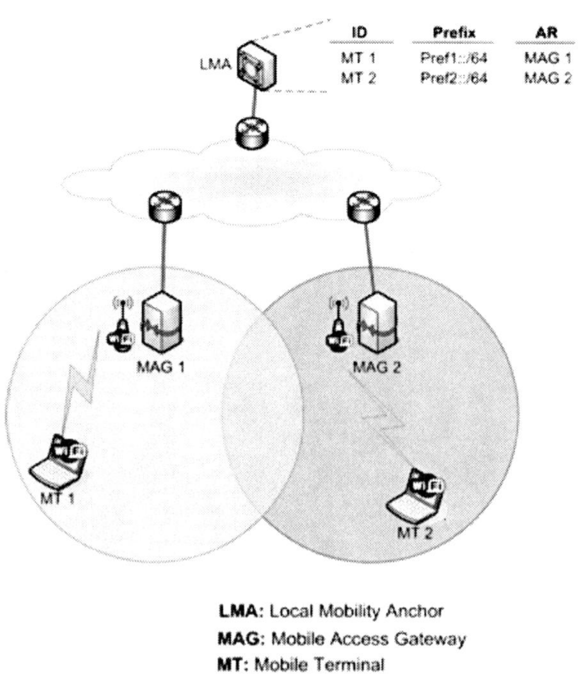

Mobility can leverage operational PMIP deployment. Our work here would actually focus on MPLS usage between the MAG (on the CGW) and the so-called Point of Attachment (PoA) (on the CAP) (i.e. be complementary).

Some proposed extensions to L3 mobility management protocol at IETF are based on approaches that distinguish between an upper hierarchical layer where wide area (macro) mobility is realized by Mobile IPv6 (MIPv6) (Johnson, 2004), and a lower level of local (micro) mobility where all nodes maintain a soft-state routing cache, mapping the source address of received packets to the node forwarding them (this enables to learn a route to the originating node). The soft state information has to be renewed periodically.

Besides Hierarchical MIPv6 (Soliman, 2005) which is specified in an RFC, there are other micro mobility protocols that were proposed at the IETF, but did not reach RFC status, namely Cellular IP (CIP) and HAWAII (Handoff-Aware Wireless Access Internet Infrastructure). Cellular IP (Valko, 1999), (Campbell, 2000) was designed to complement and interwork with global IP mobility solutions, such as MIP or PMIP, and supports paging (location management across cell borders for idle nodes) and a number of (hard and semi-soft) handoff techniques. Both CIP functions are integrated with hop-by-hop routing along bi-directional paths. Cellular IP requires each node to use exactly one CIP gateway (towards the global IP network) at a time. This is due to the Gateway's role to serve as a node's Foreign Agent which has to relay its traffic in both directions up- and downlink. It is also required to make uplink routing unambiguous. This is not suitable for mesh access networks.

Generally, the mobility management solution that we propose also distinguishes between macro- and micro-mobility. It uses standard PMIP in the operator core network (i.e. between LMA and MAG) and a PMIP-like but mesh-tailored solution in the mesh access network (between MAG and PoA). For the signaling between PoA and mobile terminal, we use IEEE 802.21, which was already introduced earlier. Details about our mobility solution will be explained in a later section.

ARCHITECTURE

General Overview

In order to enable carrier-grade services over a wireless mesh network comprising of heterogeneous wireless technologies, one of the most fundamental contributions which CARMEN makes is the introduction of a technology-independent link (control) interface. Using this interface, higher-layer modules like Routing, Self-Configuration, etc. can interact with the individual interfaces (i.e. their drivers) in a homogeneous way. The technology-independent interface is also called Abstract Interface in this chapter or *AI* for short. The AI can be used to obtain current parameters of interfaces like the currently used transmission power or modulation and coding scheme (MCS), but also to set these and other parameters, allocate resources in the underlying MAC layers, or obtain measurements from the individual radios. Figure 4 depicts the general architecture of a CARMEN node. Details of the AI will be presented later.

The AI is designed as an interface for the control and management plane of the network. The data plane in CARMEN is based on MPLS.

Generally, user traffic in CARMEN is transported in so-called *pipes*. A pipe is an end-to-end tunnel through the wireless mesh network associated with a specific QoS allocation (e.g. required bandwidth or delay). Pipes are usually dimensioned to be able to accommodate multiple user flows of a certain QoS class. This has the advantage that route discovery and resource allocation has to be done only once per pipe. Once a pipe p between a source node s and a destination node d is constructed and its resources in each intermediate node reserved, new flows between s and d (belonging to p's traffic class) can sim-

Figure 4. CARMEN node architecture

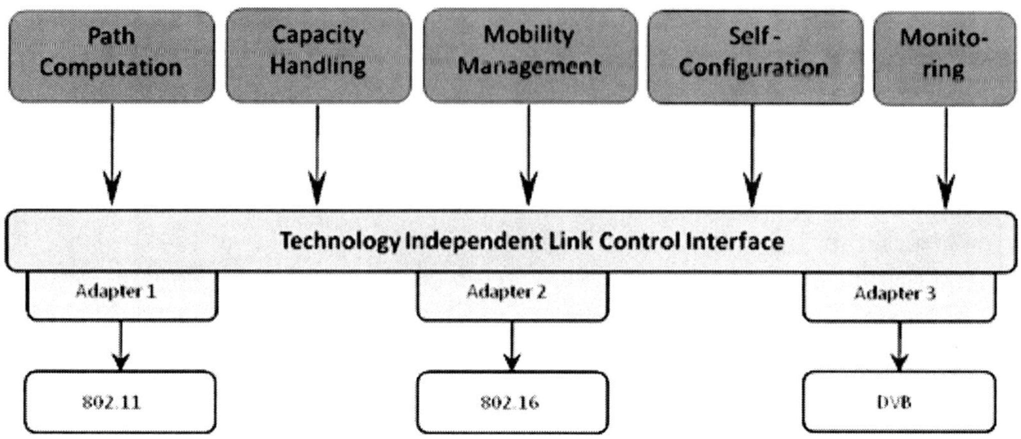

ply be forwarded along the pipe without having to consult routing again. Also, doing explicit resource reservations for pipes instead of more opportunistic forwarding (as in the case of standard IP-routing) allows controlled traffic engineering and admission control of user traffic. This prevents the network from becoming congested.

In each CARMEN node, according to Figure 4, the five software modules routing, capacity handling, self configuration, mobility management and monitoring are running. In this chapter, these modules will generally be called *mesh modules*. Routing is responsible for computing and constructing path (i.e. MPLS tunnels) within the mesh network. Capacity handling keeps track of the available capacities in the network and decides which capacities to assign to new pipes, whether to increase existing pipes or construct new ones in case the existing ones are not sufficient anymore, etc. Self configuration is responsible for bootstrapping the network and continuously maintaining the network in terms of interface (de-)activation, channel/transmission power/MCS assignment, neighbor selection, etc. Mobility management handles flow-to-pipe bindings, IP address configuration and maintenance as well as handover execution of user terminals. Monitoring takes care of performing neighborhood scanning and obtaining measurement statistics about link qualities. These modules will be described later in more detail.

Self Configuration and Monitoring Overview

The main role of Self Configuration is to provide the intelligence and configuration mechanisms to autonomously perform network setup and configuration, in particular node discovery and initial topology formation. Self-configuration also provides dynamic adaptation and optimization during regular network operation as well as during network failure. These SCF functionalities are dependent on accurate and timely monitoring functionality which can provide the SCF with the required network measurements to make informed configuration decisions and react to changes in the networks condition.

SCF and Monitoring Architecture

Although the SCF is a module in its own right, it has very tight integration with the measurement and monitoring modules, in particular Monitoring

Storage (MoMs). For this reason, these modules and their interaction with the each other will be described together in this section.

- **SCF:** Within the CARMEN system there are two variants of the SCF; a CGW SCF and an SCF module for CAP and CMP nodes. The CGW SCF is significantly more complex than the non-CGW SCF; the SCF on the CGW can be seen as the centralized entity that is responsible for computing radio configuration parameters for each node under the control of the gateway. Specifically, when computing radio configurations, the SCF must not only consider neighboring node radio configurations but also the interference between multiple radios on the same node. The SCF communicates with the MoMs module located in each CARMEN node to obtain the required information (Radio conditions, loss rates, neighbor lists etc.) to perform configuration of each node.
- **Measurement:** As shown in Figure 5 the measurement architecture incorporates four main modules, the lower layer measurement module, neighborhood scanner and Monitoring aggregation module (MoMa), and the higher layer monitoring storage module. Each of these is present on every CARMEN node. The measurement module is responsible for collecting raw measurement data from each radio interface on a frame per frame basis. Due to the high granularity of the of the lower layer measurements that this produces it is not feasible to send this vast quantity of information to the centralized SCF and thereby occupy large network resources. Rather, this information is passed locally to the MoMa which performs aggregation, processing and filtering of the raw measurements; this results in higher level metrics which are sent to MoMs for storage and forwarding to any requesting module such as the centralized SCF. The neighborhood scanner is responsible for continually monitoring the radio environment and detecting new neighboring CARMEN nodes. In particular, the neighborhood scanner plays a vital role during bootstrap and configuration of each CARMEN node. During this phase it must scan all interfaces of the bootstrapping node searching for other active CARMEN nodes and also provide interference measurements between each pair of co-located radio interfaces in the same node.

Although the CARMEN architecture incorporates heterogeneous technologies, the most difficult technologies on which to perform automatic configuration are those operating in unlicensed spectrum. The difficulty of self-configuration is further exacerbated when using low cost hardware and when co-locating multiple radios on the same mesh node.

For these reasons the CARMEN Self Configuration Function was developed to perform automated configuration for IEEE 802.11 radios. In order to maximize the system capacity the Industrial, Scientific and Medical (ISM) spectrum at 5GHz (802.11a) was used for the mesh network backhaul and the 2.4GHz band for user access; this helps to mitigate interference between mesh backhaul traffic and user access. Another advantage of using the higher 5GHz band is the larger number of available non-overlapping channels, which make it more suitable when multiple radios are co-located on the same mesh node.

However, although channels in the 802.11a band are said to be non-overlapping there are still significant interference issues between neighboring antennas, even beyond channel separations of one. In order to demonstrate this problem, to figure out the impact of this interference and the required channel separations to mitigate it, a comprehensive set of experiments were performed on a dual radio mesh node with varying antenna separations and varying channel separations over all possible

Figure 5. SCF and MoMs architecture

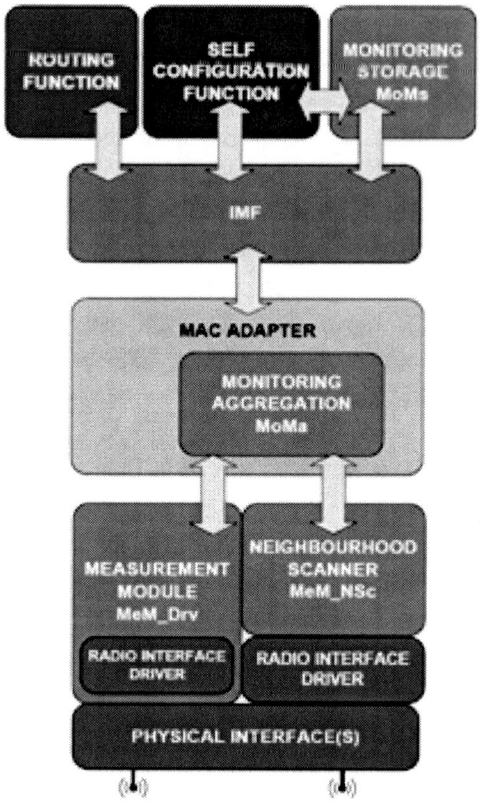

physical layer data rates. Each channel separation corresponds to a distance of 20MHz between the centre frequencies of each adjacent channel and each radio transmitting at the maximum transmit power of 15dbm to a third mesh node. In order to saturate each link, UDP traffic was generated using IPerf such that each radio was always trying to transmit at the maximum possible rate.

Figure 6, 7, 8, and 9 show the results of these experiments. Each figure shows the aggregated throughput for both radios with a fixed antenna separation and operating at the maximum transmission power (Tx Power) with different channel separations over the range of available physical layer data rates. The aggregated throughput is the sum of the throughputs for both links; assuming no interference the maximum possible aggregate throughput would be approximately 82Mbps (twice the throughput of a single UDP connection).

The general trend of these results shows that the higher the channel separation, the better the aggregated throughput of the system is. However, in each experiment the maximal expected aggregate throughput of approximately 82Mbps was not achieved until a channel separation of 7 (e.g. Operating on Channels 36 and 64) This clearly demonstrates there are other interference factors that affect the system performance and not simply the ACI effect, since the side-band of a 20 MHz channel would only affect the adjacent channel.

These results also show that the higher the physical antenna separation between co-located antennas, the higher the achievable throughput. It should also be noted that during these experiments higher antenna separations also resulted in higher link stabilities. Both of these factors are of critical importance for achieving a mesh network capable of carrying carrier grade traffic. Due to space constraints, only a small subset of the experiments performed can be presented here; further details can be found in (Robitzsch, 2009) and (Robitzsch, 2010). However, the presented results demonstrate the importance of an intelligent, interference aware radio configuration mechanism for wireless mesh networks.

Radio Configuration

Deploying such a multi-radio 802.11 Wireless Mesh Network (WMN) with stable and reliable links would require individual manual configuration of each Mesh Node (MN) based on the radio environment and the level of interference from co-located radios; this requirement greatly reduces the cost-effectiveness of such deployments. For this reason a self-configuration framework which can negate the need for manual measurements, radio planning and configuration of radio parameters such as channel, Transmission Power (TxPower), Modulation Coding Scheme (MCS) is required. As shown by (Nachtigall, 2008) the use multi-radio equipped 802.11a MNs for backbone commu-

Figure 6. Aggregated throughput for antenna separation of 20cm with Tx power of 15dbm

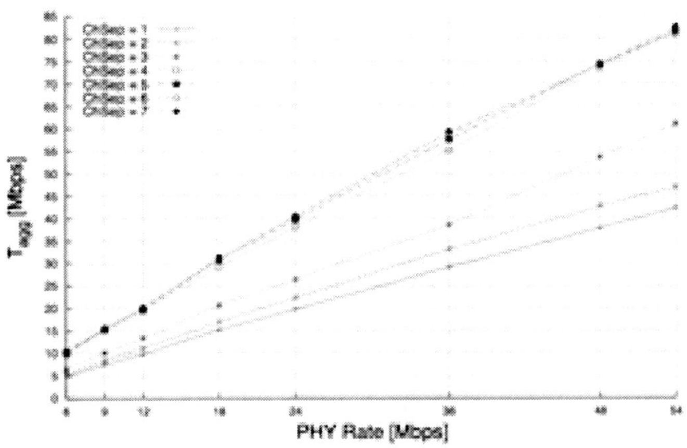

nication can greatly increase the overall WMN capacity. However, multi-radio equipped MNs suffer from significant levels of Adjacent Channel Interference (ACI); as investigated by(Congdon, 2010), (Angelakis, 2008) and(Cheng, 2006), as well as Inter-Channel Interference (ICI) described in previous work (Robitzsch, 2009). The SCF must consider each of these affects in order to compute the radio parameters for each node to minimize interference while maximizing system capacity.

In order to minimize complexity and maximize efficiency of the SCF, the implemented radio configuration takes some offline planning assumptions into account.

Firstly, it is assumed that the node density has been planned such that there is a fair and realistic relationship between the total number of available channels and the number of nodes within a range of one another. This assumption is also taken as an input parameter into the decision on the size of a Link Group (*A link group is simply a set of interfaces that share the same radio resources; this is explained in detail in the next section*). Another assumption that impacts the link group

Figure 7. Aggregated throuhput for antenna separation of 10cm with Tx power of 15dbm

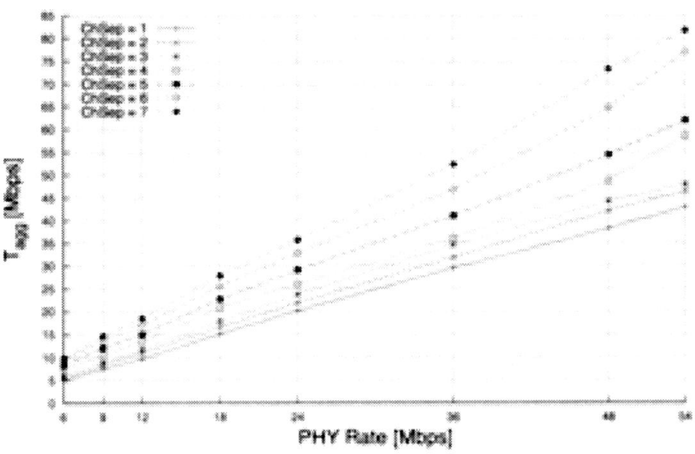

Figure 8. Aggregated throughout for antenna seperation of 30cm with Tx power of 15dbm

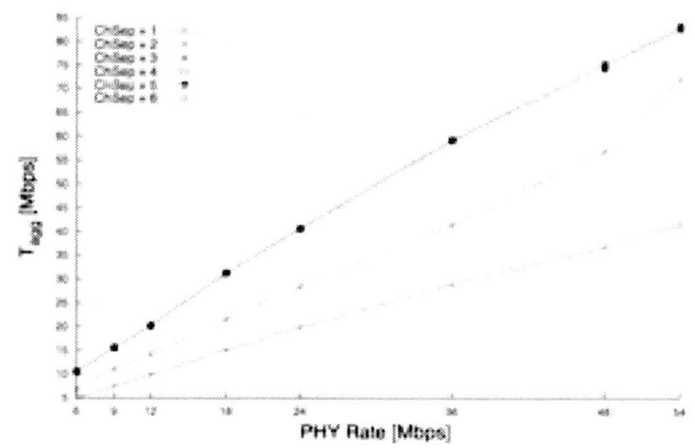

size, channel assignment and radio configuration settings is that a large number of interfaces sharing the same physical medium will lower the overall performance within the Link-Group; this is due to the uncoordinated CSMA/CA procedure used in 802.11. Therefore, the SCF always attempts to minimize the size of each link group with the optimal result being the creation of point-to-point links between neighbors. Radio configuration is performed in a per-node fashion as the nodes bootstrap.

The SCF on the CGW will instruct the new node to set the Tx Power to the maximum possible level on all available 802.11 radio interfaces in the node. Based on the level of interference between the radios in each node, as described earlier, the minimal channel separation required between the co-located radios can be calculated in order to mitigate Adjacent Channel Interference and Inter-Channel Interference.

Based on both external users and other nodes operating the CARMEN mesh, the number of available channels may not be sufficient to completely mitigate these interferences. In this case the Tx Power is decreased such that at least one potential neighbor can be reached. Once the minimum channel separation is known the channel selection is computed using an Edge-coloring ap-

Figure 9. Aggregated throughput for antenna separation of 40cm with Tx power of 15dbm

proach. If there are not enough available channels the assumption of creating point-to-point links is changed to create a link group of three radio interfaces or more.

Radio Abstraction Layer

Architectural Components

While Figure 4 showed a high-level representation of the AI, Figure 10 gives a more detailed depiction. As can be seen, the AI is internally more than a simple API exposing technology-independent event or command primitives. Rather, a significant part of the AI is defined by the so-called *Interface Management Function* (*IMF*). The main features of the IMF architecture are

- Inter Process Communication (IPC) down- and up-calls between mesh modules (routing, self-configuration, etc.) and IMF
- IPC down- and up-calls between IMF and Interface Adapters
- Communication between different mesh modules via the IPC Mechanisms offered by the IMF
- Collaboration of the IMF Message Dispatcher and the IMF Node Manager Function
- Support for remote event subscription/reception

The CARMEN IMF borrows concepts for linking modules inside a protocol stack based on adjacent layer protocols as this is described in the OSI Open Systems Connection Model (Chakraborty, 1999), (Smith, 1998). The IMF building blocks include Service Access Points (SAPs) to define service primitives issued by CARMEN mesh modules. Generally, two types of primitives which can be exchanged between mesh modules and wireless interfaces (via IMF) can be distinguished. These are event and command primitives. Examples for event primitives are LINK_UP (a new link to a neighboring node has been established), LINK_PARAMETER_CHANGE (the QoS pa-

Figure 10. IMF architecture and SAPs

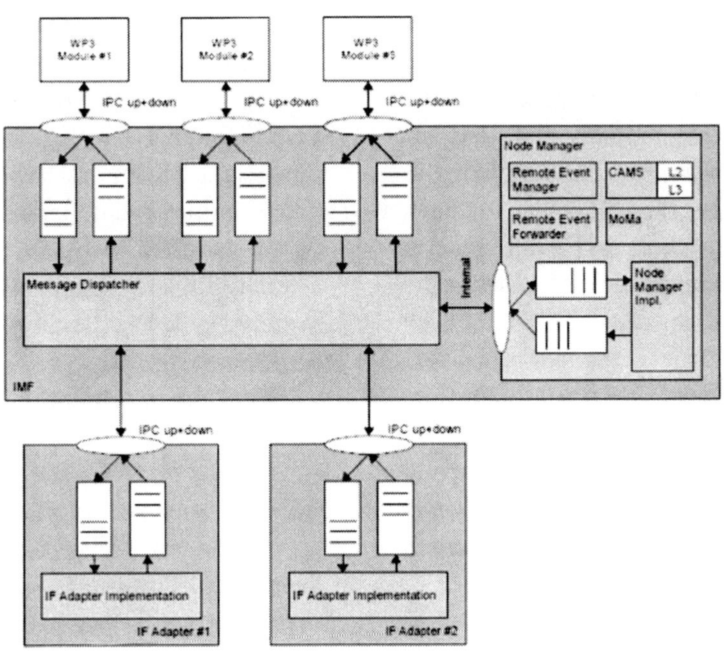

rameters of an existing link have changed), or NEW_FLOW (a new user flow has been detected at an interface). Examples for command primitives are ALLOCATE_RESOURCES (reserve QoS resources like capacity in the MAC layer of an interface), or SET_RADIO_PARAMETERS (to configure the radio parameters of a radio card like MCS, channel or power).

Actually, it has been identified that some of the required primitives are already included in the IEEE 802.21 standard (IEEE 802.21, 2010). This standard also defines a media-independent interface, but is restricted in its scope to handovers across heterogeneous networks. Thus, the only expected user (mesh module in CARMEN terminology) of the IEEE 802.21 interface is mobility management. Hence, primitives beyond handover needs are missing and the interface management capabilities of IEEE 802.21 are not as generic as CARMEN's.

While event primitives are normally unidirectional by nature (i.e. going only upwards from the interface to mesh modules), command primitives are usually bidirectional and involve a request and a response part. Therefore, when bidirectional service primitives are invoked or received by mesh modules, they need to be accompanied by transaction IDs (correlators) for allowing mesh modules to relate responses with requests. This will be done by the mesh modules and transparently forwarded through the AI and included in responses. Furthermore, the IMF needs to attach SAP IDs to allow response and event service primitives to be forwarded to the originator of the request. Each mesh module has its distinct SAP ID which is assigned to it by the IMF upon IPC attachment. The validity of SAP IDs is for the time of attachment of the mesh modules. SAP IDs are not reused for a long time in order to avoid wrong assignments of responses, indications, and events to requests in cases where a more dynamic mesh node setup is used in the future.

The IMF Node Manager Function is a node wide information store for information pertaining to the node interface management (e.g. number and IDs of interfaces). This ensures the correct referencing of interfaces in the IPC communication with mesh modules. The Node Manager also hosts other node-wide functions such as Remote Event Managers, CAMS messaging service, and Monitoring aggregators (MoMa).

The Remote Event Manager is responsible for sending, relaying and receiving remote events and commands. In CARMEN, events can not only be received locally in a node. Rather, they can also be forwarded to arbitrary other nodes and received by those nodes' mesh modules. Likewise, commands cannot only be sent to local interfaces via the IMF, but also to interfaces of remote nodes. The Remote Event Manager is responsible for the remote communication. Generally, in order to receive events, mesh modules have to subscribe to them in the IMF by specifying the event type as well as the originating node (be it the local or a remote one). Then, once an event is received at the IMF, the latter will propagate it to all mesh modules which had subscribed for it previously.

In order to propagate events to other nodes, layer-2 or layer-3 transport is employed depending on whether the target node is a direct neighbor of the originating node or not. An entity called *CARMEN Messaging Service* (*CAMS*) (cf. Figure 10) is responsible for the transport of remote events or commands. Basically, CAMS is similar to the transport mechanisms defined in IEEE 802.21 where remote event signaling is also supported via the so-called MIH_NET_SAP.

Finally, the IMF encompasses an entity called *Monitoring Module Aggregator* (*MoMa*). In the technology-specific Interface Adapter, there is a Measurement Module (MeM) for obtaining measurement data in real-time about link stability, error rates, RSSI and SNR values, etc. The MoMa queries and aggregates this data in order to compute more expressive link and interface metrics out of sequences of measurement values. These metrics in turn can be queried my mesh modules for cross-layer optimization of protocols.

Mapping Technology-Specifics to Technology Independence

One of the most critical and difficult tasks of the radio abstraction layer is to map the technology-independent primitives to technology-specific ones and vice versa. This goes far beyond syntactical transformations, since, generally, it cannot be assumed that there is a one-to-one mapping for each event and command in each target technology. A fundamental problem in carrier-grade wireless mesh networks is, for example, the radio resource management and admission control. The admission control needs to have up-to-date information about the available resources in the network, in order to decide whether a new flow can be admitted to the network or not. However, the amount of available resources changes dynamically because of wireless link fluctuations and added or removed pipes (which have dedicated resources allocated to them). This means that admission control (as well as capacity-aware routing) needs a technology-independent link state model of the network topology. The link information includes at least the available (residual) link capacity, but could also include loss, delay or jitter.

In the following, we will present such a technology-independent link capacity model. This is to serve as an example for the general mapping complexity of the radio abstraction layer.

In our model, *logical links* are the technology-independent representation of radio links as exposed by the CARMEN AI to all higher layer functions. We use the term *logical*, because there need not be a one-to-one correspondence to physical links below the AI. We assume logical links are directional and point-to-point, as

- routing and capacity handling need to be aware if a technology only supports unidirectional communication (e.g. DVB-T) or if up- and downlink capacities are asymmetric (e.g. WiMAX),
- a point-to-point model simplifies link modeling and link QoS description (and thus the AI), and
- this model makes the mesh topology appear like a wired network, which enables the reuse of protocols designed for wired networks for benchmarking purposes.

A difference between a "wired" and a "wireless" point-to-point link is that in the former case a transmission is only received by the other end-point of the link, whereas in the latter case the transmission is often received by multiple (intentional or unintentional) receivers. This needs to be modeled, because it means the throughput capacity of a logical point-to-point link may depend on the traffic or resource allocation of some other logical point-to-point links. Here, this is modeled by the concept of a *logical link group* that describes the set of logical links that share the capacity available to the group as a whole.

More formally, we can model the CARMEN mesh network as a directed graph (=digraph) $G := (V;E)$, where V is a set of vertices (CARMEN Mesh Points) and E a set of directed edges (the logical links). A directed edge $e \in E$ is an ordered pair of vertices, i.e. $e := (v_1, v_2)\ ;\ v_1, v_2 \in V$, and $p(e) := v_1$ is called the predecessor of e and $s(e) :=$ v_2 the successor of e.

A logical link group is a set of logical links and each logical link in E belongs to exactly one logical link group. In other words, E can be partitioned into a set of logical link groups Λ for which

$$\bigcup_{L \in \Lambda} L = E \wedge \forall L_i, L_j \in \Lambda, L_i \neq L_j : L_i \cap L_j = \varnothing$$

Let $L(e)$ denote the logical link group that logical link e belongs to and $C(L)$ the *capacity* of a group L. We measure the capacity of a logical link group in a technology-independent manner in terms of "scheduling units per second" ($\left[\frac{SU}{s}\right]$).

These scheduling units could, for example, be

TXOPs in IEEE 802.11 or time-frequency-slots in OFDMA, or could be just symbols and they can vary from link group to link group. The capacity should already be the net capacity (i.e. discounting for protocol overhead). Note that in one link group, as here defined, only one radio technology can be employed at any given time.

Further, let *c(e)* denote the current cost of transmitting to the receiver *s(e)* expressed in terms of "scheduling units per bit" ($\left[\frac{SU}{b}\right]$). This cost can be seen as the price paid, and deducted from the total wealth *C(L)* available for that group, to transmit at link *e* at that particular instant. It includes the efficiency of the modulation and coding scheme used to match the current radio link *e*'s quality.

Finally, let's assume *F* is a set of flows already admitted to the network by admission control. Let L_f be the subset of links of logical link group *L* ($L_f \subseteq L$) that flow *f* needs to use to reach the intended receivers of a given transmission. L_f can then be used, for example, to model a multicast transmission directed to those receivers. Note that in this model a unicast transmission can be considered as a special case where the cardinality of L_f is 1. The traffic demand imposed by flow *f* will be defined as r_f (and measured in "bits per second" ($\left[\frac{b}{s}\right]$). The *effective cost* of a flow to a particular set of receivers, assuming a particular admission control allocation, $c_{eff}(f)$, can then be computed and it depends on whether the underlying physical technology supports native multicast or whether multicast has to be emulated by separate unicast transmissions to each receiver. In the first case, the effective cost is equal to the highest cost of any point-to-point link in L_f, because the modulation and coding needs to support the weakest multicast receiver. In the second case, the effective cost is equal to the sum of costs of all point-to-point links in L_f. Therefore, we have:

$$c_{eff}(f) = \begin{cases} \max\limits_{e \in L_f}\{c_e\} & \textit{native multicast} \\ \sum_{e \in L_f} c_e & \textit{emulated multicast} \end{cases}$$

Given the information defined above, routing and capacity handling can then test in advance and remotely, how an admission control or routing decision would influence the remaining capacity in the network by computing a link group *L*'s remaining capacity *C*(L)* and available transmission rate for link *e*, *r*(e)*, at cost *c(e)* as:

$$C^*(L) = C(L) - \sum_{f \in F} c_{eff}(f) \cdot r_f$$

$$r^*(e) = \frac{C^*(L)}{c(e)}$$

A discussion how these values can be computed for specific technologies such as Wi-Fi or WiMAX must be omitted due to space limitations here. However, concrete mappings have been done for Wi-Fi, WiMAX and a non-standardized Soft-TDMA MAC.

Routing and Capacity Handling

This section provides an overview of the CARMEN routing function, the Capacity Handling Function (CHF) and the Forwarding Function (FF).

Routing Overview

Routing in a heterogeneous resource-constraint wireless mesh network requires sophisticated algorithms and mechanisms to efficiently manage the available resources and provide the high levels of guaranteed QoS required by end users. The CARMEN approach aims at reducing the CAPEX required to deploy carrier-grade networks by providing carrier-grade network availability and reliability utilizing lower cost "off the shelf" equipment.

For this reason CARMEN incorporates both an advanced admission control mechanism and a centralized QoS aware routing mechanism that is capable of providing the high levels of guaranteed QoS required by carrier grade services. The CARMEN routing function is responsible for determining how to populate and manage forwarding tables for different traffic classes and is responsible for the capacity management in the network. Unlike other mesh networks the CARMEN network must deliver carrier grade performance with guaranteed levels of QoS across heterogeneous access networks. Furthermore, to allow for triple-play service provisioning, the CARMEN routing architecture supports unicast as well as 1-to-N multicast path computation and setup. In order to support Mobile Terminal mobility and QoS differentiation while maintaining scalability, individual flows are aggregated into pipes, which are configured between any two CARMEN nodes. The routing mechanism must meet these requirements and therefore must have support for multipath routes, the ability to guarantee resources, and it must consider how traffic flows across different paths impact one another.

To achieve these requirements the CARMEN routing architecture must maintain real time up-to-date state information about the available links, their committed resources and QoS requirements and to efficiently route traffic across the network. Considering the relatively static (non-mobile) nature of a CARMEN mesh network and to reduce complexity in mesh points a centralized routing mechanism was opted for. The centralized approach allows the gateway to have global knowledge of its routing area thereby allowing it to make informed and more capacity-optimal decisions.

Figure 11 shows the performance of our centralized routing. The routing algorithm is a Linear Program that basically tries to maximize the number of admissible flows in the network by evaluating all possible candidate solutions and selecting the best one. A novelty of our LP formulation is that it includes constraints that model capacity inter-dependencies of different links such as links sharing the same spectrum. We compare our routing approach with previous, more distributed approaches such as ETT based, ETX based or shortest path based routing with a varying ratio of nodes that act as source nodes (i.e. produce traffic).

Obviously, the debate on whether to adopt a centralized or a distributed routing approach has

Figure 11. Capacity gain through centralized link group aware routing

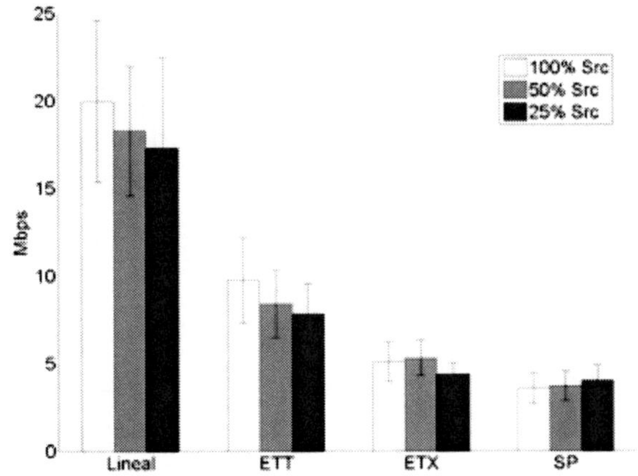

existed for a long time. While classical backhaul networks typically adopt a centralized management approach, research in the wireless mesh community generally focuses on distributed approaches. However, the convergence time of standard distributed routing protocols such as OLSR or OSPF suggests adopting a centralized approach. Our simulation results based on grid topologies in Figure 12 show that OLSR or OSPF variants (PMP, MANET, Broadcast) need in the order of 7 seconds and more to converge after link failures in the network. This is certainly too much for efficient carrier-grade failure recovery.

Unlike a decentralized routing mechanism such as OSPF link state updates in the centralized approach do not require the flooding of messages in to the network, rather updates are transmitted directly to the default gateway of each node. This greatly reduces the overhead required for performing updates, particularly if there are a high number of pipes being setup. It also negates any problems that may be experienced with multiple nodes attempting to perform pipe setup simultaneously.

The routing architecture builds upon existing technologies such as MPLS-based Traffic Engineering and PCE-based path computation. CARMEN has incorporated adaptations to address the more volatile nature of the wireless technologies supported and to handle with the uni-directionality of links. Specifically, Pipe setup, teardown and modification are performed using RSVP-TE to establish MPLS labels at each node. When the routing function of a mesh node receives a pipe request it must forward the request to its default gateway. Once the gateway has computed a route it will reply, providing the path through the network over which the pipe must be constructed. The node can then establish the pipe over the route defined by the gateway using the optional Explicit Route Object (ERO) of RSVP-TE. This allows the pipe to be constructed over any route across the network and allows multiple pipes with different routes between the same endpoints.

Routing Architecture

Due to a centralized routing solution being used in the CARMEN network, the routing module being used in each node type has some slight differences, with most of the complexity existing in the routing module of each CGW. As can be seen from Figure 13 each CARMEN routing module contains a management element and a path setup element, however only a CGW contains a routing database and the Pipe Computation Element.

Figure 12. Convergence time for distributed routing protocols

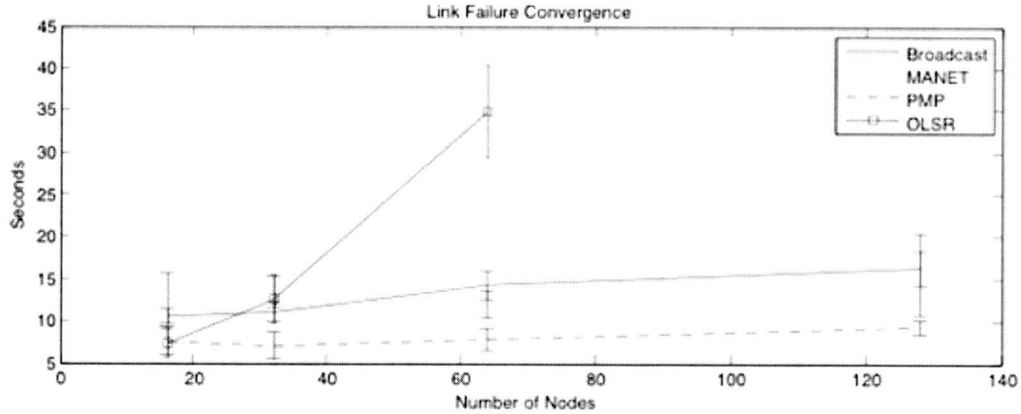

The functionality of each of the routing components is summarized as follows:

- **Management:** The Management component is responsible for maintaining the system state and acting as the interface to receive and send primitives to the IMF. Essentially the Management block contains the event handlers for each received primitive.
- **Routing Database:** The routing database is present only in each CGW and must maintain an up to date logical link view of the network topology, including installed pipes and allocated and available resources. The data in the routing database is used by the PiCE to compute optimal flow distribution in the network and to serve pipe requests from the CHF. For this reason it is critical that the database always has accurate and up to date knowledge about link states within the network. This is achieved by having the routing management component subscribed to receive link events from all links within the routing area of the CGW and maintaining data of all installed pipes.
- **PiCE:** The Pipe Computation Element (PiCE) does not maintain any state information but rather utilizes the data in the routing database. When a request is received by the management component an up to date view of the network topology is passed to the PiCE, which computes a path through the network that can meet the requested traffic requirements. If a path can be successfully computed it is returned to the management component as a set of links through the network. It is then the responsibility of the management component to update the database and request that the pipe be installed.
- **Path Setup:** The Path Setup component is present in each CMP and CGW; it is responsible for the installation and teardown of CARMEN pipes between two nodes in the network as instructed by the Management component in the CGW. Each CARMEN pipe is represented as an MPLS LSP; in order to setup a pipe the Path Setup component installs the MPLS forwarding state and resource allocations at each node through which the pipe traverses.

Figure 13. Routing architecture

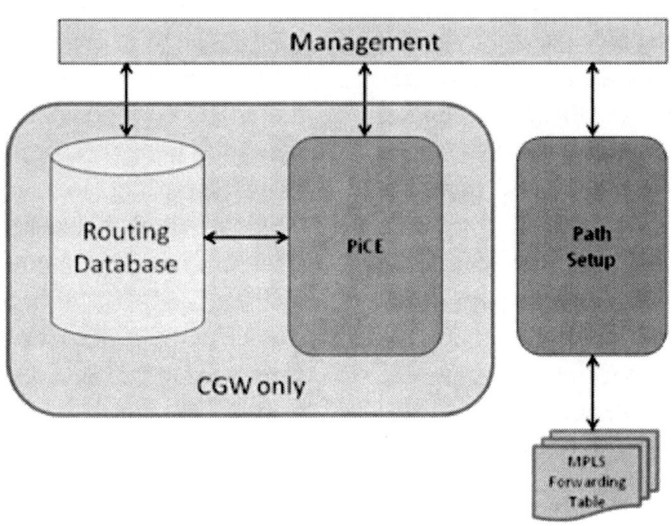

Capacity Handling

Although the Capacity handling Module (CHF) is a module in its own right and not a sub module of routing, there is a large amount of dependency and interaction between the CHF and routing. For this reason it is logical to include a description of the CHF in this section. It should be noted that the CHF module is present on every CARMEN Access Point.

In order to meet the strict QoS constraints required by carrier grade services explicit resource management techniques like admission control are needed to efficiently utilize network resources. It is the responsibility of the CHF to implement this functionality and act as the interface between the Mobility Management Function and the Routing Module. The CHF receives new flow requests from the mobility management module and must make pipe based admission control decisions for each new flow and to request new pipes from routing when required.

The CHF module has two primary functions; these are admission control and pipe initiation and maintenance.

Admission Control

CARMEN subscriber flows of the same traffic class and between common ingress and egress points are aggregated into the same pipe assuming that the pipe has available capacity. A flow request originates from the local Mobility Management Function (MMF) and then, on reception of this request the CHF checks if the flow can be admitted into an existing pipe, based upon the resources available. If so the CHF informs the MMF of the label to which the flow should be assigned; otherwise the CHF must request a new pipe from routing.

In order to be able to accurately assign new flows to existing pipes without invoking routing, admission control maintains a record of the available and allocated resources on all pipes which ingress or egress on the local CARMEN Access Point.

Pipe Initiation and Maintenance

Pipe Initiation and Maintenance is responsible for providing the pipe data rates satisfying predefined QoS requirements in cooperation with the Routing Function. On receipt of a new flow request from the MMF for which no available pipe exists, the CHF must request a new pipe from the routing function. This triggers the routing module to set up a new pipe or modify an existing pipe to meet the capacity and QoS requirements of the newly requested flow.

The Routing Function also signals the CHF when a certain pipe capacity or QoS has changed. This happens due to degraded or failed wireless links. Pipe Initiation and Maintenance then tries to redistribute the flows of this pipe to other pipes of the same traffic class and with the same endpoint, otherwise it will request that the Routing Function set up of new pipe/pipes and/or terminates excess flows. Note that this recovery process is automated. While this may seem self-understanding for the wireless mesh networks community, this is actually a big step for network operators, who are used to static route placement and repair.

Forwarding Function

This function is integrated into all CARMEN nodes and is responsible for forwarding packets (deciding their next hop) based on information in its forwarding table. Entries in the forwarding tables are established, removed and modified on request from the Routing Module. Forwarding inside a CARMEN network is done on a per-pipe basis, thus all packets carried by a given pipe are forwarded in the same way, with packets assigned to pipes based on their class of service. Packets are forwarded differently depending on the class of pipe to which they are assigned thereby providing different classes of service to different user flows.

In order to avoid an overly complex solution the CARMEN specific forwarding functionality is implemented on top of existing implementations of the forwarding module in current Linux operating systems. This implies that the standard interfaces and protocols between layer 3 and layer 2 are left unmodified; including the IPv4/IPv6 addressing schemes, the standard address resolution protocols (ARP/NDP) and the control message protocols (ICMP/ICMPv6) and are therefore still part of the protocol stack in each CARMEN node.

As stated earlier, traffic in the CARMEN mesh is transported through pipes into which traffic with the same QoS constraints and endpoint is aggregated. Although in the current implementation of pipe traffic forwarding is done using MPLS, the architecture is general enough to support other tunneling schemes such as IP in IP encapsulation just.

The CARMEN Routing Function ensures the forwarding tables are built such that traffic aggregates will be forwarded to the next-hop based on the required QoS constraints. For this purpose the aggregates use MPLS labels which are also use as the pipe identifiers; these indicate to layer 2 the type of treatment a packet should receive in order to meet the QoS requirements. At intermediary mesh points, incoming MPLS packets are decapsulated from the MAC frames, passed through the look up process and delivered to the appropriate interface toward the next hop node. The outgoing interfaces encapsulate the packet into the specific data link technology. The MAC layer will map the pipe IDs (which in the current implementation correspond to the MPLS labels) to traffic classes and schedule the packets according to their priorities.

Mobility Management

This section provides an overview of the CARMEN Mobility Management Function (MMF), as well as presenting general architecture aspects and relevant evaluation results

Overview

Mobility Management has the tasks of maintaining radio link layer connectivity of a given QoS between UTs and their points of attachment (PoAs) to the network as well as re-establishing the UTs' network layer connectivity after a change of PoA. A change of PoA may become necessary for various reasons such as:

- Degradation of access link quality below a given threshold due to movement of the UT or channel fading
- A UT's demand for a change of radio technology (e.g. to obtain a higher data rate or to conserve energy by choosing a more efficient radio technology)
- The need to load balance data traffic across PoAs for capacity reasons

In a carrier-grade mesh networking context, when backhaul is provided via multiple wireless hops, Mobility Management needs to further take into consideration the availability of capacity on the pipes between CAPs and CGWs as well as failures of backhaul links. To allow for an autonomous operation of the CARMEN Mesh in case of intermitted connectivity to the core network and to improve the scalability of Mobility Management, a hierarchical approach is considered, in which a micro-mobility solution handles mobility inside the CARMEN Mesh and a macro-mobility solution handles mobility across gateways to the core network. The latter should integrate well with existing macro-mobility solutions of the core network operator (e.g. PMIPv6).

Since PMIPv6 is the protocol that CARMEN adopts for mobility management in the core operator's network, and PMIPv6 only provides limited multi-homing support to multi-mode mobile devices, PMIPv6 extensions were also designed that allow the UT to use multiple interfaces simultaneously for flow mobility. These extensions were proposed at the IETF in the form of drafts.

As a result of the carrier-grade mesh requirements, as well as the fact that we consider heterogeneous mesh networks, the CARMEN Mobility Management solution

- Supports micro-mobility operation despite possibly intermitted connectivity between CARMEN Mesh and the core network
- Is network-centric (i.e. handover decisions are performed by the CARMEN mesh). To perform this task, information gathered by UTs is used
- Considers both access link and mesh capacity in its management decision
- Is based on a handover technology that supports heterogeneous technologies (i.e. IEEE 802.21)
- Supports seamless handovers for IEEE 802.21-based UTs
- Incurs only low signaling overhead

Furthermore, our mobility management solution

- Provides basic support for UTs without IEEE 802.21 functionality (e.g. using technology-specific solutions to enable interaction between UT and network during handover such as IEEE 802.16e specific mechanisms or IEEE 802.11k and IEEE 802.11v)
- Decentralizes HO decisions.

The CARMEN micro-mobility solution requires several supporting functions

- An Access Status Input provisioning that reports the status of the access links from candidate PoAs
- A Handover Preparation function that collects the information required to prepare a handover (i.e. to create list of candidate APs and the correspondent resources)
- A Handover Target Decision function that decides the target AP to which a UT should be attached and
- A Handover Execution function that performs all the procedures

The scenarios considered might be composed of up to thousands of nodes. Due to scalability and performance concerns, each CGW manages a certain amount of CAPs. CAPs are assigned to CGWs in such a way that the distance – in number of hops – between a CAP and its CGW is less than a certain threshold. This helps achieving a certain level of performance in terms of throughput and delay. The set of CAPs and associated CGW, together with the CMPs that interconnect them can be considered a different *Localized Mobility Domain* (*LMD*). Support for a UT moving from a CAP belonging to one LMD to a CAP belonging to different LMD (that is, to a CAP managed by a different CGW), is provided by the hierarchical scheme adopted, in which a micro-mobility solution handles mobility inside the LMD and a macro-mobility solution handles mobility across gateways to the core network. Standard PMIPv6 is adopted as the macro-mobility solution, whereas an extended PMIPv6-alike solution is used as the micro-mobility domain.

Depicted in Figure 14 is an example of a CARMEN mesh network attached through several CGWs to an operator's core network. The CARMEN network uses several CAPs to provide access to UTs. The CMP's found inside the mesh network are transparent to the mobility management and thus the mesh path from CAP to CGW is seen as one single hop, from the mobility management point of view.

At the top level of the hierarchy we assume the operator runs standard PMIPv6 which deals with handovers between CGWs of the same CARMEN mesh network and it is also the point of anchoring of the UT's Home Network Prefix (HNP), keeping track of the corresponding CGW and establishing a bidirectional IPv6 over IPv6 tunnel between the Local Mobility Anchor (LMA) and the CGW, like in standard PMIPv6. This presents particular advantages from the point of view of an operator that already runs PMIPv6 in its core network, to which a CARMEN mesh network can be attached through several CGWs.

Below this level, there are the CGWs which have Mobile Access Gateway (MAG) functionality towards the core network and enhanced LMA functionalities towards the mesh network. The CGWs acts as the mid-level in the hierarchy and must anchor the location of the UT regarding to which CAP it is attached in the edges of the mesh and also the respective LMA located in the core.

Finally, at the bottom level of the hierarchy the CAPs run enhanced MAG capabilities.

Mobility Management Architecture

The architecture of the mobility related modules consists of three main parts. The standard Proxy Mobile IPv6 (PMIPv6) protocol running in the core, the Mobility Management Function (MMF) running at the CGWs and CARMEN Access Points (CAPs) and an IEEE 802.21 stack running in the access part of the mesh (CAP and User Terminal, UT).

A detailed view of the MMF interactions with other modules, as well as with other mobility related modules is shown in Figure 15. It also depicts which parts of the software need to be running on different nodes.

The IMF provides the interface to communicate remotely and locally with other mesh modules. From the IMF, the MMF must process and dispatch any arriving messages from other mesh modules and send any pending messages to be delivered by the IMF to other CARMEN modules. The MMF at the CAP can also receive events from the IEEE 802.21 sub-module which assists in the handover procedure.

Apart from the protocol interface differences (PMIPv6 at the CGW and IEEE 802.21 at the

Figure 14. Hierarchical approach solution based on PMIPv6

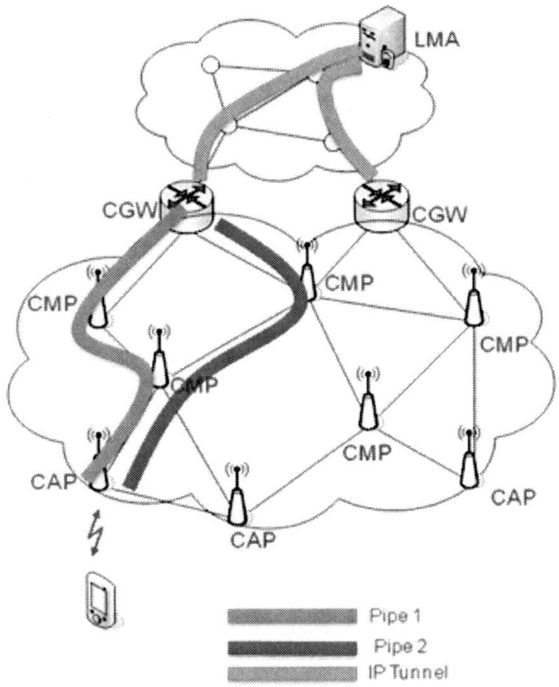

CAP), the depicted generic architecture can be used on both the CGW and the CAP, with other small implementation differences. Namely, the internal implementation of the state machine must be different in order to handle the different messages to process.

Depicted in Figure 16 are the internal details of the functionality implemented for the MMF that runs at the CAP. Again, the IMF is used for the remote and local communication with other mesh modules. The IMF at the CAP is extended in order to support delivery of standard IEEE 802.21 primitives towards the access network (handover preparation and execution) and also other CAPs (resource query and reservation).

Also shown in Figure 16 is the Binding List, which consists of a conceptual structure that contains all the information relevant to tracking the User Terminal's (UT) movements as well as associated flows. Every time the UT changes point

Figure 15. MMF interactions

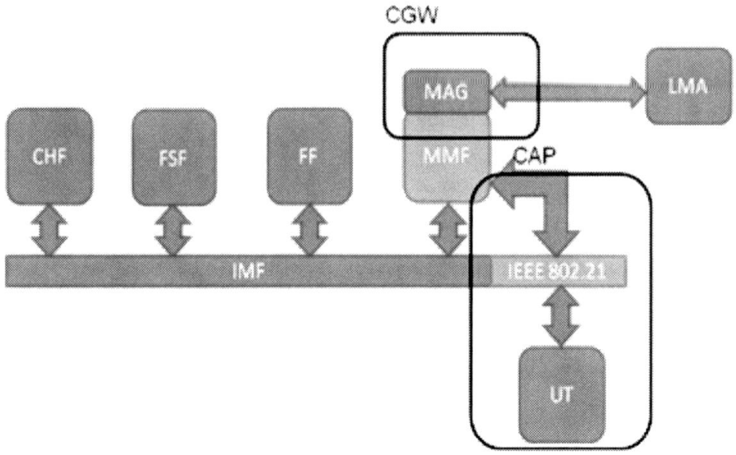

of attachment or initiates/terminates new flows this list is updated with the new information and the relevant states are also updated.

The architecture at the CGW is similar to the one found at the CAP, but without the 802.21 interface and new one for the communication towards the LMA.

Depicted in Figure 17 is the result of an experiment that considers an MN which has an IEEE 802.11a/b/g card as one of its interfaces, while the second interface is a standard 3G modem, both being able to access the mesh. The MN starts a TCP video flow in the 3G interface and this flow is switched to the WLAN and 3G back and forth. As shown in the figure, the bandwidth requirements of the video are quite low; hence the video does not suffer from congestion while being transmitted / received at any of the interfaces. This scenario

Figure 16. CAP mobility internal functions

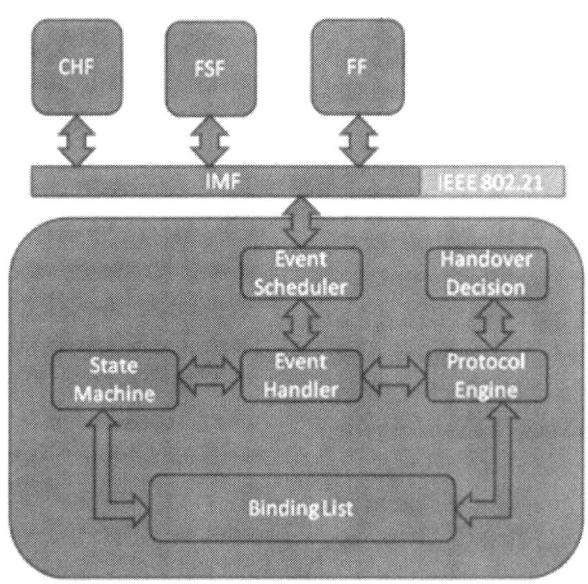

was selected since we want to assess the impact of changing the underlying technology to a standard traffic without QoS constraints. Observed results show that the handover between both technologies is almost transparent from the viewpoint of the flow performance. In the case of WLAN to 3G handover, we find that for each handover, some retransmissions occur, as the bandwidth of the 3G interface is lower than the WLAN one, and its delay is higher. This decrease in the performance would be hardly noticeable due to the low requirements of the traffic being used and the fact that the TCP pace is recovered quickly. For the case of the 3G to WLAN handover, we find the inverse behavior, observing an increase in the speed of the sequence number growth. Observed results show that our design does not impose any penalty in the performance of the flow apart from the effect of changing the characteristics of the underlying technology, which is known to affect the TCP performance. Nevertheless, the flow handover itself is seamless and transparent for the involved communications peers.

Summing up, the following concepts make our mobility solution specifically suitable for heterogeneous wireless mesh networks:

- The smooth and tight integration with IEEE 802.21 as the underlying handover mechanism
- The fact that not only the access, but also the mesh backhaul capacity toward the CGW is considered in the handover decision (this allows better load-balancing in resource-constrained wireless networks)
- The clear separation between macro- and micro-level mobility (while the latter is mesh-tailored, the former is well-established in operator networks)

FUTURE RESEARCH DIRECTIONS

While current wireless mobile backhaul networks are almost exclusively tree topologies (since no routing is required in that case), several factors make it likely that a migration towards mesh topologies will happen in the future. First, backhaul networks are already under heavy stress nowadays and investments to increase backhaul capacity are inevitable. Second, with the imminent rollout of next-generation access networks such as LTE, user traffic will again multiply, which will add to the

Figure 17. TCP sequence number and throughput for inter-technology (WLAN->3G) handover

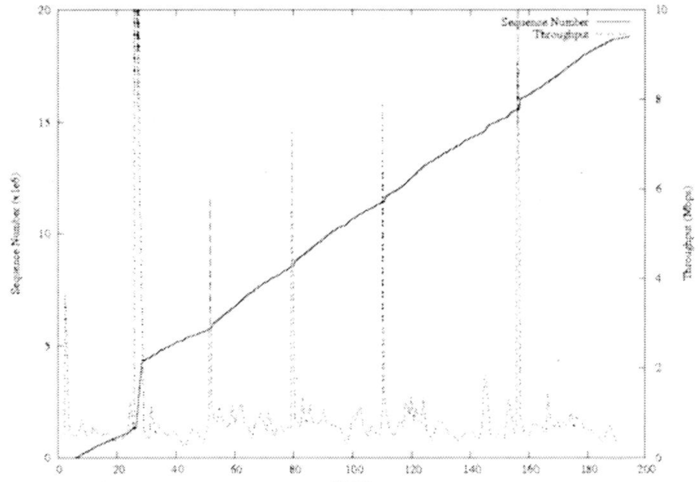

load on backhauls dramatically. Third, the fundamental advantages of mesh over tree networks (such as load balancing and resilience gains) are more and more understood by network operators.

A next step to evolve our solution would be to study integration of our architecture and mechanisms for heterogeneous wireless mesh backhauls into the overall mobile backhaul context spanning access, wireless last mile, wired middle mile/metro network. The goal would be to understand the mechanisms that are required and potentially established in these different domains, find commonalities and identify possible problems and pitfalls when integrating them. Ideally, there would be a homogeneous end-to-end solution encompassing the whole backhaul.

Our solution has been evaluated in several test networks, one of which is depicted in Figure 18. It has 6 mesh nodes and 6 logical link groups. Logical link group LG3 comprises three links, namely the ones mutually connecting CAP, CMP2 and CMP4. The wireless technologies used in the network are standard WiFi, on the one hand, and a proprietary TDMA MAC, on the other hand. Experiments have shown promising results in all testbeds. We could verify that our radio abstraction layer indeed works in a heterogeneous wireless mesh network. Routing, that operates on top of the abstracted logical topology made possible by the abstraction layer was able to place about four times as many flows in the network as standard shortest path routing. This result also demonstrates that resource management actually performs well even in the presence of link groups which include multiple links sharing the same spectrum (such as LG3). As opposed to the convergence time of distributed routing protocols, which are in the domain of 5 seconds or above, our routing exhibited a link failure reaction time of about 100 ms, which is an impressive responsiveness. Mobility Management achieved handover latencies of about 200 ms, which is a very satisfactory value, given that a considerable share of this latency is due to the operating system and driver latency, which is out-of-control of our solutions.

Currently, it is planned to extend one of the test networks and to gather experience with run-

Figure 18. CARMEN test network

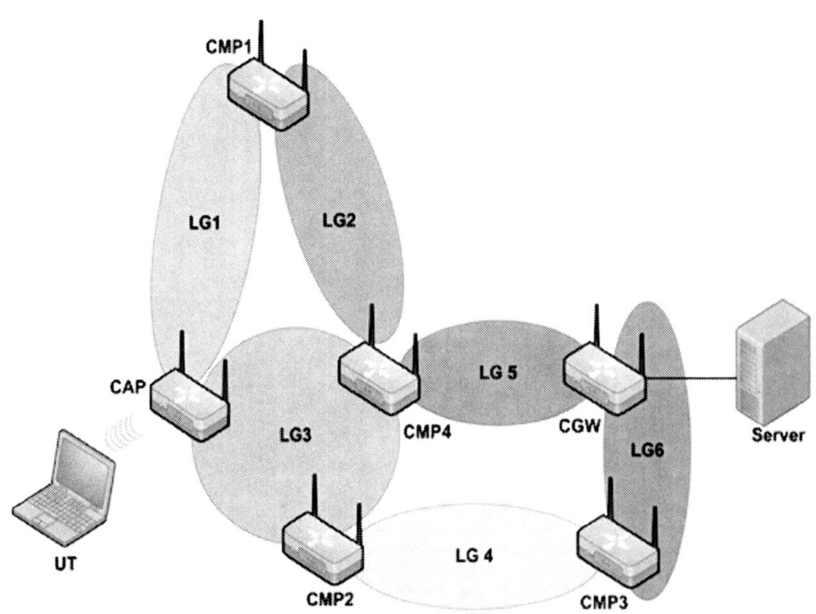

ning and optimizing our solutions in larger heterogeneous wireless mesh networks.

CONCLUSION

Mesh topologies in the wireless last mile backhaul come with increased complexity for an operator. First, meshes require routing instead of simple switching in trees. Second, cell site and frequency planning must be done carefully to maximize the usability of each wireless link, as installing such links in backhauls is very costly and meshes have more links than trees by design. Third, wireless links are more unstable compared to wired ones; this means that there must be failure recovery strategies, which should be self-organized to not unduly keep alarming the network management system in the operator premises.

In this chapter, we have presented solutions for these problems. Our routing solution is centralized, based on abstracted logical topological network views made possible by the radio abstraction layer, and has demonstrated superior performance in an actual test network. It can also recover from link failures by shifting affected flows to other links. Our self-configuration solution allows bootstrapping and reconfiguring of heterogeneous wireless networks. Our mobility management solution also supports heterogeneous networks since it is based on IEEE 802.21. Besides, it makes handover decisions based on both access and mesh load, thereby contributing to optimal load balancing.

By proposing a complete architecture, a radio abstraction layer, a self-configuration, routing and mobility management solution, we have gone far beyond the current state-of-the-art in the area of heterogeneous wireless (backhaul) networks. Our radio abstraction layer is a very generic approach which facilitates operation of heterogeneous wireless networks in many different use cases.

It must be said that the architecture and mechanisms which we presented are not restricted to pure mobile backhaul networks, they can also be used in more "community-style" or "best-effort" mesh networks, in order to increase their performance towards carrier-grade quality.

REFERENCES

Angelakis, V., Papadakis, S., Siris, V., & Traganitis, A. (2008). Adjacent channel interference in 802.11a: Modeling and testbed validation. *IEEE Radio and Wireless Symposium*, (pp. 591–594). Orlando, FL: IEEE Press.

Bahl, P., Chandra, R., & Dunagan, J. (2004). SSCH: Slotted seeded channel hopping for capacity improvement in IEEE 802.11 ad-hoc wireless networks. In *Proceedings from ACM MobiCom* (pp. 216–230). Philadelphia, PA: ACM Press.

Banerjee, S., Patra, S., Roy, K., & Vidyarthi, D. (2006). Improved genetic algorithm for channel allocation with channel borrowing in mobile computing. *IEEE Transactions on Mobile Computing*, *5*(7), 1536–1233.

Beckmann, D., & Killat, U. (1999). A new strategy for the application of genetic algorithms to the channel assignment problem. *IEEE Transactions on Vehicular Technology*, *48*(4), 1261–1269. doi:10.1109/25.775374

Campbell, A., Gomez, J., Wan, C. Y., Kim, S., Turanyi, Z., & Valko, A. (2000). *Cellular IP, internet-draft* (work in progress). Draft-ietf-mobileip-cellularip-00, (work in progress).

Chakraborty, G., & Chakraborty, B. A. (1999). Genetic algorithm approach to solve channel assignment problem in cellular radio networks. In *Proceedings of IEEE Workshop Soft Computing Methods in Industrial Applications*, (pp. 34–39). Kuusamo, Finland: IEEE Press.

Cheng, C., Hsiao, P., Kung, H., & Vlah, D. (2006). *Adjacent channel interference in dual-radio 802.11a nodes and its impact on multi-hop networking. IEEE Globecom* (pp. 1–6). San Francisco, CA: IEEE Press.

Congdon, P., & Lane, B. (2010). *802.1AB - Station and media access control connectivity discovery*. Retrieved from http://www.ieee802.org/1/pages/802.1ab.html

Funabiki, N., & Takefuji, Y. (1992). A neural network parallel algorithm for channel assignment problems in cellular radio networks. *IEEE Transactions on Vehicular Technology*, *41*(4), 430–437. doi:10.1109/25.182594

Hamida, E. B. (2008). Neighbor discovery in multi-hop wireless networks: Evaluation and dimensioning with interference consideration. *Discrete Mathematics and Theoretical Computer Science*, *10*(2), 87–114.

Ho, C., Obraczka, K., Tsudik, G., & Viswanath, K. (1999). Flooding for reliable multicast in multi-hop ad hoc networks. *Proceedings of the International Workshop on Discrete Algorithms and Methods for Mobile Computing and Communication (DIALM)*, (pp. 64–71). Seattle, WA: ACM Press.

IEEE. Computer Society. (2009). *IEEE standard for local and metropolitan area networks - Part 21*. Media Independent Handover Services. Retrieved October 3rd from http://standards.ieee.org/getieee802/802.21.html

IETF. (2001). *Multi protocol label switching (MPLS)*. RFC3031 (Proposed Standard). Retrieved from http://www.ietf.org/rfc/rfc3031.txt

IETF. (2008). *RFC 5213: Proxy mobile IPv6*. Retrieved on October 3rd from http://www.ietf.org/rfc/rfc5213.txt?number=5213

Johnson, D., Perkins, C., & Arkko, J. (2004). *Mobility support in IPv6*. Internet Engineering Task Force, RFC 3775 (Proposed Standard). Retrieved from http://www.ietf.org/rfc/rfc3775.txt

Kunz, D. (1991). Channel assignment for cellular radio using neural networks. *IEEE Transactions on Vehicular Technology*, *40*(1), 188–193. doi:10.1109/25.69987

Kunz, D., Duque-Anton, M., & Ruber, B. (1993). Channel assignment for cellular radio using simulated annealing. *IEEE Transactions on Vehicular Technology*, *42*(1), 14–21. doi:10.1109/25.192382

Kyasanur, P., & Vaidya, N. (2005). Routing and interface assignment in multi-channel multi-interface wireless networks. In *Proceedings IEEE Conference on Wireless Communications and Networking Conference*, (pp. 2051–56). New Orleans, LA: IEEE Press.

Kyasanur, P., & Vaidya, N. (2006). Routing and link-layer protocols for multi-channel multi-interface ad hoc wireless networks. *SIGMOBILE Mobile Computing and Communications Review*, *10*(1), 31–43. doi:10.1145/1119759.1119762

Li, G., & Zhu, J. (2007). Performance analysis on utilizing directional antenna in high-density WLAN. *Proceedings of the Sixth International Conference on Networking* (ICN'07). Sainte-Luce, Martinique: IEEE Press.

Li, N., & Hou, J. C. (2006). Localized fault-tolerant topology control in wireless ad hoc networks. *IEEE Transactions on Parallel and Distributed Systems*, *17*(4), 307–320. doi:10.1109/TPDS.2006.51

Luo, H., & Shankaranarayanan, N. (2004). A distributed dynamic channel allocation technique for throughput improvement in a dense WLAN environment. In *Proceedings of IEEE International Conference on Acoustics, Speech, and Signal Processing*, vol. 5, (pp. 345-348). Montreal, Canada: IEEE Press.

Marina, M., & Das, S. (2005). A topology control approach for utilizing multiple channels in multi-radio wireless mesh networks. *2nd International Conference on Broadnets*, (pp. 381–90). Boston, MA: IEEE Press.

Mathar, R., & Mattfeldt, J. (1993). Channel assignment in cellular radio networks. *IEEE Transactions on Vehicular Technology, 42*(4), 647–656. doi:10.1109/25.260746

Mishra, A., Banerjee, S., & Arbaugh, W. (2005). Weighted colouring based channel assignment for WLANs. *ACM SIGMOBILE Mobile Computing and Communications Review, 9*(3), 19–31. doi:10.1145/1094549.1094554

Mishra, A., Shrivastava, V., Agarwal, D., Banerjee, S., & Ganguly, S. (2006). Distributed channel management in uncoordinated wireless environments. In *Proceedings of the ACM Annual International Conference on Mobile Computing and Networking*. Los Angeles, CA: ACM Press.

Nachtigall, J., Zubow, A., & Redlich, J. (2008). The impact of adjacent channel interference in multi- radio systems using IEEE 802.11. *International Wireless Communications and Mobile Computing Conference*, (pp. 874–881). Crete, Greece: IEEE Press.

Papagiannaki, K., Mhatre, V., & Baccelli, F. (2007). *Interference mitigation through power control in high density 802.11 WLANs. Proceedings of IEEE INFOCOM 2007*. Anchorage, AK: IEEE Press.

Riihijarvi, J., Petrova, M., & Mahonen, P. (2004). Automatic channel allocation for small wireless local area networks using graph colouring algorithm approach. In *Proceedings of IEEE International Symposium on Personal, Indoor, and Mobile Radio Communications*, (pp. 536–539). Barcelona, Spain: IEEE Press.

Riihijärvi, J., Petrova, M., & Mähönen, P. (2005). Frequency allocation for WLANs using graph colouring techniques. In *Proceedings of 2nd Annual Conference on Wireless On-Demand Network Systems and Services*, (pp.19–21). St. Moritz, Switzerland: IEEE Press.

Robitzsch, S., Fitzpatrick, J., Murphy, S., & Murphy, L. (2010). Behind-the-scenes of IEEE 802.11a based multi-radiomesh networks: A measurement driven evaluation of inter-channel interference. *International Journal on Advances in Internet Technology, 3*(1-2), 65–76.

Robitzsch, S., Niephaus, C., Fitzpatrick, J., & Kretschmer, M. (2009). Measurements and evaluations for an IEEE 802.11a based carrier-grade multi-radio wireless mesh network deployment. In *2009 Fifth International Conference on Wireless and Mobile Communications*, (pp. 272–278). Cannes, France: IEEE Press.

Skalli, H., Ghosh, S., Das, S., Lenzini, L., & Conti, M. (2007). Channel assignment strategies for multiradio wireless mesh networks: Issues and solutions. *IEEE Communications Magazine, 45*(11), 86–95. doi:10.1109/MCOM.2007.4378326

Smith, K. (1998). A genetic algorithm for the channel assignment problem. In *Proceedings of Global Telecommunications Conference*, vol. 4, (pp. 2013–2018). Sydney, Australia: IEEE Press.

Smith, K., & Palaniswami, M. (1997). Static and dynamic channel assignment using neural networks. *IEEE Journals on Selected Areas Communications, 15*(2), 238–249. doi:10.1109/49.552073

So, J., & Vaidya, N. (2004). Multi-channel MAC for ad hoc networks: Handling multi-channel hidden terminals using a single transceiver. In *Proceedings ACM Mobihoc*, (pp. 222–33). Tokyo, Japan: ACM Press.

Soliman, H., Castelluccia, C., Malki, K., & Bellier, L. (2005). *Hierarchical mobile IPv6 mobility management (HMIPv6)*. Internet Engineering Task Force, RFC 4140 (Experimental). From http://www.ietf.org/rfc/rfc4140.txt

Sparl, P., & Zerovnik, J. (2004). 2-local 5/4-competitive algorithm for multi-colouring triangle-free hexagonal graphs. *Information Processing Letters*, *90*(5), 239–246. doi:10.1016/j.ipl.2004.02.017

Sudeep, K., & Vishwanathan, S. (2005). A technique for multi-colouring triangle-free hexagonal graphs. *Discrete Mathematics*, *300*(1-3), 256–259. doi:10.1016/j.disc.2005.06.002

Valko, A. G. (1999). Cellular IP: A new approach to internet host mobility. *Mobile Computing and Communications Review*, *29*(1), 50–65.

Wang, X., & Lim, A. O. (2008). IEEE 802.11s mesh networks: Framework and challenges. *Ad Hoc Networks*, *6*(6), 970–984. doi:10.1016/j.adhoc.2007.09.003

KEY TERMS AND DEFINITIONS

Capacity-Aware Routing: A routing algorithm that optimizes path placement in a way that optimally exploits the available capacities in the network, which could imply deviations from shortest path towards longer paths for the sake of congestion avoidance..

Carrier-Grade Requirements: Quality level that meets carrier demands on aspects such as reliability, predictability, controllability, integration to existing infrastructure, standards compliance, quad-play services support, etc.

Capacity Handling: A function in the network that performs admission control and radio resource management (i.e. determines the bandwidths of new pipes, adapts bandwidths of existing pipes, and steers the flow-to-pipe association).

Heterogeneous Wireless Network: A network consisting of links with potentially more than a single wireless technology (e.g. a network whose links are partly IEEE 802.11 links and partly IEEE 802.16 links).

Media-Independence Layer: A layer that is located between the network and the link layer, comprising an API (i.e. a set of primitives) that allow sending link control commands (e.g. to configure the transmission power of the wireless interface) and receiving link state events (e.g. a notification about a newly detected neighbor) in a technology-independent way, as well as models (e.g. a capacity model) that capture link properties (such as capacity) in a manner that is independent of the wireless technology.

Pipe: An end-to-end tunnel from the access point through the wireless network to the gateway, along a defined path with a certain bandwidth, accommodating multiple user flows of the same traffic class.

Self Configuration: While generally a term with a broader meaning, in the context of this chapter a function in the network that is responsible for topology discovery and formation in a way that is as automated as possible, including monitoring and configuring/optimizing the radio parameters of all wireless interfaces in the network.

ENDNOTE

1 The research leading to these results has received funding from the European Community's Seventh Framework Programme (FP7/2007-2013) under grant agreement no. 214994 (CARMEN).

Section 2
QoS Provisioning Solutions

Chapter 9
Toward Intelligent Fuzzy QoS Model in Wireless Ad Hoc Networks

Lyes Khoukhi
University of Technology of Troyes, France

Ali El-Masri
University of Technology of Troyes, France

Dominique Gaiti
University of Technology of Troyes, France

ABSTRACT

Wireless mobile ad hoc networks, also called MANETs, hold the key to the future generation of wireless communication, promising adaptive communication any time and anywhere. Ad hoc networks constitute a best alternative for developing countries, and everywhere communications infrastructure does not exist. This kind of network is very flexible and suitable for applications such as temporary information sharing in conferences, military actions, and disaster rescues. Numerous challenges must be overcome to realize the practical benefits of wireless ad hoc technology. These include mobility management, security, and of principal interest here, Quality of Service (QoS) issues. QoS provisioning is becoming a critical issue in designing wireless ad hoc networks due to the necessity of providing real time services in such networks. These services are typically delay-sensitive and have high bandwidth requirements.

On the other hand, the advance in the field of artificial intelligence and the progress in its application in telecommunications, and the growing scientific interest in ad hoc networks are reasons to think about coexistence and interoperability among these two areas. In the past few years, some intelligent methods have been successfully applied in the wireless networks with the aim to obtain more adaptive and flexible QoS models over the existing models. This chapter focuses on the cross layer QoS model (i.e., frameworks for achieving QoS over multiple layers). It presents two model-categories of QoS provisioning in MANETs: "conventional QoS models" that are based on traffic engineering mechanisms and QoS optimization techniques, and "intelligent QoS models" that exploit the efficiency of intelligent learning

DOI: 10.4018/978-1-4666-0017-1.ch009

heories such as Fuzzy Logic control in wireless networks. Then, it proposes an intelligent fuzzy logic QoS model to deal with the buffer management and traffic regulation issues in MANET. The fuzzy logic theory control promises to be an efficient solution and may constitute a best alternative to the conventional methods, based on the complex optimization and overloaded traffic engineering mechanisms.

INTRODUCTION

The interest in wireless Mobile Ad hoc Networks (MANETs) has grown immensely over the last 10 years. Conventional wireless networks require as prerequisites some form of fixed network infrastructure and centralized administration for their operation. In contrast, the so-called MANETs, consisting of a collection of wireless nodes, all of which may be mobile, dynamically create a wireless network among themselves without using any such infrastructure or administrative support (Haas et al., 1999). The developing world, where a higher proportion of people live in areas with limited infrastructure, could also benefit from MANET technology. In fact, the "*One Laptop-Per-Child*" project (laptop.org et al., 2008) is enabling the establishment of what are the largest real-world MANET-like networks to date (Lajos et al., 2009). Often-suggested applications of ad hoc networks include disaster recovery, highly mobile vehicle-to-vehicle networks (VANETs), and battlefield communications. Numerous challenges must be overcome to realize the practical benefits of ad hoc networking. These include mobility and power management, medium access, security, and, of principal interest here, Quality of Service (QoS) issue.

QoS requirements in MANETs mainly result from the rising popularity of end-to-end delay- and bandwidth- real time multimedia services. Different real time applications have various QoS requirements expressed in terms of end-to-end QoS metrics (e.g., delay, available bandwidth, jitter, loss rate). The ad hoc network is thereby required to provide better services than original best effort service, such as differentiated services and guaranteed services, for end-to-end user/applications. Providing QoS assurances to support real time traffic in MANET is difficult due to the unreliable wireless channel, the lack of centralized control, contention for channel access and node mobility. Various solutions have been proposed to provide QoS over ad hoc networks from different layers in the network protocol stack. However, the layered concept was primarily created for wired networks, and multi hop wireless networks operate better with layered design because of their dynamic nature, infrastructure-less architecture, and time varying unstable links and topology (Zhang et al., 2008).

In this chapter, we attempt a brief description to the new but rapidly growing area of research on QoS support in wireless mobile ad hoc networks. We focus mainly on the cross layer QoS models (i.e., frameworks for achieving QoS over multiple layers), thus we do not consider the works based on a single layer (e.g., network layer). We present two kinds of QoS models; "*conventional QoS models*" which are completely based either on traffic engineering approach or mathematical optimization modeling, and "*intelligent QoS models*" that exploit the efficiency of intelligent learning theories such as Fuzzy Logic control in wireless networks. Then, we propose a fuzzy cross layer QoS model for traffic regulation and buffer management in wireless ad hoc networks, named FuzzyCCG. The use of fuzzy logic theory in our proposal is justified by the fact that this theory is well adapted to systems characterized by imprecise states, dynamic nature, and uncertain information, as in the case of wireless mobile ad hoc networks.. Furthermore, the use of fuzzy logic can add more flexibility and capability for managing both the random traffic state and the arbitrary buffer change occupancy of mobile devices due to nodes mobility and dynamic topology of ad hoc networks. This theory has been successfully applied in many industrial systems, such as embedded systems and telecommunications control.

The reminder of the chapter is organized as follows: we first present a brief review of ad hoc networks and introduce some networking concepts pertinent to QoS. Following this, we list some relevant conventional QoS models that were found in the literature. Next we introduce the intelligent fuzzy logic theory and its applications in wireless networks; we describe some basic concepts related to this theory and QoS models based on this intelligent tool. Finally, we present our proposed intelligent fuzzy QoS model to support multimedia services in MANETs. The last section summaries the advantages and drawbacks of both convention and intelligent QoS models, discusses the trends in the field, and highlights potential areas of future work. Note that through this chapter, we use the terms "QoS cross layer models" and "QoS models" interchangeably.

BACKGROUND

The mobile ad hoc network is a collection of mobile nodes that are connected via wireless links without using an existing fixed infrastructure or centralized management. Such kinds of networks are infrastructure-less networks since they do not require any fixed infrastructure such as a base station for their operation. The mobile devices in MANET are free to move arbitrarily and organize themselves randomly (Figure 1).

MANETs offer unique advantages and versatility for certain environments and certain strategic applications. No pre-existing fixed infrastructure being prerequisite, they can be created and used "anytime, anywhere". In addition, such networks could be intrinsically fault-resilient, for they do not operate under the restrictions and limitations of a fixed topology. Indeed, since all nodes are allowed to be mobile, the composition of such networks is necessarily time-varying. Such perceived benefits elicited urgent attention in the early days among military, police, and rescue agencies in the use of ad hoc networks, especially under disorganized or hostile environments, including isolated scenes of natural disaster and armed conflict. In recent days, home or small office networking and collaborative comput-

Figure 1. Infrastructure-based wireless network and ad hoc network

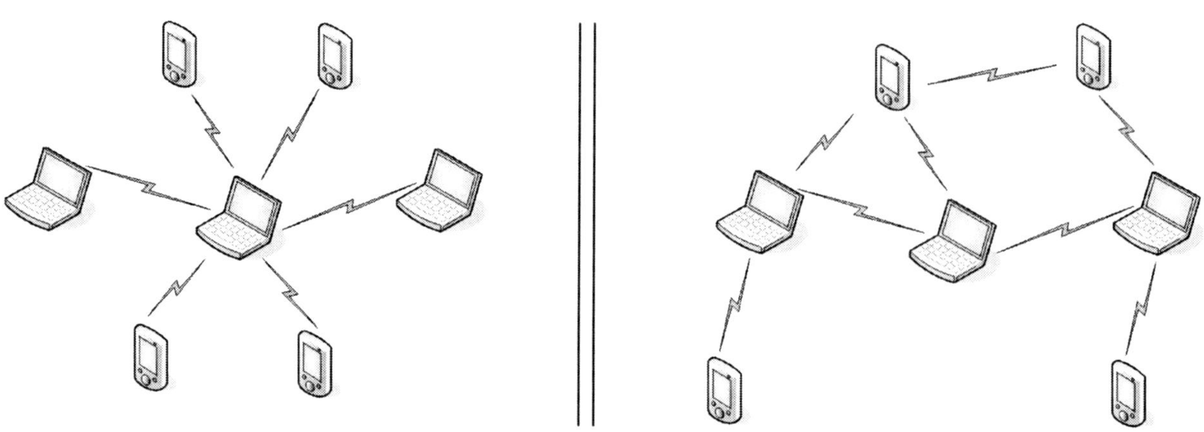

ing with laptop computers in a small area (e.g., classroom, hotels, etc.) have emerged as other foremost areas of potential application (Chakrabarti et al., 2001).

Wireless ad hoc networks are expected to deliver a broad spectrum of real-time applications such as voice and video seamlessly and efficiently for mobile users. These applications impose severe QoS constraints on timely delivery of packets. It is known (Ganz et al., 2004) that QoS is a general term with various perspectives. Different technical communities may interpret QoS in different ways. In the application communities, QoS generally refers to the quality as perceived by the user/application while in the networking community; QoS is accepted as a measure of the service quality that the network offers to the applications/users. RFC 2386 (Crawley et al., 1998) characterizes QoS as a set of service requirements to be met when transporting a packet stream from the source to its destination. In this scenario, QoS refers to an assurance by the network to offer a set of measurable service attributes to the end-to-end users/applications in terms of QoS metrics (e.g., delay, available bandwidth, jitter, and packet loss, etc.). At the same time, different types of traffic (voice, video, and data) may impose specific constraints on the QoS support due to their particular characteristics. Before the required QoS is determined, the mobile user needs to provide an accurate description of its traffic to the network so that proper network resources can be allocated (Sedoyeka et al., 2009).

It is the task of QoS mechanisms and algorithms to measure and estimate the state of the network's resources (e.g., bandwidth) and thereby to decide which application user connexions (i.e., sessions) can be admitted without providing more resources than are available and thus violating previously made guarantees. However, due to the dynamic nature of MANETs and to the constraints of multimedia applications and wireless environment, estimating the network resources and maintaining QoS guarantees during any topology updating are very challenging tasks. It is just not the connectivity that affects the QoS, but equally essential is the availability of enough resources along the previous and new routes during and after the transition (Chakrabarti et al., 2001).

QoS Metrics

This section describes a brief recap of QoS specification metrics (Table 1). A service's QoS requirements are usually derived from its traffic specification. The QoS requirements can typically be expressed using one or more of the following metrics:

Table 1. Applications QoS requirements

Performance Dimensions				
Applications	*Bandwidth*	**Sensitivity to:**		
		Delay	*Jitter*	*Loss*
VoIP	Low	High	High	Medium
Video Conference	High	High	High	Medium
Streaming Video on Demand	High	Med	Medium	Medium
Streaming Audio	Low	Medium	Medium	Medium
Client/Server Transactions	Medium	Medium	Low	High
E-mail	Low	Low	Low	High
File Transfer	Medium	Low	Low	High

- Maximum packet delay bound (s): the accumulation of the queuing and MAC delays at each node plus the propagation delay, which is relatively short (Lajos, et al., 2009).
- Maximum packet loss ratio (PLR) bound; the maximum tolerable fraction of the generated data packets that are lost en-route (Lajos et al., 2009).
- Maximum delay jitter bound: can be defined as the variance of the absolute packet delay (Wang, et al., 2005);
- Minimum average throughput (bps).

QoS Support Models in Wired Networks

Supporting QoS in wired networks can generally be obtained via the "over-provisioning" of resources and/or "traffic engineering" solutions (Demetrios, 2001) (Zeinalipour, et al., 1999). The "over-provisioning" models add abundant resources in the network so that it can provide satisfactory services to bandwidth-sensitive real time applications. Such models are easy to realize but all the users are served at the same service class. Therefore, the service may become unpredictable during peak traffic.

On the other hand, the QoS models based on "traffic engineering" classify the users/applications in service classes and assign each service class a different priority. The researchers in the literature have proposed two approaches based on traffic engineering; reservation-based and reservation-less models. The network resources, in the reservation-based models, are assigned according to the user/application's QoS request and subject to bandwidth reservation and management policy. The best examples of reservation-based models are: Integration Service (IntServ) and Asynchronous Transfer Mode (ATM). In the reservation-less models, QoS support is achieved via some strategies such as traffic classes, admission control, queuing mechanisms, and policy managers. No reservation is required by these strategies.

It is obvious that the QoS models developed for wired networks (e.g., IntServ and DiffServ) cannot entirely cope with the needs of QoS provisioning in MANETs. The IntServ/RSVP model is not suitable for ad hoc networks because of the necessity to maintain and establish the QoS information for each mobile node. This IntServ/RSVP feature requires a high capacity of processing and storage, which is undesirable for power-and-resource constrained ad hoc networks (Zhang et al., 2005). On the other hand, DiffServ seems to be an interesting model for the implementation in the internal routers (cores routers) because various flows will be in an aggregate, which facilitates the routing inside the network. However, the dynamic topology of MANETs disturbs the definition of the DiffServ architecture, in particular the notion of ingress, egress, and core routers in the network.

In order to realize the practical benefits of the multimedia QoS support in ad hoc networks, the researchers in the literature have pursued an end-to-end QoS support using a number of mechanisms and algorithms in a single layer (Network, MAC, or Transport), and, of principal interest here, QoS frameworks operating on multiple layers (i.e., cross layer models), mainly pertaining to multiple QoS layers (cross layer model). The concept of cross layer design is based on architecture where different layers can exchange information in order to improve the overall network performance (Zhang et al., 2008). As stated above, in this chapter the terms "QoS cross layer models" and "QoS models" are used interchangeably. In what follows, first we describe some challenges facing the design of QoS models, and then we summarize some important QoS models in MANET; we will present two kinds of models: conventional QoS models based on traffic engineering techniques, and intelligent models based on the theory of fuzzy logic control.

CONVENTIONAL QoS MODELS IN WIRELESS AD HOC NETWORKS

The assumptions made in the design of such models are: First, as for resource availability, each mobile device is assumed to be able to monitor (implicitly or explicitly) the available resources on each of its outgoing links. Second, as for resource reservation, a medium access protocol is supposed to be able to resolve media contention and support resource reservation at the MAC layer. Before presenting some remarkable QoS models proposed in the literature, we first discuss some challenges facing the design of both conventional and intelligent QoS models.

Challenges Facing the QoS Models

Wireless Environment Challenges

- **Node mobility:** the mobile devices in ad hoc networks may move completely independently and randomly as far as the communications protocols are concerned. This means that topology information has a limited lifetime and must be updated frequently to allow data packets to be routed to their destinations. Furthermore, Mobility can also cause QoS assurance violations without breaking routes and a QoS state must be updated with a frequency that increases with node mobility (Hanzo et al., 2007).
- **Channel contention:** even if the MAC protocol in use is not the single-channel 802.11 scheme, mobile devices in ad hoc network should communicate on a common channel. However, this leads to channel contention and interference problems, which can impact on the fraction of channel capacity available to a mobile device.
- **Unreliable wireless channel:** received signals are prone to bit errors due to interference from other transmissions, thermal noise, shadowing and multi-path fading effects (Saunders, 1999).
- **Connectivity issue:** a mobile device may lose connectivity with the rest of the group just because it has wandered off too far, or its power reserve has dropped under a certain threshold.

Computation Design Challenges

To some extent the limited device resources is an historical limitation, since wireless devices are becoming increasingly powerful and capable. However, it still holds true that such devices generally have less computational power, less memory, and a limited (battery) power supply, compared to devices such as desktop computers typically employed in wired networks (Hanzo et al., 2007). This factor has an impact on the design of QoS model, since low CPU and memory capacity limits the amount of QoS state that can be stored, necessitating more regular updates, which generate larger overhead.

Thus, one main design criteria of QoS models is related to lower algorithmic complexity to facilitate limited-bandwidth and low-power QoS solutions that can be embedded into low-cost mobile devices' microprocessors, and to extend the lifetime of the network without endangering efficient and reliable communications between mobile nodes. Usually, this is impacted by the kind of traffic (e.g., CBR voice real time traffic, rt-VBR video conferencing flows) generated by the source mobile devices, the available buffer sizes within intermediate nodes, the nodes' connectivity, QoS constraints, etc. In the wireless ad hoc environment, this is further impacted by the fact that the common medium is the wireless channel which is broadcast in nature (i.e. control and data packets can get corrupted at physical layer [due to errors and interferences] or at MAC layer [due to collision], etc.). Wireless link between mobile devices are "dynamic" in that they come and go over time (i.e. two nodes which could speak to one another suddenly cannot, and vice-versa (Bachir, et al., 2010)). Furthermore, mobile devices usually suffer from limited resources in terms of bandwidth, processing and memory capacities, and energy. Note also that the number of nodes and their density play an important part in the MANET QoS model design. A large number of mobile devices at low densities usually implies many hops at a limited degree of connectivity, which may facilitate the QoS recovery process because of a large number of routes offered in this case.

In summary, all the challenges enumerated above are potential sources of service impairment in ad hoc networks, and hence may degrade the QoS seen by users of the network. A QoS model should guarantee a successful data delivery with a good quality under a given limited memory, buffers, and processing capabilities of the mobile devices, implemented algorithm complexity, and wireless environments constraints. The overhead-heavy QoS models may have a significant impact on resource and battery life and hence may limit the usefulness of devices.

Conventional QoS Models

In this sub-section we provide a brief review of major contributions to the pool of cross layer QoS issue in MANETs, based either on optimization techniques or traffic engineering methods.

SWAN (Service Differentiation in Stateless Wireless Ad Hoc Networks) (Ahn et al., 2002) is one of the best stateless cross layer QoS approaches proposed in the literature. SWAN offers a service differentiation in wireless ad hoc networks by using distributed control algorithms. SWAN distinguishes between two traffic classes: real-time and best-effort. A classifier decides which type an incoming packet belongs to. In the case of real-time data packet, it is processed as soon as possible. Otherwise, the packet is considered as low-priority and should wait until all real-time packets are dispatched. A shaper limits the relay of low-priority traffic to reduce the contention between neighboring stations.

SWAN relies on feedback information received from the MAC layer as a measure of congestion in the network by using mechanisms of rate control and source-based admission control. The AIMD rate control algorithm is used at each node in order to perform the control of the best-effort traffic. The rate control restricts the bandwidth usage of best-effort traffic so that real-time applications can exploit the required bandwidth. The bandwidth not used by real-time applications can be exploited by the best-effort traffic. On the other side, SWAN uses sender-based admission control in order to perform the real-time traffic control (Hejmo et al., 2006).

One of the drawbacks of SWAN is how to calculate the threshold rate limiting any excessive delay that might be experienced. SWAN adopts engineering techniques that attempt to set the admission threshold rate at mobile nodes to operate under the saturation level of the wireless channel (MORGAN et al., 2003). SWAN suffers also from the problem of false admission, the common drawback of stateless approaches as we will show later on. However, SWAN remains the best example of stateless distributed QoS framework developed for wireless ad hoc networks.

A stateless distributed admission control for MANET environments (DACME) was presented in (Calafate *et al.*, 2007) (Calafate et al., 2005). DACME protocol assesses the achievable QoS on a given route by means of a set of back-to-back probe packets. In (Calafate et al., 2005) techniques are explained for estimating route capacity, end-to-end delay, and delay jitter, when DACME is operating with a single-path routing protocol. In (Calafate et al., 2007), the emphasis is not on throughput-constrained applications only, but also on a multi-path routing protocol. The authors state that the optimal operating environment for DACME is based on an EDCA MAC scheme, but it can also operate with a non-QoS-aware MAC

(Calafate et al., 2005). The estimation of delay in DACME is determined to be half of a probe/probe reply's round-trip time. The source device explicitly notifies the destination of its packet sending rate, so that the expected inter-packet interval can be calculated. A disadvantage of DACME approach is that, until traffic is actually being carried on the two routes, there is no way to predict the effect of the inter-route interference on the achievable QoS.

In (Xiang et al., 2010), Xiang *et al.* have proposed a stateless geographic multicast protocol (RSGM) which can scale to a large group size and network size and provide multicast packet transmissions in a dynamic mobile ad hoc network environment. The protocol is designed to be simple; thus, it can operate more efficiently and reliably. RSGM includes several virtual architectures for more robust and scalable membership management and packet forwarding in the presence of high network dynamics due to unstable wireless channels and frequent node movements. Both the data packets and control messages are transmitted along tree-like paths; however, different from other treebased protocols, there is no need to explicitly create and maintain a tree structure. RSGM makes use of position information to support reliable packet forwarding. Instead of addressing only a specific part of the problem, RSGM introduces a zone-based scheme to handle the group membership management, and takes advantage of the membership management structure to efficiently track the locations of all the group members without resorting to any external location server (Xiang et al., 2010).

In (Yang et al., 2005), the authors proposed the Contention-aware Admission Control Protocol (CACP) to support QoS in MANET. CACP is the first work to introduce the concept of c-neighborhood available bandwidth. It requires that a node must have enough local and c-neighborhood available bandwidth to successfully admit a flow. CACP depends on source routing to build the "c-neighbor set" to calculate bandwidth consumption. However, CACP does not deal well with route failures. While searching for a new route, it reduces the data rate of affected sessions. This implies that CACP can only support applications with elastic throughput requirements.

In (Wang et al., 2006), Wang *et al.* proposed a new model named Opportunistic Medium Access and Auto Rate (OMAR), to efficiently utilize the shared medium in 802.11-based wireless ad hoc networks. In OMAR, each mobile device with a certain number of links is enabled to form a cluster and function as the cluster head to coordinate multi-user communications locally. In each cycle of data transmission, the cluster head initiates medium access, and then the cluster members in a distributed way make medium access decisions based on the observed channel conditions, where the proposed scheme can guarantee only the user with the best normalized instantaneous channel quality wins the channel.

Lee *et al.* have presented in (Lee et al., 2000) an in-band signalling in ad hoc networks (INSIGNIA) model. In-band signalling refers to the carrying of control information in data packets headers, as opposed to in separate control packets. With INSIGNIA, applications can specify a basic level of QoS, the minimum throughput they require and an enhanced level, the maximum they can benefit from. INSIGNIA admits all user sessions but it is the task of the admission controller to decide whether the admission is on a best-effort, basic QoS or enhanced QoS basis. A flow's data packets carry resource reservation requests to intermediate nodes. Nodes reserve resources and mark packet headers according to their residual resources and the QoS they can provide. The destination notifies the source of the lowest common level of service provided by the nodes on the route. The source notifies all nodes of the level of service the session will then use, so that unused resources can be freed. Reservations are soft state and so they are also erased if not periodically refreshed by data packets. Source node adapts its sending rate according to the information received by the intermediate and destination nodes. INSIGNIA can adapt quickly to ad hoc network dynamics. However, it requires the maintaining of per-flow state information at each mobile node, which can be seen as compromising the scalability of a whole network.

Yang *et al.* have proposed in (Yang et al., 2007), (Yang et al., 2004) a unique capacity allocation model for a network handling sessions with heterogeneous priorities, named Multi-Priority Admission and Rate Control (MPARC). MPARC can utilise any mobile ad hoc routing protocol for route discovery. Also, existing signalling protocols, such as INSIGNIA are suggested for use in reserving and freeing resources along a route. The model assumes that different classes of data are differentiated by their minimum contention windows and frame sizes. MPARC aims at ensuring that no admitted session degrades the throughput of previously- admitted sessions that have an equal or higher priority. However, the throughput of lower-priority sessions may be degraded.

FQMM (Flexible Quality of Service Model for Mobile Ad Hoc Networks) (Xiao et al., 2000) was the first QoS model proposed for wireless ad hoc networks. This model combines the advantages of both solutions implemented in the internet: IntServ and DiffServ. FQMM tries to preserve the per-flow granularity for a small portion of traffic in MANETs, given that a large amount of the traffic belongs to per-aggregate of flows, that is, per-class granularity. FQMM defines three types of nodes as in DiffServ: *ingress node* (mobile node that sends data), *interior nodes* (nodes that forward data and control messages), and *egress node* (destination node). Since nodes are free to move resulting in topology changes, a single host may have multiple roles. An adaptive traffic conditioner is placed at the ingress nodes where the traffic originates. It includes several components: a traffic profile, meter, marker and dropper. The traffic conditioner is responsible for marking the traffic streams and discarding packets according to the traffic profile.

FQMM offers a good solution for small- and medium-size ad hoc network with fewer than 50 nodes, but it is not suitable for large networks, which means that it suffers from the problem of scalability. Due to bandwidth limitation, FQMM tries to preserve the per-flow granularity for a small portion of traffic types in a MANETs, given that these form a small percentage of the total traffic load (Xiao et al., 2003). Since the states of per-flow granularity come from only small portion of the traffic, the scalability problem as in IntServ is expected to be significant. On the other side, this model does not consider large mobility scenarios.

Due to the dynamic and unpredictable nature of ad hoc networks, most QoS models described above do not ensure scalability. The designers of these protocols have assumed that QoS assurances cannot be upheld in the face of high mobility, as we will show in a further section.

INTELLIGENT FUZZY LOGIC QoS MODELS IN WIRELESS NETWORKS

Wireless communications is a rapidly evolving industry, continually challenging researchers for new tools and techniques in order to meet the demands of ever higher performance and efficiency. We believe that the wireless communications area is in need of innovative approaches and novel algorithms based on intelligent control theories such as fuzzy logic and neural networks to address the growing complexity in every one of its different aspects; performance, QoS, etc. Intelligence control is applied when dealing with classes of unknown or partially known nonlinear dynamic systems such as the case of ad hoc networks. In such kind of networks, the devices are too mobile, the topology is dynamic, and the traffic capacity is unpredictable. Hence, the intelligent fuzzy logic theory seems well suitable for application in ad hoc environment.

Why Use Intelligent Fuzzy Logic?

The development of fuzzy logic in the industrial systems was motivated in large measure by the need for a conceptual frame-work which can address the issue of uncertainty and lexical imprecision. The essential characteristics of fuzzy logic are as follows (Zadeh, 1996) (Salmani et al., 2008) (Lee, 1990):

- In fuzzy logic, everything is a matter of degree.
- In fuzzy logic, exact reasoning is viewed as a limiting case of approximate reasoning.
- Any logical system can be fuzzi□ed (i.e., Boolean logic is a subset of Fuzzy logic).
- In fuzzy logic, knowledge is interpreted as a collection of elastic or, equivalently, fuzzy constraint on a collection of variables.
- Inference is viewed as a process of propagation of elastic constraints.
- It does not matter if the variables are independent and normally distributed often required by traditional statistical theory.

There are two main characteristics of fuzzy systems that give them better performance for specific applications

- Fuzzy systems are suitable for uncertain or approximate reasoning, especially for the system with a mathematical model that is difficult to derive.
- Fuzzy logic allows decision making with estimated values under incomplete or uncertain information.

Fuzzy Logic: Basic Concepts

Fuzzy logic theory (Zadeh, 1996) was first introduced by L. Zadeh of UC/Berkeley in the 1960s as a tool for modeling the uncertainty of natural language, and has been commonly employed for supporting intelligent systems. Fuzzy logic provides an inference mean that enables approximate human reasoning capabilities to be applied to knowledge-based systems. It was speci□cally invented to mathematically represent uncertainty and vagueness, and to provide formalized concepts for dealing with the imprecision intrinsic to many problems (Fullr, 1999) (Salmani *et al.*, 2008). This technology has proven efficiency in a various applications such as decision support and intelligent control, especially where a system is difficult to be characterized.

Fuzzy logic' foundations have become firmer, its applications have grown in number and variety, and its influence within the sciences, especially-computer, mathematical, and physical sciences has become more visible and more substantive (Zadeh, 1996). Fuzzy logic technology has achieved impressive success in diverse engineering applications ranging from mass market consumer products to sophisticated decision and control problems.

The importance of fuzzy logic derives from the fact that most manners of human reasoning and especially common sense reasoning are approximate in nature. Bellman and Zadeh write: "Much of the decision making in the real world takes place in an environment in which the goals, the constraints and the consequences of possible actions are not known precisely" (Bellman *et al.*, 1970) (Tomsovic *et al.*, 2000). This "imprecision" represents the core of fuzzy logic and fuzzy sets theory. This later was proposed as a generalization of classical set theory.

Fuzzy Inference Systems

Fuzzy inference is the process of formulating the mapping from a given input to an output using fuzzy logic. The mapping then provides a basis from which decisions can be made, or patterns discerned. Usually, and because of its multidisciplinary nature, fuzzy inference system is associated with a number of names, such as fuzzy modeling, fuzzy expert system, fuzzy logic controller, fuzzy-rule-based system, fuzzy associative memory, and simply (and ambiguously) fuzzy system (Mamdani et al., 1975). Mamdani's method (Sugeno,1985) and Sugeno-type (Giupponi et al., 2008) are the most commonly used fuzzy inference systems.

Basically, a fuzzy inference system consists of three steps: fuzzification, rules evaluation, and defuzzification:

1. Fuzzification is the process of changing a real scalar value into a fuzzy value. This process takes the inputs and determines the degree to which they belong to each of the appropriate fuzzy sets via membership functions.
2. The second step performs reasoning on the input data by following predefined fuzzy rules. Once the input data are processed by rules evaluation, the third step is activated.
3. Defuzzification step takes the task of converting back these input data into crisp values using a defuzzification method. The *Centroid Calculation* is one of the most popular defuzzification methods, which returns the center of area under the curve. Other methods are: Bisector, Middle of Maximum, Smallest of Maximum, and Largest of Maximum (Mamdani et al., 1975).

Innovative Models in Wireless Networks Using Fuzzy Logic

In recent years, the applications of intelligence control techniques have been developed quickly in wireless communication systems. The intelligent fuzzy logic control theory is used in these systems to model situations that take into consideration the uncertainty and imprecision of events; and to resolve problems for which precise knowledge of all constraints is insufficient or difficult to obtain. Schemes furnished with fuzzy logic theory have been shown to be useful in complex conditions with respect to non-linear and time-variant wireless communication systems. Hereafter, we briefly describe some works related to this issue.

Several works in the literature have proposed the use of fuzzy logic in the multi-hop cognitive radio networks (CRNs). Fuzzy logic is suitable in such environment characterized by uncertain, incomplete, and heterogeneous information. The authors in (Giupponi et al., 2008) proposed a fuzzy-based approach to handle the spectrum handoff in wireless cognitive networks. The spectrum handoff in CRNs means the switching to different frequency channel due to the fact that the aggregate interference caused at a primary user exceeds a certain threshold. Simulation results show that the fuzzy-based approach outperforms a solution based on fixed thresholds in terms of spectrum handoff rate and interference temperature at primary users.

Transmission power and QoS control models based on fuzzy logic have been developed in (Merentitis *et al.*, 2010) (Le *et al.*, 2007) (Tabakovic *et al.*, 2009) in order to allow a cognitive secondary user to satisfy the required transmission rate and quality, while minimizing interference to primary users and other concurrent secondary users. The authors is (Merentitis et al., 2010) considered important features such as the interference level at primary users caused by secondary users, the distance between primary and secondary users, and the received power difference of secondary users at base stations. Simulations results have shown that the fuzzy model can achieve lower power consumption and achieve lower outage probability compared with a conventional fixed-step model. Other features are considered in (Tabakovic *et al.*, 2009) such as number of users, mobility level, and update interference time interval. Other fuzzy-based solutions are found in the literature to handle different kinds of problems in cognitive radio networks such as channel Selection (Al-Fuqaha *et al.*, 2008), traffic modelling (Chen *et al.*, 2009), cooperative spectrum sensing (Qin *et al.*, 2009), and cross-layer configuration (Jiang *et al.*, 2009), etc.

Fuzzy logic is also used in Vehicular Ad hoc Networks (VANETs). Chen *et al.* (Chen *et al.*, 2009) have proposed the use of fuzzy logic theory to handle location errors caused by the mobility of vehicles in VANETs. A novel location prediction technique (LPVN) is presented based on many features such as vehicle position information, digital map data and traffic knowledge. Simulation results have shown that LPVN decreases the average of error prediction compared with conventional state-of-the-art models. A fuzzy-AODV-based QoS routing protocol (Fcar) is presented in (Wang *et al.*, 2009) to deal with the rapidly changing topology of VANETs. The fuzzy logic QoS control is used to make decision under multiple selection criteria to find out comparatively more stable routes. The authors proved by simulations that Fcar has its adaptability in VANETs and outperforms existing routing protocols in each of the evaluating criteria.

Wireless Mesh Networks (WMNs) is also one of the applied research area of the fuzzy logic theory. The authors in (Gomes *et al.*, 2010) have introduced an extended version of the OLSR protocol (Jacquet *et al.*, 2004), called OLSR-FLC (Fuzzy Link Cost). OLSR-FLC aims to improve the user perception, optimize the network resources, and achieve the QoS and the QoE requirements for multimedia applications. The fuzzy logic theory is used to solve the problem of using multiple metrics for routing. A fuzzy-based MAC scheme (FLSAC) is proposed in (Djahel *et al.*, 2009) to counter the unfairness on channel access in WMNs due the misbehaviour of selfish and greedy nodes. Greedy nodes may decline the proper use of MAC protocol in order to increase their bandwidth shares at the expense of the well-behaving nodes. Simulation results have proved that the fuzzy-based model has a high robustness and a quick detection of any cheater node.

Several works have also interested by the application of this theory in the context of mobile ad hoc networks. In (Oliveira *et al.*, 2004), the authors proposed a new model to investigate the use of fuzzy logic theory for assisting the TCP error detection mechanism in an ad hoc network. An elementary fuzzy logic engine was presented as an intelligent technique for discriminating packet loss due to congestion from packet loss by wireless induced errors. The results have shown that the fuzzy engine may distinguish congestion from channel error conditions, and consequently assist the TCP error detection (Oliveira *et al.*, 2004). Reznik in (Reznik *et al.*, 2004) has investigated the issues for improving the reliability and accuracy of the decisions in wireless ad hoc networks. He proposed an approach that offers a way of integrating wireless units measurement results with association information available or priori derived at aggregating nodes. This approach is used for describing both wireless units results and association information with consideration given to both Neuro-Fuzzy and probabilistic models and methods. The information sources available in the system are classified according to the model (fuzzy or probabilistic), which seems more feasible to be applied (Reznik *et al.*, 2004).

The researchers have explored the efficiency of the fuzzy logic theory control in many other kinds of networks. Edwards *et al.* (Edwards *et al.*, 1997) have reported encouraging results in a fuzzy algorithm driven hand-off operation for a cellular system where the cell size is reduced to 10–100 m. The algorithm combines the strength of the received signal with distance measurements to yield a hand-off factor, which decreases monotonically as a mobile device moves away from the base station, thus tracking the actual signal closely. Xia *et al.* has proposed in (Xia *et al.*, 2007) a fuzzy logic control based QoS management (FLC-QM) scheme for Wireless Sensors Networks (WSNs) with constrained resources and unpredictable environment. This scheme deals with the impact of unpredictable changes in traffic load on the QoS of WSANs. It utilizes a fuzzy logic controller inside each source sensor node to adapt the sampling period to the deadline miss ratio associated with data transmission from the sensor to the actuator. The deadline miss ratio is maintained at a pre-determined desired level so that the required QoS can be achieved. The FLC-QM model has the advantages of generality and simplicity.

In (Wang *et al.*, 2008) a fuzzy logic based co-operative MAC protocol (FLCMAC) is proposed to cooperate amongst network flows and dynamically adjust access probability of each low priority flow affecting the high priority flows to satisfy their QoS requirement. The simulation results have indicated that compared to the enhanced distributed channel access (EDCA) scheme of 802.11e, the FLCMAC gives better performances in terms of throughput and delay under moderate and heavy background traffic both in single-hop and multi-hop scenarios. This work addresses the problem of spatial reference estimation in mobile scenarios. Novel techniques based on fuzzy logic are introduced to enhance the performance of a tracking system. Specifically, the model-free function approximation capability of fuzzy logic is used to obtain resolution angle estimates from the spatial spectral density. These estimates are used to improve the resolution of the tracker. This work also develops a fuzzy controller for acting as an interpolative supervisor of different trackers that apply in different operating conditions of the dynamic nonlinear system. One of the features of the proposed technique is its low computational burden (Perez-Neira *et al.*, 1997).

INTELLIGENT QoS TRAFFIC ADAPTATION MODEL

As shown with the models described above, the fuzzy logic theory control promises to be an efficient solution and may constitutes a best alternative to the conventional methods based on the complex optimization and overloaded traffic engineering mechanisms.

In the aim of better exploring the efficiency of this theory in highly dynamic networks, we propose a fuzzy traffic regulation QoS model in ad hoc networks for improving the congestion control and traffic regulation, based on the feedback information received from the MAC layer. This is exploration is useful first, because of the dynamic nature of buffer occupancy and congestion at a mobile device, second, because of the uncertain nature of information in wireless ad hoc networks due to the network mobility. The proposed fuzzy regulation model combines two mechanisms; a fuzzy logic approach for best-effort traffic regulation and a fuzzy logic approach for threshold buffer management, as described in part in our previous work (Khoukhi et al., 2010). The feedback delay information received from the MAC layer is the key parameter used by our model to adjust dynamically the transmission of traffic according to the network conditions in order to assure a good utilization of resources, especially when best effort traffic coexists with real time traffic.

In order to deal with the dynamic buffer occupancy and the imprecision nature of MANET information, a fuzzy logic threshold selection mechanism is also used. Using fuzzy logic, our model investigates the ability of fuzzy thresholds to adapt to the dynamic conditions of the network over the classical inflexible thresholds. The notion of threshold is practical for discarding data packets and adapting the traffic service depending on the occupancy of buffers. As such, the threshold function has a significant influence on the performance of networks in terms of both packets average delay and throughput. Therefore, the selection of a particular threshold may be decisive to a good control of congestion. The selection in the proposed approach is based on fuzzy logic. The performances evaluation of our fuzzy model is realized under different network conditions in terms of mobility and scalability.

Best-Effort Traffic Regulation

FuzzyCCG-1 explores the efficiency of fuzzy logic to perform the regulation of best-effort traffic. The feedback delay information received from the MAC layer is used by FuzzyCCG-1 to adjust its operation. This feedback delay measure represents the packet delay measured by the IEEE 802.11 MAC which is used as an underlying layer in FuzzyCCG. The measure of the packet delays is performed as follows: at the reception of a packet by the MAC layer, the later listens to the channel and differs the access to the channel according to the CSMA/CA algorithm. When the MAC gets access to the channel, then RTS-CTS-DATA-ACK packets are exchanged. The reception of the ACK packet by the transmitter means that the packet was successfully received by the receiver. The time taken to send the packet between the transmitter and the receiver including total differed time represent the packet delay. This delay is calculated as the difference between the time that a packet is passed to the MAC layer (from the upper layer), and the time of reception of ACK packet from the receiver. Note that the packet delay can reflect well the network state: a high delay signifies that a possible situation of congestion has occurred in the network. Thus, when one or more packets have delays bigger than a certain value, the traffic regulation process should be triggered in order to reduce the traffic because there may be a situation of congestion in the network.

We have used SWAN model (Ahn et al., 2002) as a component to build FuzzyCCG model (Figure 2). This model contains some functionality such as routing scheme, admission controller, and classifier. The local available bandwidth at each node is estimated by the admission controller, which decides to admit or reject a new flow. The classifier differentiates between flows in terms of QoS requirements; best-effort flows and real-time flows, in order to delay the best-effort packets. Note that even the `admission control is performed to guarantee enough available bandwidth before accepting new flow; the congestion may be experienced in the network because of mobility of nodes. Therefore, it is of central importance to assure the traffic regulation. The classified best-effort packets are regulated using a fuzzy logic system according to the application requirements and the network state. The fuzzy regulation process is performed in three steps: fuzzification, inference rules evaluation, and defuzzification.

The regulation control process of best-effort traffic in FuzzyCCG-1 is performed in response to the feedback delay by a means of fuzzy logic system. Three steps are considered by FuzzyCCG-1:

1. In the first step (fuzzification), the feedback delay measurement received from the MAC layer represents the fuzzy input parameter for FuzzyCCG-1. The fuzzy output parameter is represented by the traffic regulation rate. These two parameters have to be converted into fuzzy sets. Note that a fuzzy set may contain elements that have different degrees of membership in various sets, whereas in an ordinary set an element needs to have a full membership in a set in order to be considered as a set member. Hence, if the delay-measurement parameter is considered in an ordinary set, it can only be either low or high and not both simultaneously. However, the delay-measurement in a fuzzy set can be classified as: not high, medium, or quite low. Thus, the membership of an element may be possible over various fuzzy set but with different degrees. The membership function represented as a line or curve, indicates how to map each input (i.e. delay-measurement) or output (i.e. regulation rate) parameters in order to obtain their membership values.
2. FuzzyCCG-1 considers the following sets: low (L), medium (M), and high (H). The threshold values of the fuzzy sets are presented in Figure 3. Let consider that the threshold for low delay-measurement is 2

Figure 2. Proposed fuzzy CCG model

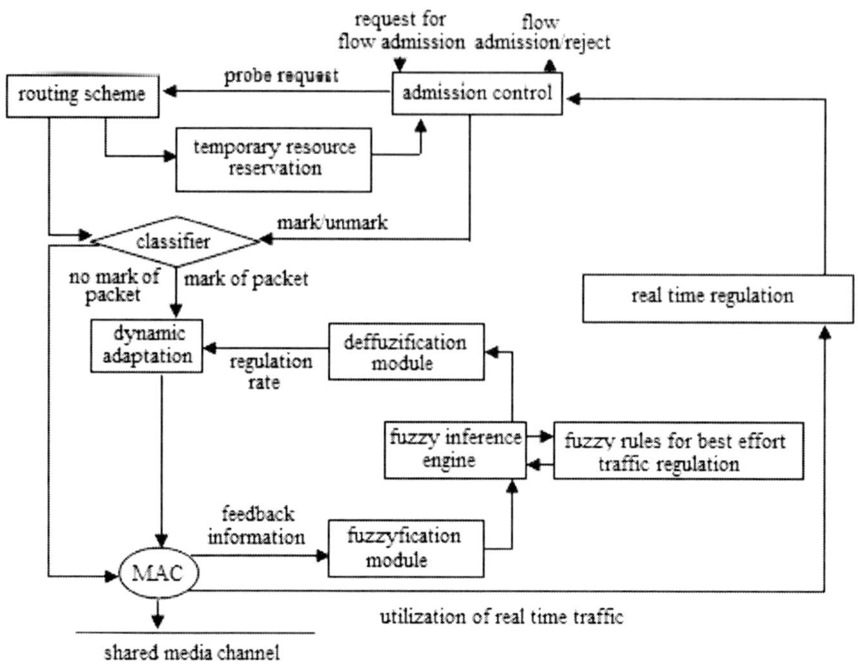

msec, for medium delay-measurement it is 3 msec, and for high delay-measurement it is 4 msec. Then, by mapping a current delay-measurement onto the graph of the membership function, the delay will be allocated a membership value in each set that is between 0 and 1. For example, in Figure 3.a if the current delay-measurement is 2.3 msec, then this value can be fuzzified into low delay with a degree of 0.7. It can also be fuzzified into medium delay with a degree of 0.3 and into a high delay with a degree of 0.

3. In the second step (inference engine), the rules controlling the behavior of the traffic in MANET are presented as a set of rules "if (…) then (…)". Throughout the rules evaluation process, these fuzzy rules are applied over the fuzzy sets using an inference engine. The inference control of the traffic regulation rate is illustrated by the following rule:

If the delay-measurement increases, then reduce the actual traffic rate

When the delay-measurement is high, this means that packets sent by the source mobile device take more time to be received by the destination device, which can be interpreted that a situation of congestion has may be occurred. Consequently, a decrease in the traffic transmission has to be performed to reduce the congestion level in the network, and the decision-making logic of the traffic regulation rate follows the delay-measurement parameter.

4. In the last step (defuzzification), the resulted decision sets are converted into numerical values. We have chosen the method of mean of maxima (MoM) (Zimmermann, 1991) to perform the defuzzification, because of its light computational complexity. In MoM, the average of the elements that reach the maximum grade in a fuzzy set is used to evaluation result.

Figure 3. Fuzzification and defuzzification process

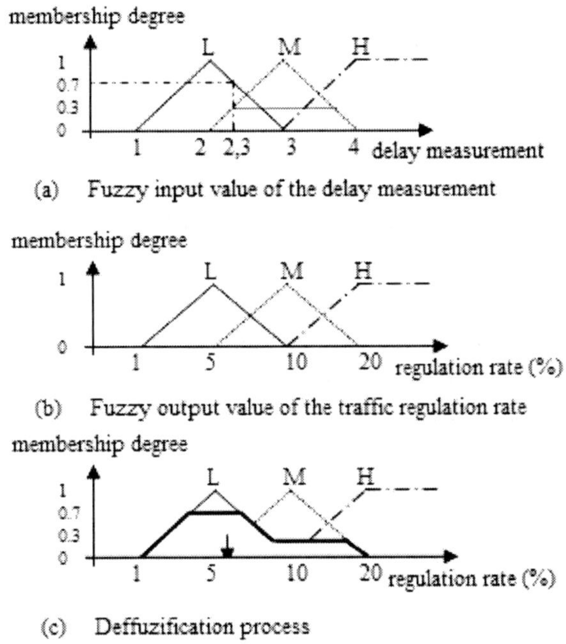

Adaptive Fuzzy Congestion System

In order to improve the reliability of the proposed model, we present in this section an adaptive fuzzy congestion system. Effective control of network congestion requires the design of new congestion control algorithms to estimate the dynamic behaviour of the traffic in node's buffer in order to send the congestion control notifications to the sources early enough to prevent packets losses caused by unexpected traffic bursts. Therefore, applying fuzzy logic control to deal with this issue may be very interesting, because of the dynamic nature of buffer occupancy and congestion at a node. It is expected that the proposed fuzzy congestion controller (Figure 4) will offer better adaptability under varying network conditions by better tuning of rules without intervention of the operator.

This system is conceived as a nonlinear controller of which the input-output relationship can be expressed by using a small number of linguistic rules or relational expressions. By monitoring

Figure 4. Fuzzy congestion system

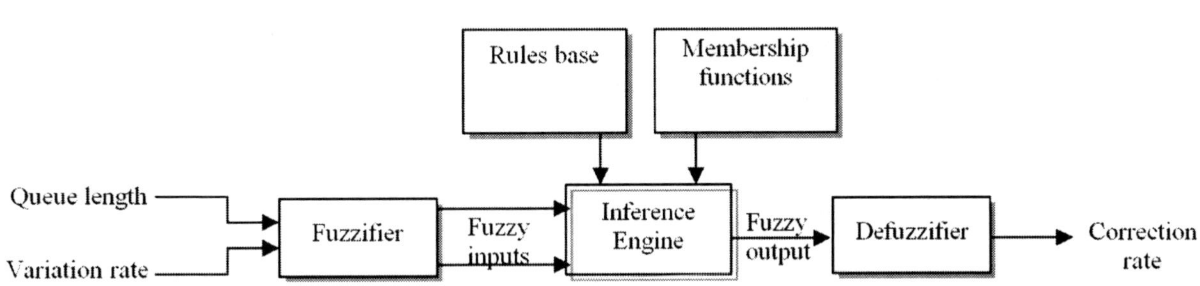

the rate of change in queue length (variation rate) in addition to the queue length, we are able to provide a measure of queue state, and by using explicit rate congestion notifications we can make source nodes more responsive to sudden changes in the network traffic volume.

Fuzzy logic promises to offer an efficient means for buffer management by using adequate thresholds that deal with the imprecise information caused by the dynamic topology of ad hoc networks. FuzzyCCG applies a fuzzy technique based on fuzzy sets theory. The later extends the classical logic set {0, 1} to use linguistic variables (e.g. full buffer, merely full buffer, empty buffer).

It is observed that the classical thresholds methods are very restrictive, because the selection of threshold is based on a single value. Thus, the utilization of a buffer may be either "poor" or "surcharged". On the one side, when the selected value is small (e.g. 30% of capacity), then the admission of new packets is possible only when the buffer occupancy is low. This means a poor utilization of the buffer; since most of incoming packets are rejected even if the buffer is almost unfilled. On the other side, when the selected value is big (e.g. 90% of capacity), problems may happen when burst traffic occurs. In a short period of time the transmission of packets generated by a bursty traffic can be very changing (from small to "near-peak" rate).

On the other side, most of events occurring in an ad hoc network are dynamic and random. Therefore manually predefining a value for threshold is not suitable. Additionally, it is important to note that the rate of packets arriving on a particular node is non static. The threshold value divides the buffer into an "admitted" part and a "no-admitted" part. Figure 5 illustrates the difference between the classical and fuzzy threshold buffers as follows, let consider that the threshold of the buffer is equal to 60%. In this scheme, the occupancy level may range from 0 to 60%. When the buffer occupancy is superior to 60%, no incoming packets is accepted in the buffer. Therefore, the change in decision making from "admit" state to "no-admit" state is performed from 60-61%. This means that a small variation in the buffer occupancy may influence the decision making of incoming packets.

FuzzyCCG extends the two-discrete states "admit" and "no-admit" of the buffer occupancy by using fuzzy logic. The aim of introducing fuzzy

Figure 5. Classical and fuzzy buffer schemes

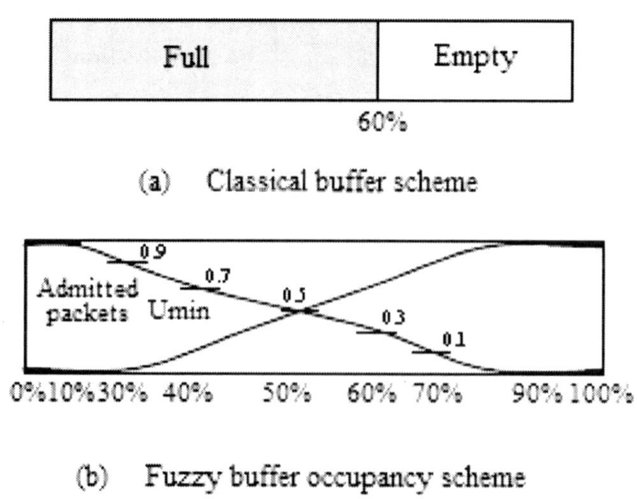

logic is to develop a more realistic representation of buffer occupancy that helps to offer an efficient decision making. Hence, the definition of "buffer occupancy" considers the two fuzzy cases of "getting full" and "not getting full", rather than "admit" and "no-admit" in the existing approaches. This fuzzy representation replaces the two-discrete sets by a continuous set membership which determine the threshold depending on the fullness of the buffer. Several membership functions may be used for that purpose: "triangular", trapezoidal", or "sigmoid" function. These functions can give a representation about the buffer fullness level. In FuzzyCCG, we used the sigmoid membership function. This choice is based on the fact that this function has the potential to represent the gradual transitions between different states of buffer occupancy.

As shown in Figure 5b, the "admit" membership function is inversely proportional to the occupancy fullness level of the buffer. Thus, when the occupancy fullness is small, the value of the "admit" membership function is big. At high fullness occupancy levels, the "admit" membership function value becomes small. When the value of the "no-admit" membership function is getting big, then only a small quantity of packets will be permitted to enter the buffer. In Figure 5b, the value of the "admit" membership function is represented by the symbol u_{adm}. The inference engine related to the buffer occupancy management is explained by the following rules:

When the value of the "admit" membership function is big, then increase the accepted incoming packets into buffer

When the value of the "admit" membership function is small, then reduce the accepted incoming packets into buffer

The rejection of packets is controlled based on the degree of fullness of the buffer, as shown in Figure 5b. For instance, when the buffer is occupied at 40%, this means that the value of u_{adm} is about 0.7 (i.e. the amount of packets admitted is about 70%). Then, about "30%" of incoming packets will be not admitted. Note that the fuzzy threshold approach covers the continuous set of values representing possible buffer occupancy (i.e. from 0 to u_{adm}). This is opposite to the classical threshold approaches that hold only one predefined single value. Hence, the threshold selection is become more flexible thanks to fuzzy logic theory.

Performance Evaluation

The presented model is implemented using the scalable ns-2 simulator. The aim is to evaluate whether the results will outperforms other existing models such as SWAN and the original model. The later refers to IEEE 802.11 wireless networks without FuzzyCCG mechanisms. Each mobile host has a transmission range of 250 meters and shares an 11 Mbps radio channel with its neighbouring nodes. The simulation considers multiple scenarios of real-time and TCP best-effort traffic. The real-time traffic is modeled as 4 voice and 4 video flows. The TCP traffic is modeled as a mixture of FTP and Web traffic. Web traffic represents micro-flows, whereas FTP traffic corresponds to macro-flows. The video and voice flows representing real-time traffic are active and monitored a duration of 100 seconds. Video traffic is modeled as 200 Kbps constant rate traffic with a packet size of 512 bytes. Voice traffic is modeled as 32 Kbps constant rate traffic with a packet size of 80 bytes. The simulation considers a multihop network of 50 mobile nodes in an area that has a rectangular shape of 1500m x 3000m.

The scalability impact of the increasing number of TCP flows on the average end-to-end delay and throughput of traffic is presented in Figure 6. Figure 6a illustrates a significant difference in terms of the average delay between FuzzyCCG and the original model. The average delay in FuzzyCCG grows slowly with the increasing number of TCP flows, and it remains between 2 and 3 msec. In contrast, the average delay in the original model grows from 7 to 31 msec as the number of TCP flows increases from 2 to 12 flows. Hence, the gain achieved by FuzzyCCG in terms of the average end-to-end delay, is by about 74-92%.

Figure 6c shows the average end-to-end delay in both FuzzyCCG and SWAN models. It is shown that the average delay is almost inferior to 3 msec in the proposed model, whereas in SWAN model the average delay is around 5 msec. This means that the achieved gain offered by FuzzyCCG in that case is about 49% in terms of average delay.

Figure 6 illustrates the impact of a growing number of TCP flows on the average throughput for the different compared models in the simulations. The average throughput of the TCP traffic in FuzzyCCG is almost the same as in the original model in most cases, as shown in Figure 6b. At lower number of TCP flows, the average throughput in the original is superior to that in FuzzyCCG. Similar results are observed in Figure 6d between FuzzyCCG and SWAN.

The previous results confirm that by adopting the FuzzyCCG mechanisms, we can achieve a reduction in terms of average end-to-end delay by about 74-92%, and 49% in comparison to respectively, the original and SWAN models, with almost the same throughput.

Figure 7 investigates the impact of mobility on the performances of FuzzyCCG. The real-time traffic is modeled in the same manner as discussed previously. The best-effort TCP flows consist of 5 web flows and 5 FTP flows. The random waypoint mobility model (Johnson et al., 1996) is implemented at each node in the network.

Figure 6. The scalability impact of the increasing number of TCP flows on the average end-to-end delay and throughput of traffic

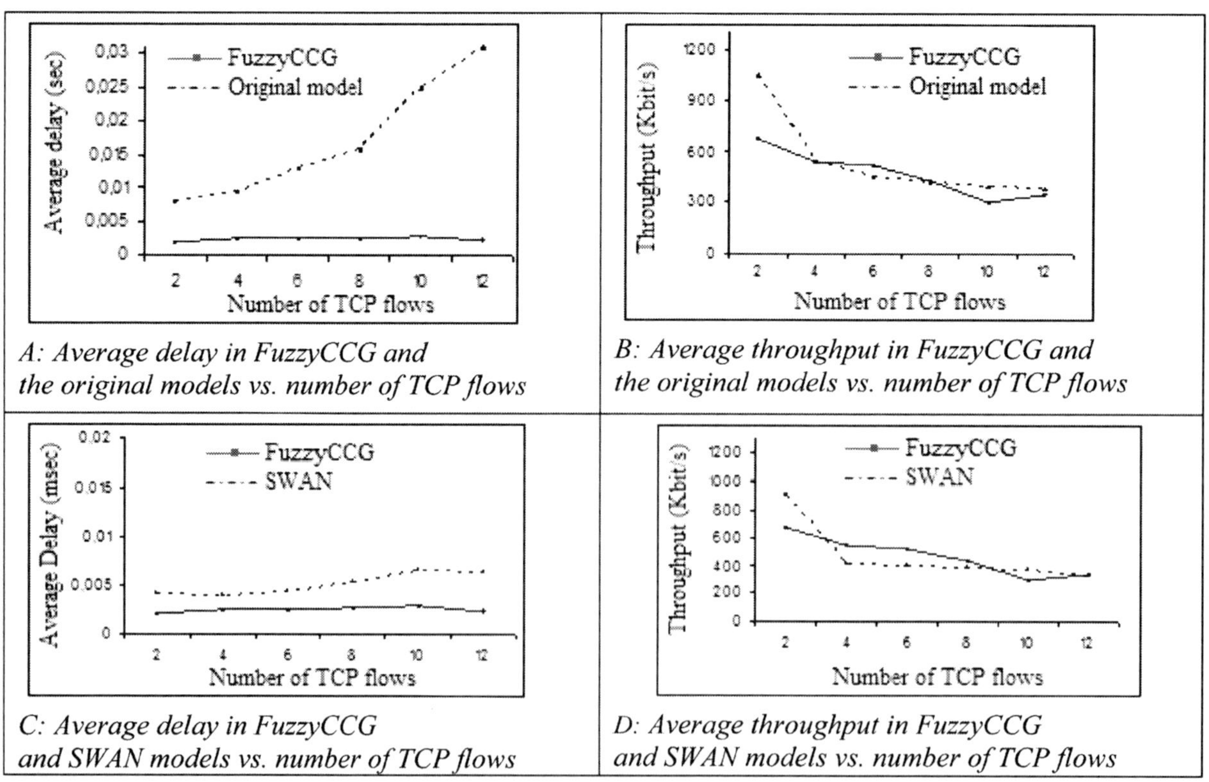

A: Average delay in FuzzyCCG and the original models vs. number of TCP flows

B: Average throughput in FuzzyCCG and the original models vs. number of TCP flows

C: Average delay in FuzzyCCG and SWAN models vs. number of TCP flows

D: Average throughput in FuzzyCCG and SWAN models vs. number of TCP flows

Figure 7. Investigations of the impact of mobility on the performances of FuzzyCCG

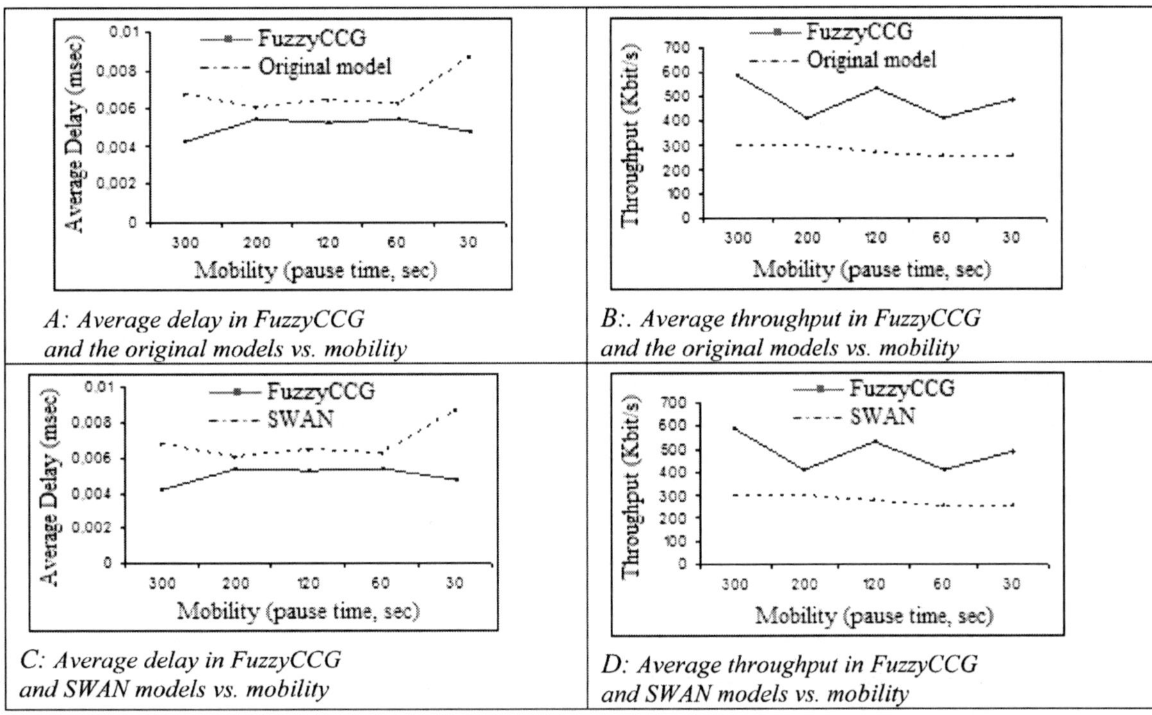

A: Average delay in FuzzyCCG and the original models vs. mobility

B:. Average throughput in FuzzyCCG and the original models vs. mobility

C: Average delay in FuzzyCCG and SWAN models vs. mobility

D: Average throughput in FuzzyCCG and SWAN models vs. mobility

We observe in Figure 7a that the average end-to-end delay in FuzzyCCG increases slowly. The average delay in the proposed model remains almost less than 5.4 msec, whereas the average delay in the original model grows from 25 to 38 msec. This means that the proposed FuzzyCCG achieves a reduction in terms of average delay by about 79-87%. On the other hand, it is observed in the Figure 7b that the throughput of TCP best-effort traffic decreases slowly in the original model as the mobility increases. The average throughput in FuzzyCCG is superior to that of the original model by about 33% for different mobility scenarios.

Figure 7c illustrates the average end-to-end delay with different mobility scenarios in both FuzzyCCG and SWAN models. It is observed that the average delay of traffic in FuzzyCCG increases slowly as the mobility increases. For different mobility scenarios, the average delay offered by FuzzyCCG is about 10-36% better than that offered by SWAN. It is shown in Figure 7d that for different mobility scenarios, the throughput in FuzzyCCG is better than in SWAN model by about 43%. During the mobility of nodes, some flows are dropped in both SWAN and FuzzyCCG models because of the difficulty in capturing the dynamics of the environment in the ad hoc network.

CONCLUSION AND FUTURE RESEARCH DIRECTIONS

Conclusion

The last few years have witnessed a wealth of research ideas on ad hoc networking that are moving rapidly toward implemented standards. Although ad hoc networks research is a relatively new field it is gaining more popularity for various new applications. For instance, multimedia applications that open up for converged services and new applications is quickly becoming a key focus area for wireless communications. With the increase in both the bandwidth of wireless channels and the computing power of mobile devices, it is expected that video and audio services will be offered over ad hoc networks in the future. However, enabling multimedia communications over such networks is remaining a challenging task for both academic and industrial communities. Video and audio services typically require stringent bandwidth and delay guarantees. This makes the deployment of quality of service (QoS) mechanisms a vital need for the satisfaction of user's requirements. This kind of applications generates traffic at varying rates and usually requires the network to be able to support such a changing rate. Besides, in a network consisting of mobile nodes, the connection path between a source and a destination is constantly broken and has to be frequently updated.

Therefore, providing QoS guarantees is crucial for supporting disparate services envisioned for the future wireless ad hoc networks. Despite the efforts made to alleviate this issue, there still exist a number of barriers to the widespread deployment of real-time applications. The most prominent one is how to ensure the quality delivery for achieving QoS over multiple layers to support real-time traffic even in the case of mobility of devices; this is accomplished by so-named cross layer QoS frameworks or QoS models. It is important to note that the existing QoS models developed for wired networks can not be deployed directly within ad hoc networks. Difficulties with these models lie in the fact that they are not adapted to different node states and resource variation, as in ad hoc environments the available bandwidth for each node varies with time since the nodes are mobile and the medium is shared.

In this chapter, we presented two kinds of QoS models: first, *conventional models* which are based on traffic engineering mechanisms (e.g., resources reservation) and optimization techniques (multi-objective QoS optimization, etc.). Most existing QoS models (e.g., CACP, OMAR, MPARC, INSIGNIA, etc.) in ad hoc networks that are based on traffic engineering mechanisms have been carried out under the assumption that the underlying QoS architecture is reservation-based. In such assumption, mobile nodes maintain some per-flow state information (e.g., flow identity, priorities, etc.) and source nodes use explicit control and reservation notifications (e.g., route reservation messages, etc.) to indicate their QoS requirements. The advantage of this kind of models is that it allows the source nodes to guarantee the requested QoS on an end-to-end basis with a high degree of accuracy. However, these models need for complex and signaling control mechanisms to refresh and update per-flow state information.

In some other existing conventional QoS models (such as SWAN, DACME, RSGM) based on traffic engineering mechanisms, there is no session information maintained at intermediate devices in support of communications between transmitter and receiver pairs. These models have the advantage of high adaptability as shown earlier. This is largely because of the light processing achieved by the intermediate mobile devices. However, in a large network with highest mobility scenarios, this kind of models suffer from the problem of false admission. This means that source nodes may have an old view about the real state of network resources (i.e., bandwidth) because of the dynamic nature of ad hoc network. This problem may occur when multiple source nodes check the state of network simultaneously, which leads to the admission of more traffic than what an intermediate node can support. Since in a mobile ad hoc network there is no centralized QoS controller, multiple sessions in a network may send probe messages to check current bandwidth availability at the same time. This may result in a situation where multiple sessions (sharing common mobile devices) send probe messages simultaneously and observe that the QoS resources are available, but in reality they are not.

On the other hand, the QoS optimization methods are split into two parts: task optimization performed on behalf of the mobile user and resources optimization performed on behalf of the network. These tasks control a network's response to user/traffic requirements and other stimuli, such as link failures or mobile node failures. The QoS models based on optimization methods can give an accurate estimation about the network's resources, and hence can achieve the prediction of congestion. However, when the mobility of devices is high, the prediction of congestion becomes very challenging task and the computation complexity is very heavy in such kind of models.

In order to alleviate the difficulties and challenges discussed above, some researchers have proposed to explore the efficiency of the intelligent learning methods such as fuzzy logic and neural networks; this is what we referenced in this chapter by the term *intelligent QoS models*. We have discussed briefly some intelligent QoS models and mechanisms that are capable of supporting real-time traffic as well as emerging data services in an integrated fashion. These models brought about the benefits of the advances in the areas of both artificial intelligence and wireless networking. Intelligent techniques such as fuzzy logic have proved their efficiency in many industrial areas. They are used to solve complex problems, and are characterized by changing their behaviour on the light of ongoing events without the intervention of the users. The intelligent fuzzy logic theory has been successfully applied to support QoS in many kinds of wireless networks WMN, VANET, WSN, as we have previously explained. The presented intelligent QoS models (FLCMAC, OLSR-FLC, LPVN, Fcar) have proven to be scalable and reliable. The validity and usefulness of these models are proved by simulations. The most important practical implication is that the fuzzy control can operate without difficulties in a high dynamic network.

After having described both conventional and intelligent QoS models, we have presented an intelligent fuzzy logic QoS model for MANETs. Applying the fuzzy logic theory to deal with QoS paradigm in MANETs is very interesting, because of the dynamic nature of buffer occupancy and congestion at a mobile device. In the proposed model, we have explored the efficiency of this theory improving the traffic regulation and congestion control based on the feedback delay information received from the MAC layer.

Future Research Directions

The obtained results indicated that the proposed model can support real-time traffic with low and stable delay for different network conditions, which is a remarkable benefit for delay- and jitter-sensitive real-time applications. These promising results show that the usage of fuzzy logic can address some of the challenges of wireless ad hoc networks such as the dynamics of topology, scalability and congestion aspects. Another advantage of using fuzzy logic to support future multimedia services is that it can offer better adaptability under varying network conditions by better tuning of rules without intervention of the operator. The intelligent fuzzy logic control is a robust artificial intelligence technology that can handle the dynamic nature of various processes.

Initial research has demonstrated the potential of fuzzy control mechanisms. Future research should mainly focus on further evaluating and optimizing the performance of intelligent QoS models. The following tasks should be conducted:

1. Extend the experiments of small scale ad hoc network for selecting a large ad hoc configuration (thousands of mobile devices). This extension would give more precision about the efficiency of the application of fuzzy theory to support QoS multimedia services.

2. Conduct experiments with other intelligent learning methods such as neural networks and genetic algorithm to compare the performances of these methods under different network conditions in terms of different QoS metrics (delay, packets loss, etc.).
3. Look at the practicality of combining multiple intelligent methods over different layers, for example using neural networks to estimate the occupancy of buffer, genetic algorithm to select the best path for data delivery, and fuzzy logic to perform the traffic regulation

REFERENCES

Ahn, G., Campbell, A. T., Veres, A., & Sun, L. H. (2002). SWAN: Service differentiation in stateless wireless ad hoc networks. *Proceeding of IEEE International Conference on Computer Communications (INFOCOM)*, (pp. 457–466). Orlando, FL: IEEE Press.

Al-Fuqaha, A., Khan, B., Rayes, A., Guizani, M., Awwad, O., & Ben Brahim, G. (2008). Opportunistic channel selection strategy for better QoS in cooperative networks with cognitive radio capabilities. *IEEE Journal on Selected Areas in Communications*, *26*(1), 156–167. doi:10.1109/JSAC.2008.080114

Bachir, A., Dohler, M., Watteyne, T., & Leung, K. K. (2010). MAC essentials for wireless sensor networks. *IEEE Communications Surveys & Tutorials*, *12*(2), 222–248. doi:10.1109/SURV.2010.020510.00058

Bellman, R. E., & Zadeh, L. A. (1970). Decision-making in a fuzzy environment. *Management Science Journal*, *17*(4), 141–164.

Calafate, C., Cano, J., Manzoni, P., & Malumbres, M. (2005). A QoS architecture for MANETs supporting real-time peer-to-peer multimedia applications. *Proceeding of IEEE International Symposium on Multimedia*, (pp. 193–200). Irvine, CA: IEEE Press.

Calafate, C., Oliver, J., Cano, J., Manzoni, P., & Malumbres, M. (2007). A distributed admission control system for MANET environments supporting multipath routing protocols. *Microprocessors and Microsystems*, *31*(4), 236–251. doi:10.1016/j.micpro.2007.01.002

Chakrabarti, S., & Mishra, A. (2001). QoS issues in ad hoc wireless networks. *IEEE Communications Magazine*, *39*(2), 142–148. doi:10.1109/35.900643

Chen, G., Song, M., Zhang, Y., Song, J., Teng, Y., & Wang, L. (2009). Fuzzy logic based traffic model considering cross-layer for future cognitive network. *Proceeding of Joint Conference on Pervasive Computing (JCPC)*, (pp. 627–632). Tamkang, Taiwan: IEEE Press.

Chen, L., Wei, S., & Shi, L. (2009). Research on location prediction of vehicular networks. *Proceeding of IEEE International Conference on Computer Science and Information Technology (ICCSIT)*, (pp. 558–561). Beijing, China: IEEE Press.

Crawley, E., Nair, R., Rajagopalan, B., & Sandick, H. (1998). *A framework for QoS-based routing in the internet*. RFC 2386. Retrieved from http://www.ietf.org/rfc/rfc2386.txt

Demetrios, Z. (2001). *A glance at quality of services in mobile ad-hoc networks*. Burns College of Engineering (Data File). Retrieved from http://www.cs.ucr.edu/ csyiazti/cs260.html

Djahel, S., & Nait-Abdesselam, F. (2009). FLSAC: A new scheme to defend against greedy behavior in wireless mesh networks. *International Journal of Communication Systems*, *22*(10), 1245–1266. doi:10.1002/dac.1027

Edwards, G., & Sankar, R. (1997). A predictive fuzzy algorithm for high performance microcellular handoff. *Proceedings of IEEE Global Communications Conference (GLOBECOM)*, (pp. 987–990). IEEE Press.

Fullr, R. (1999). *Introduction to neuro-fuzzy systems. Advances in Soft Computing Series*. Springer-Verlag.

Ganz, A., Ganz, Z., & Wongthavarawat, K. (2004). *Multimedia wireless networks: Technologies, standards, and QoS*. Upper Saddle River, NJ: Prentice Hall.

Giupponi, L., & Perez-Neira, A. I. (2008). Fuzzy-based spectrum handoff in cognitive radio networks. *Proceeding of IEEE International Conference on Cognitive Oriented Wireless Networks and Communications (CrownCom)*, (pp. 1–6). Singapore: IEEE Press.

Gomes, R. Jr, Cerqueira, W. A., & Abelem, A. (2010). A QoE fuzzy routing protocol for wireless mesh networks. *Lecture Notes in Computer Science*, 6157.

Haas, Z. J., Gerla, M, Johnson, D. B., Perkins, C. E., Pursley, M. B., Steenstrup, M., … Hayes, J. F. (1999). Guest editorial on wireless ad hoc networks. *IEEE Journal on Selected Areas in Communications – Special Issue on Wireless Networks, 17*(8): 1329–1332.

Hanzo, L., & Tafazolli, R. (2007). A survey of QoS routing solutions for mobile ad hoc networks. *IEEE Communications Surveys & Tutorials*, *9*(2), 50–70. doi:10.1109/COMST.2007.382407

Hejmo, M., Mark, B., Zouridaki, C., & Thomas, R. (2006). Design and analysis of a denial-of-service resistant quality-of-service signaling protocol for MANETs. *IEEE Transactions on Vehicular Technology*, *55*(3), 743–751. doi:10.1109/TVT.2006.873834

Jacquet, P., Muhlethaler, P., Clausen, T., Laouiti, A., Qayyum, A., & Viennot, L. (2004). Optimized link state routing protocol for ad hoc networking. *Proceedings of IEEE International Multi Topic Conference (INMIC)*. IEEE Press.

Jiang, H., Bao, Y., Li, Q., & Huang, Y. (2009). Multi-objective optimization of cross-layer configuration for cognitive wireless network. *Proceeding of IEEE international Conference on Dependable, Autonomic and secure computing (DASC)*, (pp. 379–383). Chengdu, China: IEEE Press.

Johnson, D. B., & Maltz, D. A. (1996). Dynamic source routing in ad hoc wireless networks. In Imielinski, T., & Korth, H. (Eds.), *Mobile computing* (*Vol. 353*, pp. 153–181). Kluwer Academic Publishers. doi:10.1007/978-0-585-29603-6_5

Khoukhi, L., & Cherkaoui, S. (2010). Intelligent QoS management for multimedia services support in wireless mobile ad hoc networks. *Computer Networks and ISDN Systems*, *54*(10), 1692–1706.

Lajos, H., & Rahim, T. (2009). Admission control schemes for 802.11-based multi-hop mobile ad hoc networks: A survey. *IEEE Communications Surveys & Tutorials*, *11*(4), 78–108. doi:10.1109/SURV.2009.090406

Laptop.org. (2008). Retrieved 18th November, 2008, from http://www.laptop.org

Le, H. S. T., & Liang, Q. (2007). An efficient power control scheme for cognitive radios. *Proceeding of IEEE Wireless Communications and Networking Conference (WCNC)*, (pp. 2559–2563). Hong Kong, China: IEE Press.

LEE, C. C. (1999). Fuzzy logic in control systems: Fuzzy logic controller - part I and II. *IEEE Transactions on Systems, Man, and Cybernetics*, *20*(2), 404–435.

Lee, S., Ahn, G., Zhang, X., & Campbell, A. (2000). INSIGNIA: An IP-based quality of service framework for mobile ad hoc networks. *Journal of Parallel and Distributed Computing*, *60*(4), 374–406. doi:10.1006/jpdc.1999.1613

Mamdani, E. H., & Assilian, S. (1975). An experiment in linguistic synthesis with a fuzzy logic controller. *International Journal of Man-Machine Studies*, *7*(1), 1–13. doi:10.1016/S0020-7373(75)80002-2

Merentitis, A., & Triantafyllopoulou, D. (2010). Transmission power regulation in cooperative cognitive radio systems under uncertainties. *Proceeding of IEEE International Symposium on Wireless Pervasive Computing (ISWPC)*, (pp. 134–139). Modena, Italy: IEEE Press.

Morgan, Y. L., & Kunz, T. (2003). PYLON: An architectural framework for ad-hoc QoS interconnectivity with access domains. *Proceeding of Annual Hawaii International Conference on System Sciences* (pp. 309–318). Big Island, HI: Univ. of Hawaii Press.

Oliveira, R., & Braun, T. (2004). A delay-based approach using fuzzy logic to improve TCP error detection in ad hoc networks. *Proc. of IEEE Wireless Communications and Networking Conference (WCNC)*, *5*(1), 1660–1665.

Perez-Neira, A., Lagunas, M. A., & Bas, J. (1997). Fuzzy logic for robust detection in wireless communications. *Proc. of IEEE International Symposium on Personal Indoor and Mobile Radio Communications (PIMRC)*, (pp. 1145–1149). IEEE Press.

Perkins, D., & Hughes, H. D. (2002). A survey on quality-of-service support for mobile ad hoc networks. *Journal of Wireless Communications and Mobile Computing*, *2*(5), 503–513. doi:10.1002/wcm.73

Qin, Q., Zhimin, Z., & Caili, G. (2009). A study of data fusion and decision algorithms based on cooperative spectrum sensing. *Proceeding of International Conference on Fuzzy Systems and Knowledge Discovery (FSKD)*, (pp. 76–80). Tianjin, China: IEEE Press.

Reznik, L., & Kreinovich, V. (2004). Fuzzy and probabilistic models of association information in sensor networks. *Proceedings of IEEE International Conference on Fuzzy Systems*, (pp. 185–189). IEEE Press.

Salmani, D., & Akbari, M. (2008). Fuzzy and research paradigms relationship: A mutual contribution. *Journal of Academic Research*, *6*(2).

Saunders, S. (1999). *Antennas and propagation for wireless communication systems: Concept and design*. New York, NY: John Wiley and Sons.

Sedoyeka, E., Hunaiti, Z., & Tairo, D. (2009). Analysis of QoS requirements in developing countries. *International Journal of Computing and ICT Research*, *3*(1), 18–31.

Sugeno, M. (1985). *Industrial applications of fuzzy control*. New York, NY: Elsevier.

Tabakovic, Z., Grgic, S., & Grgic, M. (2009). Fuzzy logic power control in cognitive radio. *Proceeding of IEEE International Conference on Systems, Signals, and Image Processing (IWSSIP)*, (pp. 1–6). Chalkida, Greece: IEEE Press.

Tomsovic, K., & Chow, M. K. (2000). *Contributors to tutorial on fuzzy logic applications in power systems*. IEEE-PES Winter Meeting in Singapore. Singapore: IEEE Press.

Tsoukalas, L. H., & Uhrig, R. E. (1993). *Fuzzy and neural approaches in engineering*. New York, NY: John Wiley & Sons.

Wang, J., Zhai, H., Fang, Y., Shea, J. M., & Wu, D. (2006). OMAR: Utilizing multiuser diversity in wireless. *IEEE Transactions on Mobile Computing*, *5*(12), 1764–1779. doi:10.1109/TMC.2006.186

Wang, M., & Kuo, G. S. (2005). An application-aware QoS routing scheme with improved stability for multimedia applications in mobile ad hoc networks. *Proceedings of IEEE Vehicular Technology Conference (VTC)*, (pp. 1901–1905). Stockholm, Sweden: IEEE Press.

Wang, X. B., Yang, Y. L., & An, J. W. (2009). Multi-metric routing decisions in VANET. *Proceeding of IEEE international Conference on Dependable Autonomic and Secure Computing (DASC)*, (pp. 551–556). Chengdu, China: IEEE Press.

Wang, Z.-X., Xia, H.-L., & Ding, W. (2008). A fuzzy logic cooperative MAC for MANET. *Journal of China Universities of Posts and Telecommunications*, *15*(1), 55–60. doi:10.1016/S1005-8885(08)60062-X

Xia, F., Zhao, W. H., Sun, Y. X., & Tian, Y. C. (2007). Fuzzy logic control based QoS management in wireless sensor/actuator networks. *Sensors (Basel, Switzerland)*, *12*(7), 3179–3191. doi:10.3390/s7123179

Xiang, X., Wang, X., & Yang, Y. (2010). Stateless multicasting in mobile ad hoc networks. *IEEE Transactions on Computers*, *59*(8), 1076–1090. doi:10.1109/TC.2010.102

Xiao, H., Khee Chaing, C., & Winston, S. K. G. (2003). *Quality of service models for ad hoc wireless networks. In 2003 Handbook of Ad Hoc Wireless Networks*. Boca Raton, FL: CRC Press.

Xiao, H., Seah, W. K. G., Lo, A., & Chaing, A. K. (2000). Flexible QoS model for mobile ad-hoc networks. *Proc. of IEEE Vehicular Technology Conference,* vol. 1, (pp. 445-449). IEEE Press.

Yang, Y., & Kravets, R. (2004). Throughput guarantees for multi-priority traffic in ad hoc networks. *Proceeding of IEEE International Conference on Mobile Ad-Hoc and Sensor Systems*, (pp. 379–388). Ft. Lauderdale, FL: IEEE Press.

Yang, Y., & Kravets, R. (2005). Contention-aware admission control for ad hoc networks. *IEEE Transactions on Mobile Computing*, *4*(4), 363–377. doi:10.1109/TMC.2005.52

Yang, Y., & Kravets, R. (2007). Throughput guarantees for multi-priority traffic in ad hoc networks. *Ad Hoc Networks*, *5*(2), 228–253. doi:10.1016/j.adhoc.2005.11.002

Zadeh, L. A. (1996). Fuzzy logic = computing with words. *IEEE Transactions on Fuzzy Systems*, *4*(2), 103–111. doi:10.1109/91.493904

Zeinalipour, D., Aristeidou, S., & Kazeli, S. (2009). *IP quality of services* (in Greek). Burns College of Engineering (Data File). Retrieved from http://www.cs.ucr.edu/ csyiazti/downloads/papers/ipqos/ papers/ip-qos.pdf

Zhang, B., & Mouftah, H. T. (2005). QoS routing for wireless ad hoc networks: Problems, algorithms, and protocols. *IEEE Communications Magazine*, *43*(10), 110–115. doi:10.1109/MCOM.2005.1522133

Zhang, Q., & Zhang, Y. (2008). Cross-layer design for QoS support in multihop wireless networks. *Proceedings of the IEEE*, *96*(1), 64–76. doi:10.1109/JPROC.2007.909930

Zimmermann, H. J. (1991). *Fuzzy sets theory–And its applications*. Boston, MA: Kluwer Academic Publishers.

KEY TERMS AND DEFINITIONS

Cross-Layer Model: Is a framework for achieving QoS over one or multiple layers.

Fuzzy Logic Theory: it provides an inference mean that enables approximate human reasoning capabilities to be applied to knowledge-based systems.

Mobile Ad Hoc Networks (MANET): Is a collection of mobile nodes that are connected via wireless links without using an existing fixed infrastructure or centralized management.

QoS Routing: It addresses the best and simple way to find a path through the network that is capable of supporting the requested level of QoS.

Quality of Service (QoS): Is a set of service requirements to be met when transporting traffic packets from the source to its destination.

Signalization Model: It defines control messages that deal with the changes in the resources availability of the network.

Traffic Regulation Process: The adaptation of transmission rates according to networks conditions.

Chapter 10

Packet Scheduling in Home and Business Femtocell Networks

Volkan Sevindik
Airvana Inc., USA

Oguz Bayat
Airvana Inc., USA

ABSTRACT

This chapter presents the performance of the existing macrocell schedulers in the home femtocell environment and provides a scheduling solution for the traffic differentiation problem in the business femtocell environment. The traffic classification concept is applied to Best Effort (BE) traffic in the femtocell network, and a Grade of Service (GoS) metric is used as a scheduler parameter to classify the traffic in terms of total delivered throughput. CDMA2000 1xEVDO Rev, a femtocell product, is used to conduct two different throughput experiments in order to collect real Data Rate Request (DRC) values. For both existing and proposed scheduling methods, the first experiment consists of 4 BE users, and the second experiment consists of 4 BE and 2 Voice over IP (VoIP) users. Collected DRC values are used in computer simulations to assess the performance of both the existing and the proposed scheduling methods. Average delivered throughput, throughput fairness, BE traffic differentiation error, and VoIP packet delay bound error are used as key indicators to show the performances of both scheduling techniques.

INTRODUCTION

Femtocell is a solution to indoor data capacity and indoor coverage problems of 3G/4G wireless communication systems. In this chapter, CDMA2000 1xEVDO based *femtocell* is considered for performance investigation of existing macrocell packet scheduling methods in home *femtocell* environment (Yavuz, 2006; Black, 2002; Bhushan, 2006; Ching, 2003; Parry, 2002). This chapter also proposes new scheduling methods for *BE* traffic differentiation for business *femtocell* environment. Scheduler is an essential part of the wireless radio resource management module in *femtocell* product.

The task of scheduler is to dedicate required amount of resources to users based on their network related parameters. The most commonly used conventional scheduling methods are Round Robin (*RR*), maximum data rate request (*maxDRC*), proportional fairness (*PF*), average DRC (*aveDRC*), exponent DRC (*expDRC*), and *maxDRC-PF*. The performances of these schedulers were investigated under two different traffic

DOI: 10.4018/978-1-4666-0017-1.ch010

scenarios which are 4 *BE* only and 4 *BE* + 2 *VoIP*. Average delivered throughput, throughput fairness, and *VoIP* delay bound error are key performance indicators (KPIs) to assess and compare the schedulers' performances.

Scheduler is also used to provide differentiated services to users in the network. *BE* traffic is the most common type traffic created by applications such as web surfing, file download/upload, and email access. These traffic types will cause disproportional traffic creation and network resource usage among users. Hence, resource usage level of different *BE* users will be different from each other although all *BE* users have the same subscription level and payment. Therefore, network operators wanted to classify *BE* users into different traffic groups in order to control their network usage and charge them accordingly. In order to achieve traffic classification at scheduler level, we have introduced *GoS* parameter into scheduler's decision metric (*DM*), which is used to make packet scheduling decisions. New scheduling methods are classified as *GoS* based schedulers and named as *GoS-RR*, *GoS-maxDRC*, *GoS-PF*, *GoS-aveDRC*, *GoS-expDRC*, and *GoS-maxDRC-PF*.

In order to assess and compare the performances of proposed *GoS* schedulers in business *femtocell* environment, average delivered throughput, *BE* traffic differentiation error, and *VoIP* delay bound error values are used as KPIs. As a result of real *femtocell* experiments and simulations, the appropriate scheduling methods are recommended for home and business *femtocell* environment.

BACKGROUND

Femtocell is a miniature base station providing 3G/4G wireless broad connection to users at home or office. *Femtocells* are mainly used to provide coverage and high quality data connection to users at indoor locations using existing broadband internet connections. The basic characteristics of *femtocells* will be:

- **Low-power:** *Femtocell* will only transmit 10-100 miliwatts, or lower power levels than Wi-Fi access points
- **Low-cost:** Mobile operators will be offering *femtocells* as a part of subscription package
- **User-installed:** *Femtocell* will be installed by users without any configuration
- **Compatible with existing mobile devices:** *Femtocell* will be supporting both existing and next generation mobile devices and handsets.
- **Broadband Connection:** *Femtocells* will connect to operators' networks using wired broadband internet connection such as DSL, cable or fiber optics.
- **Based on cellular network standards:** *Femtocells* will be built using existing UMTS and CDMA standards. And they will also support LTE, WiMAX, and UMB.
- **Deployed in operator-owned spectrum:** *Femtocells* will operate in licensed frequency spectrum owned by network operators. They can also share spectrum with the macro cell network.

The main advantages of *femtocells* for users are;

- **Increase indoor coverage:** *Femtocells* will provide coverage up to 200 meters. In most home and office environments, this means five bars of coverage.
- **High performance data:** *Femtocells* will support 4-6 concurrent active data or voice calls, which means that users will be experiencing high data rates than they do in macrocell environment.
- **Improved multimedia experience:** *Femtocell* will deliver high quality of multimedia experience with high rate data connection.
- **High voice quality:** *Femtocells* will support a new generation of voice quality, which is called High Definition Voice (HDVo).

- **Converged mobile VoIP services:** Femtocell will able to connect existing land line phones as well as mobile phones using VoIP technologies.
- **Enhanced emergency services:** *Femtocells* have highly accurate location information of themselves. By using this location information, emergency services will locate callers easily.

Increasing demand for wireless data forced network operators to plan for increased network capacity and coverage. However, new site acquisition is expensive, and more sites will require more backhaul connections, which cause extra cost. Current cellular wireless network standards use high frequencies, and high frequency signals do not penetrate walls easily. Therefore, network operators cannot provide superior indoor wireless service. Moreover, with each new site, existing legacy mobile phones has to be supported as well. All of these crucial problems make *femtocells* attractive technology for wireless network operators. Basically, there are 5 main advantages of *femtocells* for network operators. These are:

- **Increased network capacity:** Each *femtocell* will offload 4-6 calls from *macrocell* for a roughly a third of each day.
- **Lower capital costs:** Increasing number of network users will not force operator to set up new macrocells anymore. Therefore, operators will reduce their capital spending for infrastructure.
- **Lower backhaul costs:** Backhauling the user traffic to operator's core network is handled by the DSL, cable or fiber access lines at home or office.
- **Expanded revenue opportunities:** Improved indoor coverage and superior data quality will open doors to new multimedia applications and services. Network operators will monetize these new opportunities by controlling, delivering and storing the user data.
- **Increased customer stickiness and conversion:** Excellent in home and office coverage, operators will reduce customer churn and attract more users to their family plans.

Even though *femtocells* have significant advantages for both users and network operators, there is a number of challenges still need to be addressed to make *femtocells* successful as high-volume products. Because each represents an area of innovation, many will become significant aspects of competition and differentiation among femto access point suppliers.

- **Radio interference mitigation:** In macrocell environment, RF coverage can be manually tuned and optimized by RF engineers. However, *femtocells* are plug and play devices, and they have to operate when they are turned on. There is no manual RF optimization. Therefore, *femtocells* need to mitigate potential interference by dynamically adjusting their power.
- **Automatic system selection:** When a user moves into *femtocell* coverage from macrocell coverage, the handset must correctly select the network to operate on. Since network operators will have different pricing plans for users connecting through *femtocells*, network selection is very crucial task.
- **Active macro-femto handoffs:** When an authorized user of a femto cell moves in or out of coverage of femto cell, the handset must correctly hand off between macrocell and *femtocell* networks.
- **Quality of Service, and Experience:** Service quality levels have to be supported regardless of the interference or the number of active users in *femtocell* environment. From user's point of view, the main purpose of *femtocell* is to provide superior service quality levels. Therefore *femtocell* should monitor all quality levels continu-

ously, and make necessary adjustments to deliver high quality of experience to all users in *femtozone*.

- **Security:** A *femtocell* must identify and authenticate itself to operator's network as being valid, so a hacker cannot spoof the network into thinking his *femtocell* is someone else's and fraudulently use another customer's services.
- **Scalability:** Each *femtocell* will communicate with operator's core network using IPsec tunnels. With millions of *femtocells* deployed in a network, operators will require large scale security gateways at the edge of their core networks to handle millions of *femtocell* originated IPsec tunnels.
- **Access Control:** Since *femtocells* use the broadband internet connection owned by the user for backhaul, consumers should be given the option to control the use of their *femtocells*. In particular, a *femtocell* owner should be allowed to easily add and delete authorized users and to exclude unauthorized users near the *femtocell*, such as a person walking by or a neighbor in a multi-tenant building.
- **Activation and Management:** This is a critical area that can determine the success or failure of many products or service. Since most *femtocell* products will be purchased in a way similar to mobile phones, ease of customer installation and activation is paramount. Customer must able to open the box, power up and plug the *femtocell* into the network, call a number for activation, and have a one-touch means of registering handsets.

Traffic Differentiation in CDMA2000 1xEVDO Standard

Figure 1 shows the general architecture of a typical *femtocell* network. *Femtocell* uses the existing broadband cable connection to receive and deliver data. *Femtocell* traffic has to go through the internet service provider's network like all user generated traffic. After service provider, traffic will reach the *femtocell* gateway in network operator's network. Depending on the type and *QoS* level of connection, gateway server will make appropriate connection with outside world to deliver voice or data packets with required *QoS*. In order to achieve traffic differentiation in *femtocell* network, different traffic classes with different requirements are defined. *BE* traffic differentiation concept is called *IUBE* feature in 1xEVDO standard, and *IUBE* feature is considered as a *QoS* for *BE* users. Users are differentiated in terms of total delivered throughput value, which means that femtocell will deliver more throughput to users in high *BE* class. In a CDMA2000 1xEVDO *femtocell* network, access terminal (*AT*) reports instantaneous channel information back to *femtocell* using DRC feedback mechanism (Sevindik, 2010; Esteves, 2003; Kyungtae, 2005; Qi, 2003; Ching, 2003; Tinnakornsrisuphap, 2004). Higher values of requested DRC imply good wireless channel conditions and vice versa. Scheduler

Table 1. DRC values for CDMA 2000 1xEVDO Rev. A

Data Rate (kbps)	Slots	Packet Length (byte)	SNIR(dB)
38.4	16	128	-12.5
76.8	8	128	-9.5
102.6	6	128	-8.5
153.6	4	128	-6.5
204.8	3	128	-5.7
307.2	2	128	-4.0
614.4	1	128	-1.0
921.6	2	192	1.3
1228.8	1	256	3.0
1536	2	640	6.6
1843.2	1	384	7.2
2457.6	1	512	9.5
3072	1	640	13.8

Figure 1. Femtocell network architecture

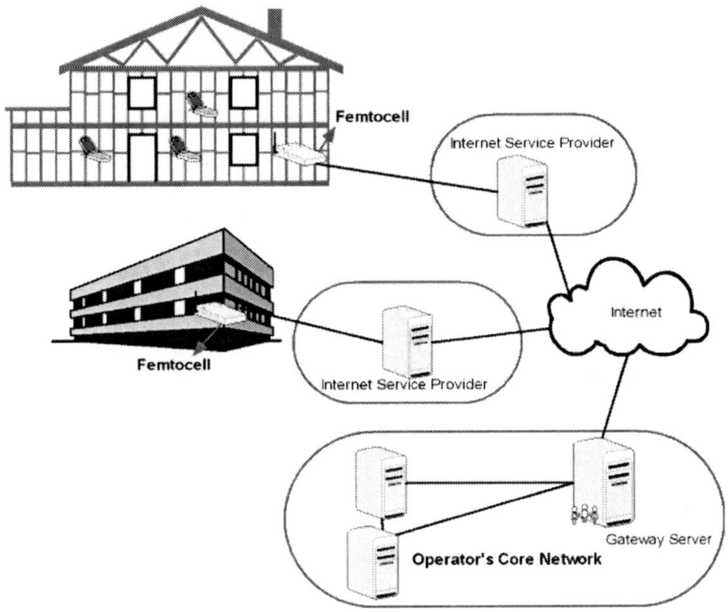

uses DRC value of each user to calculate *DM* of that user, and it transmits the packet to the user having largest *DM.* DRC values for Revision A (Rev. A) of CDMA2000 1xEVDO standard are given in Table 1

Radio Resource Management in Femtocell Environment

Radio resource management is defined as the management of available network resources in order to satisfy network users' *QoS* requirements. Radio resource management mainly consists of two important parts. First part is named as *admission control*, which means to control the user admission to available network resources at the time of connection (Taewon, 2004). For instance, *femtocell* can accept or reject users if it has enough resources to satisfy the requested *quality of service* (*QoS*) of the new user without worsening the *QoS* of current users in network. Second part is called *air link resource management*, which means to dedicate adequate air link resources to achieve the user's requested *QoS* target. We mainly focused on the second part of the radio resource management process. In terms of resource management, *femtocell* will only deal with already admitted users, and try to manage airlink resources among them in order to deliver the optimum *QoS* to each user in the network or *femtozone*. In this chapter, we considered *data throughput* the most important resource to share and to manage. User traffic will be classified in terms of received total data throughput. From now on, we will use the term *femtozone* instead of *femtocell network environment*. *Femtozone* is defined as the area that femtocell provides service and coverage. *Home femtocell environment* will be called as *home femtozone*, and *business femtocell environment* will be called as *business femtozone*.

Scheduling Concept

The most important component involved in managing airlink resources is *scheduler*. *Scheduler* manages the user packet transmission process by using user related parameters. At each scheduling time, scheduler calculates the metric called *deci-*

sion metric (DM) in order to decide which user to transmit data. In CDMA2000 1xEVDO *femtocell*, the forward link transmission is implemented as TDMA to increase the data rate (Sevindik, 2010). For CDMA2000 1xEVDO *femtocell*, the scheduler's task is to select the owner of the next transmission slot by finding the user with the highest *DM*. All of the existing scheduling algorithms have been developed for 1xEVDO *macrocell* so far. Since the characteristics of macrocell and *femtocell* are very different from each other, the performance of schedulers developed for macrocell should be investigated in *femtozone*. Since indoor wireless channel conditions are experienced in *femtozone*, schedulers' performances will change significantly. By dedicating different number of slots to different users, scheduler can provide different average data throughput values to different users, and this is considered as throughput differentiation or traffic classification concept. By changing the scheduler's parameters, we can change the delivered data throughput, and throughput distribution in the *femtozone*. In CDMA2000 1xEVDO standard, traffic differentiation is applied to best effort users in the *femtozone* since each *BE* user's resource usage is very different from each other. This *BE* traffic differentiation concept is named as *IUBE* feature in CDMA2000 1xEVDO standard. *IUBE* implementations are more applicable in business *femtozone* than home *femtozone* since in business environment, there will be different types of *BE* users demanding different *QoS* levels. *BE* traffic differentiation is achieved by introducing *grade of service (GoS)* parameter into the scheduler's *DM*. *GoS* parameter is a predefined *QoS* parameter for a specific *IUBE* class in the *femtozone*. By assigning different *GoS* values for each *IUBE* class, the behavior of scheduler is forced to change for each *IUBE* class. Classes having large *GoS* values receive more average sector throughput than the classes having low *GoS* values. In EVDO standard, *GoS* values take integer values of 2^i, where i = 0, 1, 2... M-1 and M is the total number of *IUBE* classes in the *femtozone*. Users in the same *IUBE* class will have the same *GoS* value.

Scheduling BE and VoIP Traffic

Scheduler design changes when *VoIP* and *BE* users exist in the network simultaneously. Priority for *VoIP* traffic is determined by calculating the *VoIP* packet delay at the time of scheduling (Young-June, 2004). There are two types of flows in the network. *VoIP* traffic is modeled as delay sensitive (DS) flow, *BE* traffic is modeled as Throughput sensitive (*TS*) flow (Santhi, 2006). In order to classify flows, a new metric called packet metric (*PM*) is introduced. *PM* definitions for *VoIP*, *BE* are given as follows:

$$PM_q = I_q \tag{1}$$

$$PM_{BE_n^i} = \frac{GoS_i}{R_n^i} \tag{2}$$

$$PM_{TS} = \sum_{i=1}^{M}\sum_{n=1}^{N_i} \frac{PM_{BE_n^i}}{T_n^i} \tag{3}$$

$$N = \sum_{i=1}^{M} N_i \tag{4}$$

$$PM_{DS} = \sum_{q=1}^{K} PM_q \tag{5}$$

where PM_q and I_q are the decision metric and the constant assigned to for VoIP user *q* respectively. $PM_{BE_n^i}$ is the *PM* of the *n-th BE* user in *BE* class *i*. PM_{TS} is the total *PM* for *TS* traffic is the network, and T_n^i is the total number of time slots dedicated to *BE* user *n* in *i-th* BE class. *N* is the total number of *BE* users, *M* is total number of *BE* classes and N_i is the total number of users in *i-th BE* class. PM_{DS} is the *PM* for delay sensitive traffic in the network, such as *VoIP*. Since there are only *VoIP* users in the network, PM_{DS} is the *PM* for all *VoIP* users in the network, and *K* is the total number of *VoIP* users in the network.

TRAFFIC DIFFERENTIATION AND SERVICE QUALITY IN FEMTOCELL ENVIRONMENT

Wireless applications for mobile devices are increasing at a rising speed. New applications create huge amount of additional data traffic that wireless network is supposed to deliver successfully. The main task of a wireless network is to transmit data to and from mobile devices under any circumstances. However, increasing types and amount of data traffic puts lots of pressure onto the network in terms of providing and keeping adequate *QoS* and *QoE* levels, connection speeds, and coverage. That is the reason that network operators have been looking for solutions to offload data traffic from their existing networks to existing cable or fiber internet backbone using user or business specific cellular wireless devices.

Femtocell is mainly developed for this purpose, and it provides users at home or business environment with very high 3G/4G data speeds and coverage. However, *femtocell* does not solve one very important problem, and this problem is directly related to network operator's revenue. The main purpose of cellular wireless network operators is to produce profit out of their existing wireless infrastructure. In wireless market, profit usually comes from different customer segments which receive or are willing to receive various differentiated services at the time and location they need. Providing differentiated services requires specific traffic class definitions and *QoS* requirements. Even though *QoS* requirements change depending on application and network operator, the main point is to deliver promised *QoS* levels to customers under any network circumstances. In today's wireless networks, the main traffic types are internet surfing (*http*), file sharing, file upload or download traffic (*ftp*) and background traffic which are classified as *BE* traffic.

The unique characteristic of *BE* traffic is its high and widespread usage among users in the network. However, there is one important problem with the users creating *BE* traffic. All of the users pay the same amount of subscription fee regardless of their usage level. Users who are constantly consuming the network resources pay the same fee as users who rarely use any application or service. This situation is unfair for both network operators and users. Customers should pay right amount of fee based on their network usage level, and network operator should charge the *BE* users depending on their subscription level.

In wireless networks, traffic differentiation concept is mostly handled by resource management unit, which is the core unit optimizing the overall network resource usage. Scheduler is the main component of resource management unit, and it is used to classify user traffic into predefined traffic classes using some user related *QoS* parameters. In (Elliott, 2002), you can find the frequently used scheduling techniques in CDMA2000 1xEVDO networks. *RR*, *maxDRC*, *aveDRC*, *expDRC*, *PF*, *maxDRC-PF* are the main scheduling techniques used by 1xEVDO networks. All of these scheduling techniques were developed for macrocell cellular systems, which experience outdoor channel conditions. Except *RR* scheduler, the rest of these conventional scheduling techniques use channel information by incorporating *DRC* values as a parameter into their *DMs*. Thus, their performances are highly dependent on experienced channel conditions by the users in the network. In order to use these schedulers for *femtocell* environment, their performances should be investigated under indoor channel conditions. The scheduler providing the best performance in terms of throughput and fairness may not give the same performance in *femtozone*. Therefore, in order to find the optimum scheduling technique, the throughput and fairness performance of mentioned macrocell schedulers should be carefully analyzed.

Scheduling Methods

Macrocell Schedulers

Some of the existing scheduling methods are *RR*, *maxDRC*, *PF*, *aveDRC*, *expDRC* and *maxDRC-PF*. Performances of these conventional schedulers are investigated in terms of average sector throughput, throughput fairness, and *VoIP* delay bound error. These performance metrics will be called as key performance indicators in this chapter. Two different real *femtocell* throughput experiments were conducted to collect real *DRC* values in *femtozone*. Collected *DRC* values were used in computer simulations to calculate key performance indicators.

QoS is defined as guaranteeing to deliver promised service levels under any network conditions. *QoS* term is usually used and defined by network operators; because the concept of delivering the guaranteed service levels is highly dependent on available network resources. In today's wireless networks, there are myriad of user oriented mobile phone applications that create a lot of different traffic types each of which has unique *QoS* requirements. Quarantining certain level of throughput and delay are two main *QoS* requirements. Based on these requirements, there are mainly three groups of 3G networks:

1. **Best Effort Traffic:** Best effort traffic does not have any throughput and delay requirements. It is created by applications such as background traffic, file downloading/ uploading and sharing applications, web applications.
2. **Expedited Forwarding:** This traffic type is created by applications that have certain delay requirements such as *VoIP* traffic.
3. **Assured Forwarding:** This traffic is generated by applications that require minimum throughput level to function properly such as video based applications.

QoS is defined as the satisfaction level of the experienced or used service which is provided by the network or service operator. For 3G networks, service or network operator is the same entity that is responsible to provide service to its customers; but for 4G networks, network and service operators are different from each other. In 4G networks, network operator is the entity that owns the infrastructure to deliver the services or applications. However, service or application provider is the third party that provides service or application to its own customers using network operators' infrastructure. Therefore, increasing *QoE* in 4G networks is more channeling than increasing it in 3G networks because of increasing number of players in the service delivery chain. There are no certain standards or scales to measure the *QoE* of a particular user in the network. According to Sevindik (2010), user's experienced service quality level highly depends on customer support after or during service usage. Therefore, increasing customer satisfaction requires total customer focus during the lifetime of network connection of a user. *Femtocell* networks are mainly designed to increase the *QoE* levels of the users by increasing the coverage, quality of broadband wireless data connecting at physically limited areas such as home and office environment. Even though the delivered service levels will be similar for all users at home environment; this will not be true for business environment in where there will be more users than home *femtocell* environment. The more the number of users, the more diversified the expected service quality levels. Therefore, the high quality service delivery capability of the business *femtocell* product should be carefully investigated in order to increase quality of experience.

Round Robin (RR) Scheduler

RR scheduler selects the users in cyclic order in order to transmit data (Elliott, 2002). Users are polled in a pre-defined order, and only the users who have data to receive will be included in the selection process. *RR* scheduler does not use any channel or user related information in order to reach its scheduling decisions. For *RR* scheduler, it does not matter user if user is having good or bad channel conditions, as soon as it is connected to the *femtocell*, user will be included into the scheduling decisions. Therefore, it delivers the lowest throughput performance among other macrocell scheduling techniques. Because of the cyclic structure in the user selection process, throughput fairness of the algorithm is the highest among schedulers.

$$DM = \mathrm{mod}(i, N + 1) \tag{6}$$

where mod(.) represents the modular operator, *DM* is the decision metric of *RR* scheduler at each transmission time slot, *i* is the user index, *N* is the total number of users in the network.

*Maximum DRC (*maxDRC*) Scheduler*

Maximun DRC (*maxDRC*) scheduler selects the user having the best channel condition at the time of scheduling (Elliott, 2002). The best channel condition is represented by the highest requested DRC. User requesting the highest DRC will be scheduled at next transmission slot to receive the data. *maxDRC* scheduler will deliver the highest sector throughput among other scheduler algorithms. However, throughput fairness performance of *maxDRC* scheduler is very bad, because some users can request very high *DRC* values for a long period of time, which will cause other users to wait considerable amount of time to received data. *DM* is given by the following expression:

$$DM_t = \max_{n=1,2,\ldots,N} \left(DRC_t^n \right) \tag{7}$$

DM_t is the decision metric of the scheduler at time t, DRC_t^n is the requested DRC value of user n at time t, N is the total number of users in the network. If users in the network experience very different channel conditions than each other, user with the best channel condition will receive data all the time. However, as in *femtocell* environment, if the users experience similar channel conditions, all of the users will have chance to receive data which will increase the throughput fairness in the network.

Proportional Fairness (PF) Scheduler

PF scheduler is proposed to create balance between deliver throughput and throughput fairness in the network. *PF* scheduler uses additional metric in addition to DRC values of the user (Elliott, 2002). New metric is denoted as $\overline{R_t^n}$ in DM_t, and it represents averaged data rate of user *n* at time *t*.

$$DM_t = \max_{n=1,2,\ldots,N} \left(\frac{DRC_t^n}{\overline{R_t^n}} \right) \tag{8}$$

$$\overline{R_t^n}(t+1) = \left(1 - \frac{1}{t_c}\right) \overline{R_t^n}(t) + \left(\frac{1}{t_c}\right) R_t^n(t) \tag{9}$$

Equation (9) is the update procedure for average data rate of user n at time t. $\overline{R_t^n}(t+1)$ and $\overline{R_t^n}(t)$ are the average data rate values of user n at time *t* and *t+1* respectively, t_c is the time window value used in averaging procedure. Its value is usually taken as 1000 time slots. When the scheduler transmits the data to a particular user in the network, the average data value of that user starts increasing. Because of being in the denominator of scheduler's *DM* (Equation 8), increasing $\overline{R_t^n}$ will decrease the value of DM_t, which means that the served user at time t will have less change to be served at time t+1. By this way, PF scheduler breaks the dominance of a particular user in the network, and creates a balance between total sector throughput and throughput fairness.

Average DRC (aveDRC) Scheduler

Average DRC scheduler uses average value of DRC values of each user in addition to the parameters used in *PF* scheduler (Elliott, 2002). *DM* for this scheduler is given by;

$$DM_t = \max_{n=1,2,\ldots,N}\left(\frac{DRC_t^n . \overline{DRC_t^n}}{\overline{R_t^n}}\right) \tag{10}$$

$$\overline{DRC_{t+1}^n} = \left(1 - \frac{1}{(\alpha . t_c)}\right)\overline{DRC_t^n} + \left(\frac{1}{(\alpha . t_c)}\right)DRC_t^n \tag{11}$$

Definitions for DM_t, DRC_t^n, N, and $\overline{R_t^n(t)}$ stay the same. $\overline{DRC_{t+1}^n}$ and $\overline{DRC_t^n}$ is the average value of requested DRC values of use n at time t+1 and t respectively. α is the tuning parameter to control throughput and fairness characteristics of the scheduler. *aveDRC* scheduler will transmit to users having good average channel conditions (good average DRC values). Under dynamic channel conditions, α value should be kept small in order to have accurate average DRC values. For indoor channel conditions like in the case of *femtocell*, α value can be big enough to find the accurate average DRC values of users in the network.

Exponent DRC (expDRC) Scheduler

Exponent DRC (expDRC) scheduler is proposed to control the effect of DRC value on the DM of the scheduler (Elliott, 2002). DM for the scheduler is given by;

$$DM_t = \max_{n=1,2,\ldots,N}\left(\frac{(DRC_t^n)^e}{\overline{R_t^n}}\right) \tag{12}$$

Definitions for DM_t, $\overline{R_t^n}$, and N stay the same. Update procedure for $\overline{R_t^n}$ is the same as it is in PF scheduling method. The control parameter is represented by "*e.*" Higher values of "*e*" will augment the effect of small changes in DRC values on DM. When users have similar channel conditions, high "*e*" values should be used in order to exploit small changes in channel conditions which will increase multi-user diversity in the network. Therefore, in *femtocell* environment, "*e*" value is taken greater than "1."

Maximum DRC and PF (maxDRC-PF) Scheduler

This scheduling method is proposed in order to have flexibility to adjust the amount total delivered sector throughput and the degree of throughput fairness in the network (Elliott, 2002). *maxDRC-PF* scheduler combines *maxDRC* and *PF* scheduling methods using an adjustable parameter. *DM* for this scheduler is given by;

$$DM_t = \max_{n=1,2,\ldots,N}\left(a.DRC_t^n + (1-a).\frac{DRC_t^n}{\overline{R_t^n}}\right) \tag{13}$$

all definitions for variables in DM stay the same as in other schedulers. For $a = 1$, *maxDRC-PF* will reduce to *maxDRC* scheduler, and similarly for $a = 0$, *maxDRC-PF* will reduce to *PF* scheduling method. For "*a*" values between 0 and 1, *maxDRC-PF* will behave between *maxDRC* and *PF*. For high values of *a*, scheduler will provide high sector throughout, and for small values of *a* scheduler will provide high throughput fairness. For *femtocell* environment, the value of *a* should be selected in the middle of [0, 1] interval in order to have good throughput with good throughput fairness. When wireless channel conditions become worse, *maxDRC-PF* scheduler should behave more like *PF* than *maxDRC* in order to provide throughput fairness among users in the network. On the contrary, when wireless channel conditions get better, *maxDRC-PF* should behave more like *maxDRC* than *PF* in order to have high sector throughput.

GoS Based Scheduling Methods

By introducing *GoS* parameter into *DMs* of macrocell schedulers, *GoS* based scheduling methods are created. As it was mentioned before, *GoS* based schedulers are used for *BE* traffic differentiation in *femtozone* by providing different throughput values to different *BE* traffic classes. The definition of *BE* classes highly depends on throughput requirements of *BE* classes (Sevindik, 2010).

GoS-RR Scheduler

By assigning *GoS* number of time slots to each *BE* traffic class, traffic differentiation is achieved. If we denote GoS_i as value of *GoS* for *BE* class *i*, all users in that class will be given GoS_i time slots to receive data. By providing perfect time slot distribution among *BE* classes does not guarantee the perfect throughput fairness in *femtozone*; because users will experience difference channel conditions and request different DRC values from each other. Proposed *GoS-RR* scheduler provides lowest throughput among other scheduling methods, with high throughput fairness.

GoS-maxDRC Scheduler

This scheduler is created by introducing *GoS* parameter into *maxDRC* scheduler. *DM* is given by:

$$DM_i^n(t) = \max_{\substack{i=1,2,\ldots,M \\ n=1,2,\ldots,N_i}} \left(GoS_i . DRC_i^n(t) \right) \tag{14}$$

Where $DM_i^n(t)$ is the *DM* for user *n* in BE class *i* at time *t*, GoS_i is the *GoS* value for *BE* class *i*, $DRC_i^n(t)$ is the requested DRC value by user *n* in class *BE* class *i* at time *t*, *M* is the number of *BE* classes, and N_i is the number of users in *i-th* *BE* class. This scheduler does not always select the user requesting the highest DRC value because of GoS_i value in the *DM*. User with bad channel conditions can have chance to receive data, and this is the reason that *GoS-maxDRC* scheduler is not expected to deliver the highest throughput among *GoS* based schedulers in *femtozone*.

GoS-PF Scheduler

This scheduler is created in the same way as other schedulers. By introducing GoS_i value into the *DM*, traffic classification between *BE* users is achieved. *DM* for the scheduler is given by:

$$DM_i^n(t) = \max_{\substack{i=1,2,\ldots,M \\ n=1,2,\ldots,N_i}} \left(GoS_i . \frac{DRC_i^n}{\overline{R_i^n}} \right) \tag{15}$$

$$\overline{R_i^n}(t+1) = \left(1 - \frac{1}{t_c}\right) \overline{R_i^n}(t) + \left(\frac{1}{t_c}\right) R_i^n(t) \tag{16}$$

$DRC_i^n(t)$ is the requested DRC value by user *n* in *BE* class *i* at time *t*, $\overline{R_i^n}(t+1)$ and $R_i^n(t)$ are the updated average and delivered data rate for user *n* in *BE* class *i* at time *t* respectively, GoS_i is the *GoS* value associated to *i-th BE* class, t_c is the constant for averaging window, *M* is the number of *BE* classes, and N_i is the number of users in *i-th BE* class. The ratio of experienced throughput values of two different *BE* classes is expected to have the same ratio between their corresponding *GoS* values. The value for t_c is taken as 1000 time slots. However, in order to have accurate average delivered data rate value under fast changing channel conditions, its value can be smaller than

1000 time slots. In *femtozone*, channel does not change too frequently, and this is the reason t_c value is set to 1000 time slots.

GoS-aveDRC Scheduler

GoS based version of *aveDRC* is called *GoS-aveDRC*, and *DM* is given by:

$$DM_i^n(t) = \max_{\substack{i=1,2,\ldots,M \\ n=1,2,\ldots,N_i}} \left(GoS_i . \frac{DRC_i^n(t).\overline{DRC_i^n(t)}}{\overline{R_i^n(t)}} \right) \quad (17)$$

$$\overline{DRC_i^n(t+1)} = \left(1 - \frac{1}{(\alpha.t_c)}\right)\overline{DRC_i^n(t)} + \left(\frac{1}{(\alpha.t_c)}\right) DRC_i^n(t) \quad (18)$$

definitions for $DRC_i^n(t)$, $\overline{R_i^n(t)}$, M and N_i stay the same. $\overline{DRC_i^n(t+1)}$ and $\overline{DRC_i^n(t)}$ are average DRC values of user *n* in *BE* class *i* at time *t+1* and *t* respectively. The procedure for average DRC calculation is the same as *aveDRC* macrocell scheduler.

GoS-expDRC Scheduler

expDRC is converted to *GoS-expDRC* by introducing *GoS* value in *DM*, which is given by:

$$DM_i^n(t) = \max_{\substack{i=1,2,\ldots,M \\ n=1,2,\ldots,N_i}} \left(GoS_i . \frac{(DRC_i^n(t))^e}{\overline{R_i^n(t)}} \right) \quad (19)$$

definitions for $DRC_i^n(t)$, GoS_i, $\overline{R_i^n(t)}$, *M* and N_i stay the same. The control parameter is denoted by "*e*." The small changes are augmented by "*e*" are multiplied by GoS_i, and this means more emphasize on DRC changes.

GoS-maxDRC-PF Scheduler

Similar to other *GoS* based schedulers, *maxDRC-PF* scheduler is converted to *GoS-maxDRC-PF* scheduling method and *DM* is given by:

$$DM_i^n(t) = \max_{\substack{i=1,2,\ldots,M \\ n=1,2,\ldots,N_i}} \left(GoS_i . (a.DRC_i^n(t) + (1-a).\frac{DRC_i^n(t)}{\overline{R_i^n(t)}}) \right) \quad (20)$$

definitions for $DRC_i^n(t)$, GoS_i, $\overline{R_i^n(t)}$, *M* and N_i stay the same. The control parameter is denoted by "*a*," and it is used for adjusting the average sector throughput and throughput fairness characteristics of the scheduler. With small "*a*" value, *GoS-maxDRC-PF* will behave more like *GoS-PF* scheduler, and with high "*a*" value, *GoS-maxDRC-PF* will behave more like *GoS-maxDRC* scheduler. This scheduler provides the lowest traffic differentiation error among other *GoS* based schedulers, which means each user at each *BE* class is receiving the throughput proportional to the *GoS* value of the class they are in.

Performance Analysis

Key Performance Indicators

There are several performance metrics that are used to compare the performances of schedulers in femtozone. Since simulations and experiments were conducted using both *BE* and *VoIP* traffic, there are some metrics to assess the schedulers' performances for VoIP users as well. These Key Performance Indicators (KPIs);

1. **Average Sector Throughput:** This KPI is defined as the total delivered throughput to all users in the network divided by total simulation time. This KPI is calculated for each scheduling algorithm.

2. **Throughput Fairness:** Equal share of the total average sector throughput among users in the *femtozone*. This KPI is calculated for each scheduling algorithm.
3. **VoIP delay bound error:** Total number of *VoIP* packets experiencing more delay than the delay bound divided by total number of *VoIP* packets. This KPI is defined for each *VoIP* user and for each scheduling algorithm.
4. **Traffic Differentiation Error:** *BE* traffic differentiation capability of a scheduler is defined as the average value of error values between achieved throughput and *GoS* ratios for each *BE* class in the network. This KPI is the main performance indicator of traffic classification capability of each a scheduler, and it is applicable to *GoS* based schedulers.

These KPIs are used for all scheduling algorithms for performance investigation.

Throughput Fairness Index

Fairness in resource sharing is defined as the equal share of available resources among users in the network. Throughput fairness means equal distribution of total deliverable data throughput among users in the network. In order for a scheduling method to be fair, it should balance the number of slot usage among users in the network so that users receive same amount of throughput. The value of throughout fairness changes between 0 and 1, where 0 means total unfair and 1 means perfect fair situation. One famous throughput fairness index is Jains's fairness index (Jain, 1984), which is given:

$$FI_{Jain} = \frac{\left[\sum_{i=1}^{N} T_i\right]^2}{N\sum_{i=1}^{N}\left(T_i\right)^2} \tag{21}$$

where FI_{Jain} stands for fairness index calculated by Jain's approach, T_i is the throughput received by the user i, and N is the total number of users in the network.

When all users in the network experience the same channel conditions, scheduler will have the perfect throughput fairness. However, when channel conditions start to change for users in the network, the task of scheduling becomes harder because of changing DRC and R values at each decision time. So, scheduler should follow and use the changing nature of wireless channel conditions in order to increase fairness among users in the network.

We modified Jain's fairness index to calculate the fairness among *BE* traffic classes in the network. Modified Jain's Index is given by:

$$IUBE_{FI_{Jain}} = \frac{\left(\sum_{i=1}^{M}\frac{RT_i}{GoS_i} + \left[1-\sum_{i=1}^{M} RT_i\right]\right)^2}{M.\left(\sum_{i=1}^{M}\left(\frac{RT_i}{GoS_i}\right)^2 + (1-\sum_{i=1}^{M}(RT_i))^2\right)} \tag{22}$$

$$RT_i = \frac{\overline{T_i}}{\sum_{i=1}^{N}\overline{T_i}} \tag{23}$$

$$\overline{T_i} = \frac{\sum_{n=1}^{N_i} R_n}{N_i} \tag{24}$$

where M is the number of *IUBE* classes in the network, RT_i is the ratio between the average throughput value of *IUBE* class i and total sector throughput, $\overline{T_i}$ is the average throughput of *IUBE* class i, N is the total number of users in the network. R_n is the received throughput value of user n, and N_i is the total number of users in *IUBE* class i. We want the value of $IUBE_{FI_{Jain}}$ is close to 1 as much as possible in order to have fair throughput distribution among *BE* classes in the network. To give a specific example, assume that there are 4 IUBE classes in the network and GoS values for $IUBE_1$, $IUBE_2$, $IUBE_3$, $IUBE_4$ classes is 1, 2, 4, and 8 respectively. Let's assume that the total delivered network throughput is 15 units. Under perfect throughput distribution, IUBE classes 1, 2, 3, 4 will receive 1, 2, 4, and 8 units of throughput values respectively and the sum of these throughput values will lead to 15 units. Therefore, throughput ratios are $RT_1=1/15$, $RT_2=2/15$, $RT_3=4/15$, and $RT_4=8/15$ for IUBE classes 1,2,3, and 4 respectively.

$$IUBE_{FI_{Jain}} = \frac{\left(\sum_{i=1}^{M} \frac{RT_i}{GoS_i} + \left[1 - \sum_{i=1}^{M} RT_i\right]\right)^2}{M.\left(\sum_{i=1}^{M} \left(\frac{RT_i}{GoS_i}\right)^2 + \left(1 - \sum_{i=1}^{M} (RT_i)\right)^2\right)}$$
$$= \frac{\left(\left(\frac{1/15}{1} + \frac{2/15}{2} + \frac{4/15}{4} + \frac{8/15}{8}\right) - \left[1 - \left(\frac{1}{15} + \frac{2}{15} + \frac{4}{15} + \frac{8}{15}\right)\right]\right)^2}{4\left(\left(\left(\frac{1/15}{1}\right)^2 + \left(\frac{2/15}{2}\right)^2 + \left(\frac{4/15}{4}\right)^2 + \left(\frac{8/15}{8}\right)^2\right) - \left[\left(1 - \left(\frac{1}{15} + \frac{2}{15} + \frac{4}{15} + \frac{8}{15}\right)\right)^2\right]\right)}$$
$$= 1 \tag{25}$$

Fairness formula proves that under perfect throughput distribution in the network, the throughput fairness index becomes 1. This value is the upper limit and means that the scheduler provides perfect throughput distribution among IUBE classes in the network.

Traffic Model

BE traffic is modeled as full-buffer traffic. Femto base station has always data to transmit all *BE* users in the network. *VoIP* traffic is modeled using the same way in (Ji, 2007). According to model and the *VoIP* codec used, *VoIP* traffic will be created at every 20msec, which means 11 time slots in 1xEVDO standard. That is, every *VoIP* user will be creating *VoIP* packet at every 11 time slots. Therefore, each *VoIP* packet should be transmitted in 11 time slots in order to keep the communication quality in the network. In this sense, 11 time-slot transmission bound is called delay bound, and any *VoIP* packet that is transmitted after waiting more than 11 time slots will be counted as packet transmission delay error, or *VoIP* delay bound error. According to standards, *VoIP* delay bound error target is %0.2, which means that only 2 *VoIP* packets out of every 1000 *VoIP* packets are allowed to be delayed more than 11 time slots.

Experiment Methodology

This work is unique in the sense of combining real *femtocell* experiments with computer simulations. First, several *femtocell* experiments were conducted to collect real DRC values for all *ATs* (*BE*, and *VoIP*) in *femtozone*. These DRC values are used in computer simulations to analyze the performance of schedulers. In each femtocell experiment, 6 CDMA 2000 1xEVDO Rev. A *ATs* were used to represent 4 *BE* and 2 *VoIP* users. 5MB file is transferred from femtocell to each *AT* using File Transfer Protocol (FTP), and DRC values were saved in order to use in computer simulations. Collected DRC values represent the real indoor wireless channel conditions since requested DRC values are functions of received SNIR. In all simulations, time constant t_c is takes as 1000 time slots, where each slot duration is 1.67 milliseconds in CDMA 2000 1xEVDO standard. For expDRC and GoS-expDRC schedulers, the parameter '*e*' is set to 0.5. For ave-DRC and GoS-aveDRC scheduling methods, parameter 'α' is set to 0.1. For maxDRC-PF and GoS-maxDRC-PF schedulers, the parameter '*a*' is set 0.5. In all experiments and computer simulations, 4 IUBE classes were used. As it was explained, GoS_i represents the GoS value of IUBE class *i* in the network. GoS_1, GoS_2, GoS_3, GoS_4 values for $IUBE_1$, $IUBE_2$, $IUBE_3$, $IUBE_4$ classes are 1, 2, 4, and 8 respectively.

In order to simulate VoIP traffic in computer simulations, conversational VoIP traffic model is used, in which Markov model is used to model

Figure 2. Received SNIR values of User 1

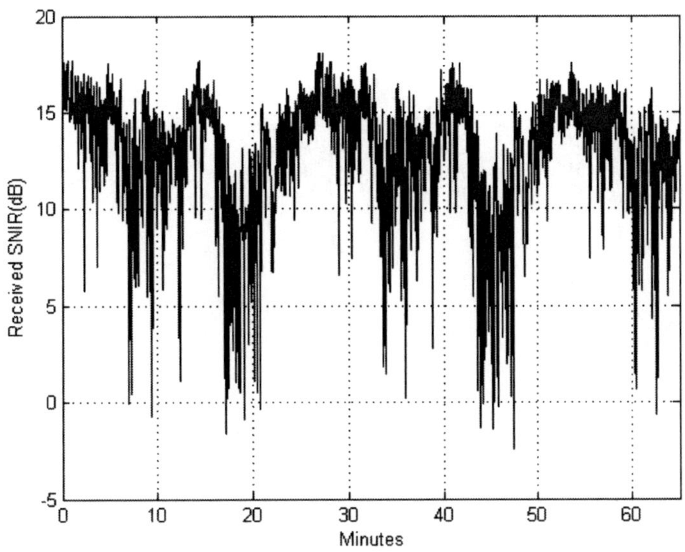

the silence, and talking period of a typical VoIP user (Ji, 2007).

For *GoS* based scheduling methods, there are 4 *BE* classes in the network, and each *BE* class has one user in it. For each *VoIP* user, *VoIP* delay bound error is calculated. When there is traffic differentiation in the *femtozone*, users in high *BE* classes will use most of the transmission time slots, and lower class *BE* and *VoIP* users will have lower chance of receiving data. Therefore, under *GoS* based scheduling methods, *VoIP* packets will experience more delay than they do under macrocell schedulers.

Channel conditions for users are given in following figures. Figure 2, 3, 4, and 5 show RF channel conditions for user 1, 2, 3, and 4 respectively.

RESULTS

Throughput Performance

Figure 6 shows the average sector throughput values for macrocell schedulers in *femtozone*. *maxDRC* scheduling method delivers the highest sector throughput with the smallest throughput fairness. *RR* and *maxDRC* schedulers provide similar sector throughput performance under slowly varying wireless channel conditions. However, varying wireless channel conditions is experienced in *femtozone*, and therefore there is a significant difference between the performances of *maxDRC* and *RR* methods. *aveDRC*, *expDRC*, *maxDRC-PF*, and *PF* schedulers deliver almost the same sector throughput values. Since *RR* scheduler does not consider channel information when making scheduling decisions, its throughput performance in the worst among schedulers.

Figure 7 shows the average sector throughput values of macrocell schedulers for 4 *BE* + 2 *VoIP* users in *femtozone*. *RR* scheduler has the lowest average sector value among scheduling methods, although it delivers the perfect throughput fairness.

PF, *maxDRC*, *expDRC*, *aveDRC*, and *maxDRC-PF* schedulers deliver the same average sector throughput. The throughput fairness value

Figure 3. Received SNIR values of User 2

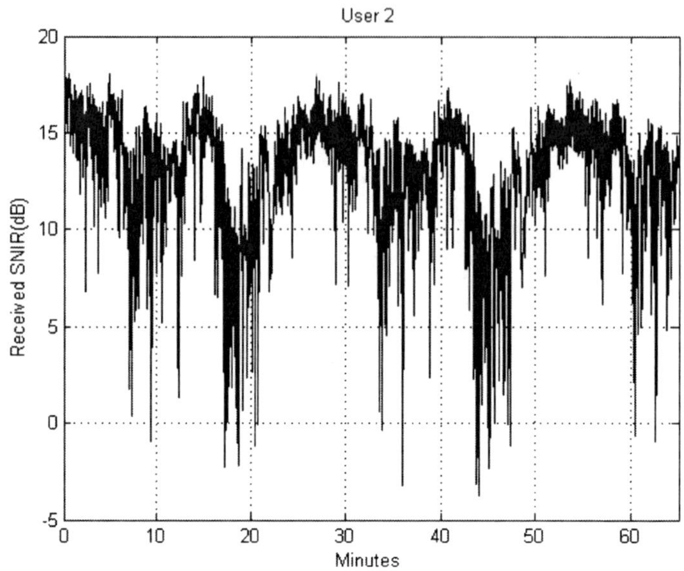

of *maxDRC* scheduler is slightly lower than other schedulers.

Figure 8 shows the sector throughput performances of *GoS* based scheduling methods. *GoS-RR* scheduler delivers the lowest sector throughput among *GoS* based scheduling methods.

GoS-PF, *GoS-aveDRC*, *GoS-expDRC*, *GoS-maxDRC-PF* has almost same throughput perfor-

Figure 4. Received SNIR values of User 3

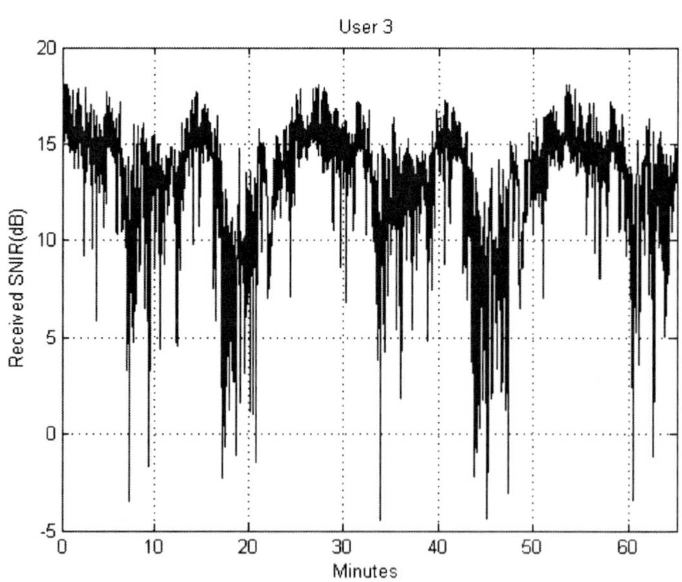

Figure 5. Received SNIR values of User 4

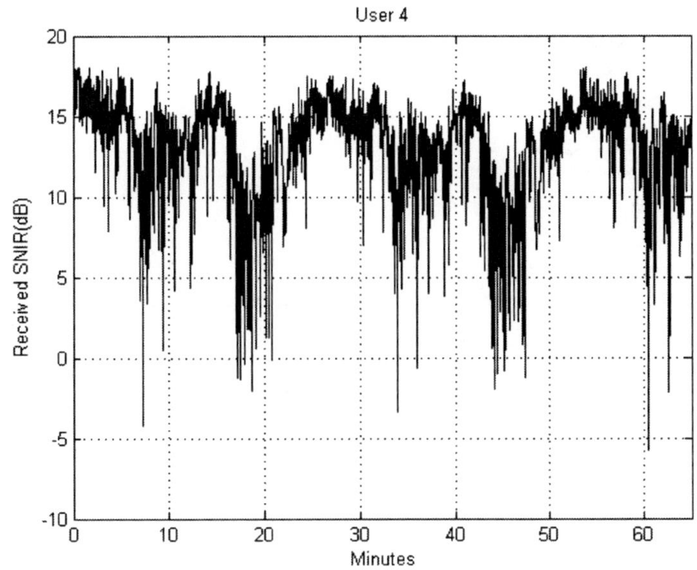

mance, which is the best among scheduling methods. *GoS-RR* has the best throughput fairness.

Figure 9 shows the performance of schedulers when there are 4 *IUBE* and 2 *VoIP* users in the network. *GoS-aveDRC* provides the highest average sector throughput (1447kbps), and *GoS-maxDRC-PF* provides the highest throughput fairness (Table 5). *GoS-aveDRC* is followed by *GoS-PF*, *GoS-expDRC*, and *GoS-maxDRC-PF* in terms of average sector throughput performance.

Figure 6. Average sector throughput values for 4 BE users in the femtozone

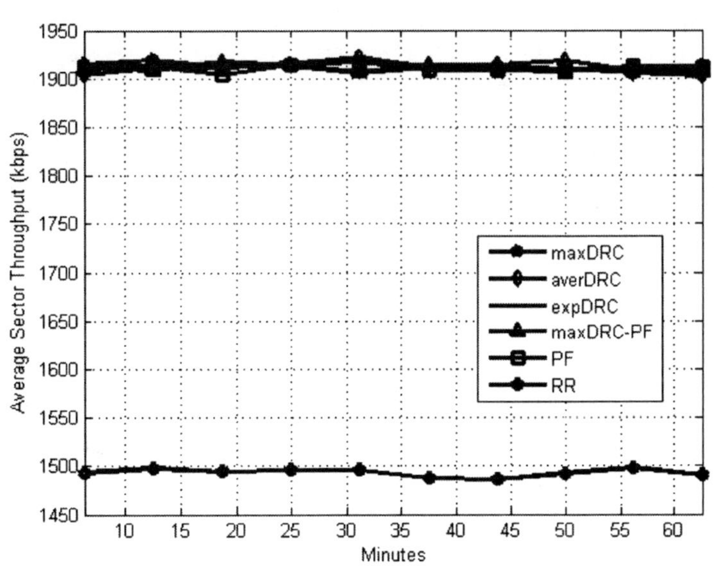

Table 2. Performance of macrocell scheduling methods

Scheduling Algorithm	User Type	Ave. Sector Throughput (kbps)	Throughput Fairness
RR	*4 BE*	1493	1
PF	*4 BE*	1910	1
maxDRC	*4 BE*	1913	0.9995
expDRC	*4 BE*	1911	1
aveDRC	*4 BE*	1911	1
maxDRC-PF	*4 BE*	1912	1

Introduced *GoS* parameter in *GoS-maxDRC* scheduler's *DM* decreases its capability of selecting the user requesting the highest *DRC* value in *femtozone*. That is the reason it delivers lower average sector throughput than other schedulers except *GoS-RR*.

In conclusion, for business *femtozone* in where there are different *BE* traffic classes, we recommend *GoS-maxDRC-PF* scheduler because of its highest throughput fairness and high sector throughput value.

Table 3. Performance of macrocell scheduling methods

Scheduling Algorithm	User Type	Ave. Sector Throughput (kbps)	Throughput Fairness
RR	4 *BE* + 2 *VoIP*	1151	1
PF	4 *BE* + 2 *VoIP*	1470	1
maxDRC	4 *BE* + 2 *VoIP*	1477	0.9995
expDRC	4 *BE* + 2 *VoIP*	1473	1
aveDRC	4 *BE* + 2 *VoIP*	1473	1
maxDRC-PF	4 *BE* + 2 *VoIP*	1474	1

VoIP Packet Delay Performance

According to ITU standards, the acceptable delay bound error is 0.2%. Table 6 shows *VoIP* packet delay bound errors for macrocell scheduling methods. In "Delay Bound Error (%)" column of the table, the first value shows the error value for the

Figure 7. Average sector throughput values for 4 BE + 2 VoIP users in the femtozone

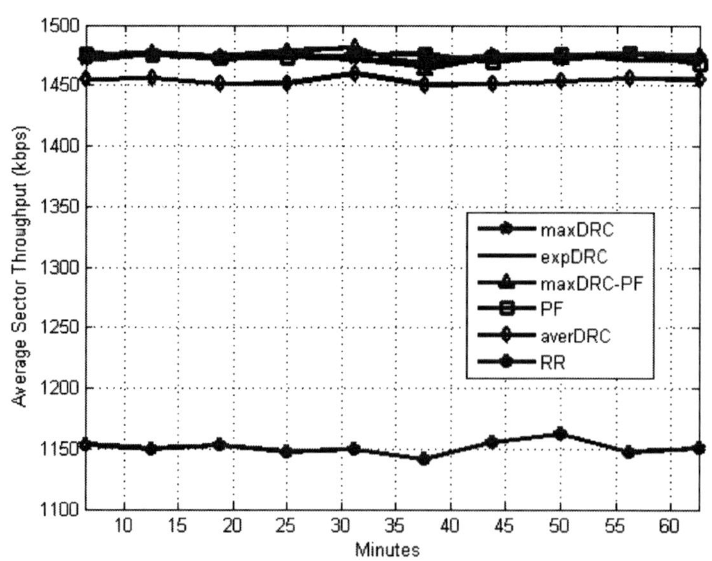

Figure 8. Average sector throughput values for 4 IUBE users in femtozone

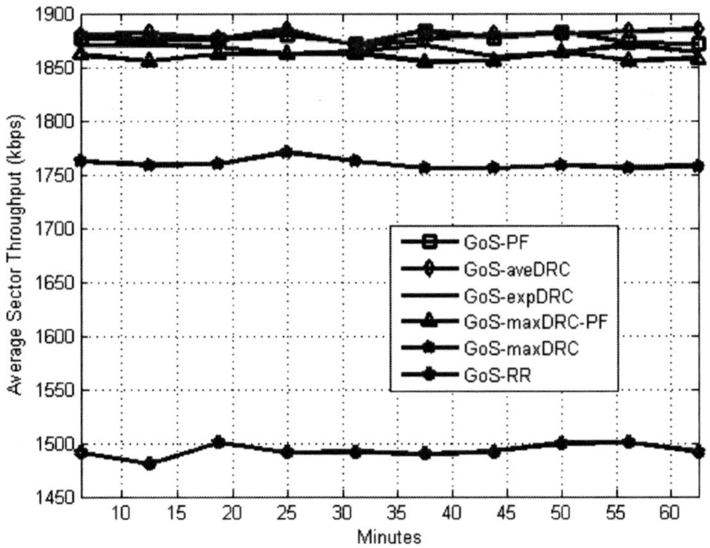

first *VoIP* user, and second values shows the error for the second *VoIP* users in the network. All of the scheduling methods satisfy delay bound error requirement. *RR*, *maxDRC*, *expDRC*, and *maxDRC-PF* schedulers have zero delay bound error.

Similarly, Table 7 shows *VoIP* packet bound error performance of *GoS* based schedulers in *femtozone*. *GoS-maxDRC* is the only scheduling method that has zero delay bound error for both *VoIP* users. Among scheduling methods that have non-zero delay bound error, *GoS-RR* has the lowest error values for the first and second *VoIP* user in the *femtozone*.

VoIP packet delay bound error is satisfied for all of the *GoS* based scheduling methods.

Table 4. Performance of GoS based scheduling algorithms

Scheduling Algorithm	User Type	Ave. Sector Throughput (kbps)	Throughput Fairness
GoS-RR	4 *IUBE*	1493	0.9980
GoS-PF	4 *IUBE*	1877	0.9879
GoS-maxDRC	4 *IUBE*	1760	0.3466
GoS-expDRC	4 *IUBE*	1867	0.9957
GoS-aveDRC	4 *IUBE*	1880	0.9883
GoS-maxDRC-PF	4 *IUBE*	1860	0.9995

Traffic Differentiation Performance

Traffic differentiation capability of a *GoS* based scheduler directly defines its potential of delivering targeted throughput values based on the type of traffic class. Table 8 shows average traffic differentiation error. Based on the results, *GoS-RR* scheduler is the best in terms of traffic discrimination.

GoS-maxDRC scheduler should not be considered for traffic differentiation because of its average error value. Other *GoS* schedulers can be used in *IUBE* traffic differentiation because of their small error value. Among these schedulers, *GoS-maxDRC-PF* scheduler method has the lowest error, and it is recommended to be used in femto base station.

Figure 9. Average sector throughput values for 4 IUBE + 2 VoIP users in the femtozone

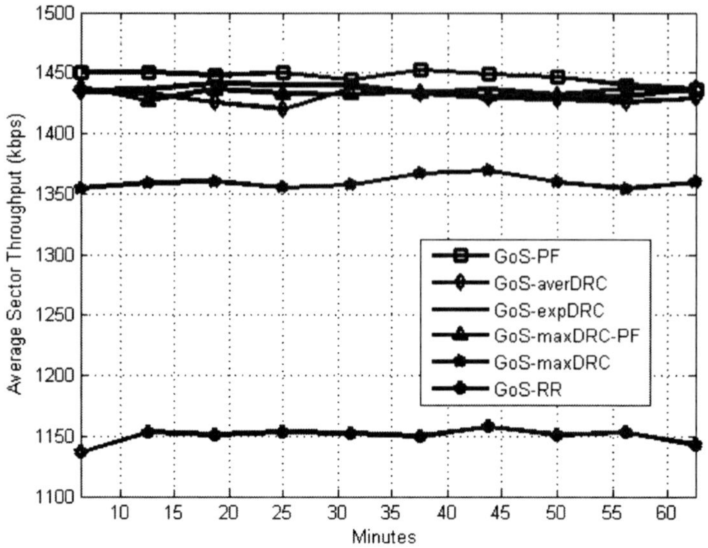

FUTURE RESEARCH DIRECTIONS

Schedulers with tunable parameters have potential of adaptivity to wireless channel conditions. *maxDRC-PF* is one of the schedulers with a control parameter called "*a*." Femto base station product can change the value of "*a*" parameter with changing indoor channel conditions. When the indoor channel conditions become worse, femto base station should decrease "*a*" value in order to increase throughput fairness among users in the network, and similarly when the wireless channel quality increases the value of "*a*" should increase in order to deliver high sector throughput.

Table 5. Performance of GoS based scheduling algorithms for 4 IUBE + 2 VoIP users in the femtozone

Scheduling Algorithm	User Type	Ave. Sector Throughput (kbps)	Throughput Fairness
GoS-RR	4 *IUBE* + 2 *VoIP*	1150	0.9970
GoS-PF	4 *IUBE* + 2 *VoIP*	1438	0.9880
GoS-max-DRC	4 *IUBE* + 2 *VoIP*	1360	0.3475
GoS-exp-DRC	4 *IUBE* + 2 *VoIP*	1437	0.9955
GoS-ave-DRC	4 *IUBE* + 2 *VoIP*	1447	0.9797
GoS-max-DRC-PF	4 *IUBE* + 2 *VoIP*	1434	0.9984

Table 6. VoIP packet delay bound error performance of macrocell schedulers

Scheduling Method	User Type	Delay Bound Error (%)
RR	4 *BE* + 2 *VoIP*	0,0
PF	4 *BE* + 2 *VoIP*	0.0198, 0.0201
maxDRC	4 *BE* + 2 *VoIP*	0,0
expDRC	4 *BE* + 2 *VoIP*	0,0
aveDRC	4 *BE* + 2 *VoIP*	0.0194, 0.02
maxDRC-PF	4 *BE* + 2 *VoIP*	0,0

Table 7. VoIP packet delay bound error performance of GoS based schedulers

Scheduling Method	User Type	Delay Bound Error (%)
GoS-RR	4 *IUBE* + 2 *VoIP*	0.0108,0.0110
GoS-PF	4 *IUBE* + 2 *VoIP*	0.0193,0.0195
GoS-maxDRC	4 *IUBE* + 2 *VoIP*	0,0
GoS-expDRC	4 *IUBE* + 2 *VoIP*	0.0191,0.0195
GoS-aveDRC	4 *IUBE* + 2 *VoIP*	0.0196,0.0193
GoS-maxDRC-PF	4 *IUBE* + 2 *VoIP*	0.0197,0.0194

Table 8. Average differentiation error

Scheduling Algorithm	User Type	Average Error (%)
GoS-maxDRC-PF	4 *IUBE* 4 *IUBE* + 2 *VoIP*	1.0903 1.5163
GoS-RR	4 *IUBE* 4 *IUBE* + 2 *VoIP*	1.5338 3.8618
GoS-expDRC	4 *IUBE* 4 *IUBE* + 2 *VoIP*	4.3336 4.4165
GoS-PF	4 *IUBE* 4 *IUBE* + 2 *VoIP*	7.1246 7.1141
GoS-aveDRC	4 *IUBE* 4 *IUBE* + 2 *VoIP*	7.0020 9.1642
GoS-maxDRC	4 *IUBE* 4 *IUBE* + 2 *VoIP*	392.25 389.5

Femto base station can also change the parameter with changing number of users in the *femtozone*. When the transmission power of femto base station increases, its network coverage will increase and the number of users in the *femtozone* will increase as well. Having more users in the *femtozone* will demand more throughput fairness, which requires *maxDRC-PF* behave more like *PF* than *maxDRC*. Therefore, small "*a*" value should be selected in order to increase throughput fairness among users. As a result, when the transmission power of femto base station increases, the tunable parameter "*a*" that *maxDRC-PF* scheduler uses should decrease.

Femto base station should have *RR*, *maxDRC*, *maxDRC-PF*, *GoS-RR*, *GoS-maxDRC-PF* scheduling algorithms implemented in it. Depending on sector throughput and throughput fairness requirements of users, femto base station can select different scheduling algorithms, and similarly if users want femto base station to provide differentiated services in *femtozone* (i.e., when they have visitors at home), they can turn on or off *BE* traffic differentiation service.

Another future research is to make femto base station transmit data based on the mobile terminal type in the *femtozone*. For instance, if a user is trying to watch video over iPad, as soon as there are available resources, the scheduler should transmit data to iPad more than the other mobile terminals. Similarly, if user wants to download a book using Amazon Kindle, user should be given the highest priority to download the requested book as fast as possible because of the promised service quality level, which is 60 seconds book download time. In essence, Femto base station should increase or decrease the priority of users depending on the information that mobile terminal provides to femto base station.

One important future research is to have dynamic traffic classification in femtocell network (Sevindik, 2011). Multimedia traffic is usually classified based on the requirements of different traffic types. However, in the future, network service providers may not have all the required types of traffic and related parameters in advance since there will be huge number of different mobile devices and applications requiring different treatments from the network. Therefore, novel traffic classification algorithms will be needed in order to determine required performance parameters of different traffic or application types running on different mobile devices.

CONCLUSION

RR scheduler delivers the lowest throughput with the highest throughput fairness. RR scheduling method does not use any channel related information in its DM, and this is the main reason that it delivers the worst throughput performance among state-of-art scheduling methods. However, in order to transmit data, RR scheduler selects one user at a time in a cyclic order and because of this nature; it delivers the highest fairness in terms of slot utilization, which does not necessarily result in perfect throughput fairness since the experienced RF channel conditions might be different for each user. But, in femtocell environment, experienced wireless channel conditions of users are very similar to each other; therefore RR scheduler delivers very high throughput fairness. *maxDRC* performs total opposite of *RR*, with the highest throughput with the lowest throughput fairness. Among state-of-art scheduling methods, maxDRC scheduler is known to be upper bound for sector throughput and to be lower bound for throughput fairness. At each time, maxDRC scheduler selects the user with the best channel condition, and its main target is to deliver the highest sector throughput regardless of the throughput fairness. maxDRC scheduler is recommended for business femtocell case, when there is a need to deliver the high sector throughput. *aveDRC*, *expDRC*, and *maxDRC-PF* schedulers deliver the same throughput performance. Among these schedulers, *maxDRC-PF* has the highest throughput fairness, and it also satisfies the *VoIP* delay bound error value defined by ITU standards. Depending on throughput and throughput fairness requirements, any of these schedulers can to be used in femtocell environment when there is no need for traffic differentiation. Specifically, maxDRC-PF scheduling is recommended for high throughput fairness, and low VoIP delay bound error.

GoS-RR scheduler delivers the lowest throughput among *GoS* based schedulers in the *femtozone* since it does not use any channel related parameter in its DM. *GoS-maxDRC* is not a preferred scheduler for *BE* traffic differentiation because of its low throughput fairness, average throughput and high traffic differentiation error. Unlike maxDRC scheduler, GoS-maxDRC scheduler is not an upper bound for sector throughput. The main reason for this is that when GoS parameter is introduced into the DM of maxDRC scheduler, the new scheduler which is named as GoS-maxDRC does not always select the user requesting the highest throughput since DRC value is multiplied by the GoS value in DM of GoS-maxDRC. Hence, there is possibility that GoS-maxDRC scheduler does not select the user with the best channel condition at the time of scheduling, and which directly affects its throughput performance, and the property of being an upper bound for the sector throughput. *GoS-PF*, *GoS-aveDRC*, *GoS-expDRC* schedulers provide almost the same throughput fairness and total throughput performance. These schedulers use average data rate and average DRC information that GoS-RR and GoS-maxDRC do not use. However, in home fetmocell environment, wireless channel quality is almost same for all the users in the network, which leads to similar average data rate and DRC values. Therefore, *GoS-PF*, *GoS-aveDRC*, *GoS-expDRC* schedulers deliver similar throughput and throughput fairness performance. In situations where wireless channel is too stationary to exploit multiuser diversity, different combination of scheduling algorithms is used. By combining GoS-PF and GoS-maxDRC scheduler, we created GoS-maxDRC-PF scheduling method in order to exploit good wireless channel conditions by providing good throughput fairness.

Among GoS based scheduling methods, *GoS-maxDRC-PF* has the highest throughput fairness, which means that it provides the lowest traffic differentiation error. It also satisfies the *VoIP* delay error bound requirement, which is %0.2 for each *VoIP* packet for each *VoIP* user in the network. Therefore, *GoS-maxDRC-PF* scheduler is the best candidate for *BE* traffic differentiation in *femtozone*. *BE* traffic differentiation concept is mostly used in business femtozone in where there are more BE users then home *femtozone*. Therefore, *GoS-maxDRC-PF* scheduler is recommended for business *femtozone*, and *maxDRC-PF* is recommended for home femtozone.

REFERENCES

Bhushan, N., Lott, C., Black, P., Attar, R., Yu-Cheun, J., & Fan, M. (2006). CDMA2000 1×EV-DO revision a: A physical layer and MAC layer overview. *IEEE Communications Magazine*, *44*(2), 37–39. doi:10.1109/MCOM.2006.1593549

Black, P. J., & Qiang, W. (2002). Link budget of CDMA2000 1xEV-DO wireless internet access system. *The 13th IEEE International Symposium on Personal, Indoor, and Mobile Radio Communications*, vol. 4, (pp. 1847- 1852). IEEE Press.

Ching, Y. H., Huei, Y. S., Vitebsky, S., Pi-Chun, C., & Nat, C. (2003). Schedulers for 1xEV-DO: Third generation wireless high-speed data systems. *Bell Labs Technical Journal*, *3*, 1710–1714.

Elliott, R. C., & Krzymien, W. A. (2002). Scheduling algorithms for the CDMA2000 packet data evolution. *IEEE Vehicular Technology Conference*, vol. 1, (pp. 304-310). IEEE Press.

Esteves, E., Gurelli, M., & Fan, M. (2003). Performance of fixed wireless access with CDMA2000 1xEV-DO. *IEEE Vehicular Tehnology Conference*, vol. 2, (pp. 836-840). IEEE Press.

Jain, R. K., Chiu, D. W., & Hawe, W. R. (1984). A quantitative measure of fairness and discrimination for resource allocation in shared computer systems. *Research Report TR-301*. Digital Equipment Corporation.

Ji, L., Xia, X., Shi, X., & Wang, Z. (2007). *Conversational model based VoIP traffic generation*. IEEE Third International Conference on Networking and Services. IEEE Press.

Kyungtae, K., Jinsung, C., Yongwoo, C., & Heonshik, S. (2005). Dynamic scheduling for scalable media transmission over CDMA2000 1xEV-DO broadcast and multicast networks. *Lecture Notes in Computer Science*, *3462*, 31–57.

Parry, R. (2002). CDMA2000 1×EV-DO for 3G communications. *IEEE Potentials*, *4*, 10–13. doi:10.1109/MP.2002.1044211

Qi, B., Ronald, R. B., Dongzhe, C., Asif, D. G., Ching-Yao, H., & Stan, V. (2003). Performance of 1xEV-DO third-generation wireless high-speed data systems. *Bell Labs Technical Journal*, *7*, 97–107. doi:10.1002/bltj.10021

Santhi, K. R., & Kumaran, G. S. (2006). Migration to 4G: Mobile IP based solutions. *IEEE International Conference on Internet and Web Applications and Services*, (p. 76). IEEE Press.

Sevindik, V. (2010). *Scheduling algorithms and QoS architectures for next generation cellular communication systems.* Doctoral dissertation. Lowell, MA: University of Massachusetts.

Sevindik, V., & Bayat, O. (2011). *Dynamic traffic group generation in LTE cellular wireless telecommunication networks*. US Patent Filled. Application #: 61/430,102. Washington, DC: USA Government.

Sevindik, V., & Bayat, O. (2011). *Multimedia content and file sharing in femtocell networks*. US Patent Filled. Application #:61/432, 629. Washington, DC: USA Government.

Sevindik, V., Bayat, O., & Weitzen, J. (2010). Traffic differentiation for BE users in CDMA 1xEVDO Networks. *International Journal of Multimedia Tools and Applications*, *55*(3), 507–523. doi:10.1007/s11042-010-0563-6

Taewon, B. K. T. F. & Kyunggi, D. (2004). A new efficient scheduling algorithm for the support of QoS in CDMA2000 1xEV-DO. *IEEE Vehicular Technology Conference*, vol. 2, (pp. 1148-1152). IEEE Press.

Tinnakornsrisuphap, P., & Lott, C. (2004). On the fairness and stability of the reverse-link MAC layer in CDMA2000 1×EV-DO. *IEEE International Conference on Communications*, vol. 1, (pp. 144-148). IEEE Press.

Yavuz, M., Diaz, S., Kapoor, R., Grob, M., Black, P., Tokgoz, Y., & Lott, C. (2006). VoIP over CDMA2000 1xEV-DO revision a. *IEEE Communications Magazine*, *44*(2), 50–57. doi:10.1109/MCOM.2006.1593550

Young-June, C., & Saewoong, B. (2004). Scheduling for VoIP service in CDMA2000 1x EV-DO. *IEEE International Conference on Communications*, vol. 3, (pp. 1495-1499). IEEE Press.

ADDITIONAL READING

Saunders, S. R., Carlaw, S., Giustina, A., Bhat, R. R., Rao, S. V., & Siegberg, R. (2009). *Femtocells: Opportunities and challenges for business and technology*. West Sussex, UK: John Wiley & Sons.

KEY TERMS AND DEFINITIONS

Best Effort (BE) Traffic: The main traffic type created by background, web and file transfer applications.

CDMA 1xEVDO: Data optimized version of CDMA2000 1x cellular wireless network standard.

Femto Base Station: Miniature personal base station used in home and business environments.

Femtozone: The network area covered by the femtocell.

Grade of Service (GoS): QoS parameter used in traffic classification.

Inter User Best Effort: Best Effort traffic differentiation feature of CDMA 1xEVDO networks.

Quality of Experience (QoE): Experienced level of quality of consumed service.

Quality of Service (QoS): The quality of delivered multimedia or voice service.

Radio Resource Management: The process of dedicating available wireless networks resources among wireless network users.

Resource Management: The process of dedicating limited resources among resource consumers.

Scheduling: The process of assigning available resources among resource consumers.

Throughput Fairness: Share of available bandwidth resources based on predefined proportions.

Chapter 11
Multimedia Broadcasting in LTE Networks

Antonios Alexiou
University of Patras, Greece

Christos Bouras
Computer Technology Institute & Press "Diophantus", Greece & University of Patras, Greece

Vasileios Kokkinos
Computer Technology Institute & Press "Diophantus", Greece & University of Patras, Greece

Andreas Papazois
Computer Technology Institute & Press "Diophantus", Greece & University of Patras, Greece

George Tsichritzis
University of Patras, Greece

ABSTRACT

Long Term Evolution (LTE) constitutes the latest step before the 4th generation (4G) of radio technologies designed to increase the capacity and speed of mobile communications. To support Multimedia Broadcast/Multicast Services (MBMS), LTE offers the functionality to transmit MBMS over Single Frequency Network (MBSFN), where a time-synchronized common waveform is transmitted from multiple cells for a given duration. In MBSFN transmissions, the achieved Spectral Efficiency (SE) is mainly determined by the Modulation and Coding Scheme (MCS) selected. This study proposes and evaluates four approaches for the selection of the MCS that will be utilized for the transmission of the MBSFN data. The evaluation of the approaches is performed for different users' distribution and from a SE perspective. Based on the SE measurement, the approach that either maximizes or achieves a target SE for the corresponding users' distribution is determined.

INTRODUCTION

Long Term Evolution (LTE) constitutes the evolution of the 3rd Generation (3G) mobile telecommunications technologies. In order to enhance 3rd Generation Partnership Project's (3GPP) radio interface, LTE utilizes Orthogonal Frequency Division Multiple Access (OFDMA) (Holma & Toskala, 2009). Moreover, 3GPP has introduced the Multimedia Broadcast/Multicast Service (MBMS) as a means to broadcast and multicast information to mobile users, with mobile

DOI: 10.4018/978-1-4666-0017-1.ch011

TV being the main service offered (3GPP, 2010b; Holma & Toskala, 2009).

In the context of LTE systems, the MBMS will evolve into e-MBMS ("e-" stands for evolved). This will be achieved through increased performance of the air interface that will include a new transmission scheme called MBMS over Single Frequency Network (MBSFN). In MBSFN operation, MBMS data are transmitted simultaneously over the air from multiple tightly time-synchronized cells. A group of those cells which are targeted to receive these data is called MBSFN area (3GPP, 2010b). Since the MBSFN transmission greatly enhances the Signal to Interference plus Noise Ratio (SINR), the MBSFN transmission mode leads to significant improvements in Spectral Efficiency (SE) in comparison to multicasting over Universal Mobile Telecommunications System (UMTS). This is extremely beneficial at the cell edge, where transmissions (which in UMTS are considered as inter-cell interference) are translated into useful signal energy and hence the received signal strength is increased, while at the same time the interference power is largely reduced (Holma, 2009).

In this study, we evaluate the performance of MBSFN in terms of SE. In general, SE refers to the data rate that can be transmitted over a given bandwidth in a communication system. Several studies, such as (Rong et al., 2008), have shown that SE is directly related to the Modulation and Coding Scheme (MCS) selected for the transmission. Additionally, the most suitable MCS is selected according to the measured SINR so as a certain Block Error Rate (BLER) target to be achieved. Taking into account the above, we focus on a dynamic user distribution, with users distributed randomly in the MBSFN area and therefore experiencing different SINRs. Based on the measured SINRs, our goal is to select the MCS which should be used by the base stations when transmitting the MBMS data. For this purpose, we consider four approaches with different goals set in each one of them. More specifically:

- The 1st approach selects the MCS that ensures that all users, even those with the lowest SINR, receive the MBSFN service (Bottom Up approach).
- The 2nd approach selects the MCS that ensures the maximum SE for all users in the MBSFN area (Top Down approach).
- The 3rd approach sets a predefined SE threshold for the area and selects the MCS that ensures that the average SE over the MBSFN area exceeds this threshold (Area-Oriented approach).
- The 4th approach selects the MCS that ensures that at least the 95% of the users receive the MBSFN service with a predefined target SE (User-Oriented approach).

The remaining of the manuscript is structured as follows: the background section presents the related work in the specific field, as well as an overview of MBSFN architecture. Afterwards, we describe the methodology for calculating the SE of the MBSFN delivery scheme in the single-user case. The four approaches for selecting the MCS of an MBSFN area as well as the evaluation results are presented subsequently. Finally, the last two sections present the conclusions and the planned next steps. For the reader's convenience, appendix A presents an alphabetical list of the acronyms used in the manuscript.

BACKGROUND

Long Term Evolution

Nowadays, mobile marketplace has enabled dramatic advances and changes in telecommunications by offering bandwidth-hungry, multimedia services that previously could only be experienced in wired networks. Although 3G technologies has significantly increased the bit rates of previous mobile technologies, the plethora of mobile applications and services poses the need for deploying a resource economic scheme.

LTE, the evolution of the 3G mobile telecommunications technologies, could constitute the solution to the explosion in demand for such applications and services. LTE supports scalable carrier bandwidths and provides downlink peak rates of at least 100 Mbps, an uplink of at least 50 Mbps and round-trip times of less than 10ms. In order to enhance 3GPP's radio interface, LTE utilizes OFDMA. This radio technology is optimized to enhance networks by enabling significant new high capacity mobile broadband applications and services, while providing cost efficient ubiquitous mobile coverage (Holma, 2009).

Moreover, LTE incorporates several key features of next generation networks. It offers low latency mobile access and implements policy enforcement and decisions at the network edge in order to improve the performance of cell-edge users. Other enhancements include the support for real-time and non-real-time applications, flexible spectrum allocations, re-use of existing cell site infrastructure and high SE performance.

In addition to the above, LTE may efficiently deliver unicast, multicast and broadcast media to the mobile users. To this direction, 3GPP has introduced the MBMS as a means to broadcast and multicast information to mobile users, with mobile TV being the main service offered. LTE infrastructure offers to MBMS an option to use an uplink channel for interaction between the service and the user, which is not a straightforward issue in usual broadcast networks (3GPP, 2010b; Holma & Toskala, 2009).

Overview of E-MBMS LTE Architecture

As depicted in Figure 1, the e-MBMS architecture is split into three domains: the User Equipment (UE) domain, the evolved UMTS Terrestrial Radio Access Network (e-UTRAN) and the Evolved Packet Core (EPC). The UE domain consists of the equipment employed by the user to access the MBSFN services. Within e-UTRAN, the evolved Node Bs (e-NBs or base stations) are the collectors of the information that has to be transmitted to users over the air-interface. The Multi-cell/multicast Coordination Entity (MCE) coordinates the transmission of synchronized signals from different cells and is responsible for the allocation

Figure 1. LTE e-MBMS flat architecture

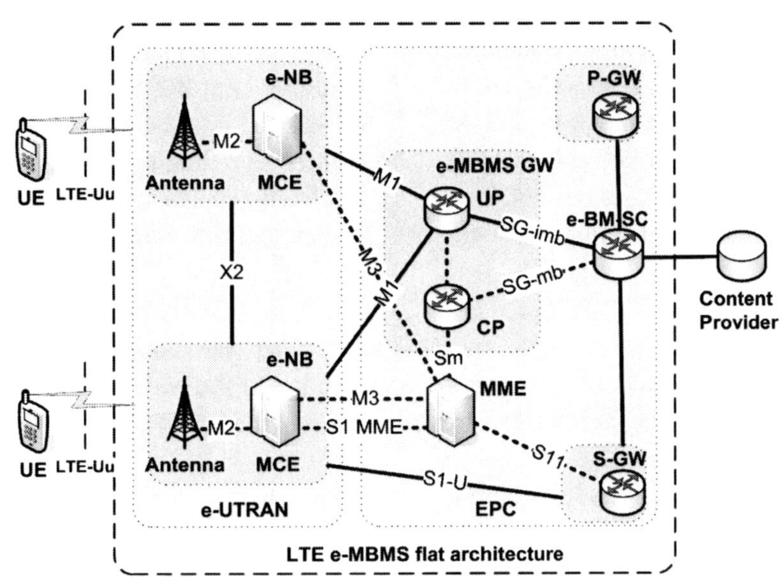

of the same radio resources, used by all e-NBs in the MBSFN area for multi-cell MBMS transmissions. Besides allocation of the time / frequency radio resources, MCE is also responsible for the radio configuration (e.g. the selection of the MCS) (3GPP, 2010c; Holma, 2009).

The EPC consists of five main nodes. These are the Mobility Management Entity (MME), the Serving Gateway (S-GW), the PDN Gateway (P-GW), the e-MBMS Gateway (e-MBMS GW) and the evolved Broadcast Multicast Service Center (e-BM-SC). The last two nodes are associated only to e-MBMS. The MME is the key control-node for the LTE access-network. All major control functionalities of LTE are executed and coordinated by this node. Among all MME functionalities the most indicative are presented below. At first, MME is responsible for the secure signaling procedure of LTE which is called Non Access Stratum (NAS) signaling. Secondly, MME controls the UE handover procedures. Furthermore, it coordinates the UE tracking, paging and polling procedures of LTE. Additionally, MME handles the UE reachabillity procedures (CONNECTED, IDLE). MME is also responsible for the authentication and authorization functions both for UEs as well as for authentication of the interconnection of LTE with external Packet Data Networks (PDNs). Another basic functionality of MME is that it controls the roaming procedures. A warning message transfer function is also implemented in MME providing in that way a more optimized selection of appropriate eNodeBs for the transmission of the data. Last but not least, MME is responsible for controlling the radio bearer management functions including dedicated bearer establishment (3GPP, 2010b; 3GPP, 2010c).

The S-GW routes and forwards user data packets, while also acting as the mobility anchor for the user plane during inter-e-NB handovers and as the anchor for mobility between LTE and other 3GPP technologies. For idle state UEs, the S-GW terminates the downlink (DL) data path and triggers paging when DL data arrives for the UE. It manages and stores UE contexts (e.g. parameters of the IP bearer service, network internal routing information). It also performs replication of the user traffic in case of lawful interception (3GPP, 2010b; 3GPP, 2010c).

The P-GW provides connectivity from the UE to external packet data networks by being the point of exit and entry of traffic for the UE. A UE may have simultaneous connectivity with more than one PGW for accessing multiple PDNs. The P-GW performs policy enforcement, packet filtering for each user, charging support, lawful Interception and packet screening. Another key role of the PGW is to act as the anchor for mobility between 3GPP and non-3GPP technologies such as WiMAX and 3GPP2 (3GPP, 2010b; 3GPP, 2010c).

The e-MBMS GW is physically located between the e-BM-SC and e-NBs and its principal functionality is to forward the MBMS packets to each e-NB transmitting the service. Furthermore, e-MBMS GW performs MBMS Session Control Signaling (Session start/stop) towards the e-UTRAN via the MME. The e-MBMS GW is logically split into two domains. The first one is related to control plane, while the other one is related to user plane. Likewise, two distinct interfaces have been defined between e-MBMS GW and e-UTRAN namely M1 for user plane and M3 for control plane. M1 interface makes use of IP multicast protocol for the delivery of packets to e-NBs. M3 interface supports the MBMS session control signaling (e.g. for session initiation and termination). The e-BM-SC is the entity that is in charge of introducing multimedia content into the LTE network. For that purpose, the e-BM-SC serves as an entry point for content providers or any other broadcast/multicast source, which is external to the network. An e-BM-SC serves all the e-MBMS GWs in a network (3GPP, 2010b; Holma, 2009).

Regarding the air (or LTE-Uu) interface, in MBSFN the transmission takes place from a time-synchronized set of e-NBs using the same resource block. The OFDM symbols in MBSFN contain a Cyclic Prefix (CP), which however is slightly longer than the CP used in conventional transmissions. This enables the UE to combine transmissions from different e-NBs located far away from each other (3GPP, 2007b). Moreover, MBSFN uses two logical channels (in downlink), namely Multicast Traffic Channel (MTCH) and Multicast Control Channel (MCCH). MTCH is a Point-to-Multipoint (PTM) downlink channel for transmitting data traffic to the UEs residing to the service area. On the other hand, MCCH is a PTM downlink channel used for transmitting MBMS control information from the network to UEs and is associated to one or several MTCHs. MCCH and MTCH are only used by UEs that receive MBMS traffic. Additionally, both MCCH and MTCH are mapped on the Multicast Channel (MCH), which is a transport channel at the Medium Access Control (MAC) layer. MCH is a broadcast channel that supports semi-static resource allocation (e.g. with a time frame of a long CP). MCH is mapped to the Physical Multicast Channel of the physical layer (3GPP, 2009; 3GPP, 2010b).

The introduction of the MBSFN feature in the set of LTE transmission techniques has triggered a set of experiments that have been performed in the context of 3GPP and have investigated the efficiency of the radio techniques that can be employed in order to provide the MBMS services (3GPP, 2007a; 3GPP, 2007c; 3GPP, 2007d). These experiments have been conducted through network simulations and have provided useful information about the relative efficiency of different approaches for the provisioning of MBMS services. Different means to provide MBMS services have been evaluated including the following techniques:

- Point-to-Point (PTP) provisioning of MBMS services: this corresponds to mapping the MBMS service to Downlink Shared Chanel (DL-SCH) and includes the possibility to apply link adaptation and Hybrid Automatic Repeat-reQuest (ARQ).
- MBSFN-based multi-cell transmission using MCH.
- PTM provisioning of MBMS services on a per-cell basis with no UE Layer 1 feedback. RAN1 has, at this stage, not made any specific assumptions whether this corresponds to DL-SCH transmission addressing multiple UEs or single-cell MCH transmission.
- PTM provisioning of MBMS services on a per-cell basis with a possibility for Hybrid ARQ UE feedback also for point-to-multipoint transmission.
- PTM provisioning of MBMS services with interference reduction by not transmitting on neighboring cells.
- PTM provisioning of MBMS services on a per-cell basis with a possibility for UE feedback, thus enabling link adaptation and Hybrid ARQ also for point-to-multipoint transmission.

The simulation experiments have been conducted for various UE densities and geometries of the UE drop locations in the examined LTE network topology. The performance evaluation results have proved that the PTP and MBSFN-based techniques provide significant benefits over the PTM techniques in terms of spectral and resource efficiency, coverage and complexity of specification (3GPP, 2007a; 3GPP, 2007d). Additionally, the MBSFN-based multi-cell technique is able to deliver the highest data rate in the central cells of the deployments (3GPP, 2007a). It should be noted that the PTM transmission techniques have been proved efficient only in cases of low UE densities (less than 1.3 UEs/sector) (3GPP, 2007d) or when the UE's subscribing to the MBMS service in question are restricted to occupy only a single cell or a very small number of cells (3GPP, 2007a).

It is obvious that the MBSFN transmission configuration augmented by single-cell PTP or PTM configurations could provide a sufficient basis for the provision of MBMS services in LTE networks by preserving optimum spectral and resource efficiency. Therefore, in the rest of this chapter we focus on the MBSFN transmission technique, which turns to be the most popular for the provision of MBMS services. More specifically we examine the very challenging issue of MCS selection during MBSFN transmission.

RELATED WORK

The performance of MBSFN has been thoroughly examined in previous research works. However, most of these works, such as (3GPP, 2007a; 3GPP, 2007b; 3GPP, 2007d), compare the performance of MBSFN transmissions with classic PTP and PTM transmissions, in which the users are served with PTP or PTM transport channels respectively and the transmissions are executed in a per-cell basis. Additionally, these works do not consider Adaptive Modulation and Coding (AMC) in order to further improve the performance of MBSFN transmission. These works performed by 3GPP have been extended by in (Alexiou et al., 2010b) where the authors focus on the MBSFN transmission scheme and evaluate techniques for the selection of the MCS that can be utilized for the transmission of the MBSFN data. The evaluation of the techniques is performed for different users' distributions and from SE perspective. Based on the SE measurements, the most suitable technique for the corresponding users' distribution is determined.

Transmission techniques, which do not adapt to the fading channel, require a fixed link margin or coding to maintain acceptable performance in deep fades. Thus, these techniques are effectively designed for the worst-case channel conditions, resulting in insufficient utilization of the full channel capacity (Goldsmith & Chua, 1998). For better utilization of the channel capacity, AMC has been proposed in a variety of publications. For example, an adaptive variable rate variable-power transmission scheme using un-coded M-ary Quadrature Amplitude Modulation (M-QAM) was proposed in (Goldsmith & Chua, 1997). This adaptive technique is more power efficient than non-adaptive modulation in fading.

Adaptive algorithms for the OFDM system are also proposed in (Wong et al., 1999). In (Wong et al., 1999) a multi-cell, multi-user OFDM system with adaptive subcarrier allocation and adaptive modulation is considered. The specific study describes an adaptive sub-carrier, bit and power allocation algorithm to maximize the total throughput of the multi-cell system in the presence of Co-Channel Interference (CCI), frequency selective Rayleigh fading and Additive White Gaussian Noise (AWGN). For the unicast system, link adaptation is possible because the Channel Status Information (CSI) can be reported to the base-station by the terminal. Our work expands (Wong et al., 1999), by focusing on the MBSFN service, which utilizes OFDM technology.

Moreover, studies such as (Ball et al., 2008; Phan et al., 2008; Rong et al., 2008; Sheng et al., 2008) have shown that SE is directly related to the MCS selected for the transmission. In (Rong et al., 2008) the authors propose an approach, which selects the lowest MCS for the MBSFN transmission that allows an expected SE target to be achieved for 95% of users. However, focusing only on the users' side may not be sufficient. Sometimes the operator's goal may be the maximization of the SE over all users of the topology or the provision of the service to all the users irrespectively of the conditions the users face. On the other hand, in (Sheng et al., 2008) an adaptive MCS based on partial feedback is proposed in order to obtain the improvement of system throughput. Our work extends and completes the above studies and, furthermore, tackles the addressed problems by proposing four approaches, each one of them fulfilling different goals in terms of SE.

Finally, a cost-based approach for the evaluation of the MBSFN delivery scheme is examined in (Alexiou et al., 2010a). For the evaluation, the packet delivery cost, cost for control procedures (synchronization, polling) and scalability of the scheme are taken into account. Based on these telecommunication cost parameters, the authors calculate the total telecommunication cost required for the transmission of the MBSFN data to mobile users of a given MBSFN service. Finally, this work estimates how many neighboring cell rings should be included in the same MBSFN area and thus transmitting in the same frequency with the cells that actually contain users, in order to achieve high SFN gains with the lowest possible cost with respect to users' distribution in the topology.

SINGLE-USER MCS SELECTION AND SE ESTIMATION

In order to select the MCS and calculate the SE in the case of a single receiver, we use the following 4-step procedure (Alexiou et al., 2010c).

Step 1: SINR Calculation

Let the MBSFN area consist of N neighboring cells. Due to multipath, the signals of the cells arrive to the receiver by M different paths, so the average SINR of a single user at a given point m is expressed as in (1) (Rong et al., 2008):

$$SINR(m) = \frac{\sum_{i=1}^{N}\sum_{j=1}^{M}\frac{w\left(\tau_i\left(m\right)+\delta_j\right)P_j}{q_i\left(m\right)}}{\sum_{i=1}^{N}\sum_{j=1}^{M}\frac{\left(1-w\left(\tau_i\left(m\right)+\delta_j\right)\right)P_j}{q_i\left(m\right)}+N_0} \tag{1}$$

with:

$$w(\tau) = \begin{Bmatrix} 1 & 0 \le \tau < T_{CP} \\ 1-\frac{\tau - T_{CP}}{T_u} & T_{CP} \le \tau < T_{CP}+T_u \\ 0 & otherwise \end{Bmatrix} \tag{2}$$

where P_j is the average power associated with the j path, $\tau_i(m)$ the propagation delay from base station i, δ_j the additional delay added by path j, $q_i(m)$ the path loss from base station i, T_{cp} the length of the cyclic prefix and T_u the length of the useful signal frame.

SINR is usually calculated in OFDMA for each subcarrier and all the SINRs are combined in order to find a non-linear average SINR (effective SINR or γ_{eff}), using the Exponential Effective SIR Mapping (EESM) (Mehlfhrer et al., 2009).

$$\gamma_{eff} = EESM\left(\gamma_i,\beta\right) = -\beta\cdot\ln\left(\frac{1}{N}\cdot\sum_{i=1}^{N}e^{-\frac{SINR_i}{\beta}}\right) \tag{3}$$

where N is the number of subcarriers and β is calibrated by means of link level simulations to fit the compression function to the AWGN (Mehlfhrer et al., 2009).

However, in 3GPP LTE systems, adjacent subcarriers allocation is considered, making subcarriers allocated to one channel experiencing similar fading conditions. All subcarriers allocated to a given channel will thus experience the same fast fading and their SINR will be equal (Rong et al., 2008).

Step 2: MCS Selection

In order to obtain the MCS that should be used for the transmission of the MBSFN data to a single user, AWGN simulations have been performed. In general, the MCS determines both the modulation alphabet and the Effective Code Rate (ECR) of the channel encoder. Figure 2 shows the BLER results for Channel Quality Indicators (CQI) 1-15 without using Hybrid Automatic Repeat Request (HARQ) and for 1.4 MHz and 5.0 MHz bandwidth. The results have been obtained from the link level simulator introduced in (Mehlfhrer et al., 2009). Each MCS is mapped to a predefined CQI value. The 15 different sets of CQIs and the corresponding MCSs are defined in (3GPP, 2010a).

In LTE networks, an acceptable BLER target value should be smaller than 10% (Mehlfhrer et al., 2009). The SINR to CQI mapping required to achieve this goal can thus be obtained by plotting the 10% BLER values over SNR of the curves in Figure 2. The 10% BLER values for each CQI are depicted in Figure 3. Using the obtained line, the γ_{eff} can be mapped to a CQI value (i.e. MCS) that should be signaled to the e-NB so as to ensure the 10% BLER target.

Step 3: Throughput Estimation

In order to estimate the achieved throughput for the selected MCS, (4) is used. In (4), *BW* is the total bandwidth offered by LTE, *e(SINR)* is the effective code rate of the selected modulation scheme and *BLER(SINR)* the block error rate (Elayoubi et al., 2008).

$$Throughput = BW \cdot e\left(SINR\right) \cdot \left(1 - BLER\left(SINR\right)\right) \quad (4)$$

Therefore, by utilizing the SINR and MCS obtained by the SINR Calculation (Step 1) and MCS Selection (Step 2) steps respectively, the achieved throughput can be calculated. Figure 4a and Figure 4b depict the relationship between the achieved throughput and the SNR for all MCSs, as calculated from (4) for the cases of 1.4 MHz and 5.0 MHz respectively.

Step 4: Single-User Spectral Efficiency

SE refers to the information rate that can be transmitted over a given bandwidth in a specific communication system. It constitutes a measure of how efficiently a limited frequency spectrum is utilized. The formula from which the SE can be obtained is:

$$SE = \frac{Throughput}{BW} \quad (5)$$

Figure 2. SNR-BLER curves obtained for: a) 1.4 MHz, b) 5.0 MHz

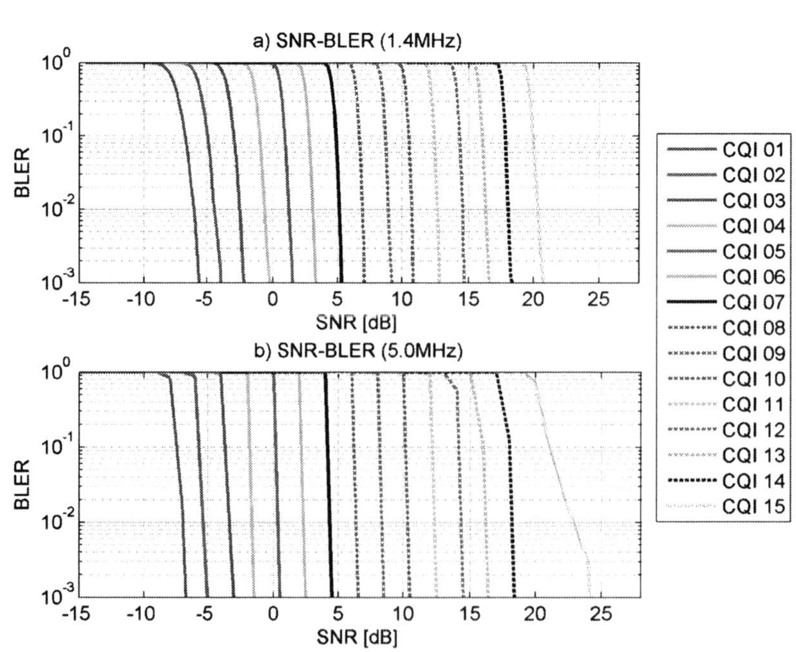

Figure 3. SINR to CQI mapping

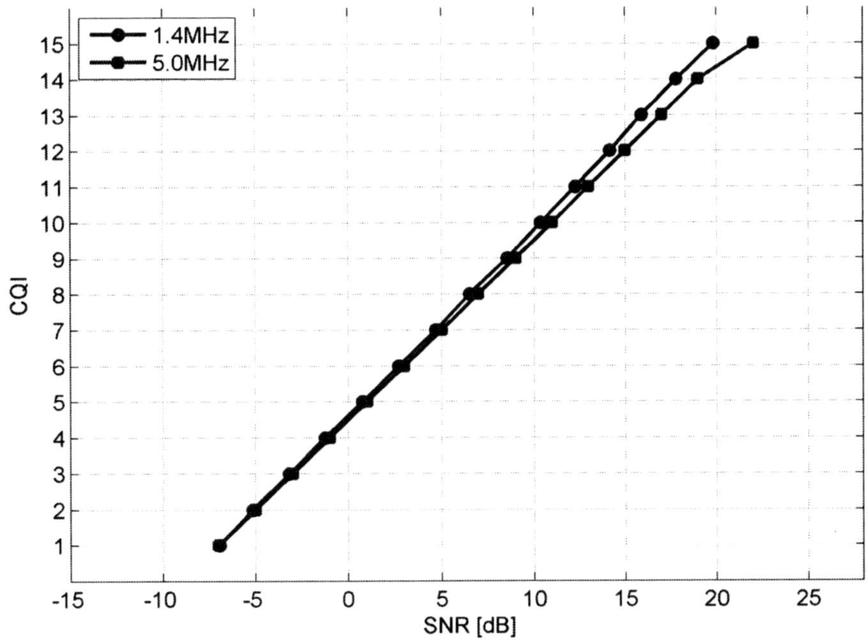

To sum up, for a single user γ_{eff} is calculated from (1), (2) and (3); while the achieved SE may be obtained from (4) and (5). If, for example, the effective SINR for a random user in the topology is 5dB and the bandwidth 5.0 MHz, from Figure 3 we obtain the equivalent CQI (CQI = 7). For the specific CQI and SINR value, the throughput as obtained from Figure 4 is 6 Mbps. Therefore, the SE as calculated from (5) is 1.2 (bit/s)/Hz.

MULTIPLE-USERS MCS SELECTION AND SE ESTIMATION

The MCS selection and the SE evaluation in the multiple-users case are deduced from the single-user approach described in the previous section. In general, when multiple users are located in the MBSFN area, the value of the total SE depends on the selected MCS. This section examines four approaches for the selection of the MCS during MBSFN transmissions.

1st Approach: Bottom Up Approach

The 1st approach ensures that all users, even those with the lowest SINR, will receive the MBSFN service. In order to achieve this goal the algorithm finds the minimum SINR and the MCS that corresponds to the minimum SINR is obtained from the MCS Selection step (Figure 3). Then, from (4) or Figure 4 the corresponding average throughput and SE are obtained. The operation of this approach indicates that all the users in the MBSFN area will uninterruptedly receive the MBMS service, irrespectively of the conditions they experience (in terms of SINR). However, the fact that the user with the minimum SINR determines the MCS indicates that users with greater SINR values will not make use of a MCS that would ensure a greater throughput. The procedure for obtaining the MCS and the SE is presented using pseudo code in Algorithm 1.

Figure 4. Throughput for all CQIs obtained for: a) 1.4 MHz, b) 5.0 MHz

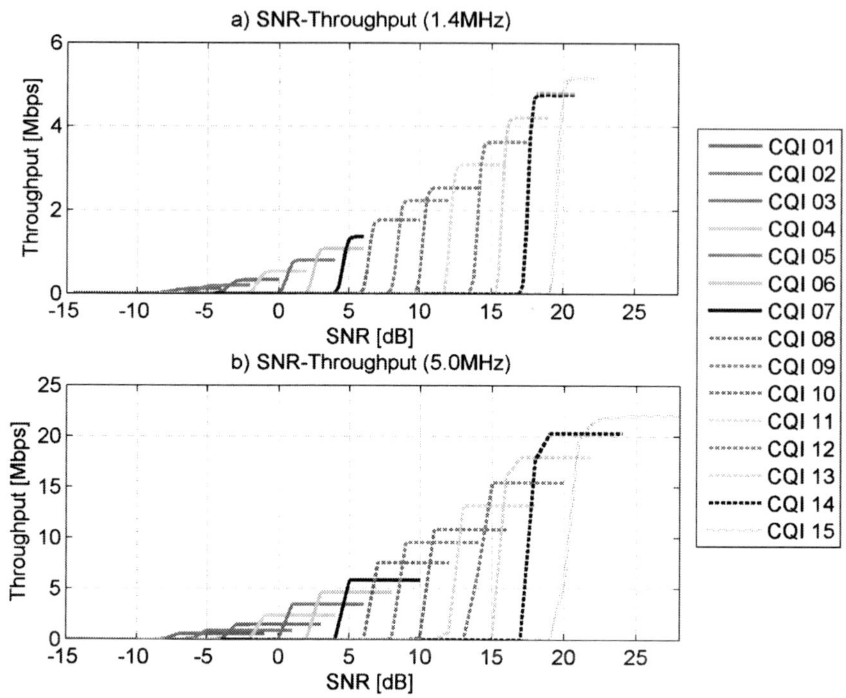

2nd Approach: Top Down Approach

The 2nd approach selects the MCS that ensures the maximum average throughput and SE over all users in the MBSFN area. At first the algorithm calculates the SINR value for each user using (1). Then, the algorithm scans all the MCSs in Figure 4. For each MCS, the algorithm calculates the per-user throughput depending on the calculated SINRs and obtains the average throughput and total SE. The MCS that ensures the maximum average throughput - and therefore the maximum total SE - is selected. Algorithm 2 presents the operation of the 2nd approach using pseudo code.

3rd Approach: Area-Oriented Approach

The goal of the 3rd approach is to find the lowest MCS that achieves a target SE for an area. This target usually equals to 1 (bit/s)/Hz (Rong et al., 2008). Initially the algorithm calculates the SINR value for each user. Then it proceeds with the scanning of the MCSs to calculate the per-user throughput. Starting from the lowest MCS, the algorithm calculates the per-user throughput and obtains the average throughput and the total SE for each MCS. If during the scanning procedure one MCS ensures that the total SE is equal or higher than the area target SE, the operation stops without scanning all the MCSs of Figure 4 and the algorithm selects this MCS for the delivery of

Algorithm 1. Pseudo code of 1st approach

```
Define MBSFN topology
% calculate the SINRs for all the users in the topology
FOR i = 1:total_users
Calculate SINR(i)
END
min_SINR = min(SINR) %find the lowest SINR
% choose the MCS that corresponds to the min SINR
selected_MCS = f_MCS (min_SINR)
% calculate the throughput for the selected MCS
throughput = f_throughput(selected_MCS, min_SINR)
   % calculate the obtained spectral efficiency
   Calculate SE
```

Algorithm 2. Pseudo code of 2nd approach

```
Define MBSFN topology
% calculate the SINRs for all the users in the topology
FOR i = 1:total_users
Calculate SINR(i)
END
% for each MCS calculate the average throughput over all users
FOR MCS = 1:15
FOR j = 1:total_users
throughput(MCS, j) = f_throughput (MCS, SINR(j))
END
avg_throughput(MCS) = average(throughput(MCS, :))
Calculate SE(MCS)
END
%find the max spectral efficiency that can be achieved
SE = max(SE(:))
```

the MBMS data. In other words, the aim of this approach is to find the lowest MCS that allows a target SE to be achieved. The scanning procedure starts from the lowest MCS in order to serve as many users as possible. If the scanning procedure starts from the highest MCS, then the SE target is achieved very quickly by utilizing a high MCS, and therefore only the users that experience high SINRs receive the MBSFN service as depicted in Figure 4. In the case the target SE cannot be achieved, this approach has identical operation with the 2nd approach (i.e. selects the MCS that ensures the maximum total SE). This procedure is presented using pseudo code in Algorithm 3.

4th Approach: User-Oriented Approach

The difference between the 4th and the 3rd approach is that in spite of defining an area-specific target SE such as the 3rd approach, the 4th approach defines a user-oriented target SE (usually equal to 1 (bit/s)/Hz (Rong et al., 2008). More specifically, the algorithm initially calculates the SINR value for each user. Then, starting from the lowest MCS, the algorithm calculates the per-user throughput and per-user SE of each MCS. If during the scanning procedure one MCS ensures that at least 95% of the users reach or exceed the

Algorithm 3. Pseudo code of 3rd approach

```
Define MBSFN topology
Define area_target_SE
% calculate the SINRs for all the users in the topology
FOR i = 1:total_users
Calculate SINR(i)
END
% scan the MCSs so as calculate the SE over the MBSFN area
FOR MCS = 1:15
FOR j = 1:total_users
throughput(MCS, j) = f_throughput (MCS, SINR(j))
END
% Calculate average throughput and spectral efficiency
avg_throughput(MCS) = average(throughput(MCS, :))
Calculate SE(MCS)
% examine if area target SE is achieved
IF SE(MCS) >= area_target_SE THEN % target is achieved
BREAK;
ELSE % target is not achieved
SE = max(SE(:))
END
END
  % obtained spectral efficiency
SE = SE(MCS)
```

target SE, the operation stops and the algorithm selects this MCS for the delivery of the MBMS data. Similar to the 3rd approach, this approach locates the lowest MCS that allows a user-specific target SE to be achieved for the 95% of the users' population. If the target SE cannot be achieved for the 95% of the users, the MCS that ensures the maximum total SE is selected. This procedure is presented using pseudo code in Algorithm 4.

PERFORMANCE EVALUATION

This section provides simulation results regarding the operation and performance of the aforementioned approaches. For the purpose of our experiments we have extended the link level simulator introduced in (Mehlfhrer et al., 2009). In particular, two different scenarios are investigated. Scenario 1 assumes that a constant number of 100 users are randomly distributed in the MBSFN area; while Scenario 2 investigates the case of variable number of users. The parameters used in the performed simulations are presented in Table 1.

Scenario 1: Predefined Number of Users

Scenario 1 attempts to make a direct comparison of the proposed approaches when the MBSFN area consists of a constant number of users. More specifically, the MBSFN area - which consists of four neighboring cells - contains 100 randomly distributed users. For comparison reasons the evaluation is performed for 1.4 MHz and 5.0 MHz bandwidth.

Let us first consider the case of 1.4 MHz bandwidth presented in Figure 5a. According to the procedure of the 1st approach, initially the users' SINRs are obtained and the lowest SINR value is selected for the determination of the MCS. In the examined scenario, the lowest SINR is -1.952dB. Therefore, from Figure 4 the CQI 3 is selected. Indeed, Figure 5a confirms that the 1st approach may provide a SE value of 0.233 (bit/s)/Hz by deploying CQI 3 for the transmission of the MBSFN data. On the other hand, the 2nd approach after the scanning procedure selects CQI 12 for the transmission of the MBSFN data. The selection of CQI 12 increases the SE drastically to 2.200 (bit/s)/Hz. As expected, this is the maximum SE that can be achieved for the specific user distribution in the case of 1.4 MHz bandwidth (Figure 5a).

Table 1. Simulation settings

Parameter	Value
Cellular layout	Hexagonal grid, 19 cell sites
Inter Site Distance (ISD)	1732 m
Carrier frequency	2.0 GHz
System bandwidth	1.4 MHz / 5.0 MHz
Channel model	3GPP Typical Urban
Propagation model	Cost Hata
Cyclic prefix / Useful signal frame length	16.67μsec / 66.67μsec
Modulation and Coding Schemes	15 different sets defined in (3GPP, 2010a)

Algorithm 4. Pseudo code of 4th approach

```
Define MBSFN topology
Define user_target_SE
% calculate the SINRs for all the users in the topology
FOR i = 1:total_users
Calculate SINR(i)
END
% scan the MCSs so as to calculate the per-user SE
FOR MCS = 1:15
FOR j = 1:total_users
% Calculate the per user throughput and spectral efficiency
throughput(MCS, j) = f_throughput (MCS, SINR(j))
SE(MCS, j) = throughput(MCS, j) / bandwidth
END
% examine if user target SE is achieved for 95% of users
IF SE(MCS, j) >= user_target_SE FOR 95% of users THEN
% target achieved
BREAK;
ELSE % target is not achieved
SE = max(SE(:, j))
END
END
  % obtained spectral efficiency
SE = SE(MCS, j)
```

Finally, the performance of the 3rd and 4th approach in Figure 5a confirms that the specific approaches have similar operation. Indeed, both approaches select CQI 8; however the 4th approach may provide a slightly increased level of SE. This is caused due to the fact that the 4th approach does not take into account the 5% of the users that experience worse network conditions (in terms of SINR). Nevertheless, it is worth mentioning that both approaches reach the target SE that was set. More specifically, the 3rd approach ensures that the total SE exceeds the SE target over the MBSFN area; while in the 4th approach the per-user SE for the 95% of the users exceeds the predefined threshold. The examination of Figure 5b that corresponds to the case of 5.0 MHz leads to similar results.

Scenario 2: Variable Number of Users

This paragraph presents simulation results concerning the operation of the proposed approaches for variable number of users. More specifically,

Figure 6 and Figure 7 examine the performance of each approach in terms of SE and selected MCS, when the users' population in the MBSFN area varies from 1 to 1000 users (for 1.4 MHz and 5.0 MHz bandwidth respectively). All the users that receive the MBMS service appear in random initial positions throughout the MBSFN area, which consists of four neighboring and tightly time-synchronized cells. The remaining simulation parameters are in accordance with Table 1.

As both figures present, the 1st approach achieves the lowest SE for the corresponding user population. On the other hand, the fact that this approach takes into account the lowest SINR in order to obtain the corresponding MCS ensures that even the users that experience low SINRs will receive the MBMS service. As a result, the users with better conditions will not receive the service with the highest possible throughput. Another disadvantage of this approach is that a potential mobility of the user with the lowest SINR could force the base station to continuously change the transmission MCS (ping-pong effect).

As depicted in Figure 6a and Figure 7a, the 2nd approach ensures the maximum SE irrespectively of the users population. This is reasonable since the 2nd approach selects the MCS that ensures the maximum average throughput and SE over all users in the topology. It is also worth mentioning, that in certain scenarios where the majority of users are distributed near the base station, the 2nd approach could achieve even higher values of SE. Indeed, the users near the base station experience high SINRs and as consequence higher values of MCS may be utilized in order for a high average throughput to be achieved. Based on the above, we conclude that the 2nd approach tends to utilize a high MCS. As stated in (Rumney, 2009), this fact has the advantage of decreasing the users' transmit power. However, the users with bad conditions will not receive the MBMS service (see Figure 4).

Figure 5. SE evaluation and CQI selection for predefined number of users for: a) 1.4 MHz, b) 5.0 MHz

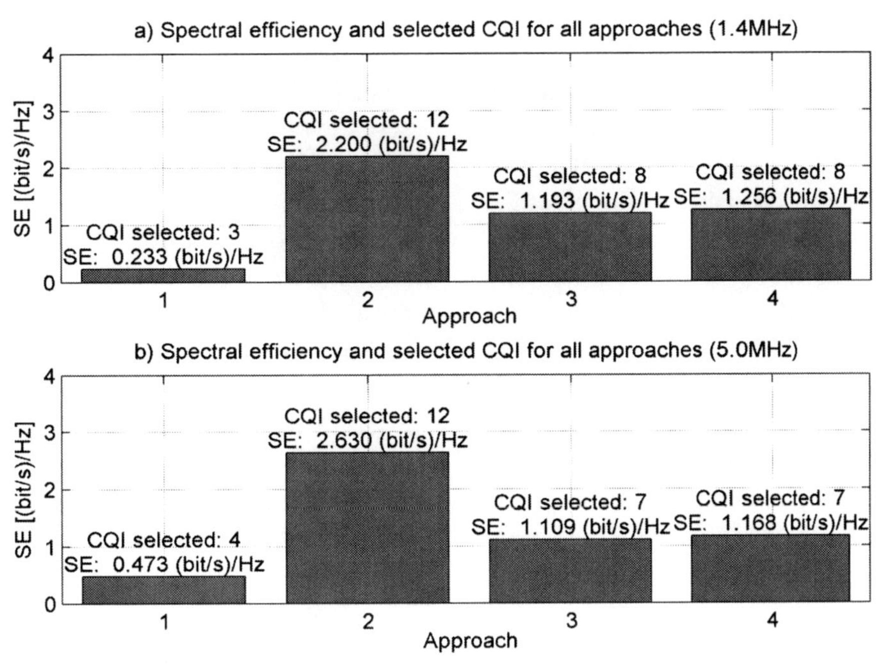

Figure 6. SE evaluation and CQI selection for variable number of users for 1.4 MHz

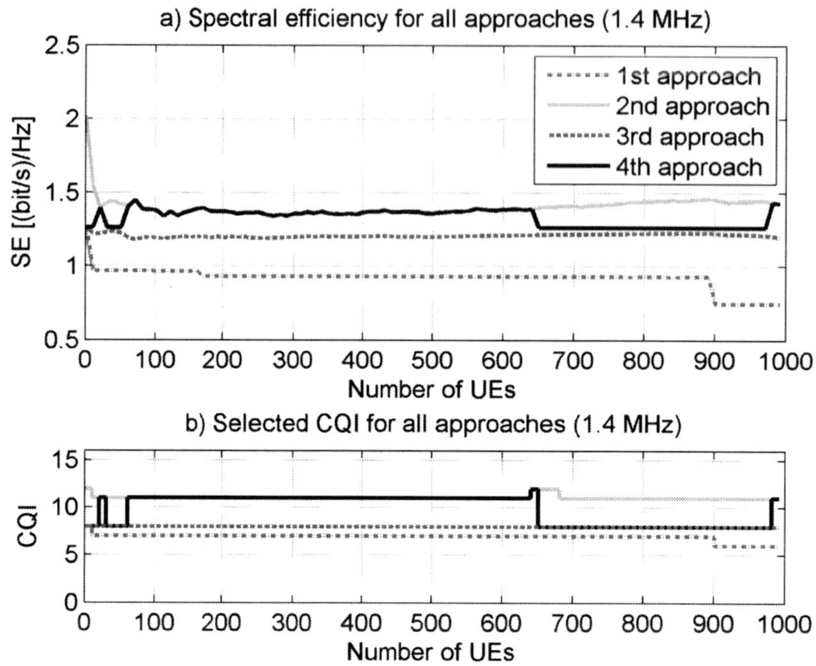

The 3rd approach selects the MCS that ensures that the average SE calculated over all users in the topology achieves the SE target. Therefore, as depicted in Figure 6b (1.4 MHz) the 3rd approach utilizes CQI 8, while in Figure 7b (5.0 MHz) the selected CQI is CQI 7. The specific MCSs achieve a SE value over the MBSFN area higher than the SE target during the whole simulation (Figure 6a and Figure 7a). One of the most important advantages of the 3rd approach is that it minimizes the ping-pong effect in MCS selection. Indeed, this approach ensures that the MCS will not necessarily change when the users' population changes. This leads to the avoidance of the ping-pong effect when new users enter the MBSFN topology or when users stop requiring the MBSFN service. However, it should be noted that the 3rd approach does not achieve the maximum possible SE, since the algorithm scans the different MCS beginning from the lowest value of MCS and stops when the selected MCS achieves the SE target.

Finally, the 4th approach selects the MCS that satisfies the SE target for the 95% of users. As depicted in Figure 6a and Figure 7a, the specific MCSs achieve a SE value higher than the per-user SE target. Moreover, the SE achieved with this approach is higher than that of the 3rd approach since the 95% of the users receive the MBSFN service with a data rate that satisfies the SE target. This implies that the remaining 5% of the users who experience bad conditions are not taken into account, in opposition to the 3rd approach in which all the users in the MBSFN area are considered for the MCS selection.

To sum up, Table 2 presents a cumulative, direct comparison between the approaches analyzed in this manuscript. The main conclusion is that the selection of the most efficient MCS is an operator dependent parameter. Therefore, the uninterrupted service provision irrespectively of the users' conditions would make the 1st approach the most efficient approach. However, this approach could not provide any guarantee for the throughput and the achieved SE. On the other hand, for maximum average throughput and maximum SE the most efficient approach would be the 2nd approach. The 3rd approach constitutes the most efficient approach when the operator targets at a specific SE value over the MBSFN area and minimizes the ping-pong effect in MCS selection (minimum MCS switching). Finally, the 4th approach achieves a predefined per-user SE target for at least the 95% of the users.

CONCLUSION

The main enhancement that the adoption of MBSFN brings in e-MBMS is the improvement of over the air SE. The achieved SE is mainly determined by the selected MCS in the physical layer. In this manuscript we proposed four different approaches for the efficient selection of the appropriate MCS and we evaluated the impact of this selection to the achieved SE. The parameters that have been taken into account in the evaluation are the number of served users and their

Table 2. Qualitative comparison of the approaches

	Performance			
Approach	**Throughput**	**Spectral Efficiency**	**Service Provision**	**MCS Switching**
1st	Minimum	Minimum	Guaranteed	Medium
2nd	Maximum	Maximum	Not Guaranteed	Medium
3rd	Medium	Target over the area	Not Guaranteed	Minimum
4th	Medium	Target per-user	Not Guaranteed	Maximum

Figure 7. SE evaluation and CQI selection for variable number of users for 5.0 MHz

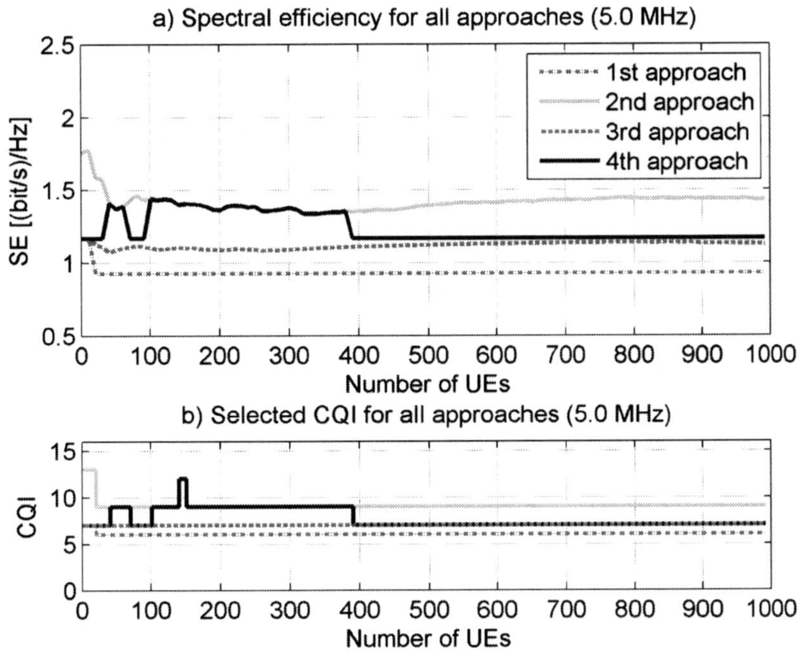

position in the topology. Based on the above two parameters, the service provider can choose the most efficient MCS selection approach for the active MBSFN sessions. The approaches cover different scenarios that could be realized in real world such as ensuring service continuity for the user with lowest SINR value and therefore for all users in the MBSFN area, selecting the MCS that maximizes the SE, selecting the MCS based on the covered area or the percentage of the users that receive the service in an acceptable quality.

In brief, we could say that the selection of the appropriate MCS is an operator dependent issue. Different operator requirements may lead to different MCS approach selection. To that end, service continuity can be secured by employing the 1st (Bottom Up) approach, while for high demanding MBMS applications, which are targeted to users that experience optimal network and link conditions, the 2nd (Top Down) approach is the most efficient one. Additionally, ping-pong effect can be regulated by employing the 3rd (Area-Oriented) approach and simultaneously, all MBMS users are treated by the approach as equal irrespectively of the network and link conditions that they experience. Finally, the 4th (User-Oriented) approach gives the ability to the network operator to predefine both the per-user target SE and the percentage of users that will be taken into account for the calculation of the achieved SE.

To conclude, it could be mentioned that the analysis presented in this manuscript underlines that the introduction of an adaptive MCS selection algorithm for MBFSN enabled LTE networks is a prerequisite for network operators in order deploy high quality broadcast networks capable of delivering high demanding real time multimedia applications to mobile users.

FUTURE RESEARCH DIRECTIONS

The step that follows this work could be the design, the implementation and the evaluation of an algorithm responsible for choosing the most efficient MCS selection approach according to operator needs each time. Our analysis indicates that approaches switching is possible to happen in real time. Furthermore, the combined usage of different approaches is also possible and could solve the particular inefficiencies that each approach has.

Another direction that we intent to investigate is the application of Forward Error Correction (FEC) for MBSFN transmissions in LTE networks. FEC is an error control method that can be used to augment or replace other methods for reliable data transmission. The main attribute of FEC schemes is that the sender adds redundant information in the messages transmitted to the receiver. This information allows the receiver to reconstruct the source data. Such schemes inevitably add a constant overhead in the transmitted data and are computationally expensive. This additional communication cost will be calculated and based on this; the efficiency of FEC use in different scenarios will be evaluated.

REFERENCES

Alexiou, A., Bouras, C., Kokkinos, V., Papazois, A., & Tsichritzis, G. (2010b). Efficient MCS selection for MBSFN transmissions over LTE Networks. *Proceedings of IFIP Wireless Days 2010*. Venice, Italy: IEEE Press.

Alexiou, A., Bouras, C., Kokkinos, V., Papazois, A., & Tsichritzis, G. (2010c). Spectral efficiency performance of MBSFN-enabled LTE networks. *Proceedings of 6th IEEE International Conference on Wireless and Mobile Computing, Networking, and Communications (WiMob 2010)*. Niagara Falls, Canada: IEEE Press.

Alexiou, A., Bouras, C., Kokkinos, V., & Tsichritzis, G. (2010a). Communication cost analysis of MBSFN in LTE. *Proceedings of 21st Annual IEEE International Symposium on Personal, Indoor, and Mobile Radio Communications (PIMRC 2010)*. Instabul, Turkey: IEEE Press.

Ball, C., Hindelang, T., Kambourov, I., & Eder, S. (2008). Spectral efficiency assessment and radio performance comparison between LTE and WiMAX. *Proceeding of IEEE International Symposium on Personal, Indoor and Mobile Radio Communications (PIMRC 2008)*. Cannes, France: IEEE Press.

Elayoubi, S., Ben Haddada, O., & Fourestie, B. (2008). Performance evaluation of frequency planning schemes in OFDMA-based networks. *IEEE Transactions on Wireless Communications*, *7*(5), 1623–1633. doi:10.1109/TWC.2008.060458.

Goldsmith, J., & Chua, G. (1997). Variable-rate variable-power MQAM for fading channels. *IEEE Transactions on Communications*, *45*(10), 1218–1230. doi:10.1109/26.634685

Goldsmith, J., & Chua, G. (1998). Adaptive coded modulation for fading channels. *IEEE Transactions on Communications*, *46*(5), 595–602. doi:10.1109/26.668727

3GPP. (2007a). *R1-070051: Performance of MBMS transmission configurations*.

3GPP. (2007b). *R1-071049: Spectral efficiency comparison of possible MBMS transmission schemes: Additional Results*.

3GPP. (2007c). *R1-071213: Response on LS on radio efficiency for delivery of broadcast/multicast services*.

3GPP. (2007d). *R1-071433: Additional results on EMBMS transmission configurations*.

3GPP. (2009). *TS 36.201, V9.0.0: Technical specification group radio access network; Evolved universal terrestrial radio access (E-UTRA); LTE physical layer - General description (release 9)*.

3GPP. (2010a). *TS 36.213, V9.1.0: Technical specification group radio access network; Evolved universal terrestrial radio access; Physical layer procedures (release 9)*.

3GPP. (2010b). *TS 36.300, V9.3.0: Technical specification group radio access network; Evolved universal terrestrial radio access (E-UTRA) and evolved universal terrestrial radio access network (E-UTRAN); Overall description; Stage 2 (release 9)*.

3GPP. (2010c). *TS 23.401 V10.2.1: General packet radio service (GPRS) enhancements for evolved universal terrestrial radio access network (E-UTRAN) access*.

Holma, H., & Toskala, A. (2009). *LTE for UMTS - OFDMA and SC-FDMA based radio access*. Chichester, UK: John Wiley & Sons.

Mehlfhrer, C., Wrulich, M., Ikuno, J., Bosanska, D., & Rupp, M. (2009). Simulating the long term evolution physical layer. *Proceedings of 17th European Signal Processing Conference (EUSIPCO 2009)*. Glasgow, Scotland: IEEE Press.

Phan Huy, D., Legouable, R., Ktenas, D., Brunel, L., & Assaad, M. (2008). Downlink B3G MIMO OFDMA link and system level performance. *Proceedings of 2008 IEEE Vehicular Technology Conference (VTC 2008-Spring)*. Marina Bay, Singapore: IEEE Press.

Rong, L., Haddada, O., & Elayoubi, S. (2008). Analytical analysis of the coverage of a MBSFN OFDMA network. *Proceedings of IEEE Global Communication Conference 2008 (Globecom 2008)*. New Orleans, LA: IEEE Press.

Rumney, M. (Ed.). (2009). *LTE and the evolution to 4G wireless: Design and measurement challenges*. Agilent Technologies.

Sheng, Y., Peng, M., & Wang, W. (2008). A novel adaptive modulation and coding strategy based on partial feedback for enhanced MBMS network. *Journal of China Universities of Posts and Telecommunications*, *15*(1), 48–54. doi:10.1016/S1005-8885(08)60061-8

Wong, C., Cheng, R., Lataief, K., & March, R. (1999). Multiuser OFDM with adaptive subcarrier, bit and power allocation. *IEEE Journal on Selected Areas in Communications*, *17*(10), 1747–1757. doi:10.1109/49.793310

ADDITIONAL READING

Adhikari, S. (2009). Critical analysis of multi-antenna systems in the LTE downlink. *Proceedings of 3rd International Conference on Internet Multimedia Systems Architecture and Applications (IMSAA 2009)*, (pp. 1-6). Bangalore, India: IEEE Press.

Alexiou, A., Bouras, C., & Kokkinos, V. (2007). Evaluation of the multicast mode of MBMS. *Proceedings of 18th Annual IEEE International Symposium on Personal, Indoor, and Mobile Radio Communications (PIMRC'07)*. Athens, Greece: IEEE Press.

Alexiou, A., Bouras, C., Kokkinos, V., Papazois, A., & Tseliou, G. (2010). Enhancing FEC application in LTE cellular networks. *Proceedings of IFIP Wireless Days 2010*. Venice, Italy: IEEE Press.

Alexiou, A., Bouras, C., Kokkinos, V., Papazois, A., & Tsichritzis, G. (2010). Adopting FEC for reliable multicasting over LTE networks. *Proceedings of 13th ACM International Conference on Modeling, Analysis, and Simulation of Wireless and Mobile Systems (MSWiM 2010)*. Bodrum, Turkey: IEEE Press.

Alexiou, A., Bouras, C., & Papazois, A. (2010). The Impact of FEC on mobile multicast power control. *Proceedings of the 6th Advanced International Conference on Telecommunications (AICT 2010)*. Barcelona, Spain: IEEE Press.

Ball, C., Mullner, L., Lienhart, J., & Winkler, H. (2009). Performance analysis of closed and open loop MIMO in LTE. *Proceedings of European Wireless Conference 2009 (EW 2009)*, (pp. 260-265). Aalborg, Denmark: IEEE Press.

Dahlman, E., Ekstrom, H., Furuskar, A., Jading, Y., Karlsson, J., Lundevall, M., & Parkvall, S. (2006). *Proceedings of 63rd Semi-Annual IEEE Vehicular Technology Conference (VTC- Spring 2006)*, (pp. 137-141). Melbourne, Australia: IEEE Press.

Dahlman, E., Parkvall, S., Skold, J., & Beming, P. (2008). *3G evolution: HSPA and LTE for mobile broadband* (2nd ed.). Elsevier Ltd.

3GPP. (2010). *TS 36.101, V9.4.0: User equipment (UE) radio transmission and reception*.

3GPP. (2010). *TS 36.104, V9.4.0: Base station (BS) radio transmission and reception*.

3GPP. (2010). *TS 36.211, V9.1.0: Physical channels and modulation*.

Kovalenko, I., Zhukov, V., & Shelkovnikov, N. (2009). Features of long term evolution technology and comparative analysis of noise stability for different kinds of modulation. *Proceedings of 19th International Crimean Conference Microwave & Telecommunication Technology (CriMiCo 2009)*, (pp. 235-236). Sevastopol, Ukraine: IEEE Press.

Krondorf, M., & Fettweis, G. (2009). Carrier frequency dependent downlink spectral efficiency of cellular LTE deployments. *Proceedings of IEEE International Conference on Communications (ICC 2009)*, (pp. 4037-4042). Dresden, Germany: IEEE Press.

Luby, M., Gasiba, T., Stockhammer, T., & Watson, M. (2007). Reliable multimedia download delivery in cellular broadcast networks. *IEEE Transactions on Broadcasting*, *53*(1), 235–246. doi:10.1109/TBC.2007.891703

Mogensen, P., Wei, N., Kovacs, Z., Frederiksen, F., Pokhariyal, A., & Pedersen, I. … Kuusela, M. (2007). LTE capacity compared to the Shannon bound. *Proceedings of 65th Semi-Annual IEEE Vehicular Technology Conference (VTC- Spring 2007)*, (pp. 1234-1238). Dublin, Ireland: IEEE Press.

Oborina, A., Moisio, M., Henttonen, T., Pernila, E., & Koivunen, V. (2008). MIMO performance evaluation in UTRAN long term evolution downlink. *Proceedings of 42nd Annual Conference on Information Sciences and Systems (CISS 2008)*, (pp. 1179-1183). Princeton, NJ: IEEE Press.

Pedersen, I., Kolding, E., Frederiksen, F., Kovacs, Z., Laselva, D., & Mogensen, E. (2009). An overview of downlink radio resource management for UTRAN long-term evolution. *IEEE Communications Magazine*, *47*(7), 86–93. doi:10.1109/MCOM.2009.5183477

Pospishny, I., Vasyuk, V., Romanchyk, S., Dovzhenko, O., Dony, O., & Shvaichenko, V. (2010). 3GPP long term evolution (LTE). *Proceedings of International Conference on Modern Problems of Radio Engineering, Telecommunications, and Computer Science (TCSET 2010)*, (p. 192). Lviv, Ukraine: IEEE Press.

Sanchez, J., Morales-Jimenez, D., Gomez, G., & Enbrambasaguas, T. (2007). Physical layer performance of long term evolution cellular technology. *Proceedings of 6th IST Mobile and Wireless Communications Summit*, (pp. 1-5). Budapest, Hungary: IEEE Press.

Sesia, S., Toufik, I., & Baker, M. (2009). *LTE - The UMTS long term evolution: From theory to practice*. Chichester, UK: John Wiley & Sons. doi:10.1002/9780470742891

Sternad, M., Svensson, T., Ottosson, T., Ahlen, A., Svensson, A., & Brunstrom, A. (2007). Towards systems beyond 3G based on adaptive OFDMA transmission. *Proceedings of the IEEE*, *95*(12), 2432–2455. doi:10.1109/JPROC.2007.907119

Taeyoung, L., Jangkeun, J., Jisun, Y., & Jitae, S. (2010). Adjustable subcarrier allocation for spectral efficiency and user fairness in 3GPP LTE downlink systems. *Proceedings of 12th International Conference on Advanced Communication Technology (ICACT 2010)*, (pp. 985-988). Gangwon-Do, South Korea: IEEE Press.

Tomazic, S., & Jakus, G. (2009). Long term evolution: Towards 4th generation of mobile telephony and beyond. *Proceedings of 9th International Conference on Telecommunication in Modern Satellite, Cable, and Broadcasting Services (TELSIKS 2009)*, (pp. 91–96). Nis, Serbia: IEEE Press.

Vieira, P., Queluz, P., & Rodrigues, A. (2008). LTE spectral efficiency using spatial multiplexing MIMO for macro-cells. *Proceedings of 2nd International Conference on Signal Processing and Communication Systems (ICSPCS 2008)*. Gold Coast, Australia: IEEE Press.

Viering, I., Buchner, C., Seidel, E., & Klein, A. (2007). Real-time network simulation of 3GPP long term evolution. *Proceedings of 2007 IEEE International Symposium on a World of Wireless, Mobile, and Multimedia Networks (WoWMoM 2007)*, (pp.1-3). Helsinki, Finland: IEEE Press.

KEY TERMS AND DEFINITIONS

AMC: A link adaptation method that raises the overall system capacity and provides the flexibility to match the MCS to the average channel conditions.

LTE: The evolution of the 3G mobile telecommunications technologies.

MBMS: A service introduced by 3GPP to broadcast and multicast information to mobile users, with mobile TV being the main feature offered.

MBSFN: A transmission scheme where data are transmitted simultaneously over the air from multiple tightly time-synchronized cells.

MBSFN Area: A group of time-synchronized cells which are targeted to receive the MBSFN data.

MCS Selection: The procedure of selecting the appropriate MCS in order to make an efficient use of the air interface.

SE: The information rate that can be transmitted over a given bandwidth in a specific communication system. It constitutes a measure of how efficiently a limited frequency spectrum is utilized.

APPENDIX A: ACRONYMS

Acronym	Explanation
3GPP	3rd Generation Partnership Project
AMC	Adaptive Modulation and Coding
AWGN	Additive White Gaussian Noise
BLER	Block Error Rate
CCI	Co-Channel Interference
CP	Cyclic Prefix
CQI	Channel Quality Indicators
CSI	Channel Status Information
e-BM-SC	evolved Broadcast Multicast Service Center
ECR	Effective Code Rate
EESM	Exponential Effective SIR Mapping
e-MBMS	evolved MBMS
e-MBMS GW	e-MBMS Gateway
e-NBs	evolved Node Bs
EPC	Evolved Packet Core
e-UTRAN	evolved UMTS Terrestrial Radio Access Network
FEC	Forward Error Correction
HARQ	Hybrid Automatic Repeat Request
ISD	Inter Site Distance
LTE	Long Term Evolution
MAC	Medium Access Control
MBMS	Multimedia Broadcast/Multicast Service
MBSFN	MBMS over Single Frequency Network
MCCH	Multicast Control Channel
MCE	Multi-cell/multicast Coordination Entity
MCH	Multicast Channel
MCS	Modulation and Coding Scheme
MME	Mobility Management Entity
M-QAM	M-ary Quadrature Amplitude Modulation
MTCH	Multicast Traffic Channel
OFDMA	Orthogonal Frequency Division Multiple Access
PTM	Point-to-Multipoint
PTP	Point-to-Point
SE	Spectral Efficiency
SINR	Signal to Interference plus Noise Ratio
UE	User Equipment
UMTS	Universal Mobile Telecommunications System

Chapter 12
Traffic Prioritization in Sensor Networks using Bandwidth Scavenging

Anthony Plummer Jr.
Johns Hopkins University Applied Physics Laboratory, USA

Mahmoud Taghizadeh
Michigan State University, USA

Subir Biswas
Michigan State University, USA

ABSTRACT

This chapter presents a history-based statistical channel access mechanism for enabling traffic prioritization in wireless sensor networks. Prioritized access is realized such that low priority non-real-time sensors can access channel bandwidth that is unused by high priority real-time traffic. The key idea is for the low priority sensor nodes to first observe and statistically model the channel usage pattern by the high priority traffic, then make advantageous probabilistic transmissions so that the non-priority traffic throughput is maximized while protecting the high-priority traffic from disruptions. The chapter details a practical whitespace measurement scheme and presents a channel history based prioritization protocol. The access mechanism is implemented in a TelosB mote-based sensor testbed in which the non-priority motes continually measure the RSSI to infer the channel usage pattern and probabilistically access the channel while different types of traffic are sent by high-priority TelosB motes.

INTRODUCTION

The concept of a Wireless Sensor Network (WSN) has provided the opportunity for many useful monitoring applications to be developed. A WSN typically consists of many small and low powered communication devices, which can be composed of a few components, including a wireless communication interface, microprocessor, small on-board memory, and sensor interface. The sensor interface allows each device to detect various environment stimuli using different sensor modules. Although these devices are individually low power and low cost, a large number of them together are capable of performing many complex applications. For example in the application of area monitoring,

DOI: 10.4018/978-1-4666-0017-1.ch012

a WSN may measure levels of sound, vibration, atmospheric pressure or temperature within a specified area. With these individual measurements, many applications can be enabled such as battlefield surveillance, animal population monitoring, or health tele-monitoring (Akyildiz, Sankarasubramaniam, & Cayirci, 2002).

With the WSN ability to monitor and report diverse information about an environment, multiple categories of data may be sent over the network. These data types can result from separate data sets generated on a single node with multiple sensor modules or multiple nodes with different sensors types. The latter case creates a heterogeneous WSN where there may be two or more sets of devices with varying capabilities.

Consider a WSN where there are two different sets of sensor devices sharing the same wireless channel. One set supports real time data applications and the other set is used for non-real time applications. Real time applications such as event surveillance typically require constant or variable rate data streams to be sent over the sensor network. Depending on the specific application, the rate of this data can be of the order of a few packets per second to tens of packets per second. The objective is to transport such real time application data packets with minimum delivery delay and packet losses at rates required by the application. Non-real time applications such as environmental monitoring need to measure levels of temperature, moisture and other ambient parameters in a time driven or event driven manner. Such applications typically require lower data rates compared to the real-time data streams and have less stringent packet delay and loss requirements. In this type of network, nodes that support real time applications must coexist with the nodes that are running non-real time application. In other words, devices with different capabilities must share the spectrum resources.

Now, if a network designer wants to ensure that the requirements of the real time traffic network are met in this heterogeneous WSN, a traffic priority structure can be used. In this case, the real time traffic is given priority access to channel over the non-real time traffic. To address this type of traffic prioritization problem in general, some issues that would need to be addressed are:

- **Maintaining a priority structure:** Support is needed to facilitate different priority levels for the various sets of network devices
- **Supporting higher priority requirements:** Data requirements of the higher priority traffic must be met
- **Limiting the effects of delaying the higher priority traffic:** When lower priority traffic has data to send, the impacts on the higher priority traffic must be minimized
- **Supporting different high priority traffic loads:** As the high priority traffic changes, the lower priority traffic should adjust its usage of bandwidth accordingly
- **Maximizing the throughput of the lower priority traffic:** Lower priority traffic throughput need to be maximized given the available channel bandwidth.

There have been a few solutions proposed to support traffic prioritization within heterogeneous sensor networks. However, some of these solutions have potential limitations in addressing the issues outlined above. In this chapter, a new and different approach to traffic prioritization that utilizes channel history-based mechanisms is developed and discussed. Channel history-based mechanisms, which we term *Bandwidth Scavenging*, may provide a suitable solution due to its ability to dynamically adjust lower priority traffic access to protect higher priority traffic. The overall objective of this chapter is to present a method of traffic prioritization through using channel history based methods.

The first part of the chapter discusses an overview of various approaches to traffic prioritization. Although traffic prioritization can be achieved through spatial or frequency separation, in this chapter, temporal based approaches are specifically highlighted since channel history-based mechanisms fall within this category. Each temporal based protocol has its advantages and drawbacks, which are highlighted within the section.

The next section develops and presents a Bandwidth Scavenging protocol called Channel History based Traffic Prioritization (CHTP). Within this section, a whitespace measurement and model are first presented that captures the traffic characteristics of the higher priority nodes. Whitespace can be defined as the idle time in between two consecutive packets in a wireless channel. Using this model, the CHTP protocol is developed for statistically identifying effective transmission opportunities for lower priority traffic while protecting higher priority traffic.

Next we introduce a prototype experimental testbed that investigates practical measurements of a sensor channel activity utilizing commonly used sensor mote hardware. Although there are some issues that would need to be addressed before an actual system deployment, the testbed provides some insights into the process of Bandwidth Scavenging.

We also present performance results of a few access strategies including CHTP under various higher priority traffic profiles and dynamic traffic cases.

We identify some research challenges involved in channel history based traffic prioritization and finally, we conclude the chapter.

APPROACHES TO TRAFFIC PRIORITIZATION

In order to protect higher priority traffic from large channel access delays due to interruptions from the lower priority traffic, an access priority structure is usually needed. Coexistence between the two types of traffic can be achieved through different techniques.

MAC Layer Role in Traffic Prioritization

Supporting traffic prioritization in wireless networks can be achieved using the Medium Access Control (MAC) layer through defining a suitable structure to control device access to the available network bandwidth. Prioritization between multiple systems can be achieved through spatial, frequency or temporal separation at the MAC layer, as reported in (Chiasserini & Rao, 2003) and (Jingli, Xiangqian, & Swami, 2009).

In spatial-based device separation, lower priority nodes define their range of communication (i.e. transmission range) such that their transmissions do not interfere with higher priority nodes' transmissions. For example, transmission power control can be used to adjust a nodes transmission power to define a suitable non-interfering transmission range (Stojmenovic, Giordano, Conti, & Basagni, 2005). Lower priority nodes can set their transmission ranges at a minimal range that still allows communication but reduces the amount of disruptions to the higher priority traffic (Mietzner, Lampe, & Schober, 2009).

Frequency-based separation involves distributing nodes into multiple channels so that their transmissions do not interfere with each other. In this case, lower priority nodes would first identify unoccupied channels, then freely access the unused channels (Ahuja, Corke, & Bok, 2008; Petty, et al., 2007). Additionally adaptive frequency hopping techniques can be used.

Another promising mechanism for traffic prioritization is to utilize the temporal separation. In this approach, channel access time is used to separate the differing priority traffics. In the following sections, a discussion on these types of techniques is presented.

TDMA Based Traffic Prioritization

Time Division Multiple Access (TDMA) protocols have been investigated as a possible method to assign priority to different traffic classes. TDMA works by dividing a channel into fixed sized time slots, which are organized into periodic frames. Each node in a network is assigned a slot or multiple slots to be used during each periodic frame. To institute priority amongst the nodes, higher priority nodes are assigned a larger portion of the available slots in a frame over the lower priority nodes. This process ensures that the higher priority nodes utilizes the channel with higher frequency and allows dedicated bandwidth to the higher priority traffic. Additionally, conflict free access is supported, which means that the high priority traffic cannot be disrupted by the lower priority traffic during channel access. However, when the higher priority nodes are not utilizing the allocated bandwidth, lower priority nodes cannot use the unused bandwidth. This can reduce the efficiency of the channel usage.

To overcome this issue, the wireless sensor network protocol presented in (Hongjun, Xun, Shen, & Hongxu, 2009) divides the TDMA frames into a broadcast period and scheduled access period. The scheduled access sections are further divided into priority groups with multiple transmission slots in each group. Within each frame, a node broadcasts their traffic priority during the broadcast period, which reserves their transmissions for the specified priority groups. During the scheduled period, nodes send packets using one of the available slots within its priority group section. Another protocol named PMAC (Lei, Hongyi, & Nian-Feng, 2009) proposes a TDMA based prioritization MAC utilizing a wide range of prefix reservation and inter-slot intervals.

TDMA based traffic prioritization provides the benefit of absolute priority for different traffic classes. A drawback of the protocol is the requirement of tight time synchronization needed to implement time slotting. Additionally, TDMA protocols can suffer from reduced efficiency of bandwidth usage due to unused scheduled slots.

Random Access Based Traffic Prioritization

Another type of temporal based traffic prioritization is utilizing random access protocols. Traffic prioritization can be achieved using the well-known Carrier Sense Multiple Access (CSMA) class of protocols by utilizing different inter-frame spacing (IFS) periods for different traffic classes. Consider an example of this approach under the IEEE 802.11 standard. In baseline 802.11 access, when a message arrives from the upper layer, the MAC layer accesses the channel with a DIFS (Distributed Coordination Function Inter-Frame Spacing) and Contention Window (CW) duration irrespective of the traffic class. The DIFS period specifies the amount of idle time that a node waits before starting a random backoff time. This back-off time is controlled by the CW process, where each node selects a random wait time between zero and a CW value.

Traffic priority has been introduced within the 802.11e standard (Mangold, Sunghyun, Hiertz, Klein, & Walke, 2003), by using traffic class specific (referred to as Access Categories or ACs) DIFS and CW values. The quantity DIFS in 802.11e is parameterized as arbitration inter-frame space (AIFS) with specified access classes (AC). Similarly, the CW is redefined with CW_{min}[AC] and CW_{max}[AC] allowing different priority traffic to be assigned with different overlapping back-off times. The addition of class-specific AIFS and CW values allow for prioritizations since nodes with smaller AIFS periods and lower CW ranges will access the channel first and more often. This core 802.11e prioritization concept has been refined in (Mishra & Sahoo, 2006) which propose mechanisms for choosing the class-specific AIFS and CW values dynamically under different network conditions. The concepts of 802.11e have also been applied to WSNs. The mechanism in (Firoze, Ju, & Kwong, 2007) is similar to 802.11e in using variable size contention windows and different inter-frame spacing for different types of sensing events.

While providing reasonable access differentiation, these approaches rely only on the instantaneous channel status (i.e. free or busy) for granting access. This leads to undesirable disruptions to the higher priority traffic as follows. Consider a situation in which lower priority traffic intends to transmit a packet and it does so after finding the channel free for the AIFS duration, which is specified for the low priority traffic. Now, in the middle of this packet transmission if a high priority node in the vicinity intends to send a packet, it needs to wait until the current packet transmission is over. This causes an undesirable delay for the higher priority traffic, which in turn will affect the application layer performance. Additionally some studies have shown that lower priority traffic can be starved in certain network conditions (Lenagala & Zeng, 2006).

Channel History Based Traffic Prioritization

A new and different approach to traffic prioritization is utilizing history-based mechanisms. In addition to the current channel state, a lower priority node uses its knowledge about the higher priority node's traffic history to access the channel. This makes the mechanism less greedy in that at times a node may not access the channel even if it is currently free. This restrained approach is based on the higher priority traffic history because the node anticipates an impending higher priority transmission. By being less greedy (compared to the prioritization random access approaches), the lower priority nodes can access the channel in a less disruptive manner. Additionally, unlike in TDMA, the lower priority nodes can adjust their transmissions depending on the traffic of the higher priority nodes.

Channel history based protocols have been mostly investigated for accessing a primary user network by a secondary user (SU) network. In this context of research, primary users (PU) can be representative of higher priority users and similarly, secondary users represent lower priority users. The authors in (Geirhofer, Tong, & Sadler, 2006) develop a methodology for formally analyzing the idle time or whitespace available within 802.11 primary traffic in infrastructure mode. The key idea is to model the whitespace as a semi-Markov process that relies on the underlying 802.11 state model involving DIFS, SIFS, DATA, and ACK transactions. The model describes the whitespace profile in terms of holding times of the idle and busy states of the channel.

The authors in (Geirhofer, Tong, & Sadler, 2008) develop an WLAN access strategy for secondary users in which the SUs utilize packet size slots for the channel access. At the beginning of each slot, a secondary user senses the channel and if the channel is free then it transmits with a specified probability that is calculated from previous measurement. The objective is to minimize primary user disruption and maximize secondary user throughput. The use of time-slotting in their approach requires time synchronization across the secondary users. In the proposed access mechanism in this chapter, the need for such secondary user time synchronization is avoided via asynchronous whitespace access based on a stochastic whitespace modeling approach.

In (Chen, Qing, & Swami, 2008; Jung & Xin, 2008), the proposed access strategies rely on specific types of PU networks. Protocols are developed in (Chen, et al., 2008) for a slotted PU network where the SUs are synchronized with the PU's slots. In this case, if a PU does not use a slot then the SU can transmit after a small duration of time in the same slot. The PUs in (Jung & Xin, 2008) are assumed to use a specific combination of time and frequency multiplexing to prevent inter-PU access collisions. In this chapter, we formulate the PU network more generally by not assuming any specific PU MAC layer such as the slotted ones in (Chen, et al., 2008). The target here is to develop an access mechanism, which can deal with unslotted as well as stochastic PU MAC layers such as 802.11, CSMA and ALOHA protocols. This assumption of a general PU access protocol broadens the applicability of the SU access approach proposed in this chapter.

The traffic profiles of higher priority users studied in literature have varied greatly. (Chen, et al., 2008; Huang, Liu, & Ding, 2009; Wang, Ji, & Liu, 2007). The authors in (Wang, et al., 2007) model the primary users' packet arrival intervals and packet durations as two separate Poisson processes. Similarly, (Huang, et al., 2008) uses an alternating ON-OFF pattern, where the duration of the idle and busy states follows an exponential distribution with means v and l respectively. In (Chen, et al., 2008), a Markov chain with known transition probabilities is used to model primary user behavior. Lastly in (Huang, et al., 2009), a Markov model with Exponential, Uniform, Weibull and Generalized Pareto distributions are considered for higher priority nodes arrival times. In line with these efforts in the literature, the access mechanism in this chapter is evaluated for a variety of higher layer traffic profiles. However, in contrast to these works, we investigate a measurement-and–model based approach for modeling higher priority nodes' traffic using the presented testbed. This approach captures the impacts of network dynamics, which may not be captured in the model-only (e.g. 2-state Markov (Chen, et al., 2008; Huang, et al., 2009)) approaches in the literature.

In the next section, we develop and discuss a channel history based protocol.

CHANNEL HISTORY BASED TRAFFIC PRIORITIZATION (CHTP)

In this section, a method to utilize history-based mechanisms is presented and detailed. With non-deterministic primary traffic patterns, a low priority node is not able to deterministically predict when a detected idle times or whitespace will end. As a result, the low priority nodes access strategy cannot be deterministic. A reasonable solution is for the low priority nodes to access a given whitespace based on a previously observed statistical whitespace profile. Once a lower priority node identifies a whitespace, it sends a packet only if the estimated chance of completing the transmission before the whitespace ends is high. The likelihood of such completion is calculated based on the historical statistics of the higher priority nodes' channel usage. We term this opportunistic access during ultra-short and non-deterministic whitespaces as *bandwidth scavenging* by the low priority nodes (Plummer, Taghizadeh, & Biswas, 2010a, 2010b). In other words, the low priority nodes scavenge the channel capacity left over by the higher priority nodes. The efficiency of whitespace access by the lower priority nodes is determined by the resulting lower priority nodes throughput, which should be maximized, and the amount of higher priority nodes traffic disruption, which should be minimized. The Channel History based Traffic Prioritization (CHTP) developed in this section attempts to accomplish these goals.

Whitespace Modeling

The available whitespace can be captured by a low priority node through detecting the received signal strength indicator (RSSI) at a given channel frequency (Geirhofer, Tong, & Sadler, 2006). This method applies energy detection to measure the higher priority nodes' signal. Methods such as matched filter detection and cyclostationary feature detection can also be used (Akyildiz, et al., 2002).

A low priority node periodically senses the channel to detect the whitespaces. Based on the channel sampling report from the hardware, the channel state is interpreted using an ON-OFF (busy-idle) model. The sensing interval of the channel is represented by T_p. The duration of whitespace is modeled using its probability density function, *w(n)*, which is the probability of a whitespace being nT_p time long. This *w(n)* is computed periodically to capture any changes in the primary user behavior. Once the *w(n)* for a measured whitespace is computed, the low priority nodes rely solely on this function for channel access.

Note that the proposed access mechanism in this chapter assumes certain degree of stationarity of the distribution *w(n)*. Meaning, the higher priority node traffic behavior remains stationary for long enough duration so that the lower priority nodes can make use of *w(n)* to make statistically meaningful access decisions. The frequency of updates to *w(n)* depends on the time scale of the higher priority traffic changes.

Functional Overview

Figure 1 shows a functional diagram of the architectural components in CHTP. The user tunable control inputs into the system are low priority node packet size (*S*), primary disruption bound (*DB*), and the sensing interval (T_p), which is a property of the low priority nodes' hardware. The lower priority node continuously computes the whitespace pdf *w(n)* with a channel-sensing interval of T_p. The access parameters are then fed into the access module, which, in addition to channel sensing, transmits packets into the channel. The objective is to satisfy a user specified primary disruption bound *DB*.

Access Based on Transmission Opportunities

A Transmission Opportunity (TO) within a whitespace is defined as a period of time that is large enough to send a single low priority node packet. Clearly, there may be multiple TOs for sending multiple low priority nodes packets

Figure 1. Functional diagram of the CHTP access strategy

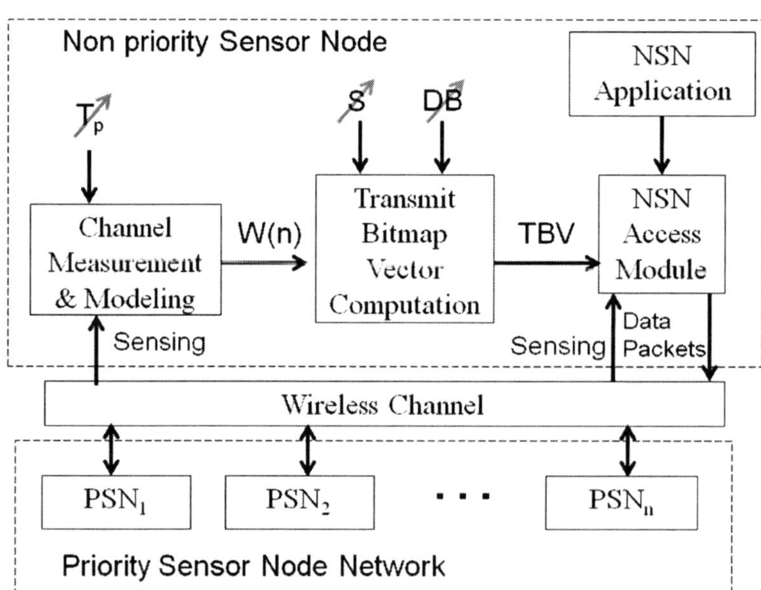

within a whitespace. However, since the duration of a whitespace is unknown in advance, there is a non-zero probability for a low priority node's transmission to spill over the end-time of the whitespace, causing a higher priority node disruption. The objective is to reduce the occurrences of such transmissions within a whitespace so that the chance of higher priority node traffic disruption is minimized. Conversely, that reduction needs to be selective so that the low priority nodes throughput can be maximized by scavenging for the legitimate TOs within the whitespace.

Consider the sample whitespace profile shown in Figure 2. Observe that the probability that a whitespace ends between [0-200] ms or [300-400] ms is zero. This means that a lower priority node can access these whitespaces without disrupting the higher priority nodes. However, utilizing whitespace durations within the ranges of [200-300] ms or [400-550] ms may cause disruption to the higher priority nodes. This is because there is a non-zero probability that a whitespace ends between these ranges. The key idea behind the CHTP strategy for lower priority nodes is to determine the best useable TOs based on the recently observed whitespace statistics. This example highlights the multiple objectives that the lower priority nodes must achieve in order to minimize the higher priority nodes disruptions and maximize its throughput.

In CHTP, a whitespace is divided into S duration TOs, where S is the duration of a low priority node packet plus control packets. As shown in Figure 2, the TOs are referred to as t_1, t_2, ..., t_m, where the quantity $t_m S$ represents the largest whitespace observed by the low priority node. Let the packet transmission decision in those TOs be represented by the bits b_1, b_2, ..., b_m, where bit b_0 indicates if the low priority node should transmit a packet (if available) in TO t_0. These m-bits (1: transmit, and 0: not transmit) form a Transmission Bitmap Vector (TBV). If the bit b_i is set to 1, then the access module in a low priority node transmits a packet in TO t_i only if a channel-sensing finds the channel to be free during that TO. In other words, a transmission can only occur if the whitespace lasted until the TO t_i. If the channel is found busy (meaning the whitespace has not lasted until the TO t_i) no transmission is made even though the b_i was set to be 1. In addition, if bit b_i was set to be 0 in the TBV, no transmission is made during the TO t_i. Once a whitespace has ended, the procedure resets for the next whitespace.

Figure 2. Transmission opportunities

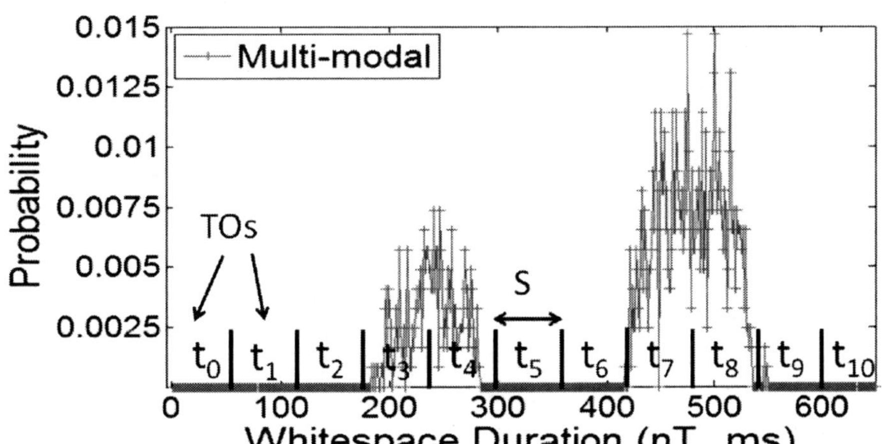

PSN Disruptions and NSN Throughput

A method must now be defined to properly select the best TOs that the lower priority nodes should utilize. Ideally, the lower priority nodes would like to use every TO to maximize its throughput but disruptions can occur since the duration of a whitespace is unknown. Therefore, each transmission opportunity has a certain gain and loss if it is chosen for use. A gain is represented by throughput for the lower priority node and a loss is represented by a possibility of disruption to the higher priority node. In this section, the throughput and disruptions are formulated for each TO.

Probability of TO Existence:

Observe that each TO can occur in every whitespace, given that a whitespace lasts long enough for a particular TO. It follows that, a TO that *exists* more often will provide greater throughput if a lower priority node uses that TO in every whitespace. To capture TO occurrences, the *probability of the existence of TO* t_i is defined as the probability that the whitespace will last at least for *iS* duration. This can be written as:

$$P_i^{existence} = 1 - W\left(iS/T_p\right) \qquad (1)$$

where *W(n)* is the cumulative distribution function (cdf) of the whitespace pdf *w(n)*, which is expressed as a multiple of the channel sensing period T_p.

Probability of TO Disruption:

A lower priority node packet transmission during a TO t_i will cause a *higher priority node traffic disruption* if the current whitespace does not last for a full *S* duration since the start of the transmission. The probability of such disruption can be expressed as the probability that a whitespace lasts anywhere between *(i-1)S* to *iS* duration, and can be written as:

$$P_i^{disrupt} = W\left(iS/T_p\right) - W\left((i-1)S/T_p\right) \qquad (2)$$

Now consider a situation in which a specific TBV $\hat{T}$ is calculated. Let δ represent a subset (of all *m* bits in $\hat{T}$) of bits that are 1. During a whitespace, after a lower priority node transmits based on the TBV $\hat{T}$, the resulting Priority Traffic Disruption (PTD) can be expressed as:

$$PTD = \sum_{\forall i \in \delta} P_i^{disrupt} \qquad (3)$$

PTD represents the probability that a Priority Traffic Disruption will occur at the end of a whitespace. Since each whitespace corresponds to a separate higher priority node's packet transmission, the quantity PTD represents the probability that a higher priority node's packet will be disrupted due to a lower priority node's packet transmission. As will be shown in the next subsection, the objective is to limit this disruption to a disruption bound *DB*.

Effective Non-priority Throughput (ENT) is defined as the number of successful lower priority node transmissions for each whitespace, which can be expressed as:

$$ENT = \sum_{\forall i \in \delta} P_i^{existence} \qquad (4)$$

Computing TBV via Solving Knapsack Problem

For a given whitespace pdf *w(n)*, the problem is to compute a Transmission Bitmap Vector $\hat{T}$ such that the ENT is maximized while limiting the PTD to a pre-defined bound *DB*. We propose to solve this problem using the 0-1 knapsack problem solution, which is summarized as follows. In the knapsack problem, there are m objects ($j=1,2,...,m$) and the jth object has a weight w_j and volume v_j. The goal is to select a subset of the objects in a knapsack so that the total weight is maximized while the volume of the knapsack does not exceed a predefined bound.

The TBV computation is posed as the knapsack problem by mapping the TO t_i to m objects. The TO existence probability $P_j^{existence}$ from Equation 1 and the disruption probability $P_j^{disrupt}$ from Equation 2 are mapped to w_j and v_j respectively. In addition, the higher priority node disruption bound DB is mapped as the knapsack volume bound, V. The mapping and solution process is formalized as follows.

The problem can be defined as:

$$\text{maximize W} = \sum_{j=1}^{m} w_j x_j \quad (5)$$

subject to:

$$\sum_{j=1}^{m} v_j x_j \leq V, \qquad x_j \in \{0,1\}. \quad (6)$$

To solve this problem, we can use dynamic programming (Toth, 1980) by defining $s[j,v]$ to be the solution for objects 1,2,…j with maximum volume v. Then define $s[j,v]$ recursively as:

$$s[j,v] = \begin{cases} 0, & \text{if j or v} = 0 \\ s[j-1,v], & \text{if v} < \text{v}_\text{j} \\ \max\{w_j + s[j-1, v-v_j], s[j-1,v]\}, & \text{if } v_j \leq v \end{cases} \quad (7)$$

The optimal solution can then be found by calculating $s[j,V]$. Under the assumption that all whitespace can be detected, the $P_j^{existence}$ and $P_j^{disrupt}$ values quantify the lower priority nodes' throughput and higher priority node disruption respectively for a given TO t_j. Hence, mapping the knapsack solution to TBV can give an optimal low priority node transmission strategy.

Sub-Optimal TBV Computation

Although providing optimal solutions, dynamic programming can be computationally intensive and may not be feasible for resource-constrained lower priority node hardware. We provide a computationally feasible heuristic solution as follows.

Initially, we compute the Density ($P_j^{existence}$ / $P_j^{disrupt}$) of all m objects (i.e. TO) and then sort them in descending order of their density. Afterwards, we assign a 1 in the Transmission Bitmap Vector $\hat{T}$ for the highest density TOs, representing good transmission opportunities. We stop assigning 1s when the Disruption Bound is reached, ensuring that the probability of disruption will meet the pre-defined bound. This scheme provides a relatively fast and simple heuristic solution to the problem. After this algorithm is completed, the $\hat{T}$ for the CHTP access strategy is created. At any given time in a whitespace, if an lower priority node has a packet to send, it will send the packet in the next TO indicated in $\hat{T}$, only if the whitespace lasts until that TO. A lower priority node following this strategy will maintain the specified DB while maximizing its throughput.

Multiple NSNs

Using CHTP, each lower priority node would independently measure, model, and calculate the access parameters (TBV) based on the lower priority nodes' point of view of the channel. This way, the disruption to the higher priority nodes in each lower priority node's environment can be bounded. The lower priority nodes can then run a traditional MAC protocol (i.e. CSMA, CSMA/CA etc.) to deal with the access contentions among themselves. From a protocol layering standpoint, it means that the inter-node contention avoidance mechanism (i.e. ALOHA, CSMA, or any other multi-access protocol) runs as a separate layer between the *NSN application* and the *NSN Access Module* in Figure 1. CHTP deals with the bandwidth scavenging part (i.e. identify transmission opportunities within the whitespaces left by the higher priority nodes) and this new layer runs within the scope of this already identified transmission opportunities.

EXPERIMENTAL TESTBED

The primary tool that channel history based methods utilize is the periodic sensing of a wireless channel. In this section, an experiment testbed is presented that consists of both a low priority sensor node network and a high priority sensor node network. The testbed provides some insights into the traffic characteristics that a low priority node would measure on a sensor channel. Note that for the remainder of the chapter, we will refer to the low priority sensor nodes as Non-priority Sensor Nodes (NSNs) and the high priority sensor nodes as Priority Sensor Nodes (PSNs).

Overview of the Experimental Testbed

The available whitespace can be measured by an NSN by detecting the received signal strength (RSSI) at a given channel frequency. To investigate this process, Crossbow TelosB sensor motes (Crossbox, 2009) are used as both the PSNs and NSNs. The TelosB mote operates at the 2.4 GHz band and has an 802.15.4 (ZigBee) compliant RF transceiver. The ZigBee based TelosB motes was used since it is designed for sensor networks as opposed to 802.11 based hardware which requires support for tight timing controls which may not be available on low power and low cost sensor hardware. Additionally, the TinyOS port provides enough driver level APIs for implementing a custom MAC layer (i.e. CSMA in this case), which is not easily possible for majority of the commercially available (802.11) network cards.

Utilizing the bandwidth scavenging concept, a NSN network attempts to access a wireless channel that is occupied by a PSN network. These two separate networks are implemented as two independent systems, namely the Priority User Network (PUN) and Non-priority User Network (NUN). The purpose of PUN is to generate various traffic patterns on a channel. While the purpose of the NUN is to first, perform measurements of the PSN channel activity.

As shown in Figure 3, the PUN and NUN is completely independent of each other, although they share an IEEE 802.15.4 Zigbee channel. The TelosB mote operates at the 2.4 GHz band with a TI MSP430 microcontroller. A Chipcon CC2420 IEEE 802.15.4 compliant RF transceiver is used with an integrated onboard antenna. The communication and programming of the motes are done using the TinyOS 2.1 operating system. For the purposes of traffic generation and statistics gathering, the motes are connected to personal computers running Ubuntu Linux 8.04 LTS. Separate computers are used for the PUN and NUN. The implementation of the computer interfaces is done using java with NetBeans as the IDE.

The communication flow of the experimental network is done using two closed paths. In the PUN path, a PSN traffic pattern is generated on the primary user computer and then sent to a PSN mote. This mote will send the traffic to another PSN mote or multiple motes. When the packets reach the last mote in the path, the packets are forwarded back to the originating PC. The NUN is designed in a similar fashion, by performing a second path using the NSN motes and NSN computer. Performance metrics for the experimental testbed are gathered on each PUN and NUN computers.

Figure 3. System implementation overview

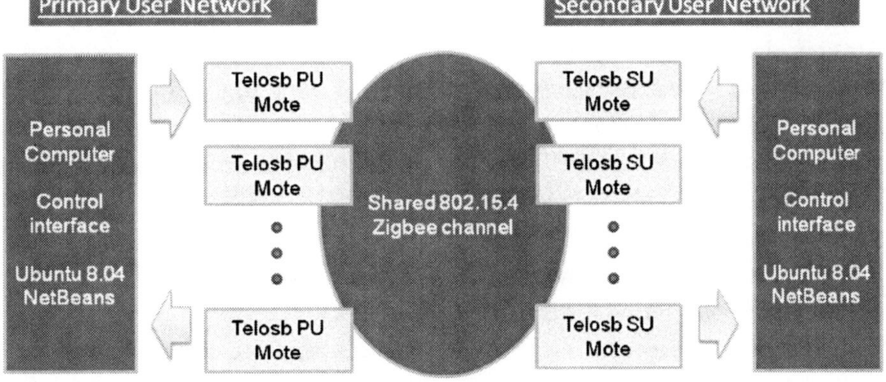

Figure 4 shows the physical setup of the experimental sensor network. Here, an NSN captures the activity on the channel using periodic RSSI measurements. The traffic on the channel is generated from the PSN Transmitter that is within transmission range of the NSN.

Priority Sensor Node Traffic Profiles

To explore the characteristics of different PSN traffic, various traffic profiles are sent by a PSN and subsequently measured by an NSN. A PSN transmitter sends messages to a PSN receiver according to the given traffic profile. Concurrently, an NSN measures the channel using an RSSI sampling program implemented on the motes. The channel is measured at 1 kHz rate with RSSI values being reported at each time step (Stanford SING, 2009). Consecutive measured RSSI values below a pre-defined threshold (-50 dBm) are considered as whitespaces.

In order to represent PSN traffic diversity, three traffic scenarios are established (i.e. sent over the PSN network):

- A uniformly distributed traffic that is sent at 7 packets per second (pps). Each packet has an added ±20% inter-packet time to vary the arrival time on the channel. This profile corresponds to TelosB based sensor network applications involving constant bit rate sensing.
- A low rate video file streamed over the channel, which represents video surveillance applications.
- A multi-modal traffic with two "peaks" of whitespaces in the pdf *w(n)*. In general, a stream with multiple peaks in its whitespace can be formed by merging multiple constant packet rate streams. This is representative of a multi-application primary traffic at different traffic generation rates.

Figure 4. TelosB sensor motes experimental setup

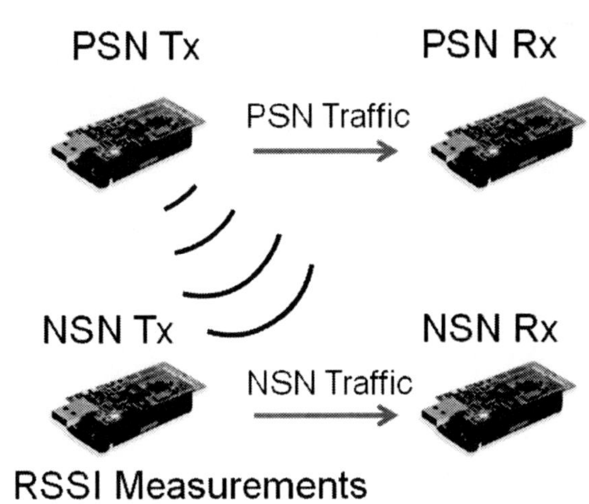

The packet size of the PSN node is set to 120 bytes, which corresponds to approximately 5 ms channel access duration given the bandwidth of the channel (250 kbps data rate). Given that the NSN senses the channel at T_p = 1 ms intervals, it is able to determine the approximate start and ending times of PSN packets (i.e. ON-OFF durations). Figures 5a-c shows the whitespace pdfs *w(n)* measured and computed by an NSN Transmitter, which is within the transmission range of the PSN Transmitter that is generating the traffic profiles as described above. In Figures 5a-c, the pdf is plotted as a function of the sensing interval time (i.e. nT_p).

The following observations can be made from the whitespace statistics shown in Figure 5. In the uniform traffic case of Figure 5a, all the whitespaces are clustered between 70 ms and 180 ms, which is the result of the 20% variation on the 7 pps traffic generation rate. Observe that the peak does not form a perfect uniform shape as expected. This result comes from the combined effects of the TinyOS operating system's scheduling jitter and the asymmetry in the hardware delays in the transmission times of the packets. For the pdf of the video stream in Figure 5b, the large peak corresponds to fixed rate video packets. The other small peaks correspond to embedded audio packets and uncorrelated video packets, which are transmitted at a slower rate. For the multi-modal pdf case in Figure 5c, the whitespace is distributed within two separate ranges. The two peaks correspond to applications that generate a traffic rate of 4 pps with 25% spread, and 2 pps with 75% spread respectively.

The analysis of whitespace in the experimental network highlights the various whitespace environments that an NSN may experience when attempting to access unused spectrum.

ACCESS STRATEGY EVALUATION

In this section, the performance of the Channel History based Traffic Prioritization (CHTP) protocol is presented. The whitespace pdfs is first extracted for the three whitespace profiles from the test-bed as discussed in Section IV-B and shown in Figure 5. Then the PTD and ENT for CHTP and comparable access strategies are evaluated under simulation with the extracted pdfs from the testbed.

Comparable Access Strategies

CHTP is compared with an inter-frame spacing based protocol (AIFS) and a Static Traffic Limit (STL) access mechanism that was presented in (Plummer & Biswas, 2009; Plummer, Taghizadeh, & Biswas, 2009). The AIFS is based on IEEE 802.11e (Mangold, et al., 2003) where priority is given between classes of traffic by waiting class-specific inter-frame spacing (IFS) before sending a packet. When the NSN has a packet to send,

Figure 5. Experimental whitespace distributions. (a) Uniform with 20% variation, (b) Low rate video stream, and (c) Multi- modal data stream

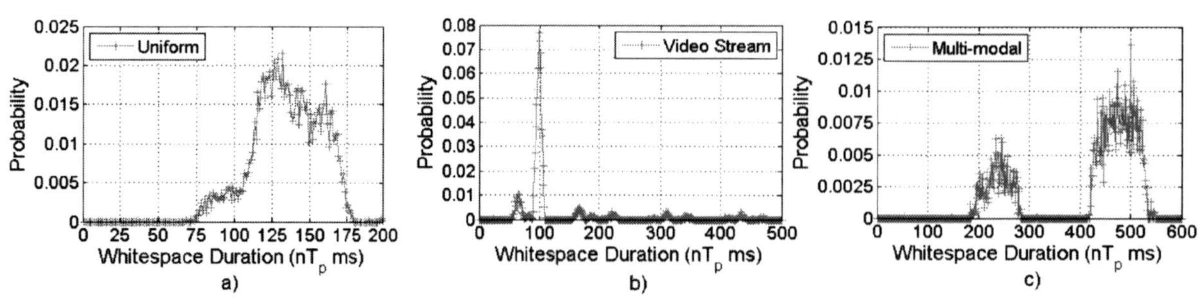

using AIFS, it waits for the specified IFS time for the channel to remain free before transmitting. In our implementation, a zero IFS was chosen for the PSNs and a non-zero IFS was chosen for the NSNs. Like in 802.11e, this arrangement gives rise to priority channel access to the PSNs. The IFS for the NSN is chosen as half the NSN packet size, which provided the best performance for the AIFS. As discussed about the limitations of 802.11e based traffic prioritization in Section I.B, AIFS relies only on the instantaneous channel status (i.e. free or busy) for granting access. As a result, it is expected that the PSN traffic will suffer from undesirable disruptions caused when the NSN traffic load is high.

Similar to the proposed CHTP, the Static Traffic Limit (STL) (Plummer & Biswas, 2009; Plummer, et al., 2009) protocol operates based on the channel usage history, and attempts to bound the amount of PSN disruption and maximize NSN throughput. In STL, there are two distinct stages. In the first stage, an NSN initially waits for the channel to become idle. Once the channel becomes idle, the NSN attempts to transmit its first packet of duration S. If the NSN has more packets to send, it continues sending until the measured value of ϵ (time elapsed since the beginning of the current whitespace) becomes so large that the expected PTD for new transmissions become equal to or larger than a pre-determined Disruption Bound (*DB*). A term J_{max} is defined, which denotes the maximum number of packets that the NSN can send before the expected disruption reaches the pre-determined *DB*. The quantity J_{max} is computed by the NSN based on the observed whitespace statistics. After J_{max} packets are sent, the low priority node abandons transmission attempts and starts channel sensing for detecting the next whitespace.

If, however, a whitespace ends before all J_{max} packets are sent by the NSN, it exits the channel after detecting that a PSN has started transmitting. Note that in spite of the NSN vacating the channel, disruption to the PSNs' traffic can still occur since the PSN must defer its transmission until the channel is vacated by the NSN. This stage essentially specifies a block of time in every whitespace that the NSN is allowed to transmit. This block is bounded by the start of a whitespace to the ending of the $J_{max}{}^{th}$ packet.

The second stage of STL corresponds to the situation in which a whitespace duration is greater than the end time of the $J_{max}{}^{th}$ packet from the NSN. No PTD will be caused in this stage because the NSN will exit the whitespace after J_{max} transmissions.

TBV for Different Traffic Profiles

Figure 6, presents the Transmission Bitmap Vectors (TBV) computed by CHTP for all three experiments corresponding to uniform, video stream and multi-modal traffic profiles. The black and while vertical bars in the diagram correspond to the ones and zeroes in the TBV. Observe how the ones in TBV, representing useable transmission opportunities for the NSNs, occupy different part of the whitespace spectrum for different traffic profiles. In the Uniform profile, most of the Transmissions Opportunities (TO) are in at the beginning of the whitespace. Similarly, in the multi-modal case, the TOs are in the beginning of the whitespaces. However, the TBV also accesses whitespaces after the first peak to maximize NSN throughput. Note that these results are for the sub-optimal protocol. In our experiments, there was little difference between the TBV optimal and sub-optimal solutions. This results from the non-integer values of $P_j^{existence}$ and $P_j^{disrupt}$. This highlights CHTP's ability in identifying the best transmission opportunities for sending an NSN packet with minimal disruption.

Performance with Uniform PSN Traffic Profile

Figure 7 shows the PSN traffic disruption and NSN throughput for all three protocols when the PSN traffic profile is uniform, causing the whitespace as shown in Figure 5(b). For all presented results, the NSN packet length is set to 3 ms, which is 0.6 times the size of the PSN packets. Additionally, the CHTP and STL strategies have been implemented with disruption bounds (*DB*) of 5% and 3%. A 5% DB means that only up to 5% of the PSN packets are allowed to be deferred by the NSN packet transmissions. The traffic load generated by the NSN is varied in order to evaluate the impacts of increasing non-priority traffic on the priority traffic. Observe that both CHTP and STL maintain the required disruption bounds (indicated as DBs in the figure) across all NSN rates, while AIFS suffers from greater priority traffic disruptions, which is above 10% for the higher NSN rates. AIFS suffers from greater disruptions since the NSN will always transmit after waiting for an IFS. Even after an IFS period, there is some probability that the PSN will return to the channel during the NSN transmission, causing disruptions for the PSN traffic.

In terms of throughput, all three strategies are able to support low NSN generation rates. However, once the specified *DB* is reached, CHTP and STL throttle the NSN transmissions in order to protect the PSN traffic. Notice that AIFS is able to achieve higher NSN throughput by always sending after waiting for an IFS, but at the expense of higher disruptions to the PSN traffic. This demonstrates how CHTP and STL are able to protect the high priority traffic by limiting the low priority traffic throughput whenever appropriate.

It is interesting to note in Figure 6 that CHTP and STL have very similar maximum NSN throughput values. CHTP achieves slightly better throughput, which is 2.2% and 1.6% more when the PSN disruption Bound (DB) is set to 5% and 3% respectively. This performance similarity is a direct result of the uniform whitespace profile. Since all of the whitespace is within a very small range (70 ms and 180 ms), in order to maintain the required *DB*, both strategies send packets until approximately the same point in a whitespace.

Performance with PSN Traffic with Multi-Modal Traffic

The protocol performance with PSN traffic from multi-modal traffic (as shown the pdf in Figure 5(c) is shown in Figure 8. Note that the relative performance of the three evaluated protocols in this case is very similar to those observed for the uniform traffic case in Figure 7. The primary difference is the protocol CHTP's ability to support higher NSN throughputs compared to STL.

As in the uniform traffic case, the CHTP and STL protocols are able to maintain the specified *DB* at the expense of limiting the NSN throughput at higher NSN load levels. However, observe that CHTP is able to support 56% and 60% higher

Figure 6. TBV of CHTP for different traffic profiles

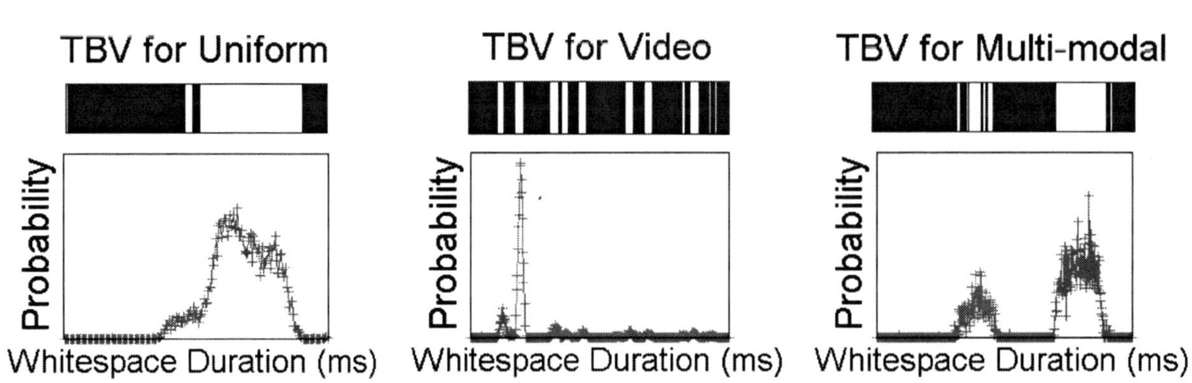

Figure 7. Access strategy comparison for uniform PSN traffic

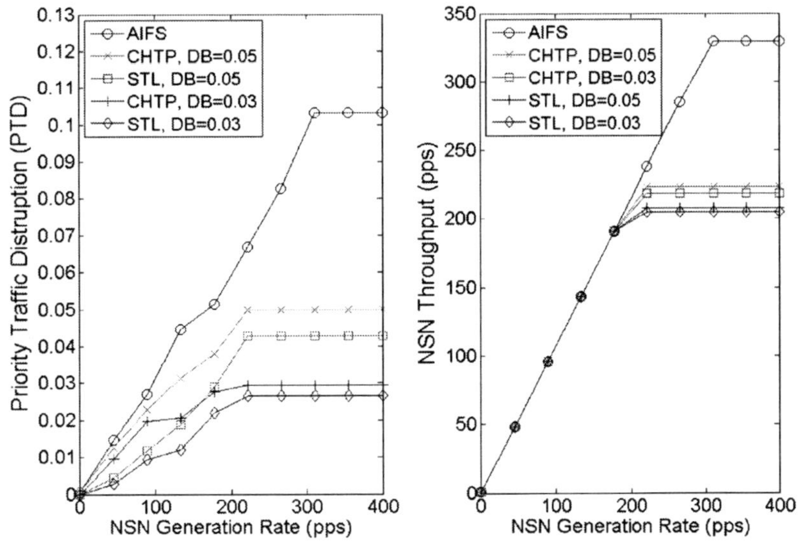

throughputs compared to STL when the DB is specified to 5% and 3% respectively. This large increase in throughput over STL results from CHTP's ability in identifying the best transmission opportunities for sending an NSN packet with minimal disruption. Notice in the *w(n)* of the multi-modal profile (Figure 5(c)) there is two large peaks at approximately 250 ms and 500 ms, with other smaller probabilities for larger whitespace. Given this whitespace compared to the uniform case, in order to maximize NSN throughput the NSN transmissions must happen at different times within a whitespace. Consequently, the Transmission Bitmap Vector (TBV) for the CHTP for this multi-modal based whitespace profile will have bits at various transmission opportunities (TOs) throughout a whitespace. Conversely, since the first peak at 250 ms contains 25% of all whitespace durations, to maintain the required DB, the STL strategy is forced to stop sending packets when the elapsed time in a whitespace reaches 250 ms. Also note that in both CHTP and STL, the throughput corresponding to *DB = 3%* is slightly lower than the *DB = 5%* case. This occurs because to maintain the lower DB, a lower number of NSN packets are sent.

Dynamic PSN Traffic Profile

The performance of CHTP and STL under a time-varying whitespace profile is shown in Figure 9. The disruption bound *DB* is set to 0.05. The PSN traffic pattern was changed using the three pdf types in Figure 5. Initially, the PSN traffic is generated with a uniformly distributed rate of 7 pps. The pattern changes after 60 seconds when the traffic profile is changed to multi-modal. Finally, at 130 seconds, the PSN traffic is changed to a video stream.

Observe that at steady state, the PTD remains close to the defined *DB* for both CHTP and STL. However, during the pdf transitions, the PTD increases past the disruption bound. Observe that the changes in pdf during traffic transitions for the different traffics are quite significant. This explains the transient peaks in the PTD during the pdf transitions in Figure 9. In terms of throughput, the protocol CHTP is able to achieve higher throughput compared to STL when the traffic changes to a video stream and to one that produces multi-modal whitespace.

Figure 8. Access strategy comparison for multi-modal whitespace

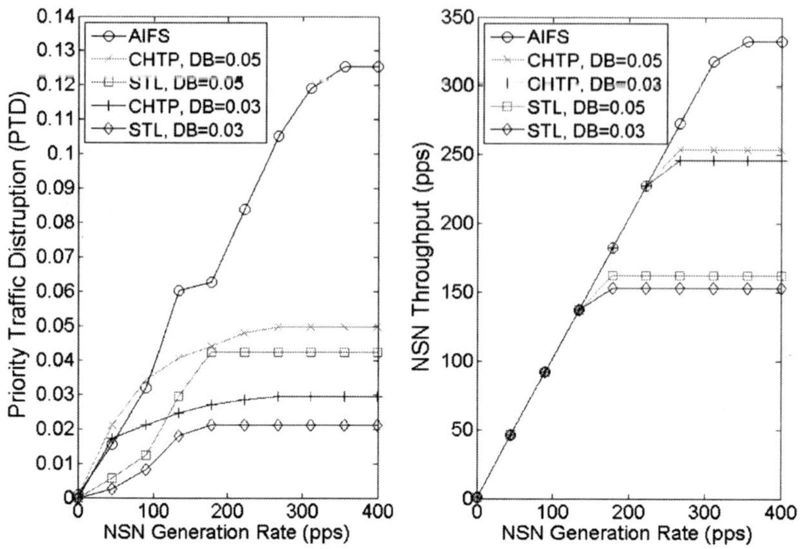

FUTURE RESEARCH DIRECTIONS

In this section, a discussion is provided of the future research directions for channel history based protocols for real time traffic protection in a Wireless Sensor Networks (WSN). Take note that this work is in the early stages of development, which means that there are many remaining issues to solve.

Multiple Non-Priority Nodes

A typical WSN utilize a large number of cooperating sensor nodes to accomplish the objectives of the network. In a heterogeneous WSN, multiple groups of sensor nodes must be managed. This is usually handled by the MAC layer. In the context of temporal based protocols similar to those discussed in this chapter, some unique issues at the MAC layer would need to be addressed.

Figure 9. Performance with time-varying w(n)

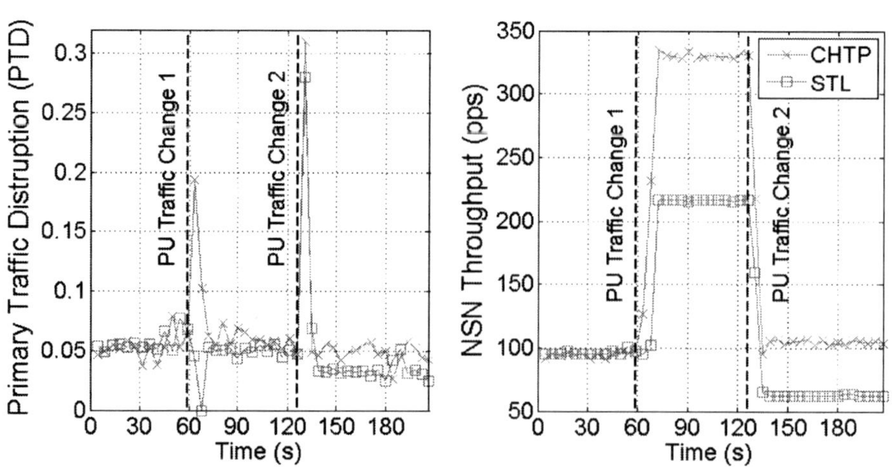

The Non-priority Sensor Nodes (NSN) goal is to protect the traffic of the Priority Sensor Nodes (PSN). The NSN first performs periodic channel sensing of the PSN traffic. Then it uses the gathered information to develop a channel access strategy that protects the PSN traffic. This process can be affected by the physical position of the NSNs with respect to the locations of the PSN. In a WSN, the locations of NSNs can vary greatly depending on the topology of the network. Therefore, each NSN transmissions can affect PSNs' traffic differently, which requires the NSNs to perform sensing at their respective physical locations. An NSN sensing report can provide information about the PSNs within its environment. Cooperation between NSNs can help mitigate the effects of low signal to noise ratio (SNR), shadowing, and hidden terminals (Unnikrishnan & Veeravalli, 2008). In cooperative sensing, NSNs at different locations sense the channel independently and send their observations to a fusion center (Gao, Yuan, Liu, Cheng, & Wang, 2009; Zhang, Wu, & Lau, 2009). The fusion center would manage access decisions for the NSN network based on the sensing reports it receives. Another approach is to distributed nodes sensing results locally to make access decisions (Liu, Yu, & Xie, 2008; Min & Shin, 2010).

Multiple NSNs attempting to access a channel must be managed to support sensing information exchange, NSN collision avoidance, and fairness. This process can be difficult because nodes channel access constrained within the duration of whitespaces. The problem is exacerbated by the fact that the timing and durations of whitespaces are not deterministic. Some researchers have suggested using similar protocols as IEEE 802.11 to manage NSN access (Q. Chen, Liang, Motani, & Wong, 2009; Zheng, Liang, Tham, & Kam, 2009). Other techniques have suggested using control channels (Eljack, Igbal, & Furong, 2009; Jia, Zhang, & Shen, 2008) or virtual control channels (Baldo, Asterjadhi, & Zorzi, 2009) to manage NSN control messages.

Energy Consumption

One concern for WSNs is the energy consumption of the sensor nodes. Energy usage is focused on because the sensor nodes are typically low power and low cost (Akyildiz, et al., 2002). Utilizing channel history based protocols for real time traffic protection can raise some issues with the energy consumption of NSNs. One issue is the energy used during the process of periodic channel sensing. In MAC protocols such as CSMA, the channel needs to be sensed when there is a message available to send. In contrast, a channel history based protocol will need to sense the channel periodically at a specific rate. Depending on the mechanics of an access protocol, the required sensing rate may be large, which may lead to higher energy usage. Another source of possible energy loss is the complexity of the access strategy algorithm. If each NSN needs to execute a complex algorithm requiring long running times or control message overheads, energy usage may be affected.

CONCLUSION

In this chapter, a channel history based traffic prioritization strategy for heterogeneous wireless sensor networks, comprising of real-time higher priority and non-real time lower priority nodes, is presented. The mechanism is based on the measurement and modeling of the idle times in a wireless channel as perceived by non-real time nodes. The strategy is general in that it does not rely on a specific traffic pattern of the real time nodes. Additionally, the access strategy provides a reasonable amount of non-real time traffic throughput while bounding the disruptions on the real-time traffic to pre-set values.

REFERENCES

Ahuja, R., Corke, R., & Bok, A. (2008). *Cognitive radio system using IEEE 802.11a over UHF TVWS.* Paper presented at the Symposium on Dynamic Spectrum Access Networks (DySPAN). IEEE Press.

Akyildiz, I. F., Su, W., Sankarasubramaniam, Y., & Cayirci, E. (2002). Wireless sensor networks: A survey. *Computer Networks*, *38*(4), 393–422. doi:10.1016/S1389-1286(01)00302-4

Baldo, N., Asterjadhi, A., & Zorzi, M. (2009). *Multi-channel medium access using a virtual network coded control channel*. Paper presented at the International Wireless Communications and Mobile Computing Conference (IWCMC). IEEE Press.

Chen, Q., Liang, Y., Motani, M., & Wong, W. (2009). *CR-CSMA: A random access MAC protocol for cognitive radio networks*. Paper presented at the IEEE 20th International Symposium on Personal, Indoor, and Mobile Radio Communications. IEEE Press.

Chen, Y., Qing, Z., & Swami, A. (2008). Joint design and separation principle for opportunistic spectrum access in the presence of sensing errors. *IEEE Transactions on Information Theory*, *54*(5). doi:10.1109/TIT.2008.920248

Chiasserini, C. F., & Rao, R. R. (2003). Coexistence mechanisms for interference mitigation in the 2.4-GHz ISM band. *IEEE Transactions on Wireless Communications*, *2*(5). doi:10.1109/TWC.2003.817417

Crossbox. (2009). *TelosB mote platform*. Retrieved September 10th, 2009, from http://www.xbow.com/Products/productdetails.aspx?sid=252

Eljack, S., Igbal, A., & Furong, W. (2009). *A multi channel cognitive MAC protocol with efficient channel reservation and collision avoidance method*. Paper presented at the Conference on Multimedia Information Networking and Security (MINES). IEEE Press.

Firoze, A. M., Ju, L. Y., & Kwong, L. M. (2007). *PR-MAC: A priority reservation MAC protocol for wireless sensor networks*. Paper presented at the International Conference on Electrical Engineering, (ICEE '07). ICEE Press.

Gao, F., Yuan, W., Liu, W., Cheng, W., & Wang, S. (2009). *Pipelined cooperative spectrum sensing in cognitive radio networks*. Paper presented at the Wireless Communications and Networking Conference (WCNC). IEEE Press

Geirhofer, S., Tong, L., & Sadler, B. M. (2006). *Dynamic spectrum access in WLAN channels: Empirical model and its stochastic analysis*. Paper presented at the ACM TAPAS. ACM Press.

Geirhofer, S., Tong, L., & Sadler, B. M. (2006). *Dynamic spectrum access in WLAN channels: Empirical model and its stochastic analysis.* Paper presented at the ACM Workshop on Technology and Policy for Accessing Spectrum (TAPAS). ACM Press.

Geirhofer, S., Tong, L., & Sadler, B. M. (2008). Cognitive medium access: Constraining interference based on experimental models. *IEEE Journal on Selected Areas in Communications*, *26*(1). doi:10.1109/JSAC.2008.080109

Hongjun, L., Xun, L., Shen, J., & Hongxu, M. (2009). *A fair multi-priority MAC protocol design of wireless sensor networks*. Paper presented at the Networks Security, Wireless Communications, and Trusted Computing, (NSWCTC '09). IEEE Press.

Huang, S., Liu, X., & Ding, Z. (2008). *On optimal sensing and transmission strategies for dynamic spectrum access*. Paper presented at the New Frontiers in Dynamic Spectrum Access Networks, (DySPAN 2008). IEEE Press.

Huang, S., Liu, X., & Ding, Z. (2009). Optimal transmission strategies for dynamic spectrum access in cognitive radio networks. *IEEE Transactions on Mobile Computing*, *8*(12), 1636–1648. doi:10.1109/TMC.2009.84

Jia, J., Zhang, Q., & Shen, X. (2008). HC-MAC: A hardware-constrained cognitive MAC for efficient spectrum management. *IEEE Journal on Selected Areas in Communications*, *26*(1). doi:10.1109/JSAC.2008.080110

Jingli, L., Xiangqian, L., & Swami, A. (2009). Collision analysis for coexistence of multiple bluetooth piconets and WLAN with dual channel transmission. *IEEE Transactions on Communications*, *57*(4), 1129–1139. doi:10.1109/TCOMM.2009.04.070103

Jung, E., & Xin, L. (2008). *Opportunistic spectrum access in heterogeneous user environments*. Paper presented at the New Frontiers in Dynamic Spectrum Access Networks, (DySPAN 2008). IEEE Press.

Lei, P., Hongyi, W., & Nian-Feng, T. (2009). *An efficient and scalable prioritized MAC protocol (PMAC) for backbone communication in wireless sensor networks*. Paper presented at the Sensor Technologies and Applications (SENSORCOMM '09). IEEE Press.

Lenagala, R., & Zeng, Q. A. (2006). Study of dynamic MAC layer parameters for starvation prevention in the IEEE 802.11e MAC layer protocol. Paper presented at the IEEE International Conference on Wireless and Mobile Computing, Networking, and Communications. IEEE Press.

Liu, Y., Yu, R., & Xie, S. (2008). *Optimal cooperative sensing scheme under time-varying channel for cognitive radio networks*. Paper presented at the DySPAN. IEEE Press.

Mangold, S., Sunghyun, C., Hiertz, G. R., Klein, O., & Walke, B. (2003). Analysis of IEEE 802.11e for QoS support in wireless LANs. *IEEE Wireless Communications*, *10*(6), 40–50. doi:10.1109/MWC.2003.1265851

Mietzner, J., Lampe, L., & Schober, R. (2009). Distributed transmit power allocation for multihop cognitive-radio systems. *IEEE Transactions on Wireless Communications*, *8*(10), 5187–5201. doi:10.1109/TWC.2009.081318

Min, A. W., & Shin, K. G. (2010). *On sensing – Access tradeoff in cognitive radio networks*. Paper presented at the Symposium on Dynamic Spectrum Access Networks (DyDPAN). IEEE Press.

Mishra, M., & Sahoo, A. (2006). *A contention window based differentiation mechanism for providing QoS in Wireless LANs*. Paper presented at the 9th International Conference on Information Technology, (ICIT '06). IEEE Press.

Petty, V. R., Rajbanshi, R., Datla, D., Frederick, W., et al. (2007). *Feasibility of dynamic spectrum access in underutilized television bands*. Paper presented at the Symposium on Dynamic Spectrum Access Networks (DySPAN). IEEE Press.

Plummer, A., & Biswas, S. (2009). Capacity scavenging in wireless networks: A comparative study. *Paper presented at the IEEE Military Communications Conference (Milcom)*. IEEE Press.

Plummer, A., Taghizadeh, M., & Biswas, S. (2009). *Measurement based capacity scavenging via whitespace modeling in wireless networks*. Paper presented at the IEEE Globecom 2009. IEEE Press.

Plummer, A., Taghizadeh, M., & Biswas, S. (2010a). Measurement based bandwidth scavenging in wireless networks. *IEEE Transactions on Mobile Computing*, *99*, 1.

Plummer, A., Taghizadeh, M., & Biswas, S. (2010b). *Traffic protection via bandwidth scavenging in heterogeneous sensor networks*. Paper presented at the IEEE Globecom. IEEE Press.

Stanford, S. I. N. G. (2009). *SourceForge.net Repository*. Retrieved October 21st, 2009, from http://tinyos.cvs.sourceforge.net/viewvc/tinyos/tinyos-2.x-contrib/stanford-sing/

Stojmenovic, I., Giordano, S., Conti, M., & Basagni, S. (2005). *Mobile ad hoc networking* (pp. 175–203). IEEE.

Toth, P. (1980). Dynamic programming algorithms for the zero-one knapsack problem. *Computing*, *25*(1), 29–45. doi:10.1007/BF02243880

Unnikrishnan, J., & Veeravalli, V. V. (2008). Cooperative sensing for primary detection in cognitive radio. *IEEE Journal of Selected Topics in Signal Processing*, *2*(1), 18–27. doi:10.1109/JSTSP.2007.914880

Wang, B., Ji, Z., & Liu, K. J. R. (2007). *Primary-prioritized Markov approach for dynamic spectrum access*. Paper presented at the Dynamic Spectrum Access Networks (DySPAN). IEEE Press.

Zhang, S., Wu, T., & Lau, V. (2009). A low-overhead energy detection based cooperative sensing protocol for cognitive radio systems. *IEEE Transactions on Wireless Communications*, *8*(11). doi:10.1109/TWC.2009.081106

Zheng, S., Liang, Y., Tham, C., & Kam, P. (2009). *Design of MAC with cooperative spectrum sensing in ad hoc cognitive radio networks*. Paper presented at the IEEE 20th International Symposium on Personal, Indoor, and Mobile Radio Communications. IEEE Press.

KEY TERMS AND DEFINITIONS

Bandwidth Scavenging: A mechanism to facilitate a low priority user to scavenge the RF channel for leftover bandwidth that is not utilized by higher priority traffic.

Effective Non-Priority Throughput (ENT): Defined as the number of successful lower priority node transmissions for each whitespace.

Non-Priority Sensor Nodes (NSN): Nodes that have a lower priority access to the channel. They must defer to the higher priority nodes.

Priority Sensor Nodes (PSN): Nodes that have priority access to the channel.

Priority Traffic Disruption (PTD): Represents the probability that a disruption will occur to a higher priority node at the end of a whitespace.

RSSI Measurement: The act of a low priority user utilizing the Received Signal Strength Indicator (RSSI) to measure the activity of higher priority users.

Traffic Prioritization: A protocol or structure that arranges various sources of network traffic into different classes based on the priority level of the traffic.

Transmission Bitmap Vector (TBV): A vector that defines which transmission opportunities that can be used within a whitespace.

Transmission Opportunity (TO): A period of time that is large enough to send a single low priority user packet within a whitespace.

Whitespace Modeling: A method to model whitespace (Idle time in between two consecutive packets in a wireless channel) occurrences within an specific area over time.

Chapter 13
Adaptive Coexistence between Cognitive and Fiber Networks

Anwer Al-Dulaimi
Brunel University, UK

Saba Al-Rubaye
Brunel University, UK

John Cosmas
Brunel University, UK

ABSTRACT

Cognitive radios are proposed as secondary users of spectrum to provision for the growth in mobile users and services. However, the dynamic changes in the wireless environment and spectrum availability are blocking the success of online communications for future cognitive mesh networks. As a solution, Cognitive Radio over Fibre (CRoF) subnet is developed through combining selected Base Stations (BSs) with the radio over fibre network. These CRoF-BSs attract the neighborhood cognitive BSs to send data through fibre whenever they are unable to formulate their own wireless links. This leads to the network splitting into many zones of services in which cognitive BSs are controlled by CRoF-BS zones. Therefore, a new paradigm for local resource sharing emerges through these architectural network modifications. In this chapter, the multi-zone structure is analyzed in order to formulate the rules of adaptation between the CRoF zones and the traditional cognitive networks.

INTRODUCTION

With the growth of wireless networks, there will be a future crisis of spectrum availability under the current spectrum allocation scheme. However, actual measurements show that current spectrum allocation makes the spectrum utilization inefficient because a wide range of the allocated frequency bands are rarely used. This gap between regulatory spectrum allocation and actual usage indicates that a new approach of spectrum usage is needed. Cognitive radio becomes an emerging idea to improve the spectrum utilization (Zhang, et al., 2005). The key enabling technology of dynamic spectrum access techniques is cognitive radio (CR) technology, which provides the capability to share the wireless channel with licensed users in an opportunistic manner. CR networks are

DOI: 10.4018/978-1-4666-0017-1.ch013

envisioned to provide high bandwidth to mobile users via heterogeneous wireless architectures and dynamic spectrum access techniques (Akyildiz, et al., 2008). However, the need for CRs is motivated by many factors. Principally, though, the need for cognition is driven by the complexity of the radio systems themselves. The existence of software defined radio (SDRs) capable of implementing a near endless number of different waveforms with different modulation schemes, power levels, error control codes, carrier frequencies, etc., means that controlling the radio becomes a problem of combinatorial optimization (MacKenzie, et al., 2009).

There are two major approaches in cognitive radio: dynamic spectrum allocation and opportunistic spectrum access. For dynamic spectrum allocation, information on spectrum occupation is used for channel allocation and planning on a long-term basis. On the other hand, with opportunistic spectrum access, instantaneous information of channel usage by a primary user is observed and utilized to grant access to secondary users to increase utilization on a short term basis (Niyato & Hossain, 2009).

Earlier research done by (Mangold, et al., 2001) shows that wireless cognitive networks are able to coexist with primary networks, and operate at the same time and location without harmful interference. The Similar goals have been formulated of the coexistence between decentralized cognitive radio and primary radio systems is Specified (Watanabe & Ishibashi, 2007) by introducing two benchmarks, the network performance with and without cognitive radio technology, with the function of detecting radio signals to avoid the primary radio system is studied. A different cooperative transmission has been investigated in (Rohokale, et al., 2010), where the cognitive cooperative diversity is a strong technique which can provide the maximum throughputs. Cooperative opportunistic large array algorithms can improve the reliability as well as the energy efficiency of the communication.

The first significant radio-domain application for such smarter radios was the autonomous sharing of pooled spectrum, which the US Federal Communication Commission (FCC) endorsed relatively soon thereafter, to encourage the development of secondary spectrum markets (Arslan, 2007). In a cognitive network, autonomous and adaptive radios select their operating parameters to achieve individual and network-wide goals. The effectiveness of these adaptations depends on the amount of knowledge about the state of the network that is available to the radios (Komali, et al., 2010). In (Chung and Tsai, 2010), the authors present an effective centralized decision approach to deal with the problem regarding how a spectrum manager should periodically re-allocate the spectrum between two networks in a single cell, while maintaining both the call blocking and dropping probabilities in their acceptable levels, and improving the spectrum efficiency simultaneously.

Additionally, (Huschke & Leaves, 2003; Leaves & Moessner, 2004) take up the idea of the dynamic allocation of spectrum among multiple wireless network systems, first discussed and investigated from a general perspective, including its major advantages and technical issues. On the other hand, resource allocation is a fundamental problem in cognitive radio networks and has been discussed a lot in the recent works (Yuan, et al., 2007; Chen, et al., 2008: Jia, et al., 2008).

Meanwhile, researchers have also been focusing on integrating cognitive radio and optical fibre systems to improve QoS in wired/wireless networks (Attar, et al., 2010) similarly to our work they investigated the feasibility of carrying a cognitive signal through RoF systems, to enhance the performance of the newly proposed Cognitive WLAN over fibre.

Our previous study (Al-Dulaimi, 2010a, 2010b) has addressed extensively a novel subnet called Cognitive Radio over Fibre (CRoF), for the opportunistic coexistence between the cognitive radio a system and fibre system. In addition, CRoF is able to use local spectrum holes efficiently and provide higher throughput to secondary users. The authors proposed as an alternative solution for the wireless links between cognitive base stations every time there is no free spectrum to communicate. In addition to participation in CRoF, we are involved in the development of the cognitive fibre system (Al-Dulaimi, 2010c). The contribution of this paper is to submit a scheme where the two networks CRN and CRoF coexist to form a new multi-zoned secondary network. At least one CRoF Base Station (BS) is placed in each zone as the wireless provider for the other CRN-BSs. CRoF-BSs control communication between different sectored zones.

In contrast to previous research, here we consider the problem of weak links in heterogeneous cognitive networks, together with adaptive fibre system selection, to maintain the connection problems.

EMERGENCE OF COGNITIVE NETWORKS

The cognitive networks that use cognitive radio applications are posing new requirements for the resource allocation schemes and the network's technical capabilities. Hence, new models are essential for carrying out network communication and the way that the network responds to the data delivery requests. All these tasks are performed in terms of opportunistic actions and are fully dependent on the agility of the cognitive network in sensing the spectrum, advanced recognition, and adaption competences. Also, the system's success is a function of extracting the performance of any cognitive station to the higher levels of network service plane and vice-versa. Hence, the attainment of a powerful cognitive network is a matter of cooperation between individual stations, and selfishness reduction more than anything else. The reason for this is that resources are available on temporary basis and any monopoly can cause the collapse of the whole secondary system. Such cases of failure in the cognitive network can only be solved by moving away the communication trajectories from the infected cells/parts. Later, these network parts can be reconnected as soon as congestion problems are fixed.

Developing strategies for network management that deal with the above cases are based on the network architecture, size, and autonomy of the network subscribers. One important thing to remember is that no packet can be shielded from dynamic changes in the cognitive network. This is due to the changes in the surrounding wireless environment and activity of the primary users. These changes take part in protocols, routing maps, and network distribution. In fact, this restless nature of the cognitive network impacts the continuity of services and the flow of data, and increases the complexity of the entire system, especially for huge sized networks.

Cognitive Network Management

The arrangement of the mobile networks architecture is managed to produce highly adaptive end-user service, and cost effective operator investments. Therefore, a clear definition is required to attain a similar network using cognitive radio systems. As a secondary user for the spectrum, it is necessary for cognitive networks to have their own architecture and system design. In other words, analysing the cognitive network performance is a matter of successful installation of equipment and free resources detection. In that case, the density of the cognitive stations' installation is decided by the locations of the free channels available.

There was been a lot of research that analyzes various arrangements of the cognitive networks, for instance, in the last few years a considerable amount has focused on ad-hoc cognitive networks, such as CRAHNs. This research focused on the dynamic allocation of wireless links using the new principle of Dynamic Spectrum Access (DSA). Therefore, secondary users are coexisting with primary networks in a dynamic wireless environment where channels are allocated using spectrum-aware operations called the Cognition Cycle. The magnet functions for this cycle are defined as (Akyildiz, et al., 2009):

- **Spectrum sensing:** A CR user can only be allocated to an unused portion of the spectrum. Therefore, a CR user should monitor the available spectrum bands and then detect spectrum holes. Spectrum sensing is a basic functionality in CR networks, and hence it is closely related to other spectrum management functions, as well as layering protocols, to provide information on spectrum availability.
- **Spectrum decision:** Once the available spectrums are identified, it is essential that the CR users select the most appropriate band according to their QoS requirements. It is important to characterize the spectrum band in terms of both radio environment and the statistical behaviours of the PUs. In order to design a decision algorithm that incorporates dynamic spectrum characteristics, we need to obtain a priori information regarding the PU activity. Furthermore, in CRAHNs, spectrum decision involves jointly undertaking spectrum selection and route formation.
- **Spectrum sharing:** Since there may be multiple CR users trying to access the spectrum, their transmissions should be coordinated to prevent collisions in overlapping portions of the spectrum. Spectrum sharing provides the capability to share the spectrum resource opportunistically with multiple CR users, which includes resource allocation to avoid interference caused to the primary network. For this, game theoretical approaches have also been used to analyse the actions of selfish CR users. Furthermore, this function necessitates a CR medium access control (MAC) protocol, which facilitates the sensing control to distribute the sensing task among the coordinating nodes, as well as spectrum access to determine the timing for transmission.
- **Spectrum mobility:** If a PU is detected in the specific portion of the spectrum in use, CR users should vacate the spectrum immediately and continue their communications in another vacant portion of the spectrum. For this, either a new spectrum must be chosen, or the affected links may be circumvented entirely. Thus, spectrum mobility necessitates a spectrum handoff scheme to detect the link failure and to switch the current transmission to a new route or a new spectrum band with minimum quality degradation. This requires collaborating with spectrum sensing, neighbour discovery in a link layer, and routing protocols. Furthermore, this functionality needs a connection management scheme to sustain the performance of upper layer protocols by mitigating the influence of spectrum switching.

The main reason for this intensive interest in cognitive ad-hoc networks is that there is no concise description for the changes that cognitive radio would bring to the current mesh networks. This problem is growing with the absence of a standard cognitive radio system definition. For example, the cognitive radio suggested by (Mitola, 2006) can adapt its transmission signals to efficiently utilize and respond to the spectrum changes. However, the application of cognitive communications to mesh networks faces many serious challenges to get successful services. One of these challenges is that cognitive mesh networks (CMNs) can transmit only if the licensed/primary networks are offline. This means that a certain cognitive operator may be pushed for long periods of silence according to the primaries' activities. This case is likely to occur at any time and site of the CMN. Such a situation can block success of the future CMN in achieving real time communications between subscribers. In addition, it raises many questions about the expected financial outcomes from such a network. The setback factors for the CMN can result also from the absence of guarantees for the stability of the wireless links in a dynamic wireless environment where primary users are active. Hence, cognitive wireless links can be interrupted at any time due to the transmissions of the primary networks. Thus, the more interrupted cognitive services are, the less time online for cognitive communications. This brings forward two new aspects: increasing the cost of accessing the spectrum holes because of the spectrum auction procedures, and increasing the complexity of running such networks. Therefore, it is very critical to analyze the expected free resources in different areas prior to any installation of CMN stations. The future analysis for spectrum usage is known as the *Spectrum Forecasting*.

The cognitive network managements can be surveyed simply as:

Distributed Network Management

The cognitive network stations are set to work on individual and independent coexistence schemes. This assumes that cognitive radios are aware of their location and capable of making observations. It should be possible for radios to collaborate to build maps of their environment. One such map could be the radio environment map discussed in (Neel, 2006). To maintain an effective working team, this form of decision making must be biased towards the accomplishment of the common team goals and the maximisation of the value and rewards for the team as a whole, rather than for a single node. The decision-making process also involves deciding whether an action or solution can be implemented by the device itself (local decision) or whether a cognitive network team approach may result in a more effective or timely solution (distributed decision) (Nolan & Doyle, 2006). Although this type of network topology allows faster attaining of local transmission opportunities faster, it results in significant time delay in peer-to-peer negotiations to establish long links.

Coordinated Network Management

The alternative way for managing the uncoordinated CRs that access the spectrum separately is the Coordinated Network Management (CNM). In this arrangement, free transmission opportunities are assigned to the network stations and users based on decisions issued by a centralized entity. There were many proposals for such a supreme spectrum entity, such as:

- (Brik, et al., 2005) designed Dynamic Spectrum Access Protocol (DSAP) that is proposed for coordinating arbitrary wireless technologies and managing access to arbitrary radio spectra. This is done by issuing clients temporary leases for parts of the radio spectrum. Spectrum allocations are performed through negotiations between the clients and the servers using special messages. The parameters in these messages include the client's (MAC) identifier, location if available, radio capabilities (e.g. supported wireless MAC protocols), destination's identifier, and the desired lease options.Then, the *RadioMap* and the set of currently active leases allow the server to determine an optimal spectrum assignment under "policy-neutral" conditions to ensure higher quality of service for a group of these nodes according to the exchanged data messages.
- (Subramanian, et al., 2007) proposed a centralized spectrum broker to coordinate access to the spectrum in a given region and assign short term spectrum leases to competing wireless service providers and/ or end users. This broker can grant spectrum leases that are for small regions (e.g. per base station) and valid for short dura-

tions (e.g. tens of minutes). Fast spectrum allocation algorithms are crucial to the design of scalable spectrum brokers that can provide such realtime spectrum access. The objects of this design were to maximise the overall demand (Max-Demand) satisfied among the various base stations, and to minimise the overall interference in the network (Min-Interference) when all the demands of the base stations are satisfied.

- The EU-FP7 ONEFIT project (Panagiotis, 2010) is supporting diversified services with vastly expanded use of wireless access for the future Internet. One of the solutions for the stressed networks comprises coordinated extensions of the infrastructure that may include network elements, and devices potentially organized in an infrastructure-less manner. The cognitive systems are managing the opportunistic networks and coordinating the infrastructure with substantial adjustability. These modifications are supported by control channels to attain the necessary cooperation between the cognitive management systems, spectrum opportunity identification/generation, spectrum allocation, routing, QoS control, and forced termination handling.

From all the above it becomes clear that the coordinated management allows better understanding of the network resources and technical capabilities. On the other hand, the scalable deployment of spectrum management entities may impact seriously on the time required for processing spectrum access requests, and cause the loss of indefinite short-slots for transmissions opportunities. This becomes more complicated for huge sized networks that require the establishment of long-distance links.

Hybrid Network Management

The design of a flexible multi-access cognitive network was studied by many to increase the reconfigurability of the future cognitive network for seamless communications. Therefore, cognitive radios were always presented as fully aware and independent systems governed by higher spectrum management entities. The edges of communications for such networks assumed the usage of mixture of mesh to ad-hoc communications to connect autonomous stations and to overcome the barriers of long-range links. One of the main projects in this field is the EU-FP7 End-to-End Efficiency (E3) project (Bogenfeld & Gaspard, 2008). This project investigates the context changes, self-adaptation, and distributed self- to deal with the heterogeneous and diverse wireless environment created by this mix of communication schemes. This type of network management may be the best in scaling resources and allocating the spectrum holes for users based on robust auction schemes. This may be done by splitting resources and users according to the transmission destinations, real-time traffic analysis and stations in action. This precise categorization of resources makes it easier to establish time-variant topologies of broadcasting. Therefore, hybrid network managements are more suitable in minimising any disruptions in cognitive network operation.

Spectrum Forecasting

Spectrum Forecasting is an emerging topic for the next generation networks. This expectation for spectrum availability resulted from long-term monitoring of the spectrum users' activities, and the already issued licenses. Initially, this analysis motivated the development of the cognitive radio, and it also motivated the anticipated selection criteria for CMN installation positions. The measurement of the channels' availability, as a function of time and place can lead to a new classification for the spectrum forecasting called: *Global and Local Forecasting Evaluations*.

For global forecasting, the spectrum channels' availability is produced by the main spectrum governing bodies, such as the FCC and the OFCOM. The reason for this is that these institutes have all the information about the spectrum licenses and users, and the time of each user's activities. Such information is highly important not just to secure the rights of the spectrum users, but also for financial issues. The more active operators are charged more for spectrum access. The spectrum management authorities are likely to monitor the licensed users for using only their channels. This is to prevent overlapping in services between adjacent channels, to assess the possibility to offer licensed channels to cognitive services, and to examine the commitment of commercial operators to the general health protection rules. In this way, the spectrum governing bodies generate spectrum reports on hourly, daily and yearly basis, which can be used to predict future spectrum usage based on former operators' behaviour. These reports and expectations cover big cities and wide sites of the states. Moreover, such reports analyse the main frequencies assigned to the principle commercial and military users. Generally, they are universal estimations for the global users and channels that are used for national services.

The other kind of spectrum forecasting is local determination for the resources. Here, the free spectrum holes are tracked for small sized areas where some channels may be accessible in some areas but not in others. Obviously, these measurements can be done by local spectrum users like cognitive radios. Therefore, efficient sensing, and the way of spreading sensors across the arena are crucial for getting comprehensive spectrum usage reports. Actually, there are two main challenges in this kind of spectrum forecasting:

Firstly, the way to collect and integrate sensors' reports from the sensing network. In other words, the positions of the installed sensors may affect their performance in getting the adequate information about the surrounding wireless environment. Therefore, new extended sensors are highly recommended to be used with the next generation networks. These sensors can track intelligently the channels not just for their usage but also for the aspect angle of the user's survival. Hence, such sensors contain a MIMO antenna and a cognition engine. Clearly, the cost of such systems would be high and cannot be delivered widely. However, there may be a solution for a new network management that solves permanently the problem of the hidden point. This kind of sensor is under development by many companies for military applications.

Secondly, as the local forecasting reports are useful only for local spectrum users, it is important to establish local spectrum management entities that spread these reports between local users. The proposals for a *Spectrum Broker* find a strong inspiration for implementation here where the broker can contact all users in its area. However, the question here is how to attain the necessary awareness to analyse the sensor's data on a real time basis. Then, how to specify the estimations that need to be sent to certain users? In fact, all these dynamic computational and geographical data are increasing considerably the complexity of the cognitive network and the future success of free channels tracking. The other important task here is to decide how to create the right image about the surrounding wireless local channels using the available data. In other words, obtained sensors' data should be enough to generate full report on the spectrum holes' occurrence in a particular area. Generally, local forecasting improves the spectrum access opportunities but increases the impacts of the management of the network architecture.

All the spectrum forecasting reports are sent to the cognitive stations according to the level they intend to access in the spectrum. Thus, two kinds of reports could be available for spectrum users according to the end-users locations: local and global forecasting reports. Therefore, global reports are provided if the packet delivery needs many hops to reach the destination. Additionally, local reports would be provided for users sharing the same area, where one to two hops may be enough for packet delivery. The last scenario suggests that both reports are provided to all stations, and hence, wireless links can be created from short connections using many channels. This is the most useful set-up, especially in a very dynamic wireless environment with an increasing number of cognitive users. These generated reports are normally saved for future comparisons to develop the management of the network based on long term notifications. Expectedly, these reports would be in XML language to simplify the way that cognitive systems talk to each other (Mahmoud, 2007). These can be exchanged between cognitive users in a cooperative way for expanding knowledge on the environment, and improving services' delivery and awareness. One application for this is that the availability of a certain channel at a certain geographical area may be expanded according to the local forecasting reports received from a distant cognitive radio sensor. Thus, forecasting reports may be received from hybrid sources to get the necessary cognition about the spectrum.

COGNITIVE RADIO OVER FIBRE SOLUTION

Restoring cognitive network services cannot be performed using the traditional procedures for link re-formulation in mobile communications. This is due to the dynamic nature of the spectrum holes' availability, auction schemes on resources, and the reconfiguration methods to access resources. Therefore, a permanent solution is necessary to guarantee the transmission of the future cognitive mesh network. Thus, the most realistic solution is to combine the CMNs with the landline networks. This expands the broadcast options for the cognitive network in order to attain reliable delivery services. The fibre network is proposed as the promising landline technology for this new hybrid cognitive-to-landline arrangement. As a matter of fact, fibre networks are re-gaining high interest for their multiple product advantages, such as high bandwidth, low power consumption, and low cost. Furthermore, many projects are currently undergoing for delivering many services via fibre networks. For example, the British Telecom (BT) 21st Century Network (21CN) programme (POSTnote, 2007), and the Japanese National Institute of Information and Communications Technology (NICT) Fiber-To-The Home (FTTH) projects (Hamanaka, et al., 2007).

The integrated connection between the fibre network and the cognitive base stations can be done easily and in a cost effective manner. This combination is named Cognitive Radio over Fibre (CRoF). This CRoF system is using the radio over fibre as the land line connection between the new CRoF base stations. Radio-over-fibre (RoF) technology gives a lot of advantages, such as the complicated signal processing at the Central Base Station (CBS) localising, therefore the overall system is cost effective and the Remote Base Station (RBS) is very simple, passive and compact (therefore it is transparent to air and its maintenance is easy). The system is very cost-effective because of localization of signal processing in CBS and also because the Base Station is simple. The reliability of the system is high due to the simple and passive structure of the RBS. This system can easily serve densely populated areas, such as shopping malls and airports, dead-zone area and highways can be covered efficiently and economically. The system can also support multiple wireless standards. Because of the high bandwidth nature of optical fibre, Broadband services are more feasible using this technology (Abdolee, et al., 2007). The integration of optical and wireless systems is considered to be one of the most promising solutions for increasing the existing capacity and mobility as well as decreasing the costs in next-generation optical access networks (Jia et al., 2007). Radio-over-Fibre is a suitable technology that provides a cost-effective solution for delivering high data rate, as well as for extending the range of the existing wireless access networks. The remotely located Base Stations (BSs) are interconnected via a Central Office (CO), using a low-loss optical fibre network (Lucarz & Seeds, 2009). Frequency independent low fibre loss, high bandwidth, and increasing availability of low-cost optoelectronic and RF integrated components are the major drivers for fibre optics RF solution (Al-Raweshidy & Komaki, 2002). Several key enabling technologies for hybrid optical–wireless access networks are raised by (Jia, et al., 2007), including optical millimeter-wave (mm-wave) generation, up conversion, transmission in a downlink direction, and full-duplex operation based on wavelength reuse by using a centralized light source in an uplink direction.

The CRoF system concepts need to develop a new radio that is equipped with fibre connections. The anticipated CRoF radio is an extended version for Mitola's cognitive radio. Such extension includes extra physical portals for accessing landlines and transferring signals on-time from wireless to wired modulation forms. This operation occurs whenever the CRoF radio failed to formulate a wireless link. This process should not take long time intervals during conversions in order to keep real time broadcasting between terminal handsets. This development can be done only by very powerful software programmable transceiver units that can be adapted on demand promptly. The other major adjustment in the CRoF radio is to extend the intelligence of the cognition machine. This allows the CRoF radio to make on-time decisions and choose the efficient link between wireless and fibre connections. The extended cognition engine also helps to explore the surrounding wireless environment to select between the available short and long wireless links and fibre connections. All these transmission assessment tasks must be performed by this extended cognition engine. Actually, these modifications may impact the computational complexity of the suggested CRoF radio and increase the time of processing of various communication attempts. However, a small comparison between the pros and cons of the CRoF radio shows that the outcomes of the new radio system are much higher than setback factors. The most interesting thing about the CRoF is the high reliability in delivering the data, which is the main problem that cognitive radio is suffering from.

MULTI-ZONED SPECTRUM SHARING

The future cognitive mesh network is to be installed in coexistence with the other primary networks. Therefore, the positions of the cognitive BSs are decided by many factors such as the transmissions power, coverage range, used channels and number of expected end-users, in order to avoid interfering with primary BSs. Thus, it is presumed that not all BSs would have the accessibility to fibre connections for many reasons:

1. CRoF-BSs are to be connected to the backbone fibre networks and not fibre terminal edges or copper connection. This will assure the highest speed for data transfers across the fibre network. In addition, this prevents the decline in services, which may result from the time taken for downloading and uploading data frames.
2. The fibre connections are not trying to eliminate wireless links between the cognitive stations. This may increase the time delay during packet delivery because of the conversions between optical and wireless equipment. Fibre is used to support the network as an alternative for wireless links only when there is no way to formulate an efficient wireless link.
3. The cost of connecting fibre to the cognitive network and the modified CRoF stations is taken into consideration in choosing the installation positions and the number of CRoF-BSs.

Therefore, only selected BSs are connected to the fibre network as CRoF gates. The best strategy for installing the cognitive BSs is to place them within the known boundaries of the primary bases stations. This is to simplify the spectrum trading between the primary and secondary networks and also to make it easier in dealing with temporary licenses issued from local spectrum entities. Thus, future cognitive BSs may have the same features for broadcasting, merged with the reconfigurability required to access multi-channels as secondary spectrum users. Figure 1 shows a huge sized cognitive network with only selected cells as CRoF areas, as explained above.

The CRoF cells will attract neighboring cognitive cells to use CRoF's fibre connections for data exchanges in severe wireless situations, when the cognitive BSs fail in contacting remote sites of the network (Al-Dulaimi, 2010a, 2010c). These groups of CRoF and cognitive cells that share the same wireless environment conditions split the network into zones of services. In each of these zones, CRoF acts as the master for all other cognitive cells. This division makes it easier to achieve local resources sharing between a zone's cells, by using the CRoF cell as a main hub of communications. The CRoF-BSs may use their intelligence to control the spectrum channels in the zone and behave like a Local Spectrum Broker. Then, all the zones are connected via fibre network through CRoF cells. In fact, the CRoF works as a subnet that is quite similar to what neurons do in the human's body. The CRoF exchanges data, information, transmission requests, and orders over far sites, and use CRoF-BSs as the gates to deliver all these events to the other cognitive cells on zone basis. The new resulting arrangement can be defined as a new hybrid system for dynamic spectrum access that is capable of achieving local, centralized-to-local, and global network resources management.

The advantages of multi-zoned application to future cognitive networks can be reviewed as:

1. Using the RoF to connect selected cognitive Base Stations as an alternative for wireless links to overcome any interruption in services due to the dynamic spectrum access models. These novel Base Stations are selected according to long-term monitoring for the

Figure 1. Adapting services between CRoF and CR stations

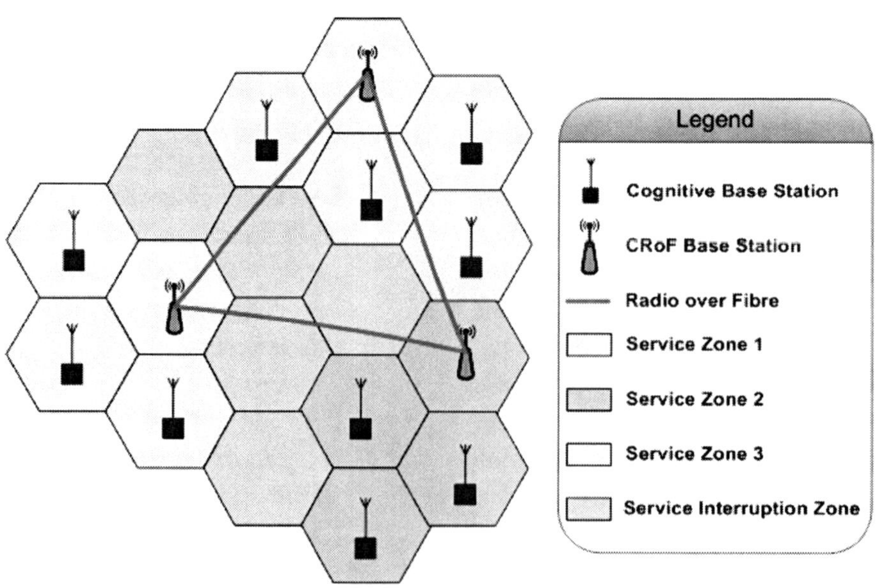

traffic requirements and the available fibre connections in the field.

2. More independent cells that are capable to reconfigure their wireless features and network interfacing managements for efficient usage of the available free resources. This requires an extended cognition engine designs for the cognitive Base Stations and new protocols for unlimited knowledge transfer between different network stations.
3. Establishing temporal, local, flexible domains of neighboured cognitive cells. These formulations denoted as zones and requires same available band of channels at a time as a condition for any cell to join or leave the set.
4. Setting mobility management schemes that can deal with cells rather than individual mobile stations. The main goal is to create a system that can respond and act to unpredicted wireless environment changes and numbers of terminal users.
5. Reducing complexity of dealing with main spectrum entities by establishing local governing units supported by protocols. The cooperation between various cells is vital for the success of the proposed models.

This chapter investigates the criterion for the formulation of these zones, the changes in the zones' sizes and boundaries, and the possible scenarios for cognitive cell migration between these zones. This new paradigm of local resources sharing through architectural adaptation between the secondary systems is taken further by analyzing the effects of such an application on the routing maps for such a network.

CELL MIGRATION BETWEEN ZONES OF SERVICES

The zone cell groups created as a result of CRoF installation are dynamically changeable to host and lose new cognitive cells. In other words, cognitive cells may join or leave zones at any time, whenever the wireless environment situation changed. The changes in wireless environment can result from the changes in the activity of the primary users, which will change the number and time of free channels' availability, due to the change in the transmission requests of the cognitive users. Additionally, the technical barriers that limit the cognitive radio abilities from accessing certain channels may be considered as part of the changes in the zoned situations. These factors are affecting the local wireless environment and make them different from other cells' wireless situations inside a certain zone. Therefore, a cell can decide to leave a certain zone and join another one or even stay independent, according to the mutual negotiations between the cell Base Station and other zone stations, including the CRoF-BS. This section analyses briefly the possible scenarios of coexistence between various zones and the migrated cell. The objectives of the cell migration will always be to attain the most efficient spectrum utilization and the highest quality of service. The main assumption for all studied cases of the cell migration between different zones is that the coexistence is done in a cooperative way and based on dynamic network arrangement that can accommodate any real-time architectural modification.

The cognitive cell may choose to migrate from one neighbour zone to another for similarity in the wireless environment situation between the cell and the new attended zone. Firstly, we presume that the cell CRoF-BS has the necessary awareness to understand the changes between the local cell wireless environment and the global zone circumstances. This is decided mainly by the way that different cells are exchanging their knowledge. However, the best scenario is to supply information to cells on two schemes, *Global Network* and *Local Zone.* In the Global Network scheme, the information is spread by different zones to the neighbored zones and cells, and vice-versa, to make them aware of the general wireless situation inside each area. This information can be provided centrally by the Operator to all stations. The Common Control Channel (CCC) can be used as the media for data delivery. On the other hand, the Local Zone scheme may provide regulating data for cells grouped by one zone. An IP address may be given to different zones on a temporary basis to simplify the way of providing services throughout the network. One suggestion is to use the CRoF-BS IP address as the zone address. Therefore, a new, fully aware network is created by establishing multi-layered data exchanging between cells, to achieve an efficient network, responding to the various expected management events.

The procedures and possible events for cell migration are explained in the following cases:

Case 1: Cell Moving between Zones

The disengagement process for a cell leaving from one cell to the other can be performed through informing the neighbored cells or the CRoF-BS, as it is the senior of the zoned cells. The cell takes the decision to leave a zone whenever it recognizes that moving to another zone is more suitable for getting extra transmission chances and better end-user services. To manage the system of zoned cells and prevent cells from turning selfish, a new global regulation should be set, namely, that a change in wireless circumstances between different zones is the only parameter for cells' migration. Thus, cells cannot leave because they want to keep the opportunities of transmission inside their coverage areas for themselves. After informing the zone cells that a cell is intending to leave the zone, all zone cells are advised about this decision. Mostly, the CRoF-BS asks all the other zone cells to terminate their Local Zone connections with the leaving cell and re-direct their zoned transmission routes using other stations. When such an order is issued the cell becomes free and can immediately attend the other zone that it intends to join, as shown in Figure 2. Here, there are two methods for joining the new zone, either the cell had talked to the zone prior to its decision to migrate, or it intends to talk as soon as it is sent free from the old zone. In both cases the cell is able to attend the new zone only if it had the same wireless environment situation and there are no more obligations any more with the old zone.

Normally, a zone accepts the request issued by an individual cell to attend the zone as long as this application follows the etiquettes expressed above and increases the zone's abilities to attain more resources. The time taken for cell migration between two zones should be fast enough to pre-

Figure 2. Cell migrations between CRoF zones

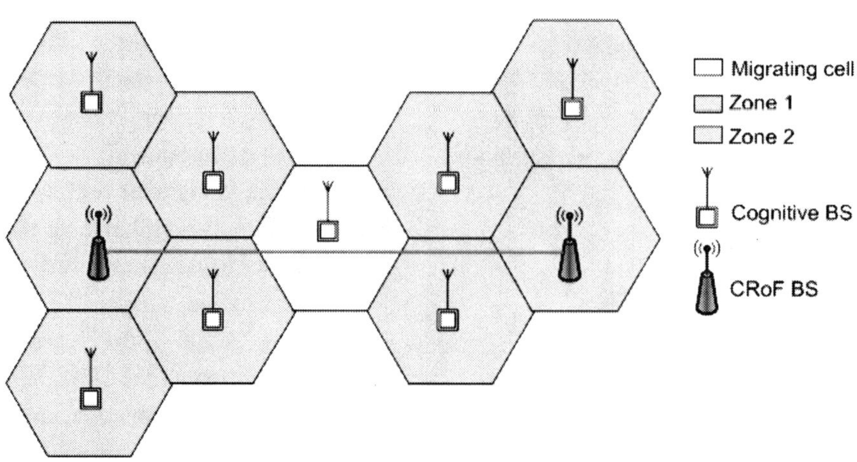

vent long time interruption in communications with the migrated cell, so it does not get trapped between two zones of services. In fact, the cell is still part of a bigger operator and network infrastructure. But the main point is that this cell had not identified itself as an independent service entity. The local communications and re-directing of the services throughout the zone is changed as soon as the new cell is unified with the service area. The time that is spent by a cell during the migration between two situations can be named as *Cell Migration Time (CMT)*. The broadcasting between the cell end-users which are situated inside the cell coverage area should not be affected during the CMT.

Case 2: Independent Cell between CRoF Zones

As the formulation of zones is based on grouping cells that have similar wireless conditions, a certain cell may stay out of these zones due to its different wireless environment, as shown in Figure 3. This independent cell can appear when the wireless environment changes at any time, which means that it has to go outside of the zone area. Additionally, the migrated cell may choose to stay independent during the CMT to achieve better services for its own end-users. In all these cases, this independent cell emerges as a new small-sized zone of service that may or may not have fibre connection. Actually, the fibre connection comes when there is more than one CRoF-BS inside a zone and this cell chooses to get independence. The last scenario may be useful for zone formulation to prevent multi- scenarios' competition on heading a group of cells constructed as one zone.

The main concern for having many independent cells is the time taken for negotiations during link formulation between the neighbored zones and this independent cell. The complexity in such a system comes from two sources: firstly, the time taken to decide to connect large zones' services and small sized zones (cells) in a very dynamic wireless environment where transmission opportunities may be lost anytime. Secondly, the identification of end-users' locations and adjacent channels' interference, in case the cells on both sides turn selfish. All this shows that the choice of a certain cognitive cell to work independently from CRoF zones may impact network performance. However, it is better for the zones when a different cell becomes independent. This will prevents any disturbance in the coverage area for having cells of unmatched resources' management and less interfacing possibilities. However, this independence should not affect at any time the overall network performance and the network life

Figure 3. Independent cell coexistences with CRoF zones

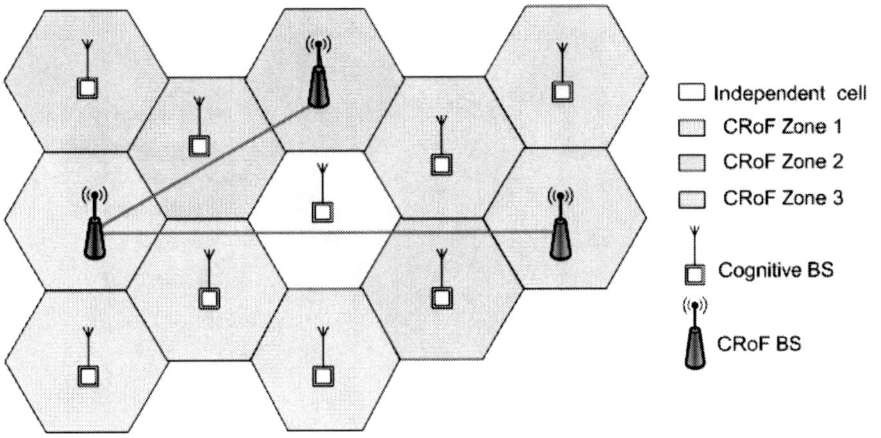

cycle. This can be achieved by fully cooperative behavior, which is guaranteed by the local spectrum governing entities, such as CRoF-BSs and the operating protocols. The same scenario is likely to occur when many networks share their infrastructure to reduce the cost and to attain better services. Therefore, it is necessary to address different management strategies through infrastructural modifications, like this scenario to host many operators using the same equipment and different frequency bands. In this way, the cognitive network is the best crucible for such management ideas and the most powerful proposed technology for real-time adaptations.

Case 3: Cell Trapped by One Zone

The unification of different cognitive cells having the same resources may lead to the isolation of a certain cell with different wireless situations, as shown in Figure 4. This cell is trapped by the same zone cells from all directions. The other assumption is that the cell leaves the zone after being part of it, for certain developments in the resources' availability in the cell area. In fact, this situation can happen when two neighbored zones choose to bond together as one zone and there is a cell in the middle of the CMT. Suddenly, this cell finds itself trapped between the boundaries of this new establishment with few other choices to take. Theoretically, the etiquettes of managing the services in a network are still the main guidance for accessing the spectrum and connecting different base stations. Practically, this is a critical situation for the trapped cell in achieving high levels of cooperation with other zone stations. To maintain seamless communications in this specific scenario, the zone has to give the zone cells that surround the trapped cell, the authority to communicate directly with the migrated cell without central approval. In this way, the zone follows the distributed access to the resources to overcome the delay resulted from requests to communicate with a small coverage zone, such as this trapped cell. There is an expected increase in the complexity of running such a network because of the long producers of having one cell stand alone between cell boundaries. However, the complexity may increase further in case the cell forced to join a zone of unmatched wireless situations. Therefore, it is preferable to obligate to the general rules of balancing services by reunifying only similar cells to a one zone formula. Even with the presumed special treatment given by the zone to this trapped distinct cell, there are many complications in achieving relaxed coexistence solutions for accessing the spectrum. One solution for this case is to find the matching channels

Figure 4. Independent cell coexistences with one zone

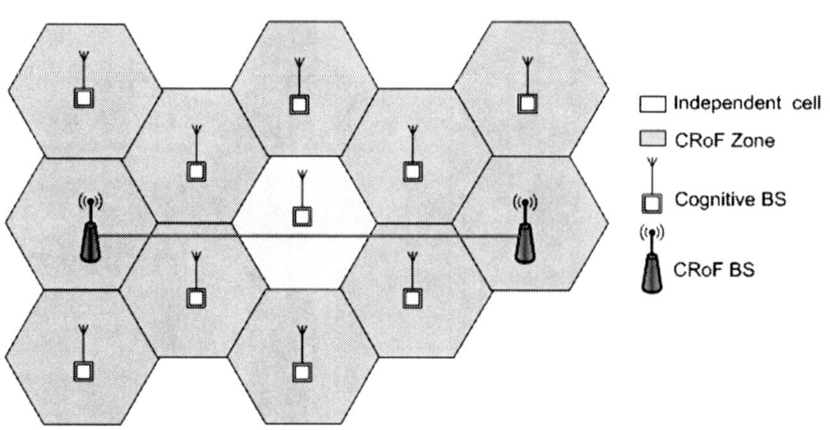

and mutual similarities in resources, and sharing them. In this way, this individual cell can re-join the zone easily, as a partially joined cell member. This eases the way of starting communications and establishing links with the main zone providers, such as CRoF cells.

As for being dynamic sets, the zone sizes, places, cell members, and shared resources may change at any time due to the activities of the primary users and the technical difficulties in adjusting transmission functions. Therefore, negotiations are undergoing between all cells at all times to achieve the highest QoS in responding to the clients' requests and spectrum developments. For such flexible arrangements, it is likely to find two CRoF cells or more in the same zone where only one should act as the senior for all other cells. This senior CRoF cell can be chosen for its position at the centre of the zone or for being the earliest CRoF station in the zone. To sum up, zones are based on optional attendance for different cognitive cells that are using CRoF cell landline connection to support their services to the cognitive end-users.

HOSTING USERS BETWEEN ZONES

In CRoF, users move between different network domains as they travel around different access points of macro cellular and micro cellular areas. The main concern for the MS in this model of networks is to re-connect smoothly and immediately to the next Access Point (AP) to avoid the interruption in services. Such an interruption is likely to occur for unpredicted time interval during the adaptation between the available channels. Hence, it is required to develop new method of roaming between various APs to reduce the unnecessary SHs. This approach is not feasible for the following reasons (Al-Dulaimi & Cosmas, 2010d):

- There is no main entity that supervises the spectrum usage and MS registration and connections in an individual basis. Although, there are many proposals about the spectrum broker supervision to the spectrum, there is no literature on any involvement of the spectrum broker in the connectivity of MS to the APs.
- The misunderstanding between SH that is happening in the same operator and the SH that happens when a MS moves between different networks. A MS within one network has the same address that can be used also to monitor the channel access history and forecasting futuristic usage along the route of movement. On the other hand, a MS that enters a foreign network needs to solicit a new care-of-address (CoA) in order to roam to the surrounding APs. Then the HA is updated with the CoA address. The time taken for getting the CoA may result in losing the MS for its channel and then driven to adapt to a new channel.

Therefore, there is a need for a new architecture to allocate the same channel to MS as long as this possible. Such modeling helps to achieve the following goals:

- Reducing the number of SH between network domains of macrocells and microcells areas. These are regular events in CRoF networks.
- Minimizing the possibility of SH failure which results from unsuccessful attaining of a new channel.
- Innovative pre-coordination method for the available resources by controlling SHs and reducing unnecessary adaptation between the accessible band channels.

There were many proposals to identify the new concepts of mobility management for the cognitive network. The existing literature aims mainly to analyse possible solutions for roaming MSs between different networks. On the other hand, some papers analyzed the effects of the dynamic spectrum changes on the mobility of the cognitive end-users. Some of these studies are described below.

In (Damljanovic, 2009), the handover control is either radio executed or network initiated. In case of user initiated handover, the proposed type of handover is Mobile Controlled Handover (MCHO). With MCHO, radio is entirely in charge of the handover decision and it decides when and where it starts the handover procedure by selecting a network with the best available service. Contrary to user initiated handover, spectrum vacation handover is better utilized by the use of the Network Controlled Handover (NCHO) and Mobile Assisted Handover (MAHO). With use of NCHO and MAHO, the cognitive network is in a better situation to "command" handover in order to avoid violation of interference temperature limits. However, a network needs continuous assistance from the cognitive radio, which is best utilized by the use of MAHO.

New entities are developed by the EUFP7-E3IIR project (Boufidis, et al., 2007), to handle mobility between 3GPP and non-3GPP access systems. The evolved packet core network consists of the Mobility Management Entity (MME), the User Plane Entity (UPE), and the Inter-Access System Anchor (IASA). The MME and UPE are collectively referred to as Access Gateway (AGW). The IASA is in charge of mobility between different access systems. It is composed of a 3GPP anchor that executes mobility between 3GPP access systems and of a System Architecture Evolution (SAE) anchor. The SAE anchor does not make any decisions regarding mobility; it just executes such decisions made by the user equipment.

When spectrum handover is needed, first an alternate channel must be determined. Then a receiver-transmitter handshake must be performed on the new channel. Only then may nodes continue their transmissions. All of these additional operations incur long delays and hence buffer overflows, which lead to packet losses, degradation in reliability, and ultimately resource waste in cognitive radio sensor networks. Therefore, a central spectrum allocation scheme that tries to minimize spectrum handover is proposed to minimize the effect of spectrum handover on various communication layers (Akan, et al., 2009).

A novel solution that ensures interoperability between several types of wireless access network is given by the developing IEEE 802.21 standard. This standard is focused on handover facilitation between different wireless networks in heterogeneous environments regardless of the type of medium. The standard names this type of vertical handover as Media Independent Handover (MIH). The goal of IEEE 802.21 is to better and ease the mobile nodes' usage by providing uninterrupted handover in heterogeneous networks. This function is responsible for communication with different terminals, networks and remote MIH functions and provides abstract services to the higher layers using a unified interface of L2.5 functionalities (Gavrilovska & Atanasovski, 2007). The 802.21 specifications are also defined by (ETSI, 2010).

To the best of our knowledge, none of the above literature analyzed the possibility of allocating the same channel within the same operator for the moving MS. This motivates to design a new architecture for SH as explained in the following section.

NEW IP MODEL

This section identifies: firstly, the new IP integrated model for the WLAN nodes, then, explains the possible scenario of application in the CRoF network.

Introducing the CAA

In order to reach a cost effective solution that reduce complexity and use the current known infrastructure to include the dynamic spectrum allocation, we allow any MS to communicate directly with its Correspondent Nodes (CN) instead of tunneling the traffic via Home Agent (HA) node whenever they can do so especially inside microcells. This utilizes efficiently local transmission opportunities in short range communications which is the main goal for the secondary systems. Additionally, two-way handshake (Solicit/Reply) is used instead of the usual four-way handshake (Solicit/Advertise, Request/Reply) to reduce the time of adaptation between channels. This is a major requirement for MSs that are moving at high speeds.

The framework solution introduces the Channel Assigning Agent (CAA) at the IP network layer to negotiate the available channels prior to any handover actions. In this way, the CAA checks the used channel at the MS side. The CAA is connected to the Access Router (AR) and from there to the main network entity named Networking Interoperating Agent (NIA) as shown in Figure 5. This model is valid for the WLAN systems and to roam MSs between various network domains. The CAA evaluates the available free channels to host any new MS in the broadcasting zone.

The main objectives of the CAA for the mobility management inside a macrocell are: Firstly, informing the corresponding node to adapt to certain channel at the time of arrival of any new MS to interface communications immediately with the new station. Secondly, the CAA is responsible to manage the allocation of the same channel for the same mobile along its journey route whenever this is possible to avoid unnecessary SHs. The last scenario is very likely to occur when mobile moves between macrocell and microcell transmission domains in the CRoF networks. This development improves the roaming of the cognitive networks for fast MSs that represents a major task with its numerous numbers of SHs.

The CAA is actually tries to keep the handover of the MS as horizontal-handover of sequence of connections to the APs along the way rather than vertical-handover where MS jumps between chan-

Figure 5. The CAA allocates the same channel for the MS that moves between two network domains

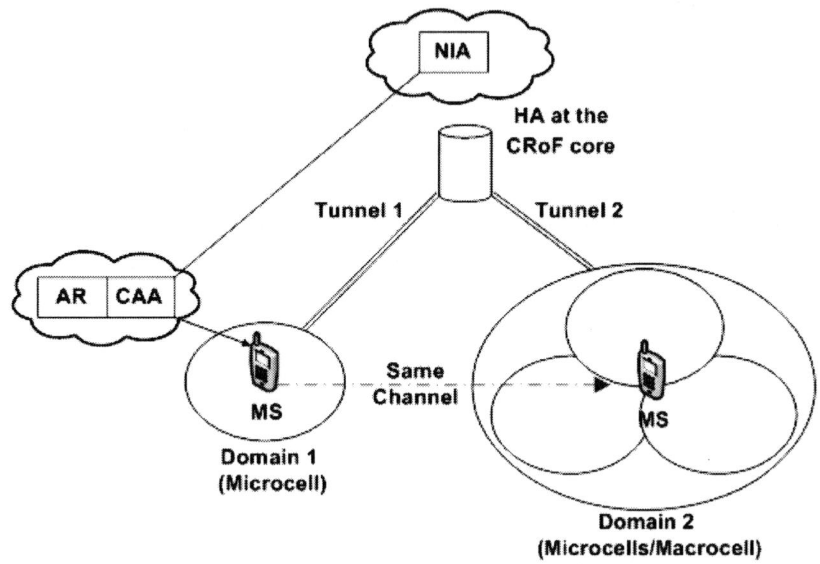

nels (Chen and Prasad, 2009). A new procedure can be suggested here for mobility management between various domains using a CAA in the CRoF networks:

- **Location identification:** A MS should be able to discover its position relevance to the surrounding APs at all times. This can be done through agent solicitation messages or through geolocational information received from the cognitive network. Such data are assumed to be available at all times to help to detect the white-time spaces as a function of location in the cognitive systems.
- **Registration:** The registration of the MS can be done as soon as the MS moves into the broadcasting area of a new AP or when the signal field strength is identified as higher than its correspondence node. Eventhough, a MS may stay connected to its correspondence node with another nearest AP to reduce the number of SH in case it cannot use the same channel with the new AP.
- **Tunneling:** The correspondence node encapsulate the packets directly to the new AP without the need to re-direct them through the HA as soon as the HA issues an order in this meaning.

When the MS movies from Domain 1 to Domain 2 in Figure 5, tunnel 1 is terminated at the time of the MS registered with the AP at Domain 2. The reduction in time delay of packets delivery using this scheme is resulted from the cancellation of the SH. However, the traditional procedures of handover and registration are $s\mu e^{-\mu t}$, then x=1/μ and $\overline{x^2} = 2/\mu^2$. Then:

$$\bar{q} = \lambda/\mu + \frac{\lambda^2/\mu^2}{(1-\lambda/\mu)} \tag{3}$$

In order to simplify the interpretation of equations, let us normalize the service time μ to 1. Now, if $\partial \bar{q}/\partial»$ and $\partial^2 \bar{q}/\partial»^2$ are evaluated, then:

$$\frac{\partial \bar{q}}{\partial \lambda} = 1 + \frac{2\lambda - \lambda^2}{(1-\lambda)^2} \tag{4}$$

$$\frac{\partial^2 \bar{q}}{\partial \lambda^2} = \frac{\left[(2-2\lambda)(1-\lambda)^2 + 2(1-\lambda)(2\lambda - \lambda^2)\right]}{(1-\lambda)^4} \tag{5}$$

Furthermore, for <μ, $\partial \bar{q}/\partial \lambda > 0$ and $\partial^2 \bar{q}/\partial \lambda^2 > 0$ showing that the $\bar{q}$ and its rate of change increases with increasing λ. In particular, $\lim_{\lambda \to \bar{\mu}} q = \infty$.

For the case of a deterministic server:

$$\bar{q} = \lambda \bar{x} + \frac{\lambda^2 \overline{x^2}}{2\left(1 - \lambda \bar{x}\right)} \tag{6}$$

which has a development similar to that of a Markovian case with the exception of certain additional constant coefficients.

Clearly, there is a close relation between the number of the mobile stations and their arrival time which result in changing the length of MSs in queue waiting for connection.

PERFORMANCE EVALUATION

Application Scenario

Spectrum handover is another source besides traditional handover for providing mobility in cognitive network architecture. Considering SH, this dynamic process of adaptation between channels may happen to fixed and moving mobile at the same time. However, splitting the effects of traditional handover schemes from the new SH is a very new topic for discussion in secondary networks. The main aim is to reduce the time of interruption in services resulted during the adaptation of the cognitive MS between channels. This new soft-spectrum handover method is performed by allowing the CRoF core to assign the same channel that a cognitive MN is using for communications during its journey inside the macrocell between macrocellular and microcellular coverage areas, whenever it is possible. This reduces service disruption and any redundant handovers. In other words, fibre network is used to establish short range communications for any MS using the same channel that was assigned to it as long as the primary user is offline. For example, the MS moving from position 2 to position 3 in Figure 6 can still use the same channel unless it is forced by the primary user to adapt to another frequency. On the other hand, we give the priority of using a certain channel to the MS that used it for the longest time. In this way, a cognitive MS may need to adapt many times and suffer many SHs for not interrupting more stable services that is given to other MSs with less SH. This means that as a cognitive MS using this scheme *the lazier the more comfortable you are while the more active the more tired you are*.

Considering the CRoF system in Figure 6, the MS experiences five different events of spectrum handover as it moves around using the trajectory shown in the figure, these handovers occurs when:

1. The MS moves into the microcell coverage area (position 1) where service capacity is much higher because of the stable and short range communications.
2. The MS leaves the microcell coverage area (position 2) to the macrocellular services area.

Figure 6. Spectrum handover events for a cognitive mobile station that moves between microcells and macrocell coverage areas

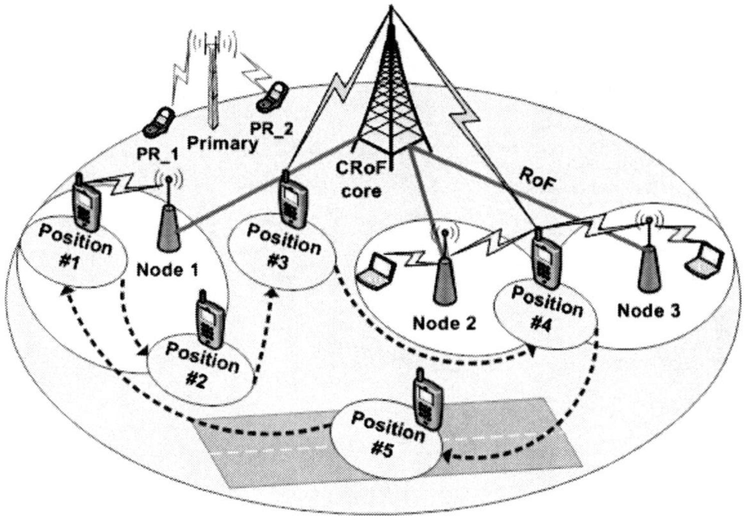

3. The MS heads directly near to the macrocell base station (position 3).
4. The MS moves to the fully covered area by three transmissions sources: the macrocell and two microcells (position 4).
5. The MS moves along the motorway (position 5) returning to its initial point (position 1).

To examine the performance of the new SH scheme using the scenario shown in Figure 6, the MS is set to use different speeds of movement along the route shown in the figure. These speeds were set according to the Table (1) that presents the vehicular speed limits in the UK and as follows:

Table 1. Simulated mobile station speed

Speed in miles per hour (Mph)	Description
2.877	Pedestrian
20	Cars speed in urban areas
30	Cars speed in urban areas & villages
40	Cars speed in non-built areas
50	Cars speed in non-built areas
70	National speed limit

An IP telephony and silence suppressed signals are generated to test the system performance. The reason to choose this kind of application is that normal phone calls are actually composed of different times of activity where the user is either talking or silent. The IP networks transmit packets only when the data and control information are in action. Therefore there is no usage for the channel if the clients are not sending anything. Thus, such application is very useful in analyzing the cognitive networks and the dynamic spectrum access models. The reason for this is that transmissions occur temporarily and when it is needed only which is the same principles of the cognitive radio systems.

The simulation setup includes two networks primary and secondary networks that coexist with each other. Primary users are transmitting along all channels while secondary systems are accessing the available band on temporarily basis whenever there are no primary activities. In order to simulate the performance of the new model at precisely, the number of the MSs moving around is set to 1, 3 and 7 respectively. In each case study, an evaluation for the system improvement with no SH is compared to the traditional case where the SH is happening along the movement route shown in Figure 6.

Figure 7. End-to end time delay as a function of the speed of mobile

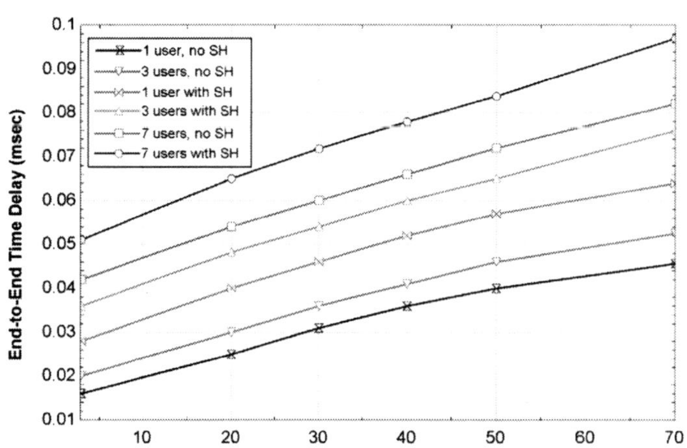

Figure 8. Throughput as a function of the speed of mobile

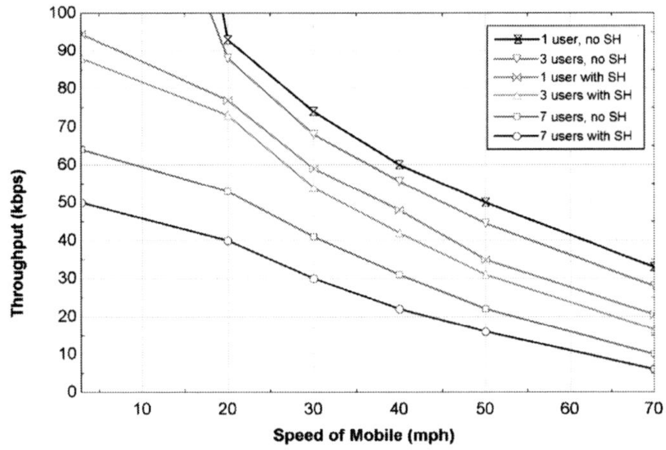

Simulation Results

In this section, the simulation results are presented to validate the proposed scheme. In Figure 7, end-to-end time delay is shown for all the simulated numbers of MSs as function for the mobile speed. The figure shows that there is a considerable time delay savings using the new scheme of CAA in compared to the traditional models of SH. The figure shows that for 1 & 3 MSs the savings in time delay are much higher compared with the same situation for 7 users. The reason for this is that limited resources available blocks the continuity of performance improvement in the system for non SH case because there is no channels to host growing number of MSs.

The throughput in Figure 8 shows also a major improvement with the application of CAA and

Figure 9. The connectivity to the access points as a function of the mobile station location according to the measurement points in Figure 6

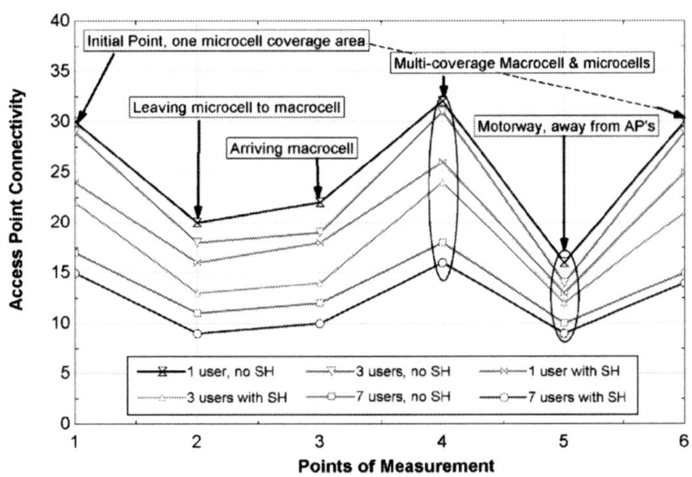

Figure 10. Traffic sent as a function of the speed of mobile

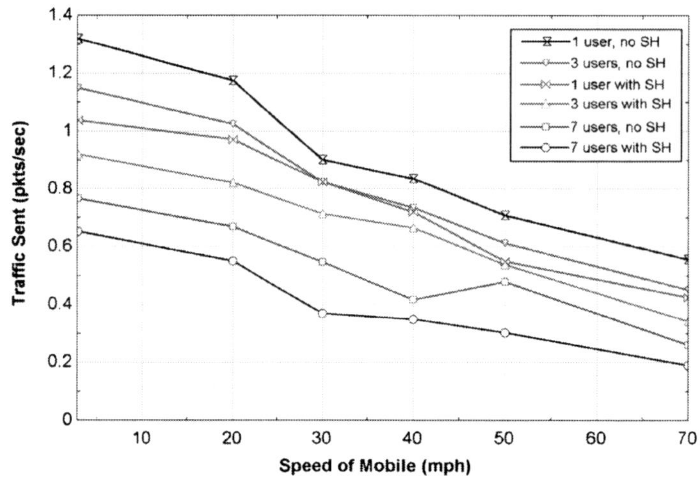

no SH events. For all simulated cases, the throughput is higher than the case for the traditional spectrum handover. The explanation for this is that the time spent in the adaptation between various channels reduces the performance of the system. This interruption time impacts the time delay in Figure 7 and the throughput in Figure 8 with the values shown in both diagrams. It can be noticed also that the performance of the simulated system is declining slowly with the increment of the MSs speed. This is because the time space opportunities are available to be accessed by the secondary users on temporarily time spaces and scattered locations in the macrocellular and micocellular areas. Therefore, the faster MSs become the less ability they become to attain white-time spaces. This raises a huge size of discussion on the credibility of the cognitive network in serving fast moving vehicles such as airplanes to landline cognitive communications.

Figure 11. Traffic sent as a function of the speed of mobile

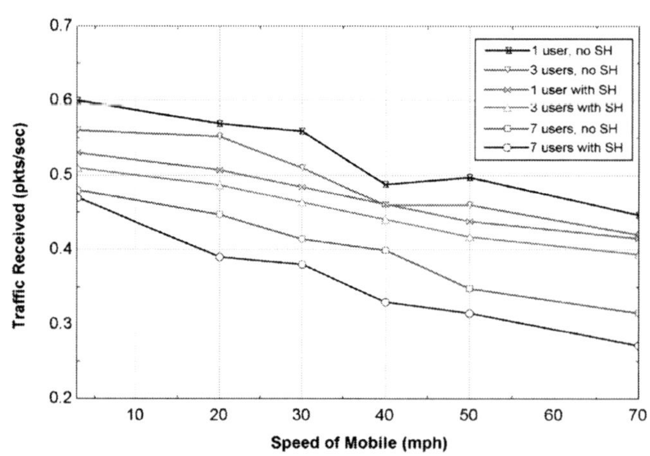

To understand further the performance of the cognitive MSs at different positions using the simulated speeds, the connectivity to the APs (Figure 9) is analyzed as a function for the measurement points illustrated in Figure 6.

Firstly, the simulated cases shows a similar performance for no SH and with SH scenarios. Secondly, there is an obvious improvement in the performance due to the CAA application. However, it can be seen that the connectivity levels between the MSs and the APs are declining as the mobile movies from the main coverage area of the microcell leaving to the macrocell (positions 1 to 2).

It can be noticed that the connectivity levels to the AP is increasing as the MSs move from towards the macrocell base station (position 2 to 3). Then, reaching the maximum value by being under the coverage of three sources: macrocell and two microcells (position 4). Afterwards, connectivity shows the lowest values as the MSs travel along the motorway away from the APs at position 5. At the end, the connectivity values raises again as the MSs return home to the initial point of their journey (position 1 or 6). Overall, interfacing the MS to the AP is a matter of MS's speed and the real effective coverage area of the AP.

The traffic analysation shows similar performance for both traffic sent and the traffic received as shown in Figures 10 & 11 respectively. In both cases, the scenarios for MSs without any SH that are using the new proposed scheme of keeping the same channel of transmission along the movement trajectory show a better performance than traditional SH scenarios. In addition, the traffic in both cases is declining as the speed of the MSs is incasing.

In summary, the CAA application for reserving channels to the cognitive MS improves the performance in accessing the spectrum and interfacing to the new APs.

CONCLUSION

This chapter proposes a comprehensive discussion for developing the infrastructural managements of the future cognitive mesh networks. The radio over fibre is analysed as a solution for the obstacles that face online cognitive communications and higher throughput transfers by cognitive base stations. These developed cognitive radios connected with fibre are proposed as a new subnet named Cognitive Radio over Fibre. The subnet stations partition the cognitive networks into different zones to ease the management and access of resources, using both wireless and fibre connections. These zones are composed of a group of cognitive cells mastered by the CRoF cell, with optional attendance. The boundaries and size of any of these zones can change at any time due to the dynamic wireless changes. Various scenarios for cells leaving and joining zones are surveyed and discussed. The aim of this proposal is to identify a new local resources sharing scheme through topology adjustments.

REFERENCES

Abdolee, R., Ngah, R., Vakilian, V., & Rahman, T. (2007). *Application of radio-over-fiber (ROF) in mobile communication*. Paper presented at the Asia-Pacific Conference on Applied Electromagnetics (APACE). APACE Press.

Akan, O. B., Karli, O. B., & Ergul, O. (2009). Cognitive radio sensor networks. *IEEE Network*, *23*(4), 34–40. doi:10.1109/MNET.2009.5191144

Akyildiz, I. F., Lee, W.-Y., & Chowdhury, K. R. (2009). CRAHNs: Cognitive radio ad hoc networks. *Ad Hoc Networks*, *7*(5), 810–836. doi:10.1016/j.adhoc.2009.01.001

Akyildiz, I. F., Lee, W.-Y., Vuran, M. C., & Mohanty, S. (2008). A survey on spectrum management in cognitive radio networks. *IEEE Communications Magazine*, *46*(4), 40–48. doi:10.1109/MCOM.2008.4481339

Al-Dulaimi, A., Al-Rubaye, S., & Al-Raweshidy, H. S. (2010a). Renovate cognitive networks under spectrum unavailability. In *Proceedings of 6th Advanced International Conference on Telecommunications*, (pp. 41–45). Spain: IEEE Press.

Al-Dulaimi, A., Al-Rubaye, S., Al-Raweshidy, H. S., Cosmas, J., & Loo, K. K. (2010b). Cognitive mesh networks: Cognitive radio over fibre for microcells applications. *IEEE Vehicular Technology Magazine.*, *5*(3), 54–60. doi:10.1109/MVT.2010.938272

Al-Dulaimi, A., Al-Rubaye, S., Al-Raweshidy, H. S., Cosmas, J., & Loo, K. K. (2010c). Multi-zones of cognitive to fibre networks. In *Proceedings of IEEE/IET 7th International Symposium on Communications Systems, Networks, and Digital Signal Processing (CSNDSP)*, (pp. 697 – 700). Newcastle upon Tyne, UK: IEEE Press.

Al-Dulaimi, A., & Cosmas, J. (2010d). Mobility management for microcells based cognitive radio over fibre networks. In *Proceedings of World Wireless Research Forum (WWRF#25) Meeting*. London, UK: WWRF Press.

Al-Raweshidy, H. S., & Komaki, S. (2002). *Radio over fiber technologies for mobile communication networks*. Boston, MA: Artech House.

Arslan, H. (Ed.). (2007). *Cognitive radio, software defined radio, and adaptive wireless systems*. Dordrecht, The Netherlands: Springer. doi:10.1007/978-1-4020-5542-3

Attar, A., Li, H., & Leung, V. (2010). Increasing TCP throughput and fairness in cognitive WLAN over fibre. In *Proceedings of 24th IEEE International Conference on Advanced Information Networking and Applications*, (pp. 1069-1076). Perth, WA: IEEE Press.

Bogenfeld, E., & Gaspard, I. (2008). *Self-x in radio access networks*. A White Paper by the EU-FP7 project End-to-End Efficiency (E3). CORDIS.

Boufidis, Z., Alonistioti, N., Feuillette, R., Holland, O., Pan, J., & Moessner, K. (2007). *End-to-end network architecture for cognitive reconfigurable mobile systems*. FP7 End-to-End Reconfigurability II (E2R II) White Paper. CORDIS.

Brik, V., Rozner, E., Banerjee, S., & Bahl, P. (2005). DSAP: A protocol for coordinated spectrum access. Paper presented at the IEEE International Symposium on New Frontiers in Dynamic Spectrum Access Networks (DySPAN). IEEE Press.

Chen, D., Zhang, Q., & Jia, W. (2008). Aggregation aware spectrum assignment in cognitive ad-hoc networks. In the *3rd International ICST Conference on Cognitive Radio Oriented Wireless Networks (CrownCom)*, (pp. 1–6). IEEE Press.

Chen, K.-C., & Prasad, R. (2009). *Cognitive radio networks*. Chichester, UK: John Wiley & Sons Ltd.doi:10.1002/9780470742020

Chung, Y., & Tsai, Z. (2010). Performance evaluation of dynamic spectrum sharing for two wireless communication networks. *IET Communications*, *4*(4), 452–462. doi:10.1049/iet-com.2009.0314

Damljanovic, Z. (2009). Mobility management strategies in heterogeneous cognitive radio networks. *Journal of Network and Systems Management*, *18*(1), 4–22. doi:10.1007/s10922-009-9146-0

ETSI. (2010). *Reconfigurable radio systems (RRS): Cognitive radio system concept. European Telecommunications Standards Institute (ETSI), (Tech. Rep. TR 102 802, V1.1.1)*. Sophia Antipolis, France: Cedex.

Gavrilovska, L., & Atanasovski, V. (2007). Interoperability in future wireless communications systems: A roadmap to 4G. *Microwave Review*, *13*(1), 19–28.

Hamanaka, T., Wang, X., Wada, N., & Kitayama, K.-I. (2007). Compound data rate and data-rate-flexible 622 Mb/s–10 Gb/s OCDMA experiments using 511-chip SSFBG and cascaded SHG-DFG-based PPLN waveguide optical thresholder. *IEEE Journal on Selected Topics in Quantum Electronics*, *13*(5), 1516–1521. doi:10.1109/JSTQE.2007.904082

Huschke, J., & Leaves, P. (2003). *Dynamic spectrum allocation algorithm including results of DSA performance simulations. DRiVE Deliverable D09*. Information Society Technologies.

Jia, J., Zhang, Q., & Shen, X. (2008). HC-MAC: A hardware-constrained cognitive MAC for efficient spectrum management. *IEEE Journal on Selected Areas in Communications*, *26*(1), 106–117. doi:10.1109/JSAC.2008.080110

Jia, Z., Yu, J., Ellinas, G., & Chang, G. K. (2007). Key enabling technologies for optical-wireless networks: Optical millimeter-wave generation, wavelength reuse, and architecture. *Journal of Lightwave Technology*, *25*(11), 3452–3471. doi:10.1109/JLT.2007.909201

Komali, R. S., Thomas, R. W., DaSilva, L. A., & MacKenzie, A. B. (2010). The price of ignorance: Distributed topology control in cognitive networks. *IEEE Transactions on Wireless Communications*, *9*(4), 1434–1445. doi:10.1109/TWC.2010.04.090400

Leaves, P., Moessner, K., Tafazolli, R., Grandblaise, D., Bourse, D., Tonjes, R., & Breveglieri, M. (2004). Dynamic spectrum allocation in composite reconfigurable wireless networks. *IEEE Communications Magazine*, *42*(5), 72–81. doi:10.1109/MCOM.2004.1299346

Lucarz, F., & Seeds, A. (2009). *Gigabit/s wireless over fibre systems for future access networks*. Retrieved September 29, 2010, from http://www.ee.ucl.ac.uk/lcs/previous/LCS2005/44.pdf

MacKenzie, A., Reed, J., Athanas, P., Bostian, C., Buehrer, R., & DaSilva, L. … da Silva, C. (2009). Cognitive radio and networking research at Virginia Tech. *Proceedings of the IEEE, 97*(4), 660-688.

Mahmoud, Q. H. (Ed.). (2007). *Cognitive networks: Towards self-aware networks*. West Sussex, UK: John Wiley & Sons Ltd.

Mangold, S., Habetha, J., Choi, S., & Ngo, C. (2001). Coexistence and interworking of IEEE 802.11a and ETSI BRAN HiperLAN/2 in MultiHop scenarios. In *Proceedings of IEEE 3rd Workshop in Wireless Local Area Networks*, (pp. 27-28). Boston, MA: IEEE Press.

Mitola, J. III. (2006). *Cognitive radio architecture: The engineering foundations of radio XML*. Hoboken, NJ: John Wiley & Sons Inc. doi:10.1002/0471773735

Neel, J. O. (2006). *Analysis and design of cognitive radio networks and distributed radio resource management algorithms*. Unpublished doctoral dissertation. Blacksburg, VA: Virginia Polytechnic Institute and State University.

Niyato, D., & Hossain, E. (2009). Cognitive radio for next-generation wireless networks: An approach to opportunistic channel selection in IEEE 802.11-based wireless mesh. *IEEE Wireless Communications*, *16*(1), 46–54. doi:10.1109/MWC.2009.4804368

Nolan, K. E., & Doyle, L. E. (2006). Principles of cognitive network teamwork. *Paper presented at the Software-Defined Radio (SDR) Forum Technical Conference*. Orlando, FL: ACM Press.

Panagiotis, D. (2010). Opportunistic networks and cognitive management systems for efficient application provision in the future Internet (OneFIT). *EU-FP7 project*. Retrieved September 29, 2010, from http://www.ict-onefit.eu/

POSTnote. (2007). *Next generation telecoms networks. Parliamentary Office of Science and Technology, Number 296*. Next Generation Networks.

Rohokale, V., Kulkarni, N., Prasad, N., & Cornean, H. (2010). Cooperative opportunistic large array approach for cognitive radio networks. *Paper presented in the 8th International Conference on Communications (COMM)*, (pp. 513–516). Bucharest, Romania: IEEE Press.

Subramanian, A. P., Gupta, H., Das, S. R., & Buddhikot, M. M. (2007). *Fast spectrum allocation in coordinated dynamic spectrum access based cellular networks*. Paper presented at New Frontiers in Dynamic Spectrum Access Networks (DySPAN). IEEE Press.

Toloo, M., & Mouftah, H. T. (1995). Speed considerations and handover mechanisms in mobile personal communication systems. In *Proceedings of the 6th IEEE International Symposium on Personal, Indoor, and Mobile Radio Communications (PMIRC'95)*, (pp. 841-845). Toronto, Canada: IEEE Press.

Watanabe, K., & Ishibashi, K. (2007). Performance of cognitive radio technologies in the presence of primary radio systems. In *Proceedings of 18th Annual IEEE International Symposium on Personal, Indoor, and Mobile Radio Communications (PIMRC'07)*, (pp. 1-5). Toronto, Canada: IEEE Press.

Yuan, Y., Bahl, P., Chandra, R., Chou, P., Ferrell, J., & Moscibroda, T. … Wu, Y. (2007). KNOWS: Kognitiv networking over white spaces. *IEEE DySPAN*, (pp. 416–427). IEEE Press.

Zhang, Q., Smit, G. J. M., Smit, L. T., Kokkeler, A., Hoeksema, F. W., & Heskamp, M. (2005). A reconfigurable platform for cognitive radio. In *Proceedings of 2nd International Conference on Mobile Technology, Applications, and Systems*, (pp. 1-5). Guangzhou, China: IEEE Press.

Chapter 14
Improving of QoS in WiFi Access Networks

Domingo Marrero
University of Las Palmas of Gran Canaria, Spain

Elsa Macías
University of Las Palmas of Gran Canaria, Spain

Alvaro Suárez
University of Las Palmas of Gran Canaria, Spain

ABSTRACT

A basic WiFi access network provides a poor connectivity and does not support acceptable Quality of Service. The standard IEEE 802.11e includes some Medium Access Control improvements. IEEE 802.11n includes some improvements alone and in combination with Wireless Media Extensions but they are not available in the most current mobile devices and do not provide in practice improved Quality of Service. Those facts provoke the user experiments multimedia service disruptions and a poor Quality of Experience. The mitigation of the adverse effects of these disruptions will be of particular interest in the coming years due to massive access to multimedia information stored in Internet servers.

The aim of this chapter is to overview some problems of WiFi access networks, to explain some solutions, and to present a combined tool (that implements these solutions) to allow the user access to multimedia information efficiently. The combined tool allows: Traffic classification to make heterogeneous bandwidth reservation, the management of the dynamic association of static mobile devices to WiFi Access Points to better balance their load sharing (number of associated mobile devices), and the management of Access Points' load sharing, guiding mobile devices to another position so that they may be associated to a low load sharing Access Point. The tool is an application's level software based on manager/agent architecture. The agents manage link level's information picked up in the mobile device and the manager regulates the access to the Internet in the Access Points. There are some works that implement solutions for the problems reviewed in this chapter, but there are not tools like the one presented here. The chapter's main contribution is the combination of these solutions to practical case studies. The experimental results presented in this chapter demonstrate that the tool improves the QoS of WiFi access networks.

DOI: 10.4018/978-1-4666-0017-1.ch014

INTRODUCTION

Nowadays, research in wireless networks is a priority in communication networks, technologies and services research areas. The traffic generated at present by home and enterprise wireless networks represents the bigger amount of traffic injected in the Fixed Internet Backbone. Moreover, the number of mobile devices grows geometrically every year. There are also an increasing growth of users and kinds of devices that use successfully wireless networks technologies. Services like mobile *Voice over Internet Protocol* (*VoIP*), *Hypertext Transfer Protocol* (*HTTP*) and *Simple Mail Transfer Protocol* (*SMTP*) are massively used. New multimedia services are a reality due to the use of mobile devices (Portable *Personal Computer* [*PC*], *Personal Digital Assistant* [*PDA*], tablets, Smartphone, e-readers...) provided with *Wireless Fidelity* (*WiFi*) and *Third Generation* (*3G*) technologies.

A mobile device in a WiFi access network can access Internet using a typical *Access Point* (*AP*), Tethering or a PC with two network interfaces (WiFi and Ethernet) in which a free operating system (hence referred to as *Linux Wireless Router* (*LWR*)) is installed. The *Service Set IDentifier* (*SSID*) of the WiFi network is defined by the AP or the LWR which identifies a particular work frequency (channel) and a method of authentication. Each mobile device that wants to use the WiFi network must be associated to that SSID. After the association process, the AP or LWR will have a record of all their associated mobile devices. Mobile devices in modern *Wireless Spontaneous Networks* (*WSN*) (Silva & Salgado, 2006) use Tethering to access Internet. In this chapter we will present some experimental results using LWRs.

The *Quality of Service* (*QoS*) of the WiFi access networks is low (Sobrinho, 1999). Different parameters affect QoS: reduced bandwidth, radio interference, limited number of available channels, overhead of security traffic and saturated spectrum... (*Jangeun, 2003*). The QoS is also affected by the *Medium Access Control* (*MAC*) (IEEE, 1999) which does not limit the amount of time of transmission of mobile devices. Therefore, the only traffic class WiFi access networks support is *best effort* which imposes severe restrictions to multimedia communication. WiFi has been improved along the last years, for example: a) *International Engineering Electric and Electronic* (*IEEE*) 802.11n takes advantage of radio reflections, fading and multipath radio signals using *Multiple Input Multiple Output* (*MIMO*) technology. b) Compliant IEEE 802.11e (Xiao, 2004; Boggia, 2004) APs add changes in the waiting time among mobile devices and introduce queues to classify and prioritize data packets (Qiang, 2005). *Vinnakote* (2006) proposes a new MAC which defines traffic classes: voice, video, best effort and prioritization of packets. The adoption of these solutions by the Market is still reduced. This is in part due to the full adoption of solutions introduces changes in the firmware or driver of the mobile device WiFi interface card. But these changes do not guarantee the compatibility with previous WiFi standards or mobile device's firmware or driver versions. The QoS of multimedia communication is typically measured combining parameters like speed, throughput, latency and jitter. The user experience is affected by aspects like automatic re-association of mobile devices to APs (Rodrigues, 2009), efficient control of APs' load sharing (Toh, 2002) and adverse effects of service disruptions.

There are several practical challenges in indoors WiFi access networks to which researchers and WiFi distributors pay little attention: a) Normally, a mobile device associates to an AP with the best *Received Signal Strength Indicator* (*RSSI*) (either manually or automatically). But this is not always the best association due to it can affect adversely the APs' load sharing and probably it can provoke interferences due to WiFi link is saturated. b) APs are not optimally allocated: they use to overlap their range areas and also their WiFi channels. In places like hotels, touristic resorts, airports, shopping centers, avenues, parks ... the APs limit the number of users and services can access to Internet. In some cases, security is turned off making the network vulnerable to attacks. c) A worst problem is unauthorized users could flood the WiFi channel. The consequence of these analyzed cases is service disruptions that annoy the users which normally abandon the Multimedia sessions. Solving these practical challenges is interesting due to hot spots loss money if the users abandon those sessions.

We propose a novel combination of well known techniques in order to solve some of the above problems: Firstly, the definition of service classes and the usage of an admission control policy (our novel proposal is to implement this policy in the mobile device). Secondly, the implementation of an efficient re-association of static mobile devices to available APs not only considering the RSSI level as is usually done. Thirdly, the implementation of an efficient re-association of moving mobile devices to APs using localization based on a coverage map. As a result of the combination of these three techniques we have implemented a combined tool that improves the Internet access in indoors WiFi networks. Its manager/agent software based architecture is able to obtain updated information about the temporal evolution of the WiFi link querying the APs. Then it distributes this information among APs and mobile devices in order them to limit the traffic they send to the WiFi channel (upstream traffic). In this way, we try to minimize service disruptions. We do not alter the MAC of mobile devices (in this way our tool is transparent to any WiFi device) and also we do not alter their particular advantages (for example the IEEE 802.11n, *Wireless Multimedia Extensions* [*WMM*] or IEEE 802.11e). The way our tool uses RSSI is independent on the problems with different WiFi interface cards vendors because we have parameterized typical cards and drivers.

This chapter is structured in six sections. In Section 2 we overview different standard techniques and our solution to improve the QoS of WiFi access networks for multimedia communication. In the Section 3 we present our combined tool that implements the above solution. In Section 4, the experimental results are described which were obtained for each one of the proposed parts of the tool in a testbed. In Section 5 we present future research work. Lastly, in Section 6 we sum up some conclusions.

NOVEL COMBINATION OF STANDARD TECHNIQUES

We review three well known techniques we have introduced in Section 1. We will present related works, the applicability of a formal model and some key features for our integrated solution.

Regulation of Traffic in the Mobile Device

In some APs, a mobile device must pass an admission control policy to access Internet multimedia services as explained by Prihandoko (2003). We do not consider typical proposals like Fang (2002) at MAC level in order not to introduce changes in the WiFi MAC. We have designed a flexible admission control policy that can learn by itself new behaviors of mobile devices. This policy runs at application level of Internet network architecture. The users must register the kind of traffic they will send in a captive web portal running in the AP. The AP uses this traffic segmentation to anticipate the reservation of the WiFi channel efficiently. Starsky (2006) proposed that an AP must regulate the maximum amount of traffic the mobile devices are authorized to send to the WiFi link. This regulation process demands an active control of the AP over the WiFi link and changes in its firmware. We propose the AP to assign the mobile devices the maximum bandwidth they can use in the WiFi channel. The mobile devices regulate itself the maximum amount of traffic they send to the WiFi link using this assignation of the

AP. The AP rejects the traffic of mobile devices in case the WiFi channel would be saturated. In the Figure 1 is shown the basic idea of our traffic regulation: *AP1* can assign the total amount of available bandwidth to the mobile device 1, but the *AP2* must share the bandwidth between mobile device 2 and 3.

A simple model of the maximum throughput *T* (the idea is only to formulate an abstract model but neither to compute an exact solution nor to specify a complete model) forwarded by an AP, taking into account *n* sessions (Web, video, SMTP, VoIP…) is:

$$T = F_1 + F_2 + \ldots + F_n \quad (1)$$

Where T represents the maximum pre-estimated throughput (Mbps) and F_1, F_2, ..., F_n are the different flows' throughput (we use interchangeably session and flow due we consider that only one flow of data packets is associated to a single session) at a given instant of time.

Assuming each flow's throughput has the form:

$$F_i = \alpha_i T \quad (2)$$

Where unknown α_i (α_i € [0, 1]) represents the *i* coefficient of reduction of flow traffic's volume for mobile device *i*, or the maximum speed permitted for F_i. The α_i coefficients can be calculated by the AP limiting the amount of traffic permitted per F_i (traffic classes that can be prioritized by a software in the AP or LWR will be discussed in Section 3). The mobile device receives its corresponding coefficient and limits the traffic it sends to the WiFi channel. Let us note that this is independent of IEEE 802.11 standards. That is, this traffic regulation technique can be applied to all WiFi standards. On the other hand, this is a complementary method to queues defined by IEEE 802.11e or WMM used in IEEE 802.11n. For example, given three flows F_1, F_2, F_3 (or three different sessions), and T equals 25 Mbps (a common value in IEEE 802.11g), α_1 could be 0,5, α_2 could be 0,25 and α_3 could be 0,25. In this case, the flow F_1 will be 12,5 Mbps, F_2 and F_3 will be 6,25 Mbps.

Figure 1. Regulation of traffic in the mobile devices

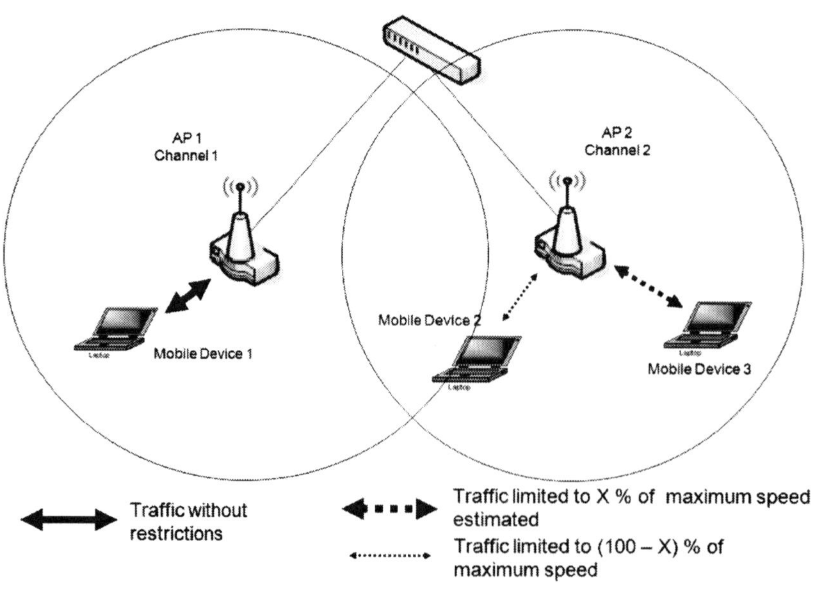

Re-Association of Static Mobile Devices to LWRs

A re-association process in a WiFi network means a mobile device is disconnected from an AP and will be connected (after some time) to a different AP (normally when a better RSSI level is detected). Typical roaming process (Liu, 2010) implies a re-association process (the re-association is a primitive of roaming protocols). We have considered a special case in which a static mobile device would be re-associated (hence referred as *static roaming*) not only considering the best RSSI level but the APs' load sharing among other parameters. This is not a simple traffic balancing process among the ports of a router; the APs control the status of the WiFi link and know the traffic assigned to mobile devices (α_i). Taking into account these parameters, the available APs self-coordinate the best balancing of the mobile devices association to APs. In Figure 2 is shown a possible scenario to explain this. The APs know their channels, associated mobile devices and services statuses. The APs interchange this information among them. The mobile device 1 detects *AP1* and *AP2*, the mobile device 2 detects *AP2*. The WiFi link defined by *AP2* would be shared by both mobile devices and no mobile device uses the WiFi link defined by *AP3*. In theory, the mobile device 1 could have better performance if it will be associated to *AP1*. If the mobile device 1 is associated or starts the Internet access using *AP2*, then *AP2* will request a static roaming of the mobile device 1 to *AP1*. The *AP3* does not participate in this process due to mobile device 1 does not detect it. In this scenario we have supposed that all APs had similar characteristics and they have processed the same number of flows and kind of traffic. A final remark is that our proposal is similar to the IEEE 802.11k but we have not analyzed the mobile devices' moving to manage the static roaming to the best AP.

Let us suppose a set of *a* APs named *APSet*: $\{AP_1, AP_2, \ldots, AP_a\}$ and let be T_j the maximum estimated throughput controlled by AP_j modeled by equation 1. The objective of our static roaming is to find the best mapping of mobile devices in the *APSet*. To do this, the *APSet* is divided in groups of sets each containing the APs that overlap their coverage areas in order to find alternatives

Figure 2. Re-association of static mobile devices to LWRs

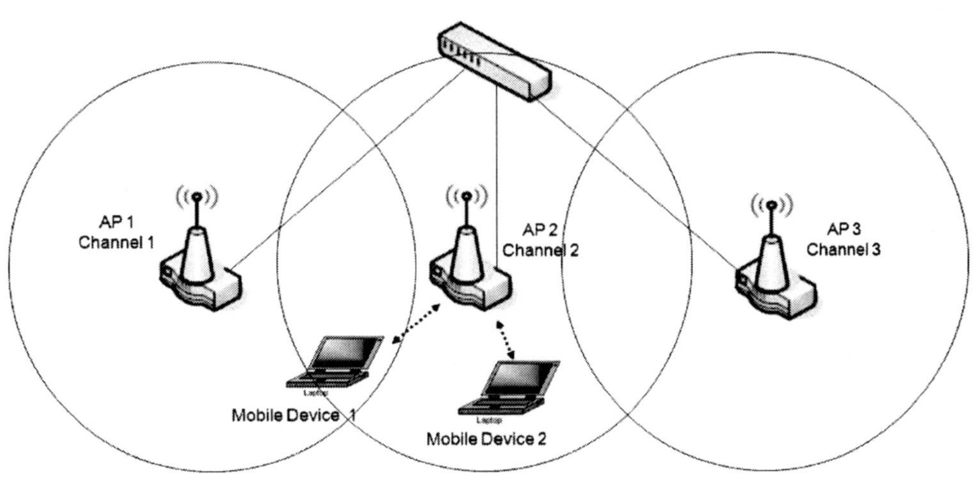

to access Internet. Then, the mapping of F_i in the *APSet* must be obtained. This mapping is a mathematic optimization problem. This is a complex problem that cannot be solved by an exact mathematic method. In general, a heuristic guided solution must be explored taking into account that this mapping varies with time. Control variables of this optimization problem are (among others): F_i, α_i, T_j and $RSSI_j$. As an example, in Figure 2, there are two groups in the *APSet*: $\{AP_1, AP_2\}$ and $\{AP_2, AP_3\}$. The best mapping of F_1 (mobile device 1), F_2 (mobile device 2), is: $F_1 \rightarrow AP1$ and $F_2 \rightarrow AP2$ independently of the value of coefficients α_1 and α_2.

Re-Association of Moving Mobile Devices to LWRs Using Localization

Normally, the WiFi users try to accommodate the QoS of their Internet connection moving the mobile device to find the best RSSI level. But not always a good result is obtained. The bad result produces the WiFi user annoys and stops the Internet access. The main idea of re-association of moving mobile devices is to transparently suggest the WiFi user a particular movement to find an appropriated coverage zone. That is, the APs localize the mobile devices and suggest them to move to a better coverage zone.

Global Positioning System (*GPS*) is an effective method of localization outdoors and it is very used in many WiFi mobile devices. Indoors walls and windows, reflections and interferences influence the localization of mobile devices which is useful in some cases (Song, 2004; Zhao, 2006). There are a lot of indoors localization methods (Akiyama, 2009; Ching, 2010). A set of RSSI values is normally used to define a coverage map for supporting efficiently the localization methods (Meng, 2008). The variability of RSSI is high in practice (it depends on several factors: wireless channel status, number of active sessions, interferences...) and it could represent a drawback. But it is the only readable coverage parameter in the WiFi interface cards. Also the location of the APs is not always correct.

We consider APs (or a Server) using a coverage map (built with RSSI values) in order to the mobile devices to scan the APs availability. Each AP calculates the position of each associated mobile device for guiding ("*guide*") it to a better location (in a better coverage area). The WiFi user receives this information and moves to a better coverage area. In general an AP could inform each associated mobile device about the best coverage area to which it could move. For example, the mobile device 2 in Figure 3 detects the availability of only *AP2* (with a very low RSSI) and receives localization information of *AP2*. Then, the user will discover the availability of *AP3* and move (guided by *AP2*) to the *AP3* coverage area.

It is important to remark that we neither are concern with a typical roaming process nor with an exact localization method. Roaming process implies a complex procedure that includes our localization information steps. We do not need an exact localization method; it is enough to indicate an approximated localization and to guide to the WiFi user to better coverage localization (maximizing the probability to access the Internet multimedia service with better conditions). Taking this into account we can formulate a new optimization problem considering localization as new control variable.

Let L_r be the localization of mobile device *r*, the main objective is to inform about neighboring localizations of L_r to the WiFi user can move to maintain or improve the mapping of the *n* F_i in the *APSet*. A second objective is avoiding the mobile device will go out of coverage. Let us take into account that in this case the solutions space is very big and finding the optimal mathematical solution is a hard complex problem. For this reason, a heuristic guided solution must be explored.

Figure 3. Re-association of mobile devices to LWRs using localization improved with a coverage map of RSSI values

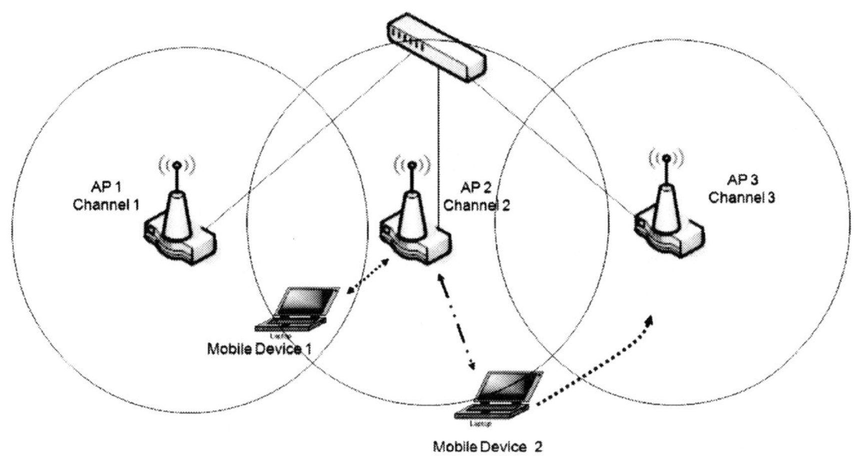

AN ALGORITHMIC INTEGRATED SOLUTION

A mathematical exact solution for an optimization problem that integrates those earlier identified is very hard or impossible. A heuristic solution can be formulated and an algorithmic solution to the above optimization problem can be specified programming the concurrent behavior of software agents (Shoham & Leyton-Brown, 2009). The main advantage is that a rapid solution can be obtained and agents can change their behavior to consider the time variability of the control parameters. Moreover, it is relatively easy to include a new control parameter. This is the reason we consider manager/agent architecture to specify an algorithmic solution (Figure 4).

For traffic regulation, an agent regulates the flow in each mobile device and it is always active. One manager is implemented in each LWR. For the admission control policy, it is necessary the user indicates the class of service and the requirements of bandwidth of the service (F_i). With this information, the manager calculates how this traffic affects to other incoming (F_j, $i \neq j$) or outgoing flows in the WiFi links. Also, the manager checks the overall availability of bandwidth and the possibility to reject flows. The manager's principal behavior is to test the WiFi channel, control the number of mobile devices connected and the services they request. Then, the manager in each LWR distributes the estimated available bandwidth between their associated mobile devices (T_i); it will communicate to each agent which regulation must apply (just in case). This bandwidth distribution depends on time (it is modified along the time) and it is optimized to parameterize different cases: number of sessions, WiFi channel status and traffic class. The communication between agent and manager uses *User Datagram Protocol* (*UDP*) to avoid flooding the WiFi link. Manager and agents communicate messages of control: activate, increase, reduce and eliminate the need of traffic regulation.

A regulation process is followed in order to adapt the solution to the environmental conditions. A sniffer detects the possibility an agent violates its declared request of bandwidth. In that case, it will communicate that event to the manager, it will lock the connection and it will send a warn-

Figure 4. The manager/agent architecture to obtain an integrated solution

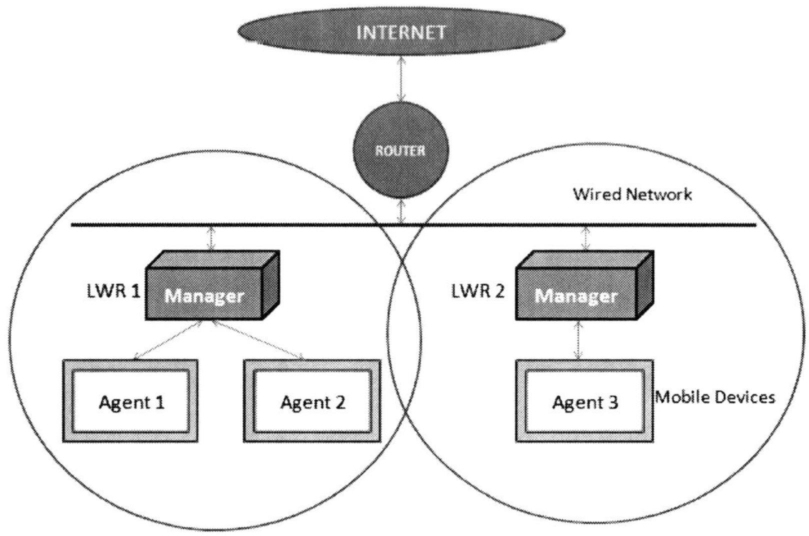

ing message to the agent. The admission lasts t s. After this t s, the access will be locked and the agent will be notified. Other events that provoke the manager locks a connection immediately are: the agent is not active, stopped or disassociated.

The mapping of static mobile devices to *APSet* is done by a three phases' process: *Initiation*, *Commitment* and *Execution*. In this process, the managers share information (using wired distribution network to assure no disruptions are produced and to reduce the control traffic injected in the WiFi channel) about their independent T_i. This process is regulated along time constantly detecting the status of the WiFi channel, using timers and doing polling, and controlling associations and disassociations of mobile devices. Periodically, the managers broadcast announcements including related information such as identification of mobile devices associated to each of them (and traffic classes F_i), T_i, their WiFi channel status, mobile devices' characteristics... Simultaneously mobile devices detect (using RSSI) several APs to which are associated. Among APs, a negotiation about the best mapping of the $n'F_i$ in the *AP'Set* is achieved (where $n' \leq n$ and the $AP'Set \subseteq APSet$). Let us note that the determination of the best static roaming is achieved by the APs (that is slightly different than in normal roaming).

Each agent informs periodically the manager about its static localization (L_r). The manager uses the *Data Base* (*DB*) of RSSI values and the current received RSSI to determine the mobile device's localization. The precision of this localization is high if three or more APs are implied but low if less than three are implied. To mitigate this problem a number we name "*grade of precision*" which defines the probability of success is used. It indicates how much reliable is the estimated L_r. The higher this number is the more accurate results are. In general, if the mobile device detects only one AP, the precision will be very low; on the contrary if there are many APs, the precision will be higher. If the manager cannot determine the mobile device's localization it will ask the agent to do a new scanning process (discovering its localization). Scanning could be done frequently in the following cases: the RSSI variability is high; the current received RSSI is not in the expected range (related to those stored in the DB). All these incidences are transmitted to the agent and then they will be shown to the user on the mobile device. The objective is the user can do corrective actions to change its location optimizing the mapping of the $n'F_i$ in the *AP'Set*.

We have not specified neither exact nor optimal solution. But, as we will present in the next sections, tuning the global regulation process an acceptable solution can be found.

THE COMBINED TOOL DESCRIPTION

We have implemented a combined tool that integrates the solution based in manager/agents. Our tool is an application level software that receives information from the low levels of Internet architecture and also from the WiFi driver (in LWRs and mobile devices) as shown in Figure 5. In this Figure we have numbered the information that the tool receives from the different levels of network architecture:

1. The tool interacts directly with the applications. That is, the tool receives input flows of typical Internet basic applications (SMTP, *File Transfer protocol* (*FTP*), video applications...). Indirectly, the tool limits the amount of information associated to the flow (F_i) the application sends to Internet applying the results of the computation of coefficients α_i of *equation 1*. Let us note that the tool does not modify the internal behavior of applications that access Internet using our tool.
2. The tool sets new sessions (ports) for establishing channels of control among the mobile devices, APs and servers. It also applies changes to the default configuration of *Transmission Control Protocol* (*TCP*) and *UDP*.
3. The tool collects information at network layer to optimize and manage the traffic in the wired and the WiFi networks using *Internet Protocol* (*IP*) family protocols like *Internet Control Message Protocol* (*ICMP*).

Figure 5. Architecture of communication and a vertical control plane to implement our combined tool

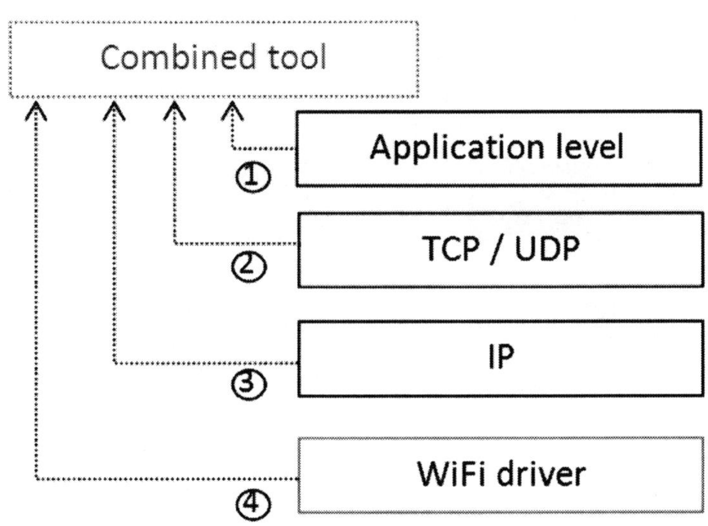

4. The tool accesses information of WiFi driver. It is important to remark we are neither interested in modifying nor accessing special parameters of particular MAC. This is because one of our design objectives is to apply our tool to any WiFi version.

Figure 6 shows the main modules of the combined tool. In the mobile device the user interface consists in the access to a *portal captive* (browser). In the LWR the tool uses a standard Web server, the manager and the sniffer.

Like traditional hotspot's WiFi admission control software, the portal captive requests to the user its identification: user and password. Our portal captive (that uses secure HTTP) also requests to the user additional information about the Internet access session. A user can declare only one kind of session, for example a FTP session, indicating the port to be used and necessary bandwidth (bits or packets per second, that is, the F_i). The user can activate, run and deactivate the session, but the user cannot issue two sessions concurrently. Once the session is declared the tool stores TCP/UDP ports for each session and user. The portal captive communicates these parameters to the manager. The manager can then estimate the available global bandwidth (T) in order to determine if it will allow the session or not. The manager can be configured to prioritize traffic classes; in general from high to low priorities the tool is initially configured to prioritize: voice, video, chat, FTP, Web, mail and others. The tool initially supposes the user specify traffic classes correctly. During the regulation process, the sniffer will check this is correct observing the amount of traffic the user sends. The manager can decline the admission control request (session rejected) or it can feed back the agent or feedback it with the final value of F_i. Moreover, the manager informs the agent about the WiFi channel status information, amount of time it will maintain the session activated and other administrative information. When this time finishes the manager will lock the session and will inform the agent.

The agents broadcast information about the number of mobile devices associated to them. Then, all agents broadcast their T_i and maximum bandwidth they can use in the particular WiFi

Figure 6. Main software modules of the combined tool

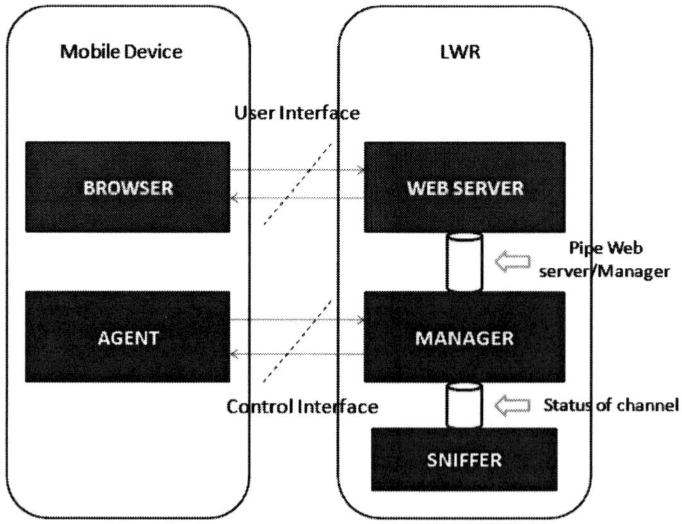

coverage area of their LWRs. An agent with a saturated T_i can request to an agent with available bandwidth to start the re-association of mobile devices. The new *APSet* is shared by all the agents. Let us note that we have to take into account several practical problems: a) a mobile device can change its IP address when it will be re-associated to a new LWR. We have not considered this case. b) Negotiation among agents to find the new *APSet* could take considerable amount of time and could reduce throughput. This depends on the number of mobile devices (n) (remember they only request one F_i), and the number of LWRs (a). The scalability of n could be done allocating the LWRs in different wired network segments. The scalability of the number of LWRs (a) is assured by our traffic regulation in the mobile device (the tool limits the amount of devices associated to LWRs). We consider only one wired network segment with less than 8 LWRs that is enough to cover a large zone (building or a large floor in a building). With these assumptions the amount of control traffic is low as shown in the experimental results. Moreover, this traffic is communicated in the wired network: The WiFi links only must support a few amount of association messages. c) The mobile device could be forced to start again its F_i. This can happen if it will change its IP address; but we do not consider that case.

In Figure 7 we show a sequence of actions achieved by two LWRs to implement static roaming.

When a mobile device is moving indoor at low speed it could experiment service disruptions (it could reach an out of coverage or very poor coverage zone). In this case the WiFi user can initiate a process of localization in order to discover its localization. The agent requests the manager's localization information. The manager first requests the agent to detect (scanning) the LWRs and others APs (using RSSI levels). The agent sends feedback to the manager with information like SSID, RSSI levels, channel's numbering… The manager seeks the received values in its DB and sends the estimated position and the *grade of precision* to the agent. Additionally, it could in-

Figure 7. Sequence of actions to realize static roaming

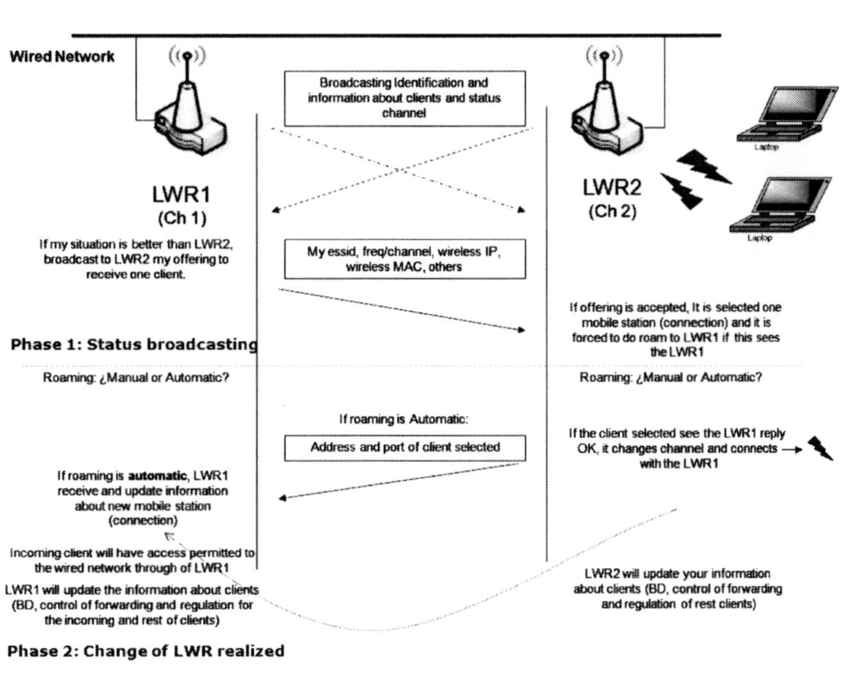

clude information about other LWRs, indication about a better localization, the need to make a particular movement, a coverage map with the localization of LWRs... The implemented localization method uses a fingerprinting method. In Figure 8 we present this process for a LWR. To cope with different definitions of RSSI the manufacturers do, the manager includes information about the manufacturer in the DB (it is made during the phase known as survey). The manager filters these RSSI values in order to systematically learn from the measurements done in different conditions (with obstacles, with different drivers, operating systems…). The *grade of precision* is updated with this filtering process. An interesting tradeoff is the time spent in scanning process (for all WiFi channels) and further localization. It is true that the scanning process could provoke insignificant service disruptions only in the case the buffer of the agent has not enough video frames (other best effort traffic will not experiment any significant problem). But it is also true that if the mobile device does not start the localization process it will be out of coverage and will experiment a service disruption. The further localization implies the manager to match RSSI values in the DB and to forward F_i from the WiFi to the Ethernet interfaces. The DB's queries and the forwarding are done in parallel by the hardware of the LWR. For this reason, it is not a significant problem. Moreover, if the mobile user issues the scanning and the localization process in a proactive manner, this mechanism will be efficient: service disruptions will be minimized and WiFi links will not be saturated with traffic which will not be received by the mobile device.

We have implemented the sniffer as an operating system's process that detects other mobile devices transmitting in the same frequency channel that affects the performance of the rest of mobile devices. Let us note that it can detect mobile devices controlled or not by the manager (the latter have not been authenticated in the portal captive). For controlled mobile devices, the manager could force them to change the WiFi frequency channel in case they are violating their declared bandwidth usage. In the case of uncontrolled mobile devices, it is impossible to avoid

Figure 8. Sequence of actions for moving mobile devices re-association

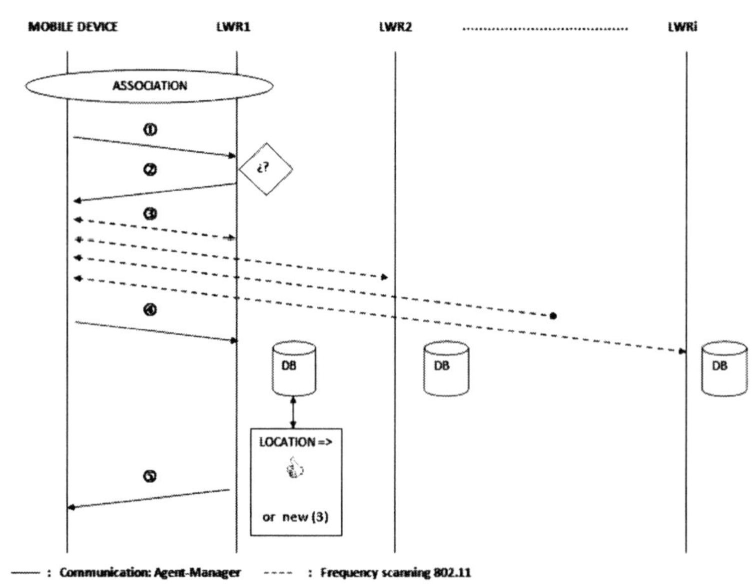

they continue transmitting in the same WiFi channel. In this case, the manager will force the registered mobile devices to change to a not saturated WiFi channel. The sniffer and the manager run concurrently and use internal operating system's inter-process communications. The process of forcing the change of WiFi channel could produce service disruptions of active sessions. For this reason, the manager must take this into account.

EXPERIMENTAL RESULTS

We have defined a testing platform for testing and applying our tool. The system software election is important. We have considered the choice of several possible platforms:

- **Router WLAN:** A Linux kernel *DD-WRT or OpenWRT* valid for Linksys router, Buffalo and others.
- **HostAP:** A software solution for converting a PC in an AP.
- **A LWR** based in any Linux Distribution (for example Ubuntu).

We have chosen the last option. The system calls and all the possibilities of open free source Linux have allowed us to access to the WiFi and Ethernet interfaces to implement: forwarding, filtering, controlling, accounting… The LWR has the following advantages: a) Use of software applications in native code, drivers and Linux kernel, b) the availability of utilities for network testing, c) the LWR can work like an AP but also like a PC (multiuser and multithread) managing several kinds of servers: Web, DHCP, proxies, email, RADIUS… and standard wireless tools while accessing the Linux kernel (*/proc*), d) easy configuration of *IP* and management of physical layer. Moreover, any old PC or new embedded PC can act as an LWR.

Two different topologies were tested:

- The mobile devices only communicate to LWR that allows the access to Internet (infrastructure mode).
- The mobile devices communicate among them (ad hoc mode) and also among them and the LWR (mixed mode).

We have tested both cases of operation modes to compare and evaluate both results. The observed performances are practically the same according to our objectives.

We have introduced new sophistication to our testing platform using two LWRs (*LWR1* and *LWR2*) and two mobile devices (portable PCs named as *M1* and *M2*). The LWRs were connected to the same wired IP subnet (switch/hub). Each of them defines a cell with the same SSID and different frequency/channel. The Figure 9 shows a diagram for this testing platform with all the resources and all the points or segments where we have done experimental tests to measure the performance of our tool. We supposed two servers (*S1* and *S2*) connected to a wired network. We have named some special traffic paths in Figure 9. For the wired segment, *TC1* (wired between *LWR1* and *LWR2*) was used to interchange information between the LWRs: basically the connectivity quality of the mobile devices in the both cells defined by the LWRs. For the WiFi links in infrastructure mode we had three paths (let us note that M1 was not in the coverage area of *LWR2*): *TI1* (between *M1* and *LWR1*), *TI2* (between *M2* and *LWR1*), *TI3* (between *M2* and *LWR2*). We have supposed *LWR1* was the only one that access to *S1* and *S2*. The two paths crossing the infrastructure links and the wired links were *TC2* and *TC3* for *S1* and *S2* respectively. Depending on the source and destination (M1, M2, S1 or S2) it could be necessary to use several network segments or links (wired or wireless). These links determined the global behavior of communication (extreme-to-extreme). We have considered that to define specific mechanisms in each link was better than extreme-to-extreme. Each link had different characteristics and problems. As an example, it has been observed that in the infrastructure mode, the traffic between M1 and M2 must cross *LWR*1. This generated two packets in the channel and the reduction of the efficiency. The path named TC2 was used for communicating M1 or M2 with S1. In this case, LWR1 could reduce the traffic in the wired network (link between LWR1 and the Switch) stopping the accessing of M1 or M2 to S1. Another way to reduce the traffic in that wired link was to force the association of M2 to LWR2.

We have used Linux Distribution Ubuntu 8.10 for *M1* and *M2,* IEEE 802.11g (IEEE 802.11g, 2003) of *Netgear* and *Dlink* with *Atheros* chipset (combined with interfaces *Ra-Link* of *Realtek*). In this way we could probe the compatibility between several WiFi interfaces. These different products and drivers (software modules) added problems because they were incompatibles among them because the manufactures do not use a common format to define the variables, objects and parameters. Also the information processing was very different. For this reason, we were forced to realize certain manual configuration in each case.

Experimental Tests of the Regulation of Traffic in the Mobile Device

We have obtained fine experimental results for traffic regulation assigning a high α_i coefficient to multimedia flows. We have done a considerable amount of tests and in all of the cases, the QoS of multimedia communications were improved when the rest of traffic were limited.

The probes have been done with a Pentium III PC with 2 GB RAM, *Dlink* IEEE 802.11b/g wireless interface card with *Atheros* chipset, *RealTek* Interface Ethernet, a 320 GB Hard Disk with Ubuntu Linux 8.10 Operating System acting as LWR. Also, we have used another PC: Pentium II 400 Mhz, 196 MB RAM *Personal*

Figure 9. Testing platform with links where we obtained experimental results

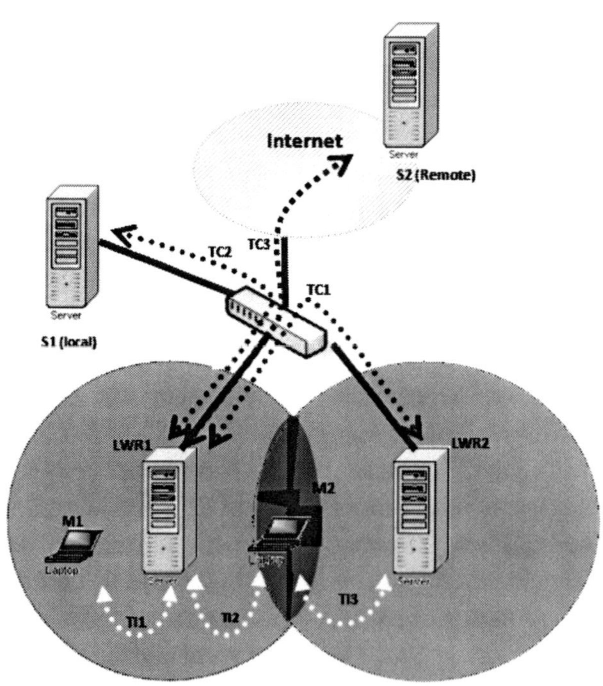

Computer Memory Card International Association (*PCMCIA*) *Compaq* IEEE 802.11b and *Fedora Core* III. In these computers were installed the managers. We have detected that the behavior was perfectly comparable with a typical AP. The mobile devices were two portables computers and one PC with Linux. The portable computers were an *Acer Aspire One Intel Atom* 1.6 Ghz, 1 GB RAM, *Atheros AR5007EG* WiFi interface card and a portable *Acer Aspire* 3 Ghz, 1 GB RAM PCMCIA *Dlink* IEEE 802.11b/g. Both used *Ubuntu 8.10 kernel 2.6.27-11*. In these mobile devices we have implemented the agents to apply the regulation with functions of traffic control of the Linux kernel (*Traffic Control* (*TC*)).

We have not considered particular properties of underlying WiFi standards for compatibility purposes. The WiFi interface card of the mobile devices that uses our tool did not modify its firmware. Also it was valid for 802.11a/b/g (IEEE 1999, 2003) because it did not include any mechanism of differentiation or traffic prioritization.

Many tests were done using tools that measure the networks performance, for example *iperf*. This application was used to inject a lot of packets in the WiFi access network. It allowed us to obtain several measurements over the capacity of the network, for example the maximum throughput. If it is jointly used with other applications, it will allow us to create many Internet access problems to those applications. We have jointly used *iperf* with traffic multimedia *Real Time Protocol* (*RTP*)/ *Real Time Streaming Protocol* (*RTSP*) with the free multimedia application named *VideoLAN Client* (*VLC*). The quality of multimedia traffic was very bad when *iperf* injected traffic. On the contrary, this was improved when *iperf* was limited with our functionality in the *LWR*.

In order to make the bandwidth distribution (assign coefficients α_i) we have calculated the available maximum bandwidth of the WiFi link without any active session. Using *iperf* we have calculated the medium value for a communication without additional sessions. Initially we defined different forms to distribute the maximum throughput. The first was to reserve 75% for multimedia traffic and 25% for other classes of traffic. Our preliminary tests were acceptable.

The empirical result according the amount of transmitted packets showed the QoS was improved. A lot of tests were done comparing communications without applying restrictions against others with restrictions (regulation). We combined many communications with different services and multimedia applications like: Web, FTP, *VLC* (video streaming) and *iperf*. We checked the manager assigned the bandwidth efficiently and the agent applied the regulation as expected. For example, the tool reduced the traffic of Web, FTP and *iperf* flows which improved the *VLC* flow.

Experimental Tests of Re-Association of Static and Moving Mobile Devices

We have used the test platform formed with two *LWR* and two portable computers to evaluate this re-association improvement as shown in Figure 5. Firstly, we have designed a simple method to classify the flows according to the number of mobile devices and services. Then, we have determined the best LWR attending to its particular conditions. The simplest situation consisted in a LWR that had not any mobile device associated and other LWR that had two mobile devices associated. In this case, only one mobile device used a multimedia service. The optimal solution could be that a mobile device did not share the WiFi channel when it issued a multimedia flow. Also, it would be interesting to group all the best effort services in the same WiFi channel.

Many tests were done satisfactorily. For example, the mobile device 1, *M1* (Figure 5), issued a video streaming session in the server *S1* and simultaneously, the mobile device 2, *M2* injected traffic with the *iperf* tool. When both mobile devices were associated to *LWR1*, the quality of video was affected (pixilation). The manager interchanged messages between LWRs and the mobile devices. Finally, it managed the re-association of *M2* to *LWR2*. After this re-association, the quality of video communication was improved because the traffic injected by *M2* did not contend with the one injected by *M1*.

The scenario for localization of moving mobile devices was a long rectangular corridor and offices on the right and on the left. There were some vertical walls in the top of the corridor that blocked the WiFi signal propagation. The width of the wall that limits the corridor with the offices is 40 cm approximately and its building's material is very dense (it reflects a lot of radio signal). We have used the two *Cisco Aironet Series 1200* APs allocated in the corridor (we have not use our LWRs). Figure 10a is a picture of the indoor and Figure 10b shows this scenario and indicates the position of the two available APs (*Cisco*). The offices 213 and 223 and their corresponding RSSI intervals for *AP1*, *AP2* and *APx* are shown in black color.

Figure 10. Indoor scenario for tests

(a)

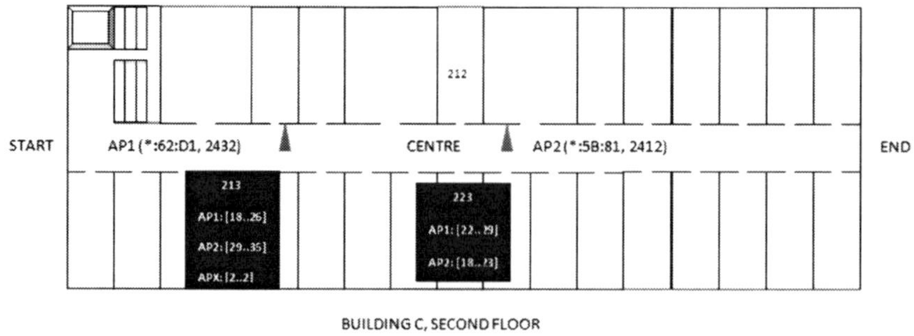

(b)

We first have made a big amount of measurements to discover problems of localization of mobile devices along the corridor and in the offices (early characterization of the indoor area). There were three distinguished facts: The covariance of different values of RSSI (in a determined indoor point) was very high (we obtained non expected results), there were shadow coverage zones inside the offices and the coverage areas of the APs were highly overlapped.

The second step we have done is theoretically characterizing the coverage areas of *AP1* and *AP2* in order to delimit the application of the localization methods. In Figure 11, we draw these areas and mark the north and south zones where a lot of signal reflections are produced. These reflections made the localization methods failed due to they produced repeated values of localization. This fact and the big overlapping area between the both APs made the exact calculation of localization were impossible. For this reason, we have decided to define the *grade of precision* metric that cannot be accurate in presence of only two APs.

The third step we have done was to build the DB of RSSI values (characterization of the coverage map for the corridor and offices). We have used an *Acer Aspire One* portable PC with *Intel Atom* 1.6 GHz. This PC was provided with an *Atheros Chipset* Wireless interface card, a *Madwifi* driver and *Ubuntu 8.10 Distribution Linux* Operating System. We have used *Linux Wireless Tools* to do the scanning (hundreds of measurements) positioning the PC in different places indoor. We have done the measurements of RSSI values at different hours of the day, different conditions (presence of users) and different services accessing the Internet in order to obtain a practical useful map of coverage. The resulting DB had a lot of out of range RSSI values, repeated values, and clearly erroneous values. A step ahead was to process these values in order to filter, group and eliminate out of range RSSI values. We could use a Kalman filter or other similar technique but it was enough to calculate the average of RSSI values for each formed group. We tested that with this new processed DB we could minimize the adverse effects of the high variability of results in different measurements. In Figure 12 we show an extract of the DB. The different columns explain: The part of the indoor area (*inicioct* for the

Figure 11. Theoretical coverage of Cisco AP1 and AP2

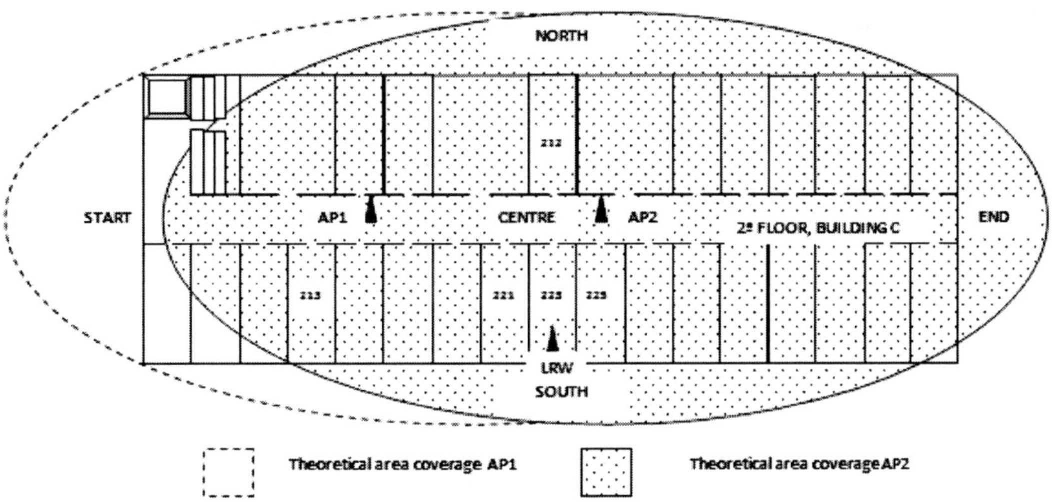

Figure 12. The extract of the DB with levels RSSI

AREA	MAC	SSID	Freq./Canal	RSSI Min, med, max	RSSI (Ref.) Min, med, max	Signal min-med-max	Noise min-med-max
....							
Inicioct	00:12:01:B5:5B:81	ULPGC	2412 1	13 21 33	70 70 70	-62 -73 -82	-95 -95 -95
Inicioct	00:12:01:B5:62:D1	ULPGC	2432 5	21 28 39	70 70 70	-56 -66 -74	-95 -95 -95
Inicioct	00:0F:24:EC:FA:41	ULPGC	2412 1	14 15 16	70 70 70	-79 -79 -81	-95 -95 -95
Inicioct	00:14:A9:75:09:11	ULPGC	2462 1	1 2 9 13	70 70 70	-82 -85 -93	-95 -95 -95
Inicioct	00:12:01:B5:61:F1	ULPGC	2442 7	32 33 36	70 70 70	-59 -61 -63	-95 -95 -95
Inicioct	00:12:01:B5:69:71	ULPGC	2422 3	23 29 38	70 70 70	-57 -65 -72	-95 -95 -95
Inicioct	00:12:01:B5:66:B1	ULPGC	2447 8	21 22 27	70 70 70	-68 -72 -74	-95 -95 -95
Inicioct	00:12:01:B5:65:21	ULPGC	2447 8	17 19 25	70 70 70	-70 -75 -78	-95 -95 -95
Inicioct	00:12:01:B5:5E:71	ULPGC	2452 9	18 21 30	70 70 70	-65 -73 -77	-95 -95 -95
Inicioct	00:12:01:B5:49:A1	ULPGC	2457 10	24 25 27	70 70 70	-68 -69 -71	-95 -95 -95
Inicioct	00:14:A9:75:05:B1	ULPGC	2422 3	9 9 9	70 70 70	-86 -86 -86	-95 -95 -95
Inicioct	00:12:01:B5:61:81	ULPGC	2452 9	6 6 6	70 70 70	-89 -89 -89	-95 -95 -95
Inicioct	00:0F:24:EC:FE:71	ULPGC	2437 6	10 11 1 2	70 70 70	-83 -84 -85	-95 -95 -95
....							
213	*00:12:01:B5:5B:81*	*ULPGC*	*2412 1*	*18 23 26*	*70 70 70*	*-69 -71 -77*	*-95 -95 -95*
213	*00:12:01:B5:62:D1*	*ULPGC*	*2432 5*	*29 31 35*	*70 70 70*	*-60 -63 -66*	*-95 -95 -95*
213	00:12:01:B5:5B:80	EDUROAM	2412 1	18 23 26	70 70 70	-69 -71 -77	-95 -95 -95
213	00:12:01:B5:62:D0	EDUROAM	2432 5	29 31 34	70 70 70	-61 -63 -66	-95 -95 -95
213	00:12:01:B5:69:71	ULPGC	2422 3	2 2 2	70 70 70	-93 -93 -93	-95 -95 -95
213	00:12:01:B5:69:70	EDUROAM	2422 3	4 4 5	70 70 70	-90 -90 -91	-95 -95 -95
223	*00:12:01:B5:5B:81*	*ULPGC*	*2412 1*	*22 27 29*	*70 70 70*	*-66 -67 -73*	*-95 -95 -95*
223	*00:12:01:B5:62:D1*	*ULPGC*	*2432 5*	*18 20 23*	*70 70 70*	*-72 -74 -77*	*-95 -95 -95*
223	00:12:01:B5:5B:80	EDUROAM	2412 1	23 27 30	70 70 70	-65 -67 -72	-95 -95 -95
223	00:12:01:B5:62:D0	EDUROAM	2432 5	19 20 23	70 70 70	-72 -74 -76	-95 -95 -95
....							

corridor or the number of an office), the MAC address, the SSID, the Frequency/Channel, the minimum (*Min*), maximum (*max*) and average (*med*) RSSI values (*RSS*) of the two Cisco APs, the RSSI reference, the level of signal and the noise. It can be observed that there were different SSIDs in different channels.

The fourth step was to study the reliability of the DB and the process of localization of the mobile devices in our indoor scenario. We have installed the DB in the LWR in order to the manager to locate a mobile device comparing the values of RSSI received with the values stored in the DB. We moved the portable PC to places under the LWR coverage (in the different places the mobile device was associated to the LWR). The localization was correct several times and in some places. But there were special cases were the localization was not correct. For example, let us consider the following values sent by the mobile device agent and received by the manager in one practical test (see Table 1).

The manager sought these values in the DB (the second row values and then the third row values) resulting the following decisions:

Table 1. Received RSSI

SSID	FREQUENCY	CHANNEL	RSSI
ULPGC	2412	1	26
ULPGC	2432	5	22

- The value of RSSI (26) was in the range 22-29 for *ULPGC* (*AP2*) in office *223* (marked in blue on Figure 12). But it was also in the range 18-26 for *AP2* in office *213* (marked in red on Figure 12). The Frequency and Channel were the same in both cases: 2412 and 1 respectively. It was clear that there was an indetermination of the location of the mobile devices (it could be in office 213 or 223). It needed the second value of RSSI (22) in order to make an accurate localization.
- The value of RSSI (22) was only found for *ULPGC* (named *AP1* due to it was in frequency 2432 and channel 5) in office *223* (marked in blue on Figure 12).
- Combining the results for both row values the manager could decide the location of

the mobile device. That is, for the second row the results were offices 213 and 223 and for the third row the results were 213. Then, with a grade of precision *p* the manager will decide the mobile device was in office 223. But it was not sure.

We also obtained the following row values (the mobile device was in the same place than the above experiment) in another practical experiment while the mobile device was in the same place (see Table 2).

- The problem was the third row values was an out of layer value. That is, it was not found in the DB. Regrettably, this happened several times. In this case, the manager could also decide the mobile device was in office 223 with a grade of precision q ($q<p$). In this case the tool tried to solve this error doing scanning processes several times and trying to minimize their reviewed adverse effects.

Table 2. Second experiment results

SSID	FREQUENCY	CHANNEL	RSSI
ULPGC	2412	1	26
ULPGC	2432	5	16

According to this variability of the RSSI values, we concluded that the location in indoors using fingerprinting based on value of RSSI was very little reliable. But in our case and for the objective that we have proposed the reliability was enough. Also we think that the results were quite dependent of distribution and number of APs or LWR available. Obviously, and mainly, it depended on the internal structure of the scenario where we needed to know the location (walls, windows, screen, stairs, etc). For this reason, we decided to implement another set of tests indoor considering other network resources for comparing purposes.

We measured the RSSI values for 2 much closed APs: one *Belkin* F6D4630 and one *Linksys* WRT54G. The mobile device that implements the survey phase was located to a distance of 2 meters of both APs. In Figure 13 are shown the results with the value of RSSI detected. The upper two

Figure 13. Levels of RSSI with other APs and a new area of location

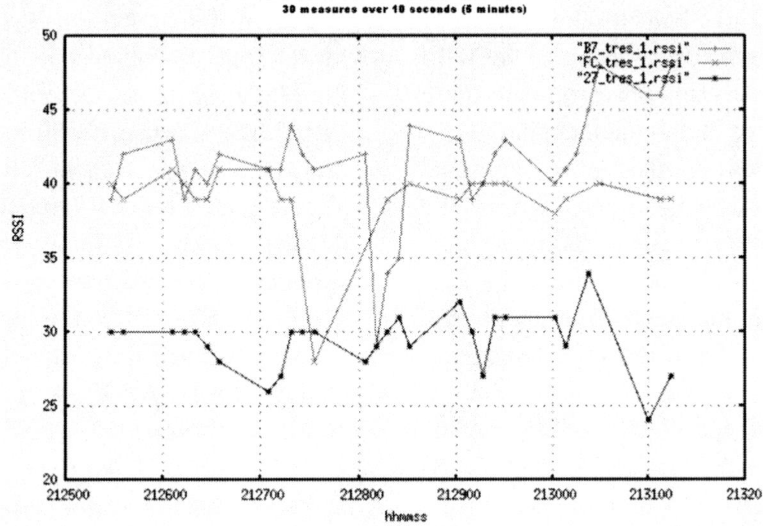

lines are the values of the two APs (RSSI around 40). The third line is an unacknowledged AP (RSSI around 30). Note that the main value is around 40 but the standard deviation is high. The scanning was done in different parts of the day during 5 minutes and each 10 seconds. This information confirmed the variability of RSSI values.

Finally, let us remark that our combined tool exhibits an acceptable behavior taking into account the commented restrictions.

FUTURE RESEARCH DIRECTIONS

We have shown that our combined tool is appropriated to improve the QoS of WiFi access networks. This aims us to continue working in this topic. We classify the future research directions in two categories:

- The improvement of different parts of the tool already implemented.
- Addition of new parts with the same objectives: to improve the QoS.

It is interesting to do some work to improve the tool in the following directions:

- To apply improved algorithms to obtain a better regulation of bandwidth in each mobile device depending on status of the WiFi link. We are analyzing other forms to share the balancing of the bandwidth adaptively. We are researching other more optimized load sharing mechanism and increase the number of simultaneous sessions per mobile device.
- To allow the manager to regulate the downstream traffic in the WiFi links from Internet.
- To select and apply filtering algorithms and to improve the processing of the RSSI values in the DB.
- To add or combine other techniques of location to improve the connectivity (time marks, header IEEE 802.11...).

The new parts of our tool on which we are working are:

- To allow the agents and manager to analyze other parameters or modify different factors, as the intensity of wireless (WiFi) signal in each AP.
- To permit a mixed system combining adhoc and infrastructure mode to obtain connectivity with wired network (network of mobile devices without nodes).
- To tune the tool to work efficiently with mesh networks.
- To design a shared multimedia buffering system among the different mobile devices.

CONCLUSION

New improvements of WiFi access networks appear continuously mainly because the number of mobile user grows exponentially. The improvements that are rapidly adopted in the Market are those related to the increasing of transmission speed. Others like MIMO improvement are timidly arriving to the Market and others like QoS at MAC level improvement have not much deployment in the Market. The next evolution of Internet is to include massive use of multimedia services which imposes new demands to WiFi access networks. That is, QoS must be improved in order to mobile devices access efficiently Internet multimedia services. It is clear that up to date WiFi access networks fail to offer this demanded QoS. The introduction of MAC level QoS requires doing changes in the AP firmware or mobile device WiFi card driver.

In this chapter we have presented a simple mathematical model for our combined solution that includes the adaptation of three different typical mechanisms: a novel regulation of traffic achieved by the mobile device (admission control policy), the static roaming of static mobile devices (the mobile devices can be re-associated to different APs in order to improve the usage of the WiFi link) and the re-association of moving mobile devices to APs using an indoor localization technique. We have combined these mechanisms in a manager/agent software based combined tool. It processes information of WiFi driver and Internet stack levels. But it does not modify the behavior of WiFi in order to clearly adapt our solution to every WiFi standards. Our tool collaborates with the applications that access Internet using our Web portal captive; but it does not modify their internal code. There are some tradeoffs that our tool must take into account. The most important disadvantage of our tool is that localization technique indoor is not robust. This produces typical localization errors. But it is true that the tool does not need an exact localization of mobile devices to produce acceptable results.

We have presented our experimental results in an indoor scenario. A lot of tests have been done to test the three parts of the tools (traffic regulation, static roaming and re-association of moving mobile devices). What was clear is that our tool improves QoS compare with the case in which it was not used. Moreover, its efficiency is acceptable.

REFERENCES

Akiyama, T., Teranishi, Y., Okamura, S., & Shimojo, S. A. (2009). Consideration of the precision improvement in WiFi positioning system. *International Conference on Complex, Intelligent, and Software Intensive Systems*, (pp. 1112-1117). IEEE Press.

Boggia, G., Camarda, P., Grieco, L. A., & Mascolo, S. (2004). Feedback-based bandwidth allocation with call admission control for providing delay guarantees in IEEE802.11e networks. *Elsevier Computer Communications*, *28*(3), 325–337.

Ching, W., Jing, T., Rue Liu, B., & Rizos, C. (2010). Uniwide Wifi based positioning system. *2010 IEEE International Symposium on Technology and Society*, (pp. 180-189). IEEE Press.

Fang, Y., & Zhang, Y. (2002). Call admission control schemes and performance analysis in wireless mobile networks. *IEEE Transactions on Vehicular Technology*, *51*(2), 371–382. doi:10.1109/25.994812

IEEE 802.11. (1999). *WG: Reference number ISO/IEC 8802-11:1999 (E) IEEE STD 802.11. International standard for information technology-Telecommunications and information exchange between systems-Local and metropolitan area networks-Specific requirements- Part 11: Wireless LAN medium access control (MAC) and physical layer (PHY) specifications*. IEEE Press.

IEEE 802.11g. (2003). *IEEE standard for information technology—Telecommunications and information exchange between systems—Local and metropolitan area networks—Specific requirements—Part 11: Wireless LAN medium access control (MAC) and physical layer (PHY) specifications—Amendment 4: Further higher-speed physical layer extension in the 2.4 GHz Band*. IEEE Press.

IEEE 802.11k. (2008), *IEEE standard for information technology telecommunications and information exchange between systems local and metropolitan area networks specific requirements part 11: Wireless LAN medium access control (MAC) and physical layer (PHY) specifications amendment 1: Radio resource measurement of wireless LANS*. IEEE Press.

Jangeun, J., Peddabachagari, P., & Sichitiu, M. (2003). Theoretical maximum throughput of IEEE 802.11 and its applications. *IEEE International Symposium on Network Computing and Applications (NCA)*, (pp. 249-256). IEEE Press.

Liu, C.-T., Wang, T.-C., & Chang, R. (2010). *Cross-layer handoff via predictive multiples pre-registrations for QoS of mobile multimedia applications*. 2010 IEEE International Conference on Wireless Communications, Networking, and Information Security. IEEE Press.

Marrero, D., Suárez, A., & Macías, E. M. (2005). Dynamic interconnection of ad hoc nodes based on the type of service to be accessed. *International Conference on Wireless Networks* (*ICWN*), (pp. 539-545). IEEE Press.

Marrero, D., Suárez, A., & Macías, E. M. (2007). *Dynamic traffic regulation for WiFi networks*. World Congress on Engineering 2007 (WCE2007). IEEE Press.

Marrero, D., Suárez, A., & Macías, E. M. (2008). An admission control and traffic regulation mechanism for infrastructure WiFi networks. *IAENG International Journal of Computer Science, 35*(1).

Marrero, D., Suárez, A., & Macías, E. M. (2009). *Aplicación multifuncional para gestión del canal en redes IEEE 802.11. VIII Jornadas de Ingeniería Telemática (JITEL2009)* (pp. 16–23). IEEE Press.

Meng, L., Zipf, A., & Winter, S. (Eds.). (2008). *Map-based mobile services design, interaction, and Usability. Lecture Notes in Geoinformation and Cartography*. Springer Verlag.

Prihandoko, F., Habaebi, M. H., & Ali, B. M. (2003). Adaptive call admission control for QoS provisioning in multimedia wireless networks. *Elsevier Computer Communications.*, *26*(14), 1560–1569.

Qiang, N. (2005). Performance analysis and enhancements for IEEE 802.11e wireless networks. *IEEE Network*, *19*(4), 21–27. doi:10.1109/MNET.2005.1470679

Rodrigues, J. C., Fraiha, S., Araujo, J., Gomes, H., Frances, C., & Cavalcante, G. (2009). Empirical study of the QoS parameters behavior of a VoIP application in Wi-Fi networks. *2009 SBMO/IEEE MTT-S International Microwave and Optoelectronics Conference* (*IMOC*), (pp. 257–261). IEEE Press.

Shoham, Y., & Leyton-Brown, K. (2009). *Multiagent systems algorithmic, game-theoretic, and logical foundations*. Cambridge, England: Cambridge University Press.

Silva, V. H., Salgado, E. I., & Quintana, F. R. (2006). AWISPA: An awareness framework for collaborative spontaneous networks. In *ASEE/IEEE Frontiers in Education Conference,* (pp. 1-6). IEEE Press.

Sobrinho, J. L., & Krishnakumar, A. S. (1999). Quality-of-service in ad hoc carrier sense multiple access wireless networks. *IEEE Journal on Selected Areas in Communications*, *17*(8), 1353–1368. doi:10.1109/49.779919

Song, L., Kotz, D., Jain, R., & Xiaoning, H. (2004). Evaluating location predictors with extensive Wi-Fi mobility data. *Twenty-Third Annual Joint Conference of the IEEE Computer and Communications Societies* (INFOCOM 2004), vol. 2, (pp. 1414–1424). IEEE Press.

Starsky, H. Y., Wong, H. Y., Lu, S., & Bharghavan, V. (2006). *Robust rate adaptation for 802.11 wireless networks*. ACM 12th Annual International Conference on Mobile Computing and Networking. Los Angeles, CA: ACM Press.

Toh, C. K., Delwar, M., & Allen, D. (2002). Evaluating the communication performance of an ad_hoc wireless network. *IEEE Transactions on Wireless Communications*, *1*(3), 402–414. doi:10.1109/TWC.2002.800539

Vinnakote, S., Naresh, S. V. S., & Pasupuleti, S. (2006). *New-MAC protocol for enhancement of QoS performance in wireless LAN*. 2006 IFIP International Conference on Wireless and Optical Communications Networks. IEEE Press.

WMM. WiFi Alliance. (2004). *Wi-Fi certified™ for WMM™ - Support for multimedia applications with quality of service in Wi-Fi® networks*. Austin, TX: WF Alliance.

Xiao, Y. (2004). IEEE: 802.11e: QoS provisioning at the MAC layer. *IEEE Wireless Communications*, *11*(3), 72–79. doi:10.1109/MWC.2004.1308952

Zhao, L., Yao, G., & Mark, J. W. (2006). Mobile positioning based on relaying capability of mobile stations in hybrid wireless networks. *IEEE Proceedings-Communications*, *153*(5), 762–770. doi:10.1049/ip-com:20050142

ADDITIONAL READING

Benvenuti, C. (2005). *Understanding Linux network internals*. O'Reilly Media.

Cooklev, T. (2004). *Wireless communication standards: A study of IEEE 802.11, 802.15, and 802.16. Standards Information Network*. IEEE Press.

Figueiras, J., & Frattasi, S. (2010). *Mobile positioning and tracking*. Wiley. doi:10.1002/9780470663035

Flickenger, R., & Weeks, R. (2005). *Wireless hacks* (2nd ed.). O'Reilly Media.

Gast, M. (2005). *802.11 wireless networks: The definitive guide* (2nd ed.). O'Reilly Media.

Geun, K., Jeong, K., & Marwan, M. (2000). Bandwidth allocation in wireless networks with guaranteed packet-loss performance. *IEEE/ACM Transactions on Networking*, *8*(3), 337–349. doi:10.1109/90.851980

Goransson, P., & Greenlaw, R. (2007). *Secure roaming in 802.11 networks (communications engineering)*. Newnes.

Guoqiang, M., & Fidan, B. (2009). *Localization algorithms and strategies for wireless sensor networks*. Hershey, PA: IGI Global.

LaRocca, J., & LaRocca, R. (2002). *IEEE 802.11 demystified*. McGraw-Hill Professional.

Muñoz, D., Bouchereau, F., Vargas, C., & Enríquez-Caldera, R. (2009). *Position location techniques and applications*. Elsevier.

O'Hara, B., & Petrick, A. (2005). *IEEE 802.11 handbook. A designer's companion*. IEEE Press.

Ohrtman, F., & Roeder, K. (2003). *WiFi handbook: Building IEEE 802.1b wireless networks*. McGraw-Hill Professional.

Pavilanskas, L. (2007). *Wireless MAC for real time applications: Analytical engineering approach to the usage of IEEE 802.11 MAC in specific conditions*. LAP Lambert Academic Publishing.

Perahia, E., & Stacey, R. (2008). *Next generation wireless LANs: Throughput, robustness, and reliability in 802.11n*. Cambridge University Press. doi:10.1017/CBO9780511541032

Reid, N. P., & Seide, R. (2002). *WiFi (802.11) 802.11 network handbook*. McGraw-Hill.

Roshan, P., & Leary, J. (2010). *IEEE 802.11 wireless LAN fundamentals*. Cisco Press.

Ross, J. (2008). *The book of wireless*. O'Reilly Media.

Saahoff, F. (2002). *QoS-aspects of mobile IPv6 clients in an IEEE 802.11 network*. VDM Verlag.

Velte, T., & Velte, A. (2005). *Cisco IEEE 802.11 wireless networking quick reference*. Cisco Press.

Walke, B. H., Mangold, S., & Berlemann, L. (2007). *IEEE 802 wireless systems: Protocols, multi-hop mesh/relaying, performance, and spectrum coexistence*. Wiley.

Weeks, R., Dumbill, E., & Jepson, B. (2004). *Linux unwired*. O'Reilly Media.

Compilation of References

3GPP TS 23.401: General packet radio service (GPRS) enhancements for evolved universal terrestrial radio access network (E-UTRAN) access.

3GPP TS 23.402: Architecture enhancements for non-3GPP accesses.

3GPP TS 29.275: Proxy mobile IPv6 (PMIPv6) based mobility and tunneling protocols.

3GPP. (2007). Requirements on 3GPP system to wireless local area network (WLAN) interworking, TS 22.243, v8.0.0. http://www.3gpp.org/specifications.

3GPP. (2007a). *R1-070051: Performance of MBMS transmission configurations.*

3GPP. (2007b). *R1-071049: Spectral efficiency comparison of possible MBMS transmission schemes: Additional Results.*

3GPP. (2007c). *R1-071213: Response on LS on radio efficiency for delivery of broadcast/multicast services.*

3GPP. (2007d). *R1-071433: Additional results on EMBMS transmission configurations.*

3GPP. (2009). *LTE Advanced.* Retrieved September 29, 2010, from http://www.3gpp.org/LTE-Advanced.

3GPP. (2009). *TS 36.201, V9.0.0: Technical specification group radio access network; Evolved universal terrestrial radio access (E-UTRA); LTE physical layer - General description (release 9).*

3GPP. (2010). *TS 36.101, V9.4.0: User equipment (UE) radio transmission and reception.*

3GPP. (2010). *TS 36.104, V9.4.0: Base station (BS) radio transmission and reception.*

3GPP. (2010). *TS 36.211, V9.1.0: Physical channels and modulation.*

3GPP. (2010a). *TS 36.213, V9.1.0: Technical specification group radio access network; Evolved universal terrestrial radio access; Physical layer procedures (release 9).*

3GPP. (2010b). *TS 36.300, V9.3.0: Technical specification group radio access network; Evolved universal terrestrial radio access (E-UTRA) and evolved universal terrestrial radio access network (E-UTRAN); Overall description; Stage 2 (release 9).*

3GPP. (2010c). *TS 23.401 V10.2.1: General packet radio service (GPRS) enhancements for evolved universal terrestrial radio access network (E-UTRAN) access.*

3GPP2. (2006). 3GPP2 to WLAN interworking, S.R0087-A, v1.0. http://www.3gpp2.org/public_html/specs/index.cfm.

Abdolee, R., Ngah, R., Vakilian, V., & Rahman, T. (2007). *Application of radio-over-fiber (ROF) in mobile communication.* Paper presented at the Asia-Pacific Conference on Applied Electromagnetics (APACE). APACE Press.

Abowd, G. D., Dey, A. K., Brown, P. J., Davies, N., Smith, M., & Steggles, P. (1999). *Towards a better understanding of context and context-awareness.* Paper presented at the 1st International Symposium on Handheld and Ubiquitous Computing.

Adhikari, S. (2009). Critical analysis of multi-antenna systems in the LTE downlink. *Proceedings of 3rd International Conference on Internet Multimedia Systems Architecture and Applications (IMSAA 2009)*, (pp. 1-6). Bangalore, India: IEEE Press.

Afanasyev, M., Chen, T., Voelker, G. M., & Snoeren, A. C. (2010). Usage patterns in an urban WiFi network. *IEEE/ACM Transactions on Networking*, *18*(5). doi:10.1109/TNET.2010.2040087

Ahmad, I., & Habibi, D. (2008a). *A proactive forward error control scheme for mobile WiMAX communication.* Paper presented at the 11th IEEE International Conference on Communication Systems, Singapore, 2008.

Ahmad, I., Habibi, D., & Rahman, Z. (2008). *An improved FEC scheme for mobile wireless communication at vehicular speeds.* Paper presented at the Telecommunication Networks and Applications Conference, 2008. ATNAC 2008. Australasian.

Ahmad, I., & Habibi, D. (2008b). Quality of service in WiMAX . In Zang, Y. (Ed.), *WiMAX network planning and optimization* (pp. 16–31). Taylor & Francis Group.

Ahn, G., Campbell, A. T., Veres, A., & Sun, L. H. (2002). SWAN: Service differentiation in stateless wireless ad hoc networks. *Proceeding of IEEE International Conference on Computer Communications (INFOCOM)*, (pp. 457–466). Orlando, FL: IEEE Press.

Ahuja, R., Corke, R., & Bok, A. (2008). *Cognitive radio system using IEEE 802.11a over UHF TVWS.* Paper presented at the Symposium on Dynamic Spectrum Access Networks (DySPAN). IEEE Press.

Akan, O. B., Karli, O. B., & Ergul, O. (2009). Cognitive radio sensor networks. *IEEE Network*, *23*(4), 34–40. doi:10.1109/MNET.2009.5191144

Akiyama, T., Teranishi, Y., Okamura, S., & Shimojo, S. A. (2009). Consideration of the precision improvement in WiFi positioning system. *International Conference on Complex, Intelligent, and Software Intensive Systems*, (pp. 1112-1117). IEEE Press.

Akyildiz, I. F., Lee, W.-Y., & Chowdhury, K. R. (2009). CRAHNs: Cognitive radio ad hoc networks. *Ad Hoc Networks*, *7*(5), 810–836. doi:10.1016/j.adhoc.2009.01.001

Akyildiz, I. F., Lee, W.-Y., Vuran, M. C., & Mohanty, S. (2008). A survey on spectrum management in cognitive radio networks. *IEEE Communications Magazine*, *46*(4), 40–48. doi:10.1109/MCOM.2008.4481339

Akyildiz, I. F., McNair, J., Ho, J. S. M., Uzunalioglu, H., & Wang, W. (1999). Mobility management in next-generation wireless systems. *Proceedings of the IEEE*, *87*(8), 1347–1384. doi:10.1109/5.775420

Akyildiz, I. F., Su, W., Sankarasubramaniam, Y., & Cayirci, E. (2002). Wireless sensor networks: A survey. *Computer Networks*, *38*(4), 393–422. doi:10.1016/S1389-1286(01)00302-4

Akyildiz, I. F., Xie, J., & Mohanty, S. (2004). A survey of mobility management in next-generation all-IP-based wireless systems. *IEEE Wireless Communications*, *11*(4), 16–28. doi:10.1109/MWC.2004.1325888

Akyildiz, I., Wang, X., & Wang, W. (2005). Wireless mesh networks: A survey. *Computer Networks*, *47*(4), 445–487. doi:10.1016/j.comnet.2004.12.001

Akyildiz, I., Xie, J., & Mohanty, S. (2004). A survey of mobility management in next-generation all-IP-based wireless systems. *IEEE Wireless Communications*, *11*(4), 16–28. doi:10.1109/MWC.2004.1325888

Al-Dulaimi, A., & Cosmas, J. (2010d). Mobility management for microcells based cognitive radio over fibre networks. In *Proceedings of World Wireless Research Forum (WWRF#25) Meeting*. London, UK: WWRF Press.

Al-Dulaimi, A., Al-Rubaye, S., & Al-Raweshidy, H. S. (2010a). Renovate cognitive networks under spectrum unavailability. In *Proceedings of 6th Advanced International Conference on Telecommunications*, (pp. 41–45). Spain: IEEE Press.

Al-Dulaimi, A., Al-Rubaye, S., Al-Raweshidy, H. S., Cosmas, J., & Loo, K. K. (2010c). Multi-zones of cognitive to fibre networks. In *Proceedings of IEEE/IET 7th International Symposium on Communications Systems, Networks, and Digital Signal Processing (CSNDSP)*, (pp. 697 – 700). Newcastle upon Tyne, UK: IEEE Press.

Al-Dulaimi, A., Al-Rubaye, S., Al-Raweshidy, H. S., Cosmas, J., & Loo, K. K. (2010b). Cognitive mesh networks: Cognitive radio over fibre for microcells applications. *IEEE Vehicular Technology Magazine.*, *5*(3), 54–60. doi:10.1109/MVT.2010.938272

Alexiou, A., Bouras, C., & Kokkinos, V. (2007). Evaluation of the multicast mode of MBMS. *Proceedings of 18th Annual IEEE International Symposium on Personal, Indoor, and Mobile Radio Communications (PIMRC '07)*. Athens, Greece: IEEE Press.

Alexiou, A., Bouras, C., & Papazois, A. (2010). The Impact of FEC on mobile multicast power control. *Proceedings of the 6th Advanced International Conference on Telecommunications (AICT 2010)*. Barcelona, Spain: IEEE Press.

Alexiou, A., Bouras, C., Kokkinos, V., & Tsichritzis, G. (2010a). Communication cost analysis of MBSFN in LTE. *Proceedings of 21st Annual IEEE International Symposium on Personal, Indoor, and Mobile Radio Communications (PIMRC 2010)*. Instabul, Turkey: IEEE Press.

Alexiou, A., Bouras, C., Kokkinos, V., Papazois, A., & Tseliou, G. (2010). Enhancing FEC application in LTE cellular networks. *Proceedings of IFIP Wireless Days 2010*. Venice, Italy: IEEE Press.

Alexiou, A., Bouras, C., Kokkinos, V., Papazois, A., & Tsichritzis, G. (2010). Adopting FEC for reliable multicasting over LTE networks. *Proceedings of 13th ACM International Conference on Modeling, Analysis, and Simulation of Wireless and Mobile Systems (MSWiM 2010)*. Bodrum, Turkey: IEEE Press.

Alexiou, A., Bouras, C., Kokkinos, V., Papazois, A., & Tsichritzis, G. (2010b). Efficient MCS selection for MBSFN transmissions over LTE Networks. *Proceedings of IFIP Wireless Days 2010*. Venice, Italy: IEEE Press.

Alexiou, A., Bouras, C., Kokkinos, V., Papazois, A., & Tsichritzis, G. (2010c). Spectral efficiency performance of MBSFN-enabled LTE networks. *Proceedings of 6th IEEE International Conference on Wireless and Mobile Computing, Networking, and Communications (WiMob 2010)*. Niagara Falls, Canada: IEEE Press.

Al-Fuqaha, A., Khan, B., Rayes, A., Guizani, M., Awwad, O., & Ben Brahim, G. (2008). Opportunistic channel selection strategy for better QoS in cooperative networks with cognitive radio capabilities. [JSAC]. *IEEE Journal on Selected Areas in Communications*, *26*(1), 156–167. doi:10.1109/JSAC.2008.080114

Al-Raweshidy, H. S., & Komaki, S. (2002). *Radio over fiber technologies for mobile communication networks*. Boston, MA: Artech House.

Angelakis, V., Papadakis, S., Siris, V., & Traganitis, A. (2008). Adjacent channel interference in 802.11a: Modeling and testbed validation. *IEEE Radio and Wireless Symposium*, (pp. 591–594). Orlando, FL: IEEE Press.

Antoniou, J., Pinto, F., Simoes, J., & Pitsillides, A. (2010). Supporting context-aware multiparty sessions in heterogeneous mobile networks. *Journal on Mobile Networks and Applications*, *15*(6), 1–14.

Ardagna, C., Cremonini, M., Capitani di Vimercati, S. D., & Samarati, P. (2008). A privacy aware access control system. *Journal of Computer Security*, *16*(4), 369–397.

Arjona, A., & Verkasalo, H. (2007). Unlicensed mobile access (UMA) handover and packet data performance analysis. In *Proceedings of the Second International Conference on Digital Telecommunications, ICDT 2007*. San Jose, CA: IEEE Press.

Arslan, H. (Ed.). (2007). *Cognitive radio, software defined radio, and adaptive wireless systems*. Dordrecht, The Netherlands: Springer. doi:10.1007/978-1-4020-5542-3

Ashley, P., Hada, S., Karjoth, G., Powers, C., & Schunter, M. (2003). *Enterprise privacy authorization language (EPAL 1.2)*. World Wide Web Consortium (W3C).

Assouma, A. D., Beaubrun, R., & Pierre, S. (2006). Mobility management in heterogeneous wireless networks. *IEEE Journal on Selected Areas in Communications*, *24*(3), 638–648. doi:10.1109/JSAC.2005.862407

Attar, A., Li, H., & Leung, V. (2010). Increasing TCP throughput and fairness in cognitive WLAN over fibre. In *Proceedings of 24th IEEE International Conference on Advanced Information Networking and Applications*, (pp. 1069-1076). Perth, WA: IEEE Press.

Awduche, D., Berger, L., Gan, D., Li, T., Srinivasan, V., & Swallow, G. (2001). *RSVP-TE: Extensions to RSVP for LSP tunnels*. RFC 3209 (Proposed Standard). Updated by RFCs 3936, 4420, 4874, 5151, 5420.

Awduche, D., Chiu, A., Elwalid, A., Widjaja, I., & Xiao, X. (2002). *Overview and principles of internet traffic engineering*. RFC 3272 (Informational). Updated by RFC 5462.

Awduche, D., Malcolm, J., Agogbua, J., O'Dell, M., & McManus, J. (1999). *Requirements for traffic engineering over MPLS. RFC 2702*. Informational.

Aydin, I., Seok, W., & Shen, C. C. (2003). Cellular SCTP: A transport-layer approach to internet mobility. In the *Proceeding of The ICCCN 2003* (285-290). Dallas, TX: IEEE Press.

Azcorra, A., Banniza, T., Chieng, D., Fitzpatrick, J., Von-Hugo, D., & Natkaniec, M. (2009). Supporting carrier grade services over wireless mesh networks: The approach of the European FP-7 STREP CARMEN. *IEEE Communications Magazine*, *47*(4), 14–16. doi:10.1109/MCOM.2009.4907397

Bachir, A., Dohler, M., Watteyne, T., & Leung, K. K. (2010). MAC essentials for wireless sensor networks. *IEEE Communications Surveys & Tutorials*, *12*(2), 222–248. doi:10.1109/SURV.2010.020510.00058

Bahl, P., Chandra, R., & Dunagan, J. (2004). SSCH: Slotted seeded channel hopping for capacity improvement in IEEE 802.11 ad-hoc wireless networks . In *Proceedings from ACM MobiCom* (pp. 216–230). Philadelphia, PA: ACM Press.

Bakre, A., & Badrinath, B. R. (1995). I-TCP: Indirect TCP for mobile hosts. In the *Proceeding of IEEE International Conference on Distributed Computing Systems* (136-143). Vancouver, Canada: IEEE Press.

Bakshi, B., Krishna, P., Vaidya, N. H., & Pradhan, D. K. (1995). *Performance of TCP over wireless networks* (Technical Report 96-014). Computer Science & Engineering. College Station, TX: TAMU Press.

Balakrishnan, H., Padmanabhan, V., Seshan, S., & Katz, R. (1997). A comparison of mechanisms for improving TCP performance over wireless links. *IEEE/ACM Transactions on Networking*, *5*(6), 756–769. doi:10.1109/90.650137

Balasubramaniam, S., & Indulska, J. (2004). Vertical handover supporting pervasive computing in future wireless networks. *Computer Communications*, *27*(8), 708–719. doi:10.1016/j.comcom.2003.10.010

Baldo, N., Asterjadhi, A., & Zorzi, M. (2009). *Multichannel medium access using a virtual network coded control channel*. Paper presented at the International Wireless Communications and Mobile Computing Conference (IWCMC). IEEE Press.

Ball, C., Hindelang, T., Kambourov, I., & Eder, S. (2008). Spectral efficiency assessment and radio performance comparison between LTE and WiMAX. *Proceeding of IEEE International Symposium on Personal, Indoor and Mobile Radio Communications (PIMRC 2008)*. Cannes, France: IEEE Press.

Ball, C., Mullner, L., Lienhart, J., & Winkler, H. (2009). Performance analysis of closed and open loop MIMO in LTE. *Proceedings of European Wireless Conference 2009 (EW2009)*, (pp. 260-265). Aalborg, Denmark: IEEE Press.

Banchs, A., Bayer, N., Chieng, D., de la Oliva, A., Gloss, B., & Kretschmer, M. … Zdarsky, F. (2008). Carmen: Delivering carrier grade services over wireless mesh networks. In *Proceeding of IEEE 19th International Symposium on Personal, Indoor, and Mobile Radio Communications PIMRC 2008*, (pp. 1-6). Cannes, France: IEEE Press.

Banerjee, S., Bhattacharjee, B., & Kommareddy, C. (2002). *Scalable application layer multicast*. Paper presented at the 2002 Conference on Applications, Technologies, Architectures, and Protocols for Computer Communications.

Banerjee, N., Wu, W., & Das, S. K. (2003). Mobility support in wireless Internet. *IEEE Wireless Communications*, *10*(5), 54–61. doi:10.1109/MWC.2003.1241101

Banerjee, S., Patra, S., Roy, K., & Vidyarthi, D. (2006). Improved genetic algorithm for channel allocation with channel borrowing in mobile computing. *IEEE Transactions on Mobile Computing*, *5*(7), 1536–1233.

Bari, F., & Leung, V. (2007). Application of ELECTRE to network selection in a heterogeneous wireless network environment. *Proceedings of IEEE Wireless Communications and Networking Conference* (WCNC'08), vol. 1 (pp. 3810-3815).

Baset, S. A., & Schulzrinne, H. (2006). An analysis of the Skype peer-to-peer internet telephony protocol. In *Proceeding of IEEE INFOCOM*, Barcelona, Spain. IEEE Press.

Bayer, N., Hock, D., Roos, A., Siebert, M., Xu, B., Rakocevic, V., & Habermann, J. (2008). VoIP performance in "meshbed": A wireless mesh networks testbed. *Vehicular Technology Conference (VTC Spring 2008)* (pp. 2218-2222). IEEE.

Beckmann, D., & Killat, U. (1999). A new strategy for the application of genetic algorithms to the channel assignment problem. *IEEE Transactions on Vehicular Technology*, *48*(4), 1261–1269. doi:10.1109/25.775374

Bellman, R. E., & Zadeh, L. A. (1970). Decision-making in a fuzzy environment. *Management Science Journal*, *17*(4), 141–164.

Benvenuti, C. (2005). *Understanding Linux network internals*. O'Reilly Media.

Berezdivin, R., Breinig, R., & Topp, R. (2002). Next-generation wireless communications concepts and technologies. *Communications Magazine, IEEE*, *40*(3), 108–116. doi:10.1109/35.989768

Berger, L., Gan, D., Swallow, G., Pan, P., Tommasi, F., & Molendini, S. (2001). *RSVP refresh overhead reduction extensions. RFC 2961*. Proposed Standard.

Berl, A., Race, N., Ishmael, J., & de Meer, H. (2010). Network virtualization in energy-efficient office environments. *Computer Networks,* 54(16). Elsevier B.V.

Berl, A., Fischer, A., & de Meer, H. (2009). Using system virtualization to create virtualized networks. *Electronic Communications of the EASST*, *17*, 1–12.

Bernaschi, M., Cacace, F., Iannello, G., & Vellucci, M. (2007). Mobility management for VoIP on heterogeneous networks: evaluation of adaptive schemes. *IEEE Transactions on Mobile Computing*, *6*(9), 1053–1047. doi:10.1109/TMC.2007.1003

Bertsekas, D. P. (1999). *Nonlinear optimization*. Belmont, MA: Athena Scientific.

Bhagwat, P., & Perkins, C. (1993). A mobile networking system based on internet protocol (IP). In *Proceeding of USENIX Symposium on Mobile and Location Independent Computing* (69-82). Cambridge, MA, USA.

Bhagwat, P., Perkins, C., & Tripathi, S. (1996). Network layer mobility: An architecture and survey. *IEEE Personal Communications*, *3*(3), 54–64. doi:10.1109/98.511765

Bhattacharyya, S. (Ed.). (2003). *RFC 3569: An overview of source-specific multicast*. SSM.

Bhushan, N., Lott, C., Black, P., Attar, R., Yu-Cheun, J., & Fan, M. (2006). CDMA2000 1×EV-DO revision a: A physical layer and MAC layer overview. *IEEE Communications Magazine*, *44*(2), 37–39. doi:10.1109/MCOM.2006.1593549

Bisnik, N., & Abouzeid, A. (2006). Delay and throughput in random access wireless mesh networks. *IEEE International Conference on Communications* (pp. 403–408). IEEE, Istanbul.

Black, P. J., & Qiang, W. (2002). Link budget of CDMA2000 1xEV-DO wireless internet access system. *The 13th IEEE International Symposium on Personal, Indoor, and Mobile Radio Communications*, vol. 4, (pp. 1847- 1852). IEEE Press.

Blake, S., Black, D., Carlson, M., Davies, E., Wang, Z., & Weiss, W. (1998). *An architecture for differentiated services*. Retrieved September 29, 2010, from http://www.ietf.org/rfc/rfc2475.txt

Bless, R., & Wehrle, R. (2004). *RFC 3754: IP multicast in differentiated services (DS) networks*.

Bloom, B. H. (1970). Space/time trade-offs in hash coding with allowable errors. *Communications of the ACM*, *13*(7), 422–426. doi:10.1145/362686.362692

Bogenfeld, E., & Gaspard, I. (2008). *Self-x in radio access networks*. A White Paper by the EU-FP7 project End-to-End Efficiency (E3). CORDIS.

Boggia, G., Camarda, P., Grieco, L. A., & Mascolo, S. (2004). Feedback-based bandwidth allocation with call admission control for providing delay guarantees in IEEE802.11e networks. *Elsevier Computer Communications*, *28*(3), 325–337.

Boivie, R., Feldman, N., Imai, Y., Livens, W., & Ooms, D. (2007). *RFC 5058: Explicit multicast (Xcast) concepts and options*.

Bolch, G., Greiner, S., de Meer, H., & Trivedi, K. S. (1998). *Queueing networks and Markov chains: Modeling and performance evaluation with computer science application* (1st ed.). Wiley Blackwell. doi:10.1002/0471200581

Bolton, W., Xiao, Y., & Guizani, M. (2007). IEEE 802.20: Mobile broadband wireless access. *IEEE Wireless Communications*, *14*(1), 84–95. doi:10.1109/MWC.2007.314554

Botta, A., Pescape, A., Ventre, G., & Karrer, R. P. (2007). High-speed wireless backbones: Measurements from MagNets. *4th International Conference Broadband Communications, Networks and Systems. (BROADNETS 2007)* (pp. 680-689). IEEE.

Boufidis, Z., Alonistioti, N., Feuillette, R., Holland, O., Pan, J., & Moessner, K. (2007). *End-to-end network architecture for cognitive reconfigurable mobile systems.* FP7 End-to-End Reconfigurability II (E2R II) White Paper. CORDIS.

Braden, R., Clark, D., & Shenker, S. (1994). *Integrated services in the internet architecture: An overview*. Retrieved September 29, 2010, from http://tools.ietf.org/html/rfc1633

Braden, R., Zhang, L., Berson, S., Herzog, S., & Jamin, S. (1997). *Resource reservation protocol (RSVP): Version 1 functional specification.* RFC 2205 (Proposed Standard). Updated by RFCs 2750, 3936, 4495.

Brik, V., Rozner, E., Banerjee, S., & Bahl, P. (2005). DSAP: A protocol for coordinated spectrum access. Paper presented at the IEEE International Symposium on New Frontiers in Dynamic Spectrum Access Networks (DySPAN). IEEE Press.

Brown, K., & Singh, S. (1996). M-UDP: UDP for mobile cellular networks. *Computer Communication Review*, *26*(5), 60–78. doi:10.1145/242896.242901

Bruno, R., Conti, M., & Gregori, E. (2008). Throughput analysis and measurements in IEEE 802.11 WLANS with TCP and UDP traffic flows. *IEEE Transactions on Mobile Computing*, *7*(2), 171–186. doi:10.1109/TMC.2007.70718

Buddhikot, M. M., Chandranmenon, G., Seungjae, H., Yui-Wah, L., Miller, S., & Salgarelli, L. (2003). Design and implementation of a WLAN/CDMA2000 interworking architecture. *Communications Magazine, IEEE*, *41*(11), 90–100. doi:10.1109/MCOM.2003.1244928

Budzisz, L., Ferrus, R., & Brunstrom, A. (2008). Towards transport-layer mobility: Evolution of SCTP multihoming. *Computer Communications*, *31*(5), 980–998. doi:10.1016/j.comcom.2007.12.014

Calafate, C., Cano, J., Manzoni, P., & Malumbres, M. (2005). A QoS architecture for MANETs supporting real-time peer-to-peer multimedia applications. *Proceeding of IEEE International Symposium on Multimedia*, (pp. 193–200). Irvine, CA: IEEE Press.

Calafate, C., Oliver, J., Cano, J., Manzoni, P., & Malumbres, M. (2007). A distributed admission control system for MANET environments supporting multipath routing protocols. *Microprocessors and Microsystems*, *31*(4), 236–251. doi:10.1016/j.micpro.2007.01.002

Campbell, A., Gomez, J., Wan, C. Y., Kim, S., Turanyi, Z., & Valko, A. (2000). *Cellular IP, internet-draft* (work in progress). Draft-ietf-mobileip-cellularip-00, (work in progress).

Capone, A., Fratta, L., & Martignon, F. (2004). Bandwidth estimation schemes for TCP over wireless networks. *IEEE Transactions on Mobile Computing*, *3*(2), 129–143. doi:10.1109/TMC.2004.5

Casals, O., Cerda, L. I., Willems, G., Blondia, C., & Van den Wijngaert, N. (2003). Performance evaluation of the post-registration method, a low latency handoff in MIPv4. In the *Proceeding of the ICC 2003*: Vol. 1 (522-526). Anchorage, AK, USA: IEEE Press.

Cassasa, M., & Beato, F. (2007). *On Parametric obligation policies: Enabling privacy-aware information lifecycle management in enterprises. Technical Report*. HP Labs, Advanced Research at HP.

Castro, M., Druschel, P., Kermarrec, A. M., & Rowstron, A. I. T. (2002). Scribe: A large-scale and decentralized application-level multicast infrastructure. *IEEE Journal on Selected Areas in Communications*, *20*(8), 1489–1499. doi:10.1109/JSAC.2002.803069

Cerqueira, E., Veloso, L., & Neto, A. Curado, I., et al. (2007). *Q3M: QoS architecture for multi-user mobile multimedia sessions in 4G systems*. Paper presented at the 10th IFIP/IEEE International Conference on Management of Multimedia and Mobile Networks and Services: Real-Time Mobile Multimedia Services.

Cerqueira, E., Veloso, L., Neto, A., Curado, M., Monteiro, E., & Mendes, P. (2006). *MMC04-3: A unifying architecture for publish-subscribe services in the next generation IP networks.* Paper presented at the Global Telecommunications Conference, 2006. GLOBECOM '06. IEEE.

Chakrabarti, S., & Mishra, A. (2001). QoS issues in ad hoc wireless networks. *IEEE Communications Magazine*, *39*(2), 142–148. doi:10.1109/35.900643

Chakraborty, G., & Chakraborty, B. A. (1999). Genetic algorithm approach to solve channel assignment problem in cellular radio networks. In *Proceedings of IEEE Workshop Soft Computing Methods in Industrial Applications*, (pp. 34–39). Kuusamo, Finland: IEEE Press.

Chalmers, D., & Sloman, M. (1999). A survey of quality of service in mobile computing environments. *IEEE Communications Tutorials and Surveys*, 2nd Quarter 1999. http://www.comsoc.org/publications.

Chandra, R., Bahl, P., & Bahl, P. (2004). MultiNet: Connecting to multiple IEEE 802.11 networks using a single wireless card. *23rd Annual Joint Conference of the IEEE Computer and Communications Societies (INFOCOM 2004)* (pp. 882- 893). IEEE.

Chaskar, H., & Madhow, U. (2003). Fair scheduling with tunable latency: A round-robin approach. *IEEE/ACM Transactions on Networking*, *11*(4), 592–601. doi:10.1109/TNET.2003.815290

Chawathe, Y. (2003). Scattercast: An adaptable broadcast distribution framework. *Multimedia Systems*, *9*(1), 104–118. doi:10.1007/s00530-002-0082-z

Chen, D., Zhang, Q., & Jia, W. (2008). Aggregation aware spectrum assignment in cognitive ad-hoc networks. In the *3rd International ICST Conference on Cognitive Radio Oriented Wireless Networks (CrownCom)*, (pp. 1–6). IEEE Press.

Chen, G., Song, M., Zhang, Y., Song, J., Teng, Y., & Wang, L. (2009). Fuzzy logic based traffic model considering cross-layer for future cognitive network. *Proceeding of Joint Conference on Pervasive Computing (JCPC)*, (pp. 627–632). Tamkang, Taiwan: IEEE Press.

Chen, L., Wei, S., & Shi, L. (2009). Research on location prediction of vehicular networks. *Proceeding of IEEE International Conference on Computer Science and Information Technology (ICCSIT)*, (pp. 558–561). Beijing, China: IEEE Press.

Chen, Q., Liang, Y., Motani, M., & Wong, W. (2009). *CR-CSMA: A random access MAC protocol for cognitive radio networks*. Paper presented at the IEEE 20th International Symposium on Personal, Indoor, and Mobile Radio Communications. IEEE Press.

Cheng, C., Hsiao, P., Kung, H., & Vlah, D. (2006). *Adjacent channel interference in dual-radio 802.11a nodes and its impact on multi-hop networking. IEEE Globecom* (pp. 1–6). San Francisco, CA: IEEE Press.

Chen, K.-C., & Prasad, R. (2009). *Cognitive radio networks*. Chichester, UK: John Wiley & Sons Ltd. doi:10.1002/9780470742020

Chen, Y., Qing, Z., & Swami, A. (2008). Joint design and separation principle for opportunistic spectrum access in the presence of sensing errors. *IEEE Transactions on Information Theory*, *54*(5). doi:10.1109/TIT.2008.920248

Chen, Y., & Yang, Y. (2007). A new 4G architecture providing multimode terminals always best connected services. *IEEE Wireless Communications*, *14*(2), 36–41. doi:10.1109/MWC.2007.358962

Chereddi, C., Kyasanur, P., & Vaidya, N. H. (2007). Net-X: A multichannel multi-interface wireless mesh implementation. *SIG-MOBILE Mobile . Computer Communication Review*, *11*(3), 84–95. doi:10.1145/1317425.1317435

Chiasserini, C. F., & Rao, R. R. (2003). Coexistence mechanisms for interference mitigation in the 2.4-GHz ISM band. *IEEE Transactions on Wireless Communications*, *2*(5). doi:10.1109/TWC.2003.817417

Chieng, D., Hugo, D., & Banchs, A. (2010). A cost sensitivity analysis for carrier-grade wireless mesh networks with Tabu optimization. In *INFOCOM IEEE Conference on Computer Communications Workshops, 2010*, (pp. 1-6). San Diego, CA: IEEE Press.

Ching, W., Jing, T., Rue Liu, B., & Rizos, C. (2010). Uniwide Wifi based positioning system. *2010 IEEE International Symposium on Technology and Society*, (pp. 180-189). IEEE Press.

Ching, Y. H., Huei, Y. S., Vitebsky, S., Pi-Chun, C., & Nat, C. (2003). Schedulers for 1xEV-DO: Third generation wireless high-speed data systems. *Bell Labs Technical Journal*, *3*, 1710–1714.

Chiron, P., Njedjou, E., Seite, P., Gosse, K., Melin, E., & Roux, P. (2008). Architectures for IP-based network-assisted mobility management across heterogeneous networks. *IEEE Wireless Communications*, *15*(2), 18–25. doi:10.1109/MWC.2008.4492974

Chiu, K.-L., Chen, Y.-S., Hwang, R.-H. (2010). Seamless session mobility scheme in heterogeneous wireless networks. *International Journal of Communication Systems*, Article first published online: 4 Nov. 2010.

Chiussi, F. M., Khotimsky, D. A., & Krishnan, S. (2002). Mobility management in third-generation all-IP networks. *IEEE Communications Magazine*, *40*(9), 124–135. doi:10.1109/MCOM.2002.1031839

Cho, Y.-B., Woo, S.-H., & Lee, S.-H. (2008). *Efficient key distribution scheme for secure routing in wireless mesh network environment* (pp. 79–82). World Scientific and Engineering Academy and Society.

Chung, Y., & Tsai, Z. (2010). Performance evaluation of dynamic spectrum sharing for two wireless communication networks. *IET Communications*, *4*(4), 452–462. doi:10.1049/iet-com.2009.0314

Cisco Visual Networking Index. (2010). *Global mobile data traffic forecast update, 2009-2014*. Retrieved September 29, 2010, from http://www.cisco.com/ en/US/ solutions/ collateral/ns341/ns525/ns537/ns705/ns827/ white_paper_c11-520862.html

Clancy, T. C. (2008). Secure handover in enterprise WLANs: CAPWAP, HOKEY, and IEEE 802.11r. *IEEE Wireless Communications*, *15*(5), 80–85. doi:10.1109/ MWC.2008.4653136

Clark, D., Sollins, K., Wroclawski, J., Katabi, D., Kulik, J., & Yang, X. (2003). *New arch: Future generation internet architecture*. Technical Report. Retrieved July, 2010, from http://www.isi.edu/newarch/iDOCS/final.finalreport.pdf

Clausen, T., & Jacquet, P. (2003). *Optimized link state routing protocol (OLSR). RFC 3626*. Experimental.

Clerckx, B., Lozano, A., Sesia, S., Rensburg, C. V., & Papadias, C. B. (2009). 3GPP LTE and LTE advanced. *EURASIP Journal on Wireless Communications and Networking*.

Congdon, P., & Lane, B. (2010). *802.1AB - Station and media access control connectivity discovery*. Retrieved from http://www.ieee802.org/1/pages/802.1ab.html

Cooklev, T. (2004). *Wireless communication standards: A study of IEEE 802.11, 802.15, and 802.16. Standards Information Network*. IEEE Press.

Coskun, H., Schieferdecker, I., & Al-Hazmi, Y. (2009). *Virtual WLAN: Going beyond virtual access points* (p. 17). ECEASST.

Costa, L. H. M. K., Fdida, S., & Duarte, O. (2001). *Hop by hop multicast routing protocol*. Paper presented at the 2001 Conference on Applications, Technologies, Architectures, and Protocols for Computer Communications.

Cottrell, R. L., Ansari, S., Khandpur, P., Gupta, R., Hughes-Jones, R., & Chen, M. … Leers, F. (2005). *Characterization and evaluation of TCP and UDP-based transport on real networks*. Third International Workshop on Protocols for Fast Long-Distance Networks, Lyon, France: IEEE Press.

Crawley, E., Nair, R., Rajagopalan, B., & Sandick, H. (1998). *A framework for QoS-based routing in the internet*. RFC 2386. Retrieved from http://www.ietf.org/ rfc/rfc2386.txt

Croker, D. (2003). *Multiple address service for transport (MAST): An extended proposal*. draft-crocker-mast-proposal-01. IETF draft (expired).

Crossbox. (2009). *TelosB mote platform*. Retrieved September 10th, 2009, from http://www.xbow.com/Products/ productdetails.aspx?sid=252

Cudak, M. (2007). *IEEE 802.16m system requirements* (IEEE 802.16m-07/002r4).

Cui, H., Yan, Q., Cai, Y., Gao, Y., & Wu, L. (2008). Heterogeneous network selection using a novel multi-attribute decision method. *Proceedings of International Conference on Communications and Networking in China* (ChinaCom'08), vol. 1, (pp. 153-157).

Dahlman, E., Ekstrom, H., Furuskar, A., Jading, Y., Karlsson, J., Lundevall, M., & Parkvall, S. (2006). *Proceedings of 63rd Semi-Annual IEEE Vehicular Technology Conference (VTC- Spring 2006)*, (pp. 137-141). Melbourne, Australia: IEEE Press.

Dahlman, E., Parkvall, S., Skold, J., & Beming, P. (2008). *3G evolution: HSPA and LTE for mobile broadband* (2nd ed.). Elsevier Ltd.

Damianou, N., Dulay, N., Lupu, E., & Sloman, M. (2001). *The ponder policy speci□cation language* (pp. 18–38). Lecture Notes in Computer Science Springer-Verlag.

Damljanovic, Z. (2009). Mobility management strategies in heterogeneous cognitive radio networks. *Journal of Network and Systems Management*, *18*(1), 4–22. doi:10.1007/s10922-009-9146-0

De la Oliva, A., Banchs, A., Soto, I., Melia, T., & Vidal, A. (2008). An overview of IEEE 802.21: media independent handover services. *IEEE Wireless Communications*, *15*(4), 96–103. doi:10.1109/MWC.2008.4599227

Deering, S. E., & Cheriton, D. R. (1990). Multicast routing in datagram internetworks and extended LANs. *ACM Transactions on Computer Systems*, *8*(2), 85–110. doi:10.1145/78952.78953

Demetrios, Z. (2001). *A glance at quality of services in mobile ad-hoc networks*. Burns College of Engineering (Data File). Retrieved from http://www.cs.ucr.edu/csyiazti/cs260.html

Devarapalli, V., Wakikawa, R., Petrescu, A., Thubert, P. (2005). *Network mobility (NEMO) basic support protocol*. IETF RFC 3963.

Dhar, J., Kiran, S., & Reddy, K. (2007). Network selection in heterogeneous wireless environment: a ranking algorithm. *Proceedings of International Conference on Wireless Communication and Sensor Networks* (WCSN'07), vol. 1 (pp. 41-44).

Dijkstra, E. W. (1959). A note on two problems in connexion with graphs. *Numerische Mathematik*, *1*(1), 269–271. doi:10.1007/BF01386390

Dilmaghani, R. B., & Rao, R. R. (2006). *Hybrid wireless mesh network deployment: A communication testbed for disaster scenarios*. 1st International Workshop on Wireless Network Testbeds, Experimental Evaluation & Characterization (WiNTECH '06). ACM.

Diot, C., Levine, B. N., Lyles, B., Kassem, H., & Balensiefen, D. (2000). Deployment issues for the IP multicast service and architecture. *IEEE Network*, *14*(1), 78–88. doi:10.1109/65.819174

Djahel, S., & Nait-Abdesselam, F. (2009). FLSAC: A new scheme to defend against greedy behavior in wireless mesh networks. *International Journal of Communication Systems*, *22*(10), 1245–1266. doi:10.1002/dac.1027

Dong, J., Curtmola, R., & Nita-Rotaru, C. (2009). Practical defenses against pollution attacks in intra-flow network coding for wireless mesh networks. *Proceedings of the Second ACM Conference on Wireless Network Security* (pp. 111-122). ACM.

Durantini, A., & Petracca, M. (2011). Performance comparison of vertical handover strategies for psdr heterogeneous networks . *IEEE Wireless Communications*, *15*(3), 54–59. doi:10.1109/MWC.2008.4547523

Duros, E., Dabbous, W., Izumiyama, H., Fujii, N., & Zhang, Y. (2001). *A link-layer tunneling mechanism for unidirectional links. RFC 3077*. Proposed Standard.

Eastwood, L., Migaldi, S., Xie, Q., & Gupta, V. (2008). Mobility using IEEE 802.21 in a heterogeneous IEEE 802.16/802.11-based, IMT-advanced (4G) network. *IEEE Wireless Communications*, *15*(2), 26–34. doi:10.1109/MWC.2008.4492975

Eddy, W. M. (2004). At what layer does mobility belong? *IEEE Communications Magazine*, *42*(10), 155–159. doi:10.1109/MCOM.2004.1341274

Edwards, G., & Sankar, R. (1997). A predictive fuzzy algorithm for high performance microcellular handoff. *Proceedings of IEEE Global Communications Conference (GLOBECOM)*, (pp. 987–990). IEEE Press.

Efstathiou, E. C., Frangoudis, P. A., & Polyzos, G. C. (2006). Stimulating participation in wireless community networks. *25th IEEE International Conference on Computer Communications* (pp.1-13).

El Hachimi, M., Abouaissa, A., Lorenz, P., & Lee, M. O. (2004). Scalable multicast provisioning in DiffServ with MPLS labeling. *Telecommunication Systems*, *27*(2), 253–272. doi:10.1023/B:TELS.0000041011.21556.da

Elayoubi, S., Ben Haddada, O., & Fourestie, B. (2008). Performance evaluation of frequency planning schemes in OFDMA-based networks. *IEEE Transactions on Wireless Communications*, *7*(5), 1623–1633. doi:10.1109/TWC.2008.060458.

Eljack, S., Igbal, A., & Furong, W. (2009). *A multi channel cognitive MAC protocol with efficient channel reservation and collision avoidance method.* Paper presented at the Conference on Multimedia Information Networking and Security (MINES). IEEE Press.

Elliott, R. C., & Krzymien, W. A. (2002). Scheduling algorithms for the CDMA2000 packet data evolution. *IEEE Vehicular Technology Conference*, vol. 1, (pp. 304-310). IEEE Press.

Emmelmann, M., Wiethoelter, S., Koepsel, A., Kappler, C., & Wolisz, A. (2007). Moving towards seamless mobility: State of art and emerging aspects in standardization bodies. *Wireless Personal Communications*, *43*(3), 803–816. doi:10.1007/s11277-007-9255-6

Eng Hwee Ong & Khan, J.Y. (2010). On optimal network selection in a dynamic multi-RAT environment. *IEEE Communications Letters*, *14*(3), 217–219. doi:10.1109/LCOMM.2010.03.092378

Eriksson, J., Agarwal, S., Bahl, P., & Padhye, J. (2006). Feasibility study of mesh networks for all-wireless offices. *4th International Conference on Mobile Systems, Applications and Services (MobiSys '06)* (pp. 69-82). New York, NY: ACM.

Esteves, E., Gurelli, M., & Fan, M. (2003). Performance of fixed wireless access with CDMA2000 1xEV-DO. *IEEE Vehicular Tehnology Conference*, vol. 2, (pp. 836-840). IEEE Press.

ETSI. (2010). *Reconfigurable radio systems (RRS): Cognitive radio system concept. European Telecommunications Standards Institute (ETSI), (Tech. Rep. TR 102 802, V1.1.1).* Sophia Antipolis, France: Cedex.

Fang, Y., & Zhang, Y. (2002). Call admission control schemes and performance analysis in wireless mobile networks. *IEEE Transactions on Vehicular Technology*, *51*(2), 371–382. doi:10.1109/25.994812

Farinacci, D., Fuller, V., Lewis, D., Meyer, D. (2010). *LISP mobile node.* Draft-Meyer-LISP-MN-03, IETF draft (work in progress).

Farinacci, D., Fuller, V., Meyer, D., & Lewis, D. (2010). *Locator/ID separation protocol (LISP).* Draft IETF-LISP-08, IETF draft (work in progress).

Farinacci, D., Li, T., Hanks, S., Meyer, D., & Traina, P. (2000). *Generic routing encapsulation (GRE).* RFC 2784 (Proposed Standard). Updated by RFC 2890.

Farrel, A., Vasseur, J.-P., & Ash, J. (2006). *A path computation element (PCE)-based architecture. RFC 4655.* Informational.

Figueiras, J., & Frattasi, S. (2010). *Mobile positioning and tracking.* Wiley. doi:10.1002/9780470663035

Firoze, A. M., Ju, L. Y., & Kwong, L. M. (2007). *PR-MAC: A priority reservation MAC protocol for wireless sensor networks.* Paper presented at the International Conference on Electrical Engineering, (ICEE '07). ICEE Press.

Flickenger, R., & Weeks, R. (2005). *Wireless hacks* (2nd ed.). O'Reilly Media.

Floyd, S., & Fall, K. (1999). Promoting the use of end-to-end congestion control in the Internet. *IEEE/ACM Transactions on Networking*, *7*(4), 458–472. doi:10.1109/90.793002

Fogelstroem, E., Jonsson, E., & Perkins, C. (2007). *Mobile IPv4 regional registration.* IETF RFC 4857.

Fu, S., Atiquzzaman, M., & Lee, Y. J. (2005). Architecture and performance of SIGMA: A seamless mobility architecture for data networks. In the *Proceeding of the ICC 2005*: Vol. 5 (3249- 3253). Seoul, Korea: IEEE Press.

Fudenberg, D., & Tirole, J. (1991). *Game theory.* Cambridge, MA: MIT Press.

Fuller, V. (2010). *LISP map server.* Draft-IETF-LISP-MS-05, IETF draft (work in progress).

Fullr, R. (1999). *Introduction to neuro-fuzzy systems. Advances in Soft Computing Series.* Springer-Verlag.

Funabiki, N., & Takefuji, Y. (1992). A neural network parallel algorithm for channel assignment problems in cellular radio networks. *IEEE Transactions on Vehicular Technology*, *41*(4), 430–437. doi:10.1109/25.182594

Gakhar, K., Gravey, A., & Leroy, A. (2005, 7-7 Oct. 2005). *IROISE: A new QoS architecture for IEEE 802.16 and IEEE 802.11e interworking.* Paper presented at the (BroadNets) 2nd International Conference on Broadband Networks.

Gallardo, R., Pineda-Rico, U., & Stevens-Navarro, E. (2009). VIKOR Method for vertical handoff decision in beyond 3G wireless networks. *Proceedings of IEEE International Conference In Automatic Control, Computing Science and Electronic Engineering* (CCE'09), vol. 1 (pp. 1-5).

Gang, W., Mizuno, M., & Havinga, P. J. M. (2002). MIRAI architecture for heterogeneous network. *Communications Magazine, IEEE, 40*(2), 126–134. doi:10.1109/35.983919

Ganz, A., Ganz, Z., & Wongthavarawat, K. (2004). *Multimedia wireless networks: Technologies, standards, and QoS*. Upper Saddle River, NJ: Prentice Hall.

Gao, F., Yuan, W., Liu, W., Cheng, W., & Wang, S. (2009). *Pipelined cooperative spectrum sensing in cognitive radio networks*. Paper presented at the Wireless Communications and Networking Conference (WCNC). IEEE Press

Gartner (2010). *Technology research & business leader insight*. Retrieved September 27, 2010, from http://www.gartner.com/it/page.jsp?id=1322414

Gast, M. (2005). *802.11 wireless networks: The definitive guide* (2nd ed.). O'Reilly Media.

Gavrilovska, L., & Atanasovski, V. (2007). Interoperability in future wireless communications systems: A roadmap to 4G. *Microwave Review, 13*(1), 19–28.

Geirhofer, S., Tong, L., & Sadler, B. M. (2006). *Dynamic spectrum access in WLAN channels: Empirical model and its stochastic analysis*. Paper presented at the ACM Workshop on Technology and Policy for Accessing Spectrum (TAPAS). ACM Press.

Geirhofer, S., Tong, L., & Sadler, B. M. (2008). Cognitive medium access: Constraining interference based on experimental models. *IEEE Journal on Selected Areas in Communications, 26*(1). doi:10.1109/JSAC.2008.080109

Gelenbe, E., Sakellari, G., & D'Arienzo, M. (2007). Controlling access to preserve QoS in a self-aware network. *International Conference on Self-Adaptive and Self-Organizing Systems* (pp. 205-213).

George, A., Kumar, A., Cavalcanti, D., & Agrawal, D. P. (2008). Protocols for mobility management in heterogeneous multi-hop wireless networks. *Pervasive and Mobile Computing, 4*(1), 92–116. doi:10.1016/j.pmcj.2007.03.002

Geun, K., Jeong, K., & Marwan, M. (2000). Bandwidth allocation in wireless networks with guaranteed packet-loss performance. *IEEE/ACM Transactions on Networking, 8*(3), 337–349. doi:10.1109/90.851980

Giupponi, L., & Perez-Neira, A. I. (2008). Fuzzy-based spectrum handoff in cognitive radio networks. *Proceeding of IEEE International Conference on Cognitive Oriented Wireless Networks and Communications (CrownCom)*, (pp. 1–6). Singapore: IEEE Press.

Goff, T., Moronski, J., Phatak, D. S., & Gupta, V. (2000). Freeze-TCP: A true end-to-end TCP enhancement mechanism for mobile environments. In the *Proceeding of IEEE International Conference on Computer Communications* (pp. 1537-1545). Tel Aviv, Israel: IEEE Press.

Goldsmith, J., & Chua, G. (1997). Variable-rate variable-power MQAM for fading channels. *IEEE Transactions on Communications, 45*(10), 1218–1230. doi:10.1109/26.634685

Goldsmith, J., & Chua, G. (1998). Adaptive coded modulation for fading channels. *IEEE Transactions on Communications, 46*(5), 595–602. doi:10.1109/26.668727

Gomes, R. Jr, Cerqueira, W. A., & Abelem, A. (2010). A QoE fuzzy routing protocol for wireless mesh networks. *Lecture Notes in Computer Science*, 6157.

Gondi, V. K. (2009). Seamless secured roaming over heterogeneous wireless networks. Doctoral dissertation, University of Evry, Evry, France.

Goransson, P., & Greenlaw, R. (2007). *Secure roaming in 802.11 networks (communications engineering)*. Newnes.

Gowrishankar, R., Raju, G. T., & Satyanarayana, P. S. (2008). Performability model of vertical handoff in wireless data networks. *Proceedings of International Conference on Wireless and Mobile Communications* (ICWMC'08), vol. 1 (pp. 309-314).

Gu, T., Wang, X., Pung, H., & Zhang, D. (2004). *An ontology-based context model in intelligent environments.*

Gulbrandsen, A., Vixie, P., & Esibov, L. (2000). *A DNS RR for specifying the location of services* (DNS SRV). IETF RFC 2782.

Gundavelli, S., et al. (2008). *Proxy mobile IPv6*. IETF RFC 5213.

Guo, Q., Zhu, J., & Xu, X. (2005). An adaptive multi-criteria vertical handoff decision algorithm for radio heterogeneous networks. *Proceedings of IEEE International Communications Conference* (ICC'05), Vol. 4, (2769-2773).

Guo, C., Guo, Z., Zhang, Q., & Zhu, W. (2004). A seamless and proactive end-to-end mobility solution for roaming across heterogeneous wireless networks. *IEEE Journal on Selected Areas in Communications*, *22*(5), 834–848. doi:10.1109/JSAC.2004.826921

Guoqiang, M., & Fidan, B. (2009). *Localization algorithms and strategies for wireless sensor networks*. Hershey, PA: IGI Global.

Gupta, V. (2009). *IEEE 802.21-2008 IEEE standard for media independent handover services*. IEEE Std. 802.21-2008, January 2009.

Gustafsson, E., & Jonsson, A. (2003). Always best connected. *IEEE Wireless Communications*, *10*(1), 49–55. doi:10.1109/MWC.2003.1182111

Haas, Z. J., & Agrawal, P. (1997). Mobile-TCP: An asymmetric transport protocol design for mobile systems. In *Proceeding of IEEE International Conference on Communications* (pp. 1054-1058), Montreal, Canada: IEEE Press.

Haas, Z. J., Gerla, M, Johnson, D. B., Perkins, C. E., Pursley, M. B., Steenstrup, M., … Hayes, J. F. (1999). Guest editorial on wireless ad hoc networks. *IEEE Journal on Selected Areas in Communications – Special Issue on Wireless Networks, 17*(8): 1329–1332.

Hamanaka, T., Wang, X., Wada, N., & Kitayama, K.-I. (2007). Compound data rate and data-rate-flexible 622 Mb/s–10 Gb/s OCDMA experiments using 511-chip SSFBG and cascaded SHG-DFG-based PPLN waveguide optical thresholder. *IEEE Journal on Selected Topics in Quantum Electronics*, *13*(5), 1516–1521. doi:10.1109/JSTQE.2007.904082

Hamida, E. B. (2008). Neighbor discovery in multi-hop wireless networks: Evaluation and dimensioning with interference consideration. *Discrete Mathematics and Theoretical Computer Science*, *10*(2), 87–114.

Han, Y. H., Jang, H., Choi, J., Park, B., & McNair, J. (2007). A cross-layering design for IPv6 fast handover support in an IEEE 802.16e wireless MAN. *IEEE Network*, *21*(6), 54–62. doi:10.1109/MNET.2007.4395111

Hanzo, L., & Tafazolli, R. (2007). A survey of QoS routing solutions for mobile ad hoc networks. *IEEE Communications Surveys & Tutorials*, *9*(2), 50–70. doi:10.1109/COMST.2007.382407

Hejmo, M., Mark, B., Zouridaki, C., & Thomas, R. (2006). Design and analysis of a denial-of-service resistant quality-of-service signaling protocol for MANETs. *IEEE Transactions on Vehicular Technology*, *55*(3), 743–751. doi:10.1109/TVT.2006.873834

Ho, C., Obraczka, K., Tsudik, G., & Viswanath, K. (1999). Flooding for reliable multicast in multi-hop ad hoc networks. *Proceedings of the International Workshop on Discrete Algorithms and Methods for Mobile Computing and Communication (DIALM)*, (pp. 64–71). Seattle, WA: ACM Press.

Holma, H., & Toskala, A. (2000). *WDCDMA for UMTS – Radio access for third generation mobile communications* [Book]. John Wiley & Sons, Ltd.

Holma, H., & Toskala, A. (2009). *LTE for UMTS - OFDMA and SC-FDMA based radio access*. Chichester, UK: John Wiley & Sons.

Hongjun, L., Xun, L., Shen, J., & Hongxu, M. (2009). *A fair multi-priority MAC protocol design of wireless sensor networks*. Paper presented at the Networks Security, Wireless Communications, and Trusted Computing, (NSWCTC '09). IEEE Press.

Hosseini, M., Ahmed, D. T., Shirmohammadi, S., & Georganas, N. D. (2007). A survey of application-layer multicast protocols. *IEEE Communications Surveys & Tutorials*, *9*(3), 58–74. doi:10.1109/COMST.2007.4317616

Hou, J., & O'Brien, D. C. (2006). Vertical handover-decision-making algorithm using fuzzy logic for the integrated radio-and-OW system. *IEEE Transactions on Wireless Communications*, *5*(1), 176–185. doi:10.1109/TWC.2006.1576541

Hsieh, H.-Y., & Sivakumar, R. (2002). pTCP: An end-to-end transport layer protocol for striped connections. In *10th IEEE International Conference on Network Protocols (ICNP 2002)*, Paris, France.

Hu, P., Portmann, M., Robinson, R., & Indulska, J. (2008). Context-aware routing in wireless mesh networks. *ACM International Conference on Context-Awareness for Self-Managing Systems* (pp. 16-23).

Huang, P. J., Tseng, Y. C., & Tsai, K. C. (2006). A fast handoff mechanism for IEEE 802.11 and IAPP networks. In the *Proceeding of 63rd IEEE Vehicular Technology Conference*: Vol. 2 (pp. 966-970). Grand Hyatt Melbourne, Melbourne, Australia: IEEE Press.

Huang, S., Liu, X., & Ding, Z. (2008). *On optimal sensing and transmission strategies for dynamic spectrum access*. Paper presented at the New Frontiers in Dynamic Spectrum Access Networks, (DySPAN 2008). IEEE Press.

Huang, C. M., Lin, C. W., & Yang, C. C. (2010). Mobility management for video streaming on heterogeneous networks. *IEEE MultiMedia*, *17*(1), 24–32. doi:10.1109/MMUL.2010.17

Huang, S., Liu, X., & Ding, Z. (2009). Optimal transmission strategies for dynamic spectrum access in cognitive radio networks. *IEEE Transactions on Mobile Computing*, *8*(12), 1636–1648. doi:10.1109/TMC.2009.84

Hui, S. K., & Yeung, K. H. (2003). Challenges in the migration to 4G mobile systems. *IEEE Communications Magazine*, *41*(12), 54–59. doi:10.1109/MCOM.2003.1252799

Huschke, J., & Leaves, P. (2003). *Dynamic spectrum allocation algorithm including results of DSA performance simulations. DRiVE Deliverable D09*. Information Society Technologies.

Hussain, S., Hamid, Z., & Khattak, N. S. (2006). Mobility management challenges and issues in 4G heterogeneous networks. In proceeding of *The First International Conference on Integrated Internet Ad Hoc and Sensor Networks* (14-19). Nice, France: ACM Press.

IEEE 802.11. (1999). *WG: Reference number ISO/IEC 8802-11:1999 (E) IEEE STD 802.11. International standard for information technology-Telecommunications and information exchange between systems-Local and metropolitan area networks-Specific requirements- Part 11: Wireless LAN medium access control (MAC) and physical layer (PHY) specifications*. IEEE Press.

IEEE 802.11g. (2003). *IEEE standard for information technology—Telecommunications and information exchange between systems—Local and metropolitan area networks—Specific requirements—Part 11: Wireless LAN medium access control (MAC) and physical layer (PHY) specifications—Amendment 4: Further higher-speed physical layer extension in the 2.4 GHz Band*. IEEE Press.

IEEE. Computer Society. (2009). *IEEE standard for local and metropolitan area networks - Part 21*. Media Independent Handover Services. Retrieved October 3rd from http://standards.ieee.org/getieee802/802.21.html

Iera, A., Molinaro, A., Campolo, C., & Amadeo, M. (2006). *An access network selection algorithm dynamically adapted to user needs and preferences*. Paper presented at the Personal, Indoor and Mobile Radio Communications, 2006 IEEE 17th International Symposium.

IETF. (2001). *Multi protocol label switching (MPLS)*. RFC3031 (Proposed Standard). Retrieved from http://www.ietf.org/rfc/rfc3031.txt

IETF. (2008). *RFC 5213: Proxy mobile IPv6*. Retrieved on October 3rd from http://www.ietf.org/rfc/rfc5213.txt?number=5213

Ioannidis, J., Duchamp, D., & Maguire, G. (1991). IP-based protocols for mobile internetworking. In *ACM Proceeding of the Conference on Communications Architecture & Protocols* (pp. 235-245). Zurich, Switzerland: ACM Press.

ITU-RReportM. 2134. (2008). *Report ITU-R M.2134: Requirements related to technical performance for IMT-advanced radio interface(s)*.

Iyengar, J. R., Amer, P. D., & Stewart, R. (2006). Concurrent multipath transfer using sctp multihoming over independent end-to-end paths. *IEEE/ACM Transactions on Networking*, *14*(5), 951–964. doi:10.1109/TNET.2006.882843

Jacquet, P., Muhlethaler, P., Clausen, T., Laouiti, A., Qayyum, A., & Viennot, L. (2004). Optimized link state routing protocol for ad hoc networking. *Proceedings of IEEE International Multi Topic Conference (INMIC)*. IEEE Press.

Jain, R. K., Chiu, D. W., & Hawe, W. R. (1984). A quantitative measure of fairness and discrimination for resource allocation in shared computer systems. *Research Report TR-301*. Digital Equipment Corporation.

Jangeun, J., Peddabachagari, P., & Sichitiu, M. (2003). Theoretical maximum throughput of IEEE 802.11 and its applications. *IEEE International Symposium on Network Computing and Applications (NCA)*, (pp. 249-256). IEEE Press.

Jannotti, J., Gifford, D. K., Johnson, K. L., Kaashoek, M. F., & James, W. O'Toole, J. (2000). *Overcast: Reliable multicasting with on overlay network.* Paper presented at the 4th Conference on Symposium on Operating System Design & Implementation.

Jesus, V., Sargento, S., & Aguiar, R. L. (2008). *Any-constraint personalized network selection.* Paper presented at the Personal, Indoor and Mobile Radio Communications, 2008. PIMRC 2008. IEEE 19th International Symposium.

Ji, L., Xia, X., Shi, X., & Wang, Z. (2007). *Conversational model based VoIP traffic generation.* IEEE Third International Conference on Networking and Services. IEEE Press.

Jia, J., Zhang, Q., & Shen, X. (2008). HC-MAC: A hardware-constrained cognitive MAC for efficient spectrum management. *IEEE Journal on Selected Areas in Communications, 26*(1). doi:10.1109/JSAC.2008.080110

Jiang, H., Bao, Y., Li, Q., & Huang, Y. (2009). Multi-objective optimization of cross-layer configuration for cognitive wireless network. *Proceeding of IEEE international Conference on Dependable, Autonomic and secure computing (DASC)*, (pp. 379–383). Chengdu, China: IEEE Press.

Jia, Z., Yu, J., Ellinas, G., & Chang, G. K. (2007). Key enabling technologies for optical-wireless networks: Optical millimeter-wave generation, wavelength reuse, and architecture. *Journal of Lightwave Technology, 25*(11), 3452–3471. doi:10.1109/JLT.2007.909201

Jingli, L., Xiangqian, L., & Swami, A. (2009). Collision analysis for coexistence of multiple bluetooth piconets and WLAN with dual channel transmission. *IEEE Transactions on Communications, 57*(4), 1129–1139. doi:10.1109/TCOMM.2009.04.070103

Joe, I., Kim, W. T., & Hong, S. (2008). A network selection algorithm considering power consumption in hybrid wireless networks. *IEICE Transactions on Communications . E (Norwalk, Conn.), 91-B*(1), 314–317.

Johnson, D., Perkins, C., & Arkko, J. (2004). *Mobility support in IPv6.* Internet Engineering Task Force, RFC 3775 (Proposed Standard). Retrieved from http://www.ietf.org/rfc/rfc3775.txt

Johnson, D. B., & Maltz, D. A. (1996). Dynamic source routing in ad hoc wireless networks . In Imielinski, T., & Korth, H. (Eds.), *Mobile computing* (*Vol. 353*, pp. 153–181). Kluwer Academic Publishers. doi:10.1007/978-0-585-29603-6_5

Jung, E., & Xin, L. (2008). *Opportunistic spectrum access in heterogeneous user environments.* Paper presented at the New Frontiers in Dynamic Spectrum Access Networks, (DySPAN 2008). IEEE Press.

Kalampoukas, L., Varma, A., & Ramakrishnan, K. K. (1998). Two-way TCP traffic over rate controlled channels: Effects and analysis. *IEEE/ACM Transactions on Networking, 6*(6), 729–743. doi:10.1109/90.748085

Karrer, R., Zerfos, P., & Piratla, N. (2006). *Magnets: A next generation access network.* Poster at IEEE International Conference on Computer Communications (INFOCOM 2006).

Kassar, M., Kervella, B., & Pujolle, G. (2008). An overview of vertical handover decision strategies in heterogeneous wireless networks. *Computer Communications, 31*, 2607–2620. doi:10.1016/j.comcom.2008.01.044

Kerry, S. J. et al. (2003). *IEEE trial-use recommended practice for multi-vendor access point interoperability via an inter-access point protocol across distribution systems supporting IEEE 802.11 operation.* IEEE Std. 802.11F-2003. July, 2003.

Kerry, S. J., et al. (2008). *IEEE standard for information technology—Telecommunications and information exchange between systems local and metropolitan area networks specific requirements—part 11: wireless LAN medium access control (MAC) and physical layer (PHY) Spec., amendment 2: fast basic service set (BSS) transition.* IEEE Std. 802.11r-2008. July 2008.

Keshava Murthy, K. S. (2008). *NextGen wireless access gateway analysis of combining WiMAX and LTE gateway functions.* Paper presented at the (IMSAA 2008), 2nd International Conference on Internet Multimedia Services Architecture and Applications.

Kewen, L., & Jing, T. (2008). *A QoS mobile multicast routing algorithm based ant colony algorithm.* Paper presented at the Computational Intelligence and Industrial Application, 2008. PACIIA '08. Pacific-Asia Workshop.

Khoukhi, L., & Cherkaoui, S. (2010). Intelligent QoS management for multimedia services support in wireless mobile ad hoc networks. *Computer Networks and ISDN Systems*, *54*(10), 1692–1706.

Kim, S., & Yeom, I. (2007). TCP-aware uplink scheduling for IEEE 802.16. *IEEE Communications Letters*, *11*(2), 146–148. doi:10.1109/LCOMM.2007.061509

Kim, Y.-S., Kwon, D.-H., & Suh, Y.-J. (2008). A client-based vertical handoff approach for seamless mobility across heterogeneous wireless networks. *Wireless Communications and Mobile Computing*, *10*(9), 1208–1220. doi:10.1002/wcm.673

Klein, D., Hartmann, M., & Menth, M. (2010). *NAT traversal for LISP mobile node*. Draft-klein-lisp-mn-nat-traversal-00, IETF draft (work in progress).

Klerer, S. M., et al. (2009). *IEEE 802.20-2008 IEEE standard for local and metropolitan area networks – Part 20: Air interface for mobile broadband wireless access systems supporting vehicular mobility – Physical and media access control layer specification*. IEEE Std. 802.20-2009, 2009.

Koh, S., J., Chang, M., J., & Lee, M. (2004). mSCTP for soft handover in transport layer. *IEEE Communications Letters, 8*(3), 189-191.

Kohler, E., Handley, M., & Floyd, S. (2006). *Datagram congestion control protocol* (DCCP). IETF RFC 4340. (. *Updated by*, *RFC5595*, RFC5596.

Komali, R. S., Thomas, R. W., DaSilva, L. A., & MacKenzie, A. B. (2010). The price of ignorance: Distributed topology control in cognitive networks. *IEEE Transactions on Wireless Communications*, *9*(4), 1434–1445. doi:10.1109/TWC.2010.04.090400

Komolafe, O., & Sventek, J. (2007). Analysis of RSVP-TE graceful restart. In *IEEE International Conference on Communications, 2007*, (pp. 2324-2329). Glasgow, Scotland. IEEE Press.

Kompella, K., & Rekhter, Y. (2003). *Signaling unnumbered links in resource reservation protocol - Traffic engineering (RSVP-TE). RFC 3477*. Proposed Standard.

Kone, V., Das, S., Zhao, B. Y., & Zheng, H. (2007). *Quorum: Quality of service routing in wireless mesh networks*. In ICST Conference on Heterogeneous Networking for Quality, Reliability, Security, and Robustness (QShine). New York, NY: ACM Press.

Koodli, R. (2005). *Fast handovers for mobile IPv6*. IETF RFC 4068 (Obsoleted by RFC5268).

Koodli, R., & Perkins, C. (2007). *Mobile IPv4 fast handovers*. IETF RFC 4988.

Kovalenko, I., Zhukov, V., & Shelkovnikov, N. (2009). Features of long term evolution technology and comparative analysis of noise stability for different kinds of modulation. *Proceedings of 19th International Crimean Conference Microwave & Telecommunication Technology (CriMiCo 2009)*, (pp. 235-236). Sevastopol, Ukraine: IEEE Press.

Kretschmer, M., & Ghinea, G. (2009). Seamless integration of unidirectional broadcast links into QoS-constrained broadband wireless mesh access networks. In *Proceedings of the 4th International Conference for Internet Technology and Secured Transactions*. London, UK: IEEE Press.

Kretschmer, M., & Ghinea, G. (2010). *An IEEE 802.21-based approach for seamless wireless mobile integration using QoS-aware paths supporting unidirectional links*. In IEEE Globecom 2010 Workshop on Seamless Wireless Mobility (SWM 2010), Miami, FL: IEEE Press.

Kretschmer, M., Niephaus, C., & Ghinea, G. (2009a). *QoS-aware flow monitoring and event creation in heterogeneous MPLS-based wireless mesh networks supporting unidirectional links*. In 9th IEEE International Conference on Communications 2009, Kuala Lumpur, Malaysia. IEEE Press.

Kretschmer, M., Niephaus, C., & Ghinea, G. (2009b). *Towards QoS provisioning in a heterogeneous carrier-grade wireless mesh access networks using unidirectional overlay cells. Lecture Notes of the Institute for Computer Sciences, Social Informatics and Telecommunications Engineering, 22*. Berlin, Germany: Springer.

Krichene, N., & Boudriga, N. (2008). Handoff management in WiMAX . In Zang, Y. (Ed.), *WIMAX network planning and optimization*. CRC Press.

Krondorf, M., & Fettweis, G. (2009). Carrier frequency dependent downlink spectral efficiency of cellular LTE deployments. *Proceedings of IEEE International Conference on Communications (ICC 2009)*, (pp. 4037-4042). Dresden, Germany: IEEE Press.

Kunz, D. (1991). Channel assignment for cellular radio using neural networks. *IEEE Transactions on Vehicular Technology*, *40*(1), 188–193. doi:10.1109/25.69987

Kunz, D., Duque-Anton, M., & Ruber, B. (1993). Channel assignment for cellular radio using simulated annealing. *IEEE Transactions on Vehicular Technology*, *42*(1), 14–21. doi:10.1109/25.192382

Kyamakya, K., & Jobmann, K. (2005). Location management in cellular networks: Classification of the most important paradigms, realistic Simulation framework, and relative performance analysis. *IEEE Transactions on Vehicular Technology*, *54*(2), 687–708. doi:10.1109/TVT.2004.842459

Kyasanur, P., & Vaidya, N. (2005). Routing and interface assignment in multi-channel multi-interface wireless networks. In *Proceedings IEEE Conference on Wireless Communications and Networking Conference*, (pp. 2051–56). New Orleans, LA: IEEE Press.

Kyasanur, P., & Vaidya, N. (2006). Routing and link-layer protocols for multi-channel multi-interface ad hoc wireless networks. *SIGMOBILE Mobile Computing and Communications Review*, *10*(1), 31–43. doi:10.1145/1119759.1119762

Kyungtae, K., Jinsung, C., Yongwoo, C., & Heonshik, S. (2005). Dynamic scheduling for scalable media transmission over CDMA2000 1xEV-DO broadcast and multicast networks. *Lecture Notes in Computer Science*, *3462*, 31–57.

Lajos, H., & Rahim, T. (2009). Admission control schemes for 802.11-based multi-hop mobile ad hoc networks: A survey. *IEEE Communications Surveys & Tutorials*, *11*(4), 78–108. doi:10.1109/SURV.2009.090406

LAN/MAN Committee. (2007). *Part 11: Wireless LAN medium access control (MAC) and physical layer (PHY) specifications* (IEEE Std 802.11-2007).

Langar, R., Bouabdallah, N., & Boutaba, R. (2009). Mobility-aware clustering algorithms with interference constraints in wireless mesh networks. *Computer Networks*, *53*(1), 25–44. doi:10.1016/j.comnet.2008.09.012

Lao, L., Cui, J. H., Gerla, M., & Chen, S. (2007). A scalable overlay multicast architecture for large-scale applications. *IEEE Transactions on Parallel and Distributed Systems*, *18*(4), 449–459. doi:10.1109/TPDS.2007.1008

Laptop.org. (2008). Retrieved 18th November, 2008, from http://www.laptop.org

LaRocca, J., & LaRocca, R. (2002). *IEEE 802.11 demystified*. McGraw-Hill Professional.

Law, C., Mehta, A. K., & Siu, K. (2003). A new Bluetooth scatternet formation protocol. *Mobile Networks and Applications*, *8*(5), 485–498. doi:10.1023/A:1025133710426

Le, H. S. T., & Liang, Q. (2007). An efficient power control scheme for cognitive radios. *Proceeding of IEEE Wireless Communications and Networking Conference (WCNC)*, (pp. 2559–2563). Hong Kong, China: IEE Press.

Leaves, P., Moessner, K., Tafazolli, R., Grandblaise, D., Bourse, D., Tonjes, R., & Breveglieri, M. (2004). Dynamic spectrum allocation in composite reconfigurable wireless networks. *IEEE Communications Magazine*, *42*(5), 72–81. doi:10.1109/MCOM.2004.1299346

Le, D., Fu, X. M., & Hogrefe, D. (2006). A review of mobility support paradigms for the Internet. *IEEE Communications Surveys & Tutorials*, *8*(1), 38–51. doi:10.1109/COMST.2006.323441

LEE, C. C. (1999). Fuzzy logic in control systems: Fuzzy logic controller - part I and II. *IEEE Transactions on Systems, Man, and Cybernetics*, *20*(2), 404–435.

Lee, C. W., Sun, Y., & Chen, M. C. (2008). HiMIP-NEMO: Combining cross-layer network mobility management and resource allocation for fast QoS-handovers. In the *Proceeding of VTC Spring 2008* (282-286). Marina Bay, Singapore: IEEE Press.

Lee, L. S., & Wang, K. (2010). Design and analysis of a network-assisted fast handover scheme for IEEE 802.16e networks. *IEEE Transactions on Vehicular Technology*, *59*(2), 869–883. doi:10.1109/TVT.2009.2037508

Lee, S., Ahn, G., & Campbell, A. T. (2001). Improving UDP and TCP performance in mobile ad hoc networks with INSIGNIA. *IEEE Communications Magazine*, *39*(6), 156–165. doi:10.1109/35.925684

Lee, S., Ahn, G., Zhang, X., & Campbell, A. (2000). INSIGNIA: An IP-based quality of service framework for mobile ad hoc networks. *Journal of Parallel and Distributed Computing*, *60*(4), 374–406. doi:10.1006/jpdc.1999.1613

Lei, P., Hongyi, W., & Nian-Feng, T. (2009). *An efficient and scalable prioritized MAC protocol (PMAC) for backbone communication in wireless sensor networks*. Paper presented at the Sensor Technologies and Applications (SENSORCOMM '09). IEEE Press.

Lenagala, R., & Zeng, Q. A. (2006). Study of dynamic MAC layer parameters for starvation prevention in the IEEE 802.11e MAC layer protocol. Paper presented at the IEEE International Conference on Wireless and Mobile Computing, Networking, and Communications. IEEE Press.

Lessmann, J., Schoeller, M., & Zdarsky, F. (2010). Rope ladder routing: Position-based multipath routing for wireless mesh networks. In *IEEE International Symposium on a World of Wireless Mobile and Multimedia Networks (WoWMoM), 2010*, (pp. 1-6). Montreal, Canada: IEEE Press.

Leung, K., Dommety, G., Narayanan, V., & Petrescu. A. (2008). *Network mobility (NEMO) extensions for mobile IPv4.* IETF RFC 5177.

Li, G., & Zhu, J. (2007). Performance analysis on utilizing directional antenna in high-density WLAN.*Proceedings of the Sixth International Conference on Networking* (ICN'07). Sainte-Luce, Martinique: IEEE Press.

Li, L. S., Tsai, Y. H., & Lin, P. C. (2010). An efficient fast handoff scheme in hierarchical mobile IPv6 for NEMO. In the *Proceeding of 2010 International Symposium on Communication Control and Automation (3CA)*: Vol. 2 (pp. 90-94). Tainan, China: IEEE Press.

Liang, Q., & Ren, Q. (2005). Energy and mobility aware geographical multipath routing for wireless sensor networks. *IEEE Wireless Communications and Networking Conference*, vol. 3, (pp. 1867-1871).

Liao, J., Wang, J., & Zhu, X. (2008). A multi-path mechanism for reliable VoIP transmission over wireless networks. *Computer Networks*, *52*, 2450–2460. doi:10.1016/j.comnet.2008.04.008

Li, N., & Hou, J. C. (2006). Localized fault-tolerant topology control in wireless ad hoc networks. *IEEE Transactions on Parallel and Distributed Systems*, *17*(4), 307–320. doi:10.1109/TPDS.2006.51

Lin, T., Wang, C., & Lin, P. (2008). A neural-network-based context-aware handoff algorithm for multimedia computing. *ACM Transactions on Multimedia Computing, Communications, and Applications*, *4*(3), 1–23. doi:10.1145/1386109.1386110

Liu, C.-T., Wang, T.-C., & Chang, R. (2010). *Cross-layer handoff via predictive multiples pre-registrations for QoS of mobile multimedia applications*. 2010 IEEE International Conference on Wireless Communications, Networking, and Information Security. IEEE Press.

Liu, Y., Yu, R., & Xie, S. (2008). *Optimal cooperative sensing scheme under time-varying channel for cognitive radio networks*. Paper presented at the DySPAN. IEEE Press.

Lombriser, C., Marin-Perianu, R., Roggen, D., Havinga, P., & Troester, G. (2009). Modeling service-oriented context processing in dynamic body area networks. *IEEE Journal on Selected Areas in Communications*, *27*(1), 49–57. doi:10.1109/JSAC.2009.090106

Luby, M., Gasiba, T., Stockhammer, T., & Watson, M. (2007). Reliable multimedia download delivery in cellular broadcast networks. *IEEE Transactions on Broadcasting*, *53*(1), 235–246. doi:10.1109/TBC.2007.891703

Lucarz, F., & Seeds, A. (2009). *Gigabit/s wireless over fibre systems for future access networks*. Retrieved September 29, 2010, from http://www.ee.ucl.ac.uk/lcs/previous/LCS2005/44.pdf

Lukas, G., & Fackroth, C. (2009). WMNSec: Security for wireless mesh networks. *Proceedings of the 2009 International Conference on Wireless Communications and Mobile Computing Connecting the World Wirelessly* (pp. 90-95). ACM.

Luo, H., & Shankaranarayanan, N. (2004). A distributed dynamic channel allocation technique for throughput improvement in a dense WLAN environment. In *Proceedings of IEEE International Conference on Acoustics, Speech, and Signal Processing*, vol. 5, (pp. 345-348). Montreal, Canada: IEEE Press.

Määttä, J., & Bräysy, T. (2009). A novel approach to fair routing in wireless mesh networks. *EURASIP Journal on Wireless Communications and Networking*, (1): 1–13. doi:10.1155/2009/379059

MacKenzie, A., Reed, J., Athanas, P., Bostian, C., Buehrer, R., & DaSilva, L. … da Silva, C. (2009). Cognitive radio and networking research at Virginia Tech. *Proceedings of the IEEE, 97*(4), 660-688.

MadWifi. (n. d.). *Multiband atheros driver for wireless fidelity*. Retrieved from http://madwifi-project.org/

Mahmoud, Q. H. (Ed.). (2007). *Cognitive networks: Towards self-aware networks*. West Sussex, UK: John Wiley & Sons Ltd.

Ma, L., Yu, F. R., & Leung, V. C. M. (2007). Performance improvements of mobile SCTP in integrated heterogeneous wireless networks. *IEEE Transactions on Wireless Communications, 6*(10), 3567–3577. doi:10.1109/TWC.2007.05403

Malki, K. E. (2007). *Low-latency handoffs in Mobile IPv4*. IETF RFC 4881.

Maltz, D. A., & Bhagwat, P. (1998). MSOCKS: An architecture for transport layer mobility. In the *Proceeding of IEEE International Conference on Computer Communications*: Vol. 3 (1037-1045). San Francisco, CA: IEEE Press.

Mamdani, E. H., & Assilian, S. (1975). An experiment in linguistic synthesis with a fuzzy logic controller. *International Journal of Man-Machine Studies, 7*(1), 1–13. doi:10.1016/S0020-7373(75)80002-2

Mangold, S., Habetha, J., Choi, S., & Ngo, C. (2001). Coexistence and interworking of IEEE 802.11a and ETSI BRAN HiperLAN/2 in MultiHop scenarios. In *Proceedings of IEEE 3rd Workshop in Wireless Local Area Networks*, (pp. 27-28). Boston, MA: IEEE Press.

Mangold, S., Sunghyun, C., Hiertz, G. R., Klein, O., & Walke, B. (2003). Analysis of IEEE 802.11e for QoS support in wireless LANs. *IEEE Wireless Communications, 10*(6), 40–50. doi:10.1109/MWC.2003.1265851

Mannie, E. (2004). *Generalized multi-protocol label switching (GMPLS) architecture. RFC 3945*. Proposed Standard.

Marina, M., & Das, S. (2005). A topology control approach for utilizing multiple channels in multi-radio wireless mesh networks. *2nd International Conference on Broadnets*, (pp. 381–90). Boston, MA: IEEE Press.

Marks, R. B. (2006). *IEEE standard for local and metropolitan area networks—Part 16: Air interface for fixed and mobile broadband wireless access systems*. IEEE Std. 802.16e-2005, Feb. 2006.

Marques, V., Costa, X. P., Aguiar, R. L., Liebsch, M., & Duarte, A. M. O. (2005). Evaluation of a mobile IPv6-based architecture supporting user mobility QoS and AAAC in heterogeneous networks. *IEEE Journal on Selected Areas in Communications, 23*(11), 2138–2151. doi:10.1109/JSAC.2005.856825

Marrero, D., Suárez, A., & Macías, E. M. (2005). Dynamic interconnection of ad hoc nodes based on the type of service to be accessed. *International Conference on Wireless Networks* (*ICWN*), (pp. 539-545). IEEE Press.

Marrero, D., Suárez, A., & Macías, E. M. (2007). *Dynamic traffic regulation for WiFi networks*. World Congress on Engineering 2007 (WCE2007). IEEE Press.

Marrero, D., Suárez, A., & Macías, E. M. (2008). An admission control and traffic regulation mechanism for infrastructure WiFi networks. *IAENG International Journal of Computer Science, 35*(1).

Marrero, D., Suárez, A., & Macías, E. M. (2009). *Aplicación multifuncional para gestión del canal en redes IEEE 802.11. VIII Jornadas de Ingeniería Telemática (JITEL2009)* (pp. 16–23). IEEE Press.

Martinez, J. P. (Ed.). (2008). *Enable efficient and operational mobility in large heterogeneous IP networks*.

Martinez-Morales, J. D., Pineda-Rico, U., & Stevens-Navarro, E. (2010). Performance comparison between MADM algorithms for vertical handoff in 4G networks. *Proceedings of IEEE International Conference In Automatic Control, Computing Science and Electronic Engineering* (CCE'10), vol. 1, (pp. 1-5).

Martini, L., Rosen, E., El-Aawar, N., & Heron, G. (2006). *Encapsulation methods for transport of ethernet over MPLS networks. RFC 4448*. Proposed Standard.

Mathar, R., & Mattfeldt, J. (1993). Channel assignment in cellular radio networks. *IEEE Transactions on Vehicular Technology, 42*(4), 647–656. doi:10.1109/25.260746

Mathieu, B., Meng, S., Galis, A., Cheng, L., Jean, K., Ocampo, R., et al. (2007). *Self-management of context-aware overlay ambient networks.* Paper presented at the Integrated Network Management, 2007. IM '07. 10th IFIP/IEEE International Symposium.

Matos, R., Sargento, S., Hummel, K., Hess, A., Tutschku, K., & de Meer, H. (2013). Context-based wireless mesh networks: A case for network virtualization. *Special Issue on Future Internet Services and Architectures: Trends and Visions, Telecommunication Systems, 52*(3). Springer.

McNair, J., & Zhu, F. (2004). Vertical handoffs in fourth-generation multinetwork environments. *IEEE Wireless Communications, 11*(3), 8–15. doi:10.1109/MWC.2004.1308935

Mehlfhrer, C., Wrulich, M., Ikuno, J., Bosanska, D., & Rupp, M. (2009). Simulating the long term evolution physical layer. *Proceedings of 17th European Signal Processing Conference (EUSIPCO 2009).* Glasgow, Scotland: IEEE Press.

Meng, L., Zipf, A., & Winter, S. (Eds.). (2008). *Map-based mobile services design, interaction, and Usability. Lecture Notes in Geoinformation and Cartography.* Springer Verlag.

Merentitis, A., & Triantafyllopoulou, D. (2010). Transmission power regulation in cooperative cognitive radio systems under uncertainties. *Proceeding of IEEE international Symposium on Wireless Pervasive Computing (ISWPC)*, (pp. 134–139). Modena, Italy: IEEE Press.

Mietzner, J., Lampe, L., & Schober, R. (2009). Distributed transmit power allocation for multihop cognitive-radio systems. *IEEE Transactions on Wireless Communications, 8*(10), 5187–5201. doi:10.1109/TWC.2009.081318

Min, A. W., & Shin, K. G. (2010). *On sensing – Access tradeoff in cognitive radio networks.* Paper presented at the Symposium on Dynamic Spectrum Access Networks (DyDPAN). IEEE Press.

Mishra, A., Shrivastava, V., Agarwal, D., Banerjee, S., & Ganguly, S. (2006). Distributed channel management in uncoordinated wireless environments. In *Proceedings of the ACM Annual International Conference on Mobile Computing and Networking.* Los Angeles, CA: ACM Press.

Mishra, M., & Sahoo, A. (2006). *A contention window based differentiation mechanism for providing QoS in Wireless LANs.* Paper presented at the 9th International Conference on Information Technology, (ICIT '06). IEEE Press.

Mishra, M., & Sivalingam, K. M. (2008). Enhancing TCP performance in AMC based broadband wireless access networks. In *Proceedings of IEEE International Conference on Communications* (pp. 2984-2989). Beijing, China: IEEE Press.

Mishra, A., Banerjee, S., & Arbaugh, W. (2005). Weighted colouring based channel assignment for WLANs. *ACM SIGMOBILE Mobile Computing and Communications Review, 9*(3), 19–31. doi:10.1145/1094549.1094554

Mishra, A., Shin, M., & Arbaugh, W. (2003). An empirical analysis of the IEEE 802.11 MAC layer handoff process. *ACM SIGCOMM Computer Communications Review, 33*(2), 93–102. doi:10.1145/956981.956990

Mitola, J. III. (2006). *Cognitive radio architecture: The engineering foundations of radio XML.* Hoboken, NJ: John Wiley & Sons Inc.doi:10.1002/0471773735

Mitrano, F., Guainella, E., Tamea, G., Castrucci, M., Di Giorgio, A., Benali, O., et al. (2007). *QoS management of multicast and broadcast services in next generation networks.* Paper presented at the Mobile and Wireless Communications Summit, 2007. 16th IST.

Mogensen, P., Wei, N., Kovacs, Z., Frederiksen, F., Pokhariyal, A., & Pedersen, I. ... Kuusela, M. (2007). LTE capacity compared to the Shannon bound. *Proceedings of 65th Semi-Annual IEEE Vehicular Technology Conference (VTC-Spring 2007)*, (pp. 1234-1238). Dublin, Ireland: IEEE Press.

Morgan, Y. L., & Kunz, T. (2003). PYLON: An architectural framework for ad-hoc QoS interconnectivity with access domains. *Proceeding of Annual Hawaii International Conference on System Sciences* (pp. 309–318). Big Island, HI: Univ. of Hawaii Press.

Moskowitz, R., Nikander, P., Jokela, P., & Henderson, T. (2008). *Host identity protocol*. IETF RFC 5201.

Muñoz, D., Bouchereau, F., Vargas, C., & Enríquez-Caldera, R. (2009). *Position location techniques and applications*. Elsevier.

Nachtigall, J., Zubow, A., & Redlich, J. (2008). The impact of adjacent channel interference in multi- radio systems using IEEE 802.11. *International Wireless Communications and Mobile Computing Conference*, (pp. 874–881). Crete, Greece: IEEE Press.

Nash, J. (1950). Equilibrium points in n-person games. *Proceedings of the National Academy of Sciences of the United States of America*, *36*(1), 48–49. doi:10.1073/pnas.36.1.48

Nasser, N., Hasswa, A., & Hassanein, H. (2006). Handoffs in fourth generation heterogeneous networks. *IEEE Communications Magazine*, *44*(10), 96–103. doi:10.1109/MCOM.2006.1710420

Neel, J. O. (2006). *Analysis and design of cognitive radio networks and distributed radio resource management algorithms*. Unpublished doctoral dissertation. Blacksburg, VA: Virginia Polytechnic Institute and State University.

Neto, A., Cerqueira, E., Rissato, A., Monteiro, E., & Mendes, P. (2007). *A resource reservation protocol supporting QoS-aware multicast trees for next generation networks*. Paper presented at the Computers and Communications, 2007. ISCC 2007. 12th IEEE Symposium.

Neves, P., Sargento, S., & Aguiar, R. L. (2006, 26-29 June 2006). *Support of real-time services over integrated 802.16 metropolitan and local area networks*. Paper presented at the (ISCC 2006), Proceedings. 11th IEEE Symposium on Computers and Communications.

Nguyen-Vuong, Q.-T., Agoulmine, N., & Ghamri-Doudane, Y. (2007). Terminal-controlled mobility management in heterogeneous wireless networks. *IEEE Communications Magazine*, *45*(4), 122–129. doi:10.1109/MCOM.2007.343621

Niyato, D., & Hossain, E. (2009). Cognitive radio for next-generation wireless networks: An approach to opportunistic channel selection in IEEE 802.11-based wireless mesh. *IEEE Wireless Communications*, *16*(1), 46–54. doi:10.1109/MWC.2009.4804368

Nogueira, J., Melo, M., Carapinha, J., & Sargento, S. (2010). *Creating virtual networks: An approach for virtual network discovery*. Workshop on the Network of the Future, IEEE Global Telecommunications Conference (IEEE GLOBECOM 2010). OASIS eXtensible Access Control Markup Language (XACML) TC. (n.d.). *OASIS: Advancing open standards for the global information society*. Retrieved September 23, 2010, from http://www.oasis-open.org/committees/tc_home.php?wg_abbrev=xacml#XACML20.

Nolan, K. E., & Doyle, L. E. (2006). Principles of cognitive network teamwork. *Paper presented at the Software-Defined Radio (SDR) Forum Technical Conference*. Orlando, FL: ACM Press.

O'Hara, B., & Petrick, A. (2005). *IEEE 802.11 handbook. A designer's companion*. IEEE Press.

Oborina, A., Moisio, M., Henttonen, T., Pernila, E., & Koivunen, V. (2008). MIMO performance evaluation in UTRAN long term evolution downlink. *Proceedings of 42nd Annual Conference on Information Sciences and Systems (CISS 2008)*, (pp. 1179-1183). Princeton, NJ: IEEE Press.

Oh, M. (2008). An adaptive routing algorithm for wireless mesh networks. *10th International Conference on Advanced Communication Technology* (pp. 2087-2091). IEEE.

Ohrtman, F., & Roeder, K. (2003). *WiFi handbook: Building IEEE 802.1b wireless networks*. McGraw-Hill Professional.

Oliveira, R., & Braun, T. (2004). A delay-based approach using fuzzy logic to improve TCP error detection in ad hoc networks. *Proc. of IEEE Wireless Communications and Networking Conference (WCNC)*, *5*(1), 1660–1665.

Opricovic, S., & Tzeng, G. H. (2004). Compromise solution by MCDM methods: A comparative analysis of VIKOR and TOPSIS. *European Journal of Operational Research*, *156*, 445–455. doi:10.1016/S0377-2217(03)00020-1

Ormond, O., Murphy, J., & Muntean, G. (2006). Utility-based intelligent network selection in beyond 3G systems. *Proceedings of IEEE International Communications Conference* (ICC'06), Vol. 4, (1831-1836).

Pack, S., Park, K., Kwon, T., & Choi, Y. (2006). SAMP: Scalable application-layer mobility protocol. *IEEE Communications Magazine*, *44*(6), 86–92. doi:10.1109/MCOM.2006.1668425

Pack, S., Shen, X., Mark, J. W., & Pan, J. (2007). Mobility management in mobile hotspots with heterogeneous multihop wireless links. *IEEE Communications Magazine*, *45*(9), 106–112. doi:10.1109/MCOM.2007.4342864

Panagiotis, D. (2010). Opportunistic networks and cognitive management systems for efficient application provision in the future Internet (OneFIT). *EU-FP7 project*. Retrieved September 29, 2010, from http://www.ict-onefit.eu/

Pan, P., Swallow, G., & Atlas, A. (2005). *Fast reroute extensions to RSVP-TE for LSP tunnels. RFC 4090*. Proposed Standard.

Papagiannaki, K., Mhatre, V., & Baccelli, F. (2007). *Interference mitigation through power control in high density 802.11 WLANs. Proceedings of IEEE INFOCOM 2007*. Anchorage, AK: IEEE Press.

Parekh, A., & Gallagher, R. (1993). A generalized processor sharing approach to flow control in integrated services networks: The single-node case. *IEEE/ACM Transactions on Networking*, *1*(3), 344–357. doi:10.1109/90.234856

Parry, R. (2002). CDMA2000 1×EV-DO for 3G communications. *IEEE Potentials*, *4*, 10–13. doi:10.1109/MP.2002.1044211

Pauli, V., Naranjo, J. D., & Seidel, E. (2010). *Heterogeneous LTE networks and inter-cell interference coordination*. Munich, Germany: Nomor Research GmbH.

Pavilanskas, L. (2007). *Wireless MAC for real time applications: Analytical engineering approach to the usage of IEEE 802.11 MAC in specific conditions*. LAP Lambert Academic Publishing.

Pedersen, I., Kolding, E., Frederiksen, F., Kovacs, Z., Laselva, D., & Mogensen, E. (2009). An overview of downlink radio resource management for UTRAN long-term evolution. *IEEE Communications Magazine*, *47*(7), 86–93. doi:10.1109/MCOM.2009.5183477

Perahia, E., & Stacey, R. (2008). *Next generation wireless LANs: Throughput, robustness, and reliability in 802.11n*. Cambridge University Press. doi:10.1017/CBO9780511541032

Perez-Neira, A., Lagunas, M. A., & Bas, J. (1997). Fuzzy logic for robust detection in wireless communications. *Proc. of IEEE International Symposium on Personal Indoor and Mobile Radio Communications (PIMRC)*, (pp. 1145–1149). IEEE Press.

Perkins, C. (Ed.). (2002). *IP mobility support for IPv4*. IETF RFC 3344.

Perkins, C., Belding-Royer, E., & Das, S. (2003). *Ad hoc on-demand distance vector (AODV) routing. RFC 3561*. Experimental.

Perkins, D., & Hughes, H. D. (2002). A survey on quality-of-service support for mobile ad hoc networks. *Journal of Wireless Communications and Mobile Computing*, *2*(5), 503–513. doi:10.1002/wcm.73

Persaud, R. (2006). *Core network mobility: Active MPLS*. Germany: Shaker Verlag.

Petty, V. R., Rajbanshi, R., Datla, D., Frederick, W., et al. (2007). *Feasibility of dynamic spectrum access in underutilized television bands*. Paper presented at the Symposium on Dynamic Spectrum Access Networks (DySPAN). IEEE Press.

Pfaff, B., Pettit, J., Koponen, T., Amidon, K., Casado, M., & Shenker, S. (2009). *Extending networking into the virtualization layer*. In ACM HotNets, New York, NY, October 2009.

Phan Huy, D., Legouable, R., Ktenas, D., Brunel, L., & Assaad, M. (2008). Downlink B3G MIMO OFDMA link and system level performance. *Proceedings of 2008 IEEE Vehicular Technology Conference (VTC 2008-Spring)*. Marina Bay, Singapore: IEEE Press.

Plummer, A., & Biswas, S. (2009). Capacity scavenging in wireless networks: A comparative study. *Paper presented at the IEEE Military Communications Conference (Milcom)*. IEEE Press.

Plummer, A., Taghizadeh, M., & Biswas, S. (2009). *Measurement based capacity scavenging via whitespace modeling in wireless networks*. Paper presented at the IEEE Globecom 2009. IEEE Press.

Plummer, A., Taghizadeh, M., & Biswas, S. (2010b). *Traffic protection via bandwidth scavenging in heterogeneous sensor networks*. Paper presented at the IEEE Globecom. IEEE Press.

Plummer, A., Taghizadeh, M., & Biswas, S. (2010a). Measurement based bandwidth scavenging in wireless networks. *IEEE Transactions on Mobile Computing*, *99*, 1.

Pollini, G. (1996). Trends in handover design. *IEEE Communications Magazine*, *34*(3), 82–90. doi:10.1109/35.486807

Pospishny, I., Vasyuk, V., Romanchyk, S., Dovzhenko, O., Dony, O., & Shvaichenko, V. (2010). 3GPP long term evolution (LTE). *Proceedings of International Conference on Modern Problems of Radio Engineering, Telecommunications, and Computer Science (TCSET 2010)*, (p. 192). Lviv, Ukraine: IEEE Press.

POSTnote. (2007). *Next generation telecoms networks. Parliamentary Office of Science and Technology, Number 296*. Next Generation Networks.

Prehofer, C., Nafisi, N., & Wei, Q. (2003). *A framework for context-aware handover decisions*. Paper presented at the Personal, Indoor and Mobile Radio Communications, PIMRC 2003.

Prihandoko, F., Habaebi, M. H., & Ali, B. M. (2003). Adaptive call admission control for QoS provisioning in multimedia wireless networks. *Elsevier Computer Communications.*, *26*(14), 1560–1569.

Qiang, N. (2005). Performance analysis and enhancements for IEEE 802.11e wireless networks. *IEEE Network*, *19*(4), 21–27. doi:10.1109/MNET.2005.1470679

Qi, B., Ronald, R. B., Dongzhe, C., Asif, D. G., Ching-Yao, H., & Stan, V. (2003). Performance of 1xEV-DO third-generation wireless high-speed data systems. *Bell Labs Technical Journal*, *7*, 97–107. doi:10.1002/bltj.10021

Qin, Q., Zhimin, Z., & Caili, G. (2009). A study of data fusion and decision algorithms based on cooperative spectrum sensing. *Proceeding of International Conference on Fuzzy Systems and Knowledge Discovery (FSKD)*, (pp. 76–80). Tianjin, China: IEEE Press.

Qingyang, S., & Jamalipour, A. (2005). *A network selection mechanism for next generation networks*. Paper presented at the IEEE International Conference on Communications (ICC).

Qingyang, S., & Jamalipour, A. (2005). Network selection in an integrated wireless LAN and UMTS environment using mathematical modeling and computing techniques. *IEEE Wireless Communications*, *12*(3), 42–48. doi:10.1109/MWC.2005.1452853

Rabbani, M. G., Kamruzzaman, J., Gondal, I., & Ahmad, I. (2011). *A new resource distribution model for improved QoS in an integrated WiMAX/WiFi architecture*. IEEE International Conference on Communications.

Ramabhadran, S., & Pasquale, J. (2003). Stratified round robin: A low complexity packet scheduler with bandwidth fairness and bounded delay. In *Proceedings of the 2003 Conference on Applications, Technologies, Architectures, and Protocols for Computer Communications. SIGCOMM '03* (pp. 239-250). Karlsruhe, Germany: ACM Press.

Ramasubramanian, V. & Mosse, D. (2008). Bra: A bidirectional routing abstraction for asymmetric mobile ad-hoc networks. *IEEE/ACM TON, 16*(1), 116-129.

Ramirez, R., & Ramos, R. (2009). A vertical handoff decision algorithm which considers the uncertainty during the decision making process. *Proceedings of IFIP International Conference on Wireless and Optical Communication Networks* (WOCN'09), vol. 1 (pp. 1-6).

Ranjan, P., La, R., & Abed, E. (2010). *Bifurcations of TCP and UDP traffic under RED*. Retrieved September 29, 2010, from www.glue.umd.edu/~priya/template.pdf

Rappaport, T. S. (1996). *Wireless communications: Principles and practice*. Prentice Hall.

Ray, S. K., & Mista, I. S. (2006). Fourth generation (4G) networks: Roadmap-migration to the future. *IETE Technical Review*, *23*(4), 253–265.

Ray, S. K., Pawlikowski, P., & Sirisena, H. (2010). Handover in mobile WiMax networks: The state of art and research issues. *IEEE Communications Surveys & Tutorials*, *12*(3), 376–399. doi:10.1109/SURV.2010.032210.00064

Reid, N. P., & Seide, R. (2002). *WiFi (802.11) 802.11 network handbook*. McGraw-Hill.

Reznik, L., & Kreinovich, V. (2004). Fuzzy and probabilistic models of association information in sensor networks. *Proceedings of IEEE International Conference on Fuzzy Systems*, (pp. 185–189). IEEE Press.

Riihijarvi, J., Petrova, M., & Mahonen, P. (2004). Automatic channel allocation for small wireless local area networks using graph colouring algorithm approach. In *Proceedings of IEEE International Symposium on Personal, Indoor, and Mobile Radio Communications*, (pp. 536–539). Barcelona, Spain: IEEE Press.

Riihijärvi, J., Petrova, M., & Mähönen, P. (2005). Frequency allocation for WLANs using graph colouring techniques. In *Proceedings of 2nd Annual Conference on Wireless On-Demand Network Systems and Services*, (pp.19–21). St. Moritz, Switzerland: IEEE Press.

Riva, O., Teittinen, V. M., Siikavirta, S., & Huovinen, L. (2008). *A next generation operator environment to turn context-aware services into a commercial reality.* Paper presented at the Mobile Data Management, 2008. MDM '08. 9th International Conference.

Robitzsch, S., Niephaus, C., Fitzpatrick, J., & Kretschmer, M. (2009). Measurements and evaluations for an IEEE 802.11a based carrier-grade multi-radio wireless mesh network deployment. In *2009 Fifth International Conference on Wireless and Mobile Communications*, (pp. 272–278). Cannes, France: IEEE Press.

Robitzsch, S., Fitzpatrick, J., Murphy, S., & Murphy, L. (2010). Behind-the-scenes of IEEE 802.11a based multi-radiomesh networks: A measurement driven evaluation of inter-channel interference. *International Journal on Advances in Internet Technology*, *3*(1-2), 65–76.

Rodrigues, J. C., Fraiha, S., Araujo, J., Gomes, H., Frances, C., & Cavalcante, G. (2009). Empirical study of the QoS parameters behavior of a VoIP application in Wi-Fi networks. *2009 SBMO/IEEE MTT-S International Microwave and Optoelectronics Conference (IMOC)*, (pp. 257–261). IEEE Press.

Rohokale, V., Kulkarni, N., Prasad, N., & Cornean, H. (2010). Cooperative opportunistic large array approach for cognitive radio networks. *Paper presented in the 8th International Conference on Communications (COMM)*, (pp. 513–516). Bucharest, Romania: IEEE Press.

Rong, L., Haddada, O., & Elayoubi, S. (2008). Analytical analysis of the coverage of a MBSFN OFDMA network. *Proceedings of IEEE Global Communication Conference 2008 (Globecom 2008)*. New Orleans, LA: IEEE Press.

Rosen, E., Tappan, D., Fedorkow, G., Rekhter, Y., Farinacci, D., Li, T., & Conta, A. (2001a). *MPLS label stack encoding*. RFC 3032 (Proposed Standard). Updated by RFCs 3443, 4182, 5332, 3270, 5129, 5462, 5586.

Rosenberg J., et al. (2002). *SIP: Session initiation protocol*. IETF RFC 3261.

Rosen, E., Viswanathan, A., & Callon, R. (2001b). *Multiprotocol label switching architecture. RFC 3031.* Proposed Standard.

Roshan, P., & Leary, J. (2010). *IEEE 802.11 wireless LAN fundamentals*. Cisco Press.

Ross, J. (2008). *The book of wireless*. O'Reilly Media.

Rowstron, A., & Druschel, P. (2001). Pastry: Scalable, decentralized object location, and routing for large-scale peer-to-peer systems. *Lecture Notes in Computer Science*, 2218.

Rumney, M. (2008). *LTE and the evolution to 4G wireless*Agilent Technologies.

Rumney, M. (Ed.). (2009). *LTE and the evolution to 4G wireless: Design and measurement challenges*. Agilent Technologies.

Saahoff, F. (2002). *QoS-aspects of mobile IPv6 clients in an IEEE 802.11 network*. VDM Verlag.

Sachs, J., Prytz, M., & Gebert, J. (2009). Multi-access management in heterogeneous networks. *Wireless Personal Communications*, *48*(1), 7–32. doi:10.1007/s11277-007-9431-8

Salber, D., Dey, A., & Abowd, G. (1999). The context toolkit: Aiding the development of context-enabled applications. In *SIGCHI Conference on Human Factors in Computing Systems* (pp. 434–441).

Sales, L. M., Almeida, H. O., Perkusich, A., & Sales, M. (2008). *An experimental evaluation of DCCP transport protocol: A focus on the fairness and hand-off over 802.11g networks. In the Proceeding of Consumer Communications and Networking Conference 2008* (pp. 1149–1153). Las Vegas, NV: IEEE Press.

Salmani, D., & Akbari, M. (2008). Fuzzy and research paradigms relationship: A mutual contribution. *Journal of Academic Research*, *6*(2).

Salsano, S., Polidoro, A., Mingardi, C., Niccolini, S., & Veltri, L. (2008). SIP-based mobility management in next generation networks. *IEEE Wireless Communications*, *15*(2), 92–99. doi:10.1109/MWC.2008.4492982

Sanchez, J., Morales-Jimenez, D., Gomez, G., & Enbrambasaguas, T. (2007). Physical layer performance of long term evolution cellular technology. *Proceedings of 6th IST Mobile and Wireless Communications Summit*, (pp. 1-5). Budapest, Hungary: IEEE Press.

Santhi, K. R., & Kumaran, G. S. (2006). Migration to 4G: Mobile IP based solutions. *IEEE International Conference on Internet and Web Applications and Services*, (p. 76). IEEE Press.

Saunders, S. (1999). *Antennas and propagation for wireless communication systems: Concept and design*. New York, NY: John Wiley and Sons.

Saunders, S. R., Carlaw, S., Giustina, A., Bhat, R. R., Rao, S. V., & Siegberg, R. (2009). *Femtocells: Opportunities and challenges for business and technology*. West Sussex, UK: John Wiley & Sons.

Schilit, B. N., & Theimer, M. M. (1994). Disseminating active map information to mobile hosts. *IEEE Network*, *8*(5), 22–32. doi:10.1109/65.313011

Sedoyeka, E., Hunaiti, Z., & Tairo, D. (2009). Analysis of QoS requirements in developing countries. *International Journal of Computing and ICT Research*, *3*(1), 18–31.

Seo, J. K., Nam, S. H., & Lee, K. G. (2007). Fast route optimization for dynamic nested NEMO. In the *Proceeding of the ICPPW 2007* (pp. 49-53). Xi'an, China: IEEE Press.

Sesia, S., Toufik, I., & Baker, M. (2009). *LTE - The UMTS long term evolution: From theory to practice*. Chichester, UK: John Wiley & Sons. doi:10.1002/9780470742891

Sevindik, V. (2010). *Scheduling algorithms and QoS architectures for next generation cellular communication systems*. Doctoral dissertation. Lowell, MA: University of Massachusetts.

Sevindik, V., & Bayat, O. (2011). *Dynamic traffic group generation in LTE cellular wireless telecommunication networks*. US Patent Filled. Application #: 61/430,102. Washington, DC: USA Government.

Sevindik, V., & Bayat, O. (2011). *Multimedia content and file sharing in femtocell networks*. US Patent Filled. Application #:61/432,629. Washington, DC: USA Government.

Sevindik, V., Bayat, O., & Weitzen, J. (2010). Traffic differentiation for BE users in CDMA 1xEVDO Networks. *International Journal of Multimedia Tools and Applications*, *55*(3), 507–523. doi:10.1007/s11042-010-0563-6

Sgora, A., & Vergados, D. (2009). Handoff prioritization and decision schemes in wireless cellular networks: a survey. *IEEE Communications Tutorials and Surveys*, 11(4), http://www.comsoc.org/publications.

Shacham, R., Schulzrinne, H., Thakolsri, S., & Kellerer, W. (2009). *Session initiation protocol (SIP) session mobility.* IETF RFC 5631.

Shah, P. A., & Yousaf, M. (2008). End-to-end mobility management solutions for TCP: An analysis. In the *Proceeding of the International Conference on Computer and Information Technology* (pp. 696-701). Venice, Italy: IEEE Press.

Shakkottai, S., Altman, E., & Kumar, A. (2007). Multihoming of users to access points in WLANs: A population game perspective. [IEEE.]. *IEEE Journal on Selected Areas in Communications*, *25*(6), 1207–1215. doi:10.1109/JSAC.2007.070814

Shao, X., Schiphorst, R., & Slump, C. H. (2009). An opportunistic error correction layer for OFDM systems. *EURASIP Journal on Wireless Communications and Networking*, 2009.

Sheng, Y., Peng, M., & Wang, W. (2008). A novel adaptive modulation and coding strategy based on partial feedback for enhanced MBMS network. *Journal of China Universities of Posts and Telecommunications*, *15*(1), 48–54. doi:10.1016/S1005-8885(08)60061-8

Shoham, Y., & Leyton-Brown, K. (2009). *Multiagent systems algorithmic, game-theoretic, and logical foundations*. Cambridge, England: Cambridge University Press.

Siddiqui, F., & Zeadally, S. (2006). Mobility management across hybrid wireless networks: Trends and challenges. *Computer Communications*, *29*(9), 1363–1385. doi:10.1016/j.comcom.2005.09.003

Sidhu, D., Nair, R., & Abdallah, S. (1991). Finding disjoint paths in networks. [ACM.]. *ACM SIGCOMM Computer Communication Review*, *21*(4), 43–51. doi:10.1145/115994.115998

Silva, V. H., Salgado, E. I., & Quintana, F. R. (2006). AWISPA: An awareness framework for collaborative spontaneous networks. In *ASEE/IEEE Frontiers in Education Conference,* (pp. 1-6). IEEE Press.

Skalli, H., Ghosh, S., Das, S., Lenzini, L., & Conti, M. (2007). Channel assignment strategies for multiradio wireless mesh networks: Issues and solutions. *IEEE Communications Magazine*, *45*(11), 86–95. doi:10.1109/MCOM.2007.4378326

Smith, K. (1998). A genetic algorithm for the channel assignment problem. In *Proceedings of Global Telecommunications Conference*, vol. 4, (pp. 2013–2018). Sydney, Australia: IEEE Press.

Smith, K., & Palaniswami, M. (1997). Static and dynamic channel assignment using neural networks. *IEEE Journals on Selected Areas Communications*, *15*(2), 238–249. doi:10.1109/49.552073

Snoeren, A. C., & Balakrishnan, H. (2000). An end-to-end approach to host mobility. In the *Proceedings of the 6th Annual International Conference on Mobile Computing and Networking* (pp.155-166). Boston, MA: ACM Press.

So, J., & Vaidya, N. (2004). Multi-channel MAC for ad hoc networks: Handling multi-channel hidden terminals using a single transceiver. In *Proceedings ACM Mobihoc*, (pp. 222–33). Tokyo, Japan: ACM Press.

Sobrinho, J. L., & Krishnakumar, A. S. (1999). Quality-of-service in ad hoc carrier sense multiple access wireless networks. *IEEE Journal on Selected Areas in Communications*, *17*(8), 1353–1368. doi:10.1109/49.779919

Soliman, H. (Ed.). (2009). *Mobile IPv6 support for dual stack hosts and routers*. IETF RFC 5555.

Soliman, H., Castelluccia, C., Malki, K., & Bellier, L. (2005). *Hierarchical mobile IPv6 mobility management (HMIPv6)*. Internet Engineering Task Force, RFC 4140 (Experimental). From http://www.ietf.org/rfc/rfc4140.txt

Song, L., Kotz, D., Jain, R., & Xiaoning, H. (2004). Evaluating location predictors with extensive Wi-Fi mobility data. *Twenty-Third Annual Joint Conference of the IEEE Computer and Communications Societies* (INFOCOM 2004), vol. 2, (pp. 1414–1424). IEEE Press.

Song, Q., & Jamalipour, A. (2005). A network selection mechanism for next generation networks. *Proceedings of IEEE International Communications Conference* (ICC'05), vol. 2, (pp. 1418-1422).

Song, Q., & Jamalipour, A. (2005). Network selection in an integrated wireless LAN and UMTS environment using mathematical modeling and computing techniques. *IEEE Wireless Communications*, *12*(3), 42–48. doi:10.1109/MWC.2005.1452853

Sparl, P., & Zerovnik, J. (2004). 2-local 5/4-competitive algorithm for multi-colouring triangle-free hexagonal graphs. *Information Processing Letters*, *90*(5), 239–246. doi:10.1016/j.ipl.2004.02.017

Stanford, S. I. N. G. (2009). *SourceForge.net Repository*. Retrieved October 21st, 2009, from http://tinyos.cvs.sourceforge.net/viewvc/tinyos/tinyos-2.x-contrib/stanford-sing/

Starsky, H. Y., Wong, H. Y., Lu, S., & Bharghavan, V. (2006). *Robust rate adaptation for 802.11 wireless networks*. ACM 12th Annual International Conference on Mobile Computing and Networking. Los Angeles, CA: ACM Press.

Staub, T., & Braun, T. (2008). *ATOM: Adaptive transport over multipaths in wireless mesh networks*. 2nd ERCIM Workshop on eMobility.

Staub, T., Ott, S., & Braun, T. (2009). Experimental evaluation of multi-path routing in a wireless mesh network inside a building. *5th Workshop on Mobile Ad-Hoc Networks - WMAN 2009 - in conjunction with the 16th Conference on Communication in Distributed Systems (KiVS)* (pp. 1-12).

Stemm, M., & Katz, R. (1998). Vertical handoffs in wireless overlay networks. *ACM Mobile Networks (MONET) . Special Issue on Mobile Networking in the Internet, 3*, 335–350.

Sternad, M., Svensson, T., Ottosson, T., Ahlen, A., Svensson, A., & Brunstrom, A. (2007). Towards systems beyond 3G based on adaptive OFDMA transmission. *Proceedings of the IEEE, 95*(12), 2432–2455. doi:10.1109/JPROC.2007.907119

Stevens-Navarro, E. (2008). *Mobility management and admission control in heterogeneous wireless networks*. PhD dissertation, The University of British Columbia, Canada.

Stevens-Navarro, E., & Wong, V. (2006). Comparison between vertical handoff decision algorithms for heterogeneous wireless networks. *Proceedings of IEEE Vehicular Technology Conference* (VTC-SPRING'06), vol. 2, (pp. 947-951).

Stevens-Navarro, E., Lin, Y., & Wong, V. (2008). An MDP-based vertical handoff decision algorithm for heterogeneous wireless networks. *IEEE Transactions on Vehicular Technology, 57*(2), 1243–1254. doi:10.1109/TVT.2007.907072

Stevens-Navarro, E., Pineda-Rico, U., & Acosta-Elias, J. (2008). Vertical handover in beyond third generation (B3G) wireless networks. *International Journal of Future Generation Communication and Networking, 1*(1), 51–58.

Stewart, R. (Ed.). (2007). *Stream control transmission protocol*. IETF RFC 4960.

Stoica, I., Morris, R., Liben-Nowell, D., Karger, D. R., Kaashoek, M. F., Dabek, F., & Balakrishnan, H. (2003, February). Chord: A scalable peer-to-peer lookup protocol for internet applications. *IEEE/ACM Transactions on Networking, 11*(1), 17–32. doi:10.1109/TNET.2002.808407

Stojmenovic, I., Giordano, S., Conti, M., & Basagni, S. (2005). *Mobile ad hoc networking* (pp. 175–203). IEEE.

Strang, T. (2005). *Ubiquitous computing mobility*. Courseware of University of Innsbruck. Retrieved September, 2010, from http://www.stiinnsbruck.at/ fileadmin/ documents/ teaching_archive/ ubicomp05/09-mobility_6.pdf

Strang, T., & Popien, C. (2004). *A context modeling survey*. Paper presented at the Workshop on Advanced Context Modelling, Reasoning and Management, UbiComp 2004 - The Sixth International Conference on Ubiquitous Computing.

Subramanian, A. P., Buddhikot, M. M., & Miller, S. (2006). Interference aware routing in multi-radio wireless mesh networks. *2nd IEEE Workshop on Wireless Mesh Networks* (pp. 55-63).

Subramanian, A. P., Gupta, H., Das, S. R., & Buddhikot, M. M. (2007). *Fast spectrum allocation in coordinated dynamic spectrum access based cellular networks*. Paper presented at New Frontiers in Dynamic Spectrum Access Networks (DySPAN). IEEE Press.

Sudeep, K., & Vishwanathan, S. (2005). A technique for multi-colouring triangle-free hexagonal graphs. *Discrete Mathematics, 300*(1-3), 256–259. doi:10.1016/j.disc.2005.06.002

Sugeno, M. (1985). *Industrial applications of fuzzy control*. New York, NY: Elsevier.

Suk Yu, H., & Kai Hau, Y. (2003). Challenges in the migration to 4G mobile systems. *IEEE Communications Magazine, 41*(12), 54–59. doi:10.1109/MCOM.2003.1252799

SuKyoung, L., Sriram, K., Kyungsoon, K., Yoon Hyuk, K., & Golmie, N. (2009). Vertical handoff decision algorithms for providing optimized performance in heterogeneous wireless networks. *IEEE Transactions on Vehicular Technology, 58*(2), 865–881. doi:10.1109/TVT.2008.925301

Tabakovic, Z., Grgic, S., & Grgic, M. (2009). Fuzzy logic power control in cognitive radio. *Proceeding of IEEE International Conference on Systems, Signals, and Image Processing (IWSSIP)*, (pp. 1–6). Chalkida, Greece: IEEE Press.

Tabassam, A. A., Trsek, H., Heiss, S., & Jasperneite, J. (2009). Fast and seamless handover for secure mobile industrial applications with 802.11r. In the *Proceeding of the LCN 2009* (pp. 750-757). Zurich, Switzerland: IEEE Press.

Taewon, B. K. T. F. & Kyunggi, D. (2004). A new efficient scheduling algorithm for the support of QoS in CDMA2000 1xEV-DO. *IEEE Vehicular Technology Conference*, vol. 2, (pp. 1148-1152). IEEE Press.

Taeyoung, L., Jangkeun, J., Jisun, Y., & Jitae, S. (2010). Adjustable subcarrier allocation for spectral efficiency and user fairness in 3GPP LTE downlink systems. *Proceedings of 12th International Conference on Advanced Communication Technology (ICACT 2010)*, (pp. 985-988). Gangwon-Do, South Korea: IEEE Press.

Taniuchi, K., Ohba, Y., & Fajardo, V. (2009). IEEE 802.21: Media independent handover: Features, applicability, and realization. *IEEE Communications Magazine*, *47*(1), 112–120. doi:10.1109/MCOM.2009.4752687

Taniuchi, K., Ohba, Y., Fajardo, V., Das, S., Tauil, M., & Yuu-Heng, C. (2009). IEEE 802.21: media independent handover: features, applicability, and realization. *IEEE Communications Magazine*, *47*(1), 112–120. doi:10.1109/MCOM.2009.4752687

Tapia, P., Liu, J., Karimli, Y., & Feuerstein, M. J. (2009). *HSPA performance and evolution* [Book]. John Wiley & Sons, Ltd.doi:10.1002/9780470742068

Tarique, M., Tepe, K. E., Adibi, S., & Erfani, S. (2009). Survey of multipath routing protocols for mobile ad hoc networks. *Journal of Network and Computer Applications*, *32*(6), 1125–1143. doi:10.1016/j.jnca.2009.07.002

Tawil, R. Demerjian, J., Pujolle, G. & Salazar, O. (2008). Processing delay reduction during the vertical handoff decision in heterogeneous wireless systems. *Proceedings of IEEE/ASC International Conference on Computer Systems and Applications* (AICCSA'08), vol. 1, (pp. 381-385).

Tawil, R. Salazar, O., & Pujolle, G. (2008). Vertical handoff decision scheme using MADM for wireless networks. *Proceedings of IEEE Wireless Communications and Networking Conference* (WCNC'08), vol. 1, (pp. 2789-2792).

Team, W. (2006). *A solution for seamless handover among multi-provider heterogeneous networks*.

Teraoka, F., Ishiyama, M., & Kunishi, M. (2003*). LIN6: A solution to multihoming and mobility in IPv6, draft-teraoka-multi6-lin6-00*. IETF draft (expired).

Teraoka, F., Yokote, Y., & Tokoro, M. (1991). A network architecture providing host migration transparency. In *ACM Proceeding of the Conference on Communications Architecture & Protocols* (pp. 209-220). Zurich, Switzerland: ACM Press.

The Cooperative Association for Internet Data Analysis. (2010). *Analyzing UDP usage in internet traffic*. Retrieved September 29, 2010, from http://www.caida.org/research/traffic-analysis/tcpudpratio/

Thomas Buchholz, M. S. (2003). Quality of context: what it is and why we need it. In *Proceedings of the 10th Workshop of the OpenView University Association: OVUA'03*.

Timm-Giel, A., Zaki, Y., Zhao, L., & Goerg, C. (2010). *A novel LTE wireless virtualization framework*. Second International ICST Conference on Mobile Networks and Management (MONAMI 2010), Sept. 2010.

Tinnakornsrisuphap, P., & Lott, C. (2004). On the fairness and stability of the reverse-link MAC layer in CDMA2000 1×EV-DO. *IEEE International Conference on Communications*, vol. 1, (pp. 144-148). IEEE Press.

Todorovic, M., & Lopez-Benitez, N. (2006). Efficiency study of TCP protocols in infrastructured wireless networks. In *ICNS '06: The International Conference on Networking and Services*, (p. 103). New York, NY: IEEE Press.

Toh, C. K., Delwar, M., & Allen, D. (2002). Evaluating the communication performance of an ad_hoc wireless network. *IEEE Transactions on Wireless Communications*, *1*(3), 402–414. doi:10.1109/TWC.2002.800539

Toloo, M., & Mouftah, H. T. (1995). Speed considerations and handover mechanisms in mobile personal communication systems. In *Proceedings of the 6th IEEE International Symposium on Personal, Indoor, and Mobile Radio Communications (PMIRC'95)*, (pp. 841-845). Toronto, Canada: IEEE Press.

Tomazic, S., & Jakus, G. (2009). Long term evolution: Towards 4th generation of mobile telephony and beyond. *Proceedings of 9th International Conference on Telecommunication in Modern Satellite, Cable, and Broadcasting Services (TELSIKS 2009)*, (pp. 91–96). Nis, Serbia: IEEE Press.

Tomsovic, K., & Chow, M. K. (2000). *Contributors to tutorial on fuzzy logic applications in power systems*. IEEE-PES Winter Meeting in Singapore. Singapore: IEEE Press.

Toth, P. (1980). Dynamic programming algorithms for the zero-one knapsack problem. *Computing*, *25*(1), 29–45. doi:10.1007/BF02243880

Touch, J. D., Shun Wang, Y., Pingali, V., Eggert, L., Zhou, R., & Finn, G. G. (2005). A global x-bone for network experiments. In *Proceeding in IEEE Tridentcom* 2005, (pp. 194-203).

Tran, P., & Boukhatem, N. (2008). Comparison of MADM decision algorithms for interface selection in heterogeneous wireless networks. *Proceedings of IEEE International Conference on Software, Telecommunications and Computer Networks* (SoftCOM'08), vol. 1, (pp. 119-124).

Tsirigos, A., & Haas, Z. J. (2001). Multipath routing in the presence of frequent topological changes. *IEEE Communications Magazine*, *39*(11), 132–138. doi:10.1109/35.965371

Tsirtsis, G., Park, V., Soliman, H. (2009). *Dual-stack mobile IPv4*. IETF RFC 5454.

Tsoukalas, L. H., & Uhrig, R. E. (1993). *Fuzzy and neural approaches in engineering*. New York, NY: John Wiley & Sons.

Unnikrishnan, J., & Veeravalli, V. V. (2008). Cooperative sensing for primary detection in cognitive radio. *IEEE Journal of Selected Topics in Signal Processing*, *2*(1), 18–27. doi:10.1109/JSTSP.2007.914880

Valko, A. G. (1999). Cellular IP: A new approach to internet host mobility. *Mobile Computing and Communications Review*, *29*(1), 50–65.

Varadarajan, S., Ngo, H. Q., & Srivastava, J. (2000). An adaptive, perception-driven error spreading scheme in continuous media streaming. In *Proceedings from 20th IEEE International Conference on Distributed Computing Systems (ICDCS'00)*, Taipei, Taiwan: IEEE Press.

Velte, T., & Velte, A. (2005). *Cisco IEEE 802.11 wireless networking quick reference*. Cisco Press.

Vieira, P., Queluz, P., & Rodrigues, A. (2008). LTE spectral efficiency using spatial multiplexing MIMO for macro-cells. *Proceedings of 2nd International Conference on Signal Processing and Communication Systems (ICSPCS 2008)*. Gold Coast, Australia: IEEE Press.

Viering, I., Buchner, C., Seidel, E., & Klein, A. (2007). Real-time network simulation of 3GPP long term evolution. *Proceedings of 2007 IEEE International Symposium on a World of Wireless, Mobile, and Multimedia Networks (WoWMoM2007)*, (pp. 1-3). Helsinki, Finland: IEEE Press.

Vinnakote, S., Naresh, S. V. S., & Pasupuleti, S. (2006). *New-MAC protocol for enhancement of QoS performance in wireless LAN*. 2006 IFIP International Conference on Wireless and Optical Communications Networks. IEEE Press.

Vixie P. (1997). *Dynamic updates in the domain name system* (DNS UPDATE). IETF RFC 2136.

Wakikawa, R. (Ed.). (2009). *Multiple care-of addresses registration*. IETF RFC 5648.

Wakikawa, R., Gundavelli, S. (2010). *IPv4 support for proxy mobile IPv6*. IETF RFC 5844.

Walke, B. H., Mangold, S., & Berlemann, L. (2007). *IEEE 802 wireless systems: Protocols, multi-hop mesh/relaying, performance, and spectrum coexistence*. Wiley.

Wang, B., Ji, Z., & Liu, K. J. R. (2007). *Primary-prioritized Markov approach for dynamic spectrum access.* Paper presented at the Dynamic Spectrum Access Networks (DySPAN). IEEE Press.

Wang, H., Katz, R., & Giese, J. (1999). Policy-enabled handoffs across heterogeneous wireless networks. Proceedings of *Second IEEE Workshop on Mobile Computing Systems and Applications (WMCSA'99)*, 1999, (pp. 51-60).

Wang, L., & Binet, D. (2009). MADM-based network selection in heterogeneous wireless networks: A simulation study. *Proceedings of International Conference on Wireless Communications, Vehicular Technology, Information Theory and Aerospace & Electronics Systems Technology* (Wireless VITAE'09), vol. 1, (pp. 559-564).

Wang, L., Jang, S., & Lee, T. Y. (2005). Redundant source routing for real-time services in ad hoc networks. *IEEE International Conference on Mobile Adhoc and Sensor Systems Conference* (pp. 7-87), IEEE.

Wang, M., & Kuo, G. S. (2005). An application-aware QoS routing scheme with improved stability for multimedia applications in mobile ad hoc networks. *Proceedings of IEEE Vehicular Technology Conference (VTC)*, (pp. 1901–1905). Stockholm, Sweden: IEEE Press.

Wang, X. B., Yang, Y. L., & An, J. W. (2009). Multi-metric routing decisions in VANET. *Proceeding of IEEE international Conference on Dependable Autonomic and Secure Computing (DASC)*, (pp. 551–556). Chengdu, China: IEEE Press.

Wang, J., Zhai, H., Fang, Y., Shea, J. M., & Wu, D. (2006). OMAR: Utilizing multiuser diversity in wireless. *IEEE Transactions on Mobile Computing*, *5*(12), 1764–1779. doi:10.1109/TMC.2006.186

Wang, X., & Lim, A. O. (2008). IEEE 802.11s mesh networks: Framework and challenges. *Ad Hoc Networks*, *6*(6), 970–984. doi:10.1016/j.adhoc.2007.09.003

Wang, Z.-X., Xia, H.-L., & Ding, W. (2008). A fuzzy logic cooperative MAC for MANET. *Journal of China Universities of Posts and Telecommunications*, *15*(1), 55–60. doi:10.1016/S1005-8885(08)60062-X

Watanabe, K., & Ishibashi, K. (2007). Performance of cognitive radio technologies in the presence of primary radio systems. In *Proceedings of 18th Annual IEEE International Symposium on Personal, Indoor, and Mobile Radio Communications (PIMRC '07)*, (pp. 1-5). Toronto, Canada: IEEE Press.

Waxman, B. M. (1988). Routing of multipoint connections. *IEEE Journal on Selected Areas in Communications*, *6*(9), 1617–1622. doi:10.1109/49.12889

Wedlund, E., & Schulzrinne, H. (1999). Mobility support using SIP. In *Proceeding of 2nd ACM International Workshop on Wireless Mobile Multimedia* (76–82). Washington, DC: ACM Press.

Weeks, R., Dumbill, E., & Jepson, B. (2004). *Linux unwired*. O'Reilly Media.

Wen, S., Fang, Y., & Sun, H. (2009). Differentiated bandwidth allocation with TCP protection in core routers. *IEEE Transactions on Parallel and Distributed Systems*, *20*(1), 34–47. doi:10.1109/TPDS.2008.71

Williams, M. H., Roussaki, I., Strimpakou, M., Yang, Y., MacKinnon, L., Dewar, R., et al. (2005). *Context-awareness and personalisation in the daidalos pervasive environment.* Paper presented at the International Conference on Pervasive Services, ICPS '05.

WiMAX Forum. (2008). WiMAX forum network architecture. *Stage 2: Architecture Tenets, Reference Models, and Reference Points (3GPP2 – WiMAX Interworking)*. Rel. 1, ver. 1.2. http://www.wimaxforum.org/resources/documents/technical.

WMM. WiFi Alliance. (2004). *Wi-Fi certified™ for WMM™ - Support for multimedia applications with quality of service in Wi-Fi® networks*. Austin, TX: WF Alliance.

Wong, C., Cheng, R., Lataief, K., & March, R. (1999). Multiuser OFDM with adaptive subcarrier, bit and power allocation. *IEEE Journal on Selected Areas in Communications*, *17*(10), 1747–1757. doi:10.1109/49.793310

Xia, F., Zhao, W. H., Sun, Y. X., & Tian, Y. C. (2007). Fuzzy logic control based QoS management in wireless sensor/actuator networks. *Sensors (Basel, Switzerland)*, *12*(7), 3179–3191. doi:10.3390/s7123179

Xiang, X., Wang, X., & Yang, Y. (2010). Stateless multicasting in mobile ad hoc networks. *IEEE Transactions on Computers*, *59*(8), 1076–1090. doi:10.1109/TC.2010.102

Xiao, H., Seah, W. K. G., Lo, A., & Chaing, A. K. (2000). Flexible QoS model for mobile ad-hoc networks. *Proc. of IEEE Vehicular Technology Conference,* vol. 1, (pp. 445-449). IEEE Press.

Xiao, H., Khee Chaing, C., & Winston, S. K. G. (2003). *Quality of service models for ad hoc wireless networks. In 2003 Handbook of Ad Hoc Wireless Networks*. Boca Raton, FL: CRC Press.

Xiao, Y. (2004). IEEE: 802.11e: QoS provisioning at the MAC layer. *IEEE Wireless Communications*, *11*(3), 72–79. doi:10.1109/MWC.2004.1308952

Xing, B., & Nalini, V. (2005). *Multi-constraint dynamic access selection in always best connected networks.* Paper presented at The Second Annual International Conference on Mobile and Ubiquitous Systems: Networking and Services, MobiQuitous 2005.

Xylomenos, G., & Polyzo, G. (2003). Wireless link layer enhancements for TCP and UDP application. In *Proceedings of IPDPS '03 the International Parallel and Distributed Processing Symposium*. Nice, France: IEEE Press.

Xylomenos, G., & Polyzos, G. (1999). TCP and UDP performance over a wireless LAN. In *Proceedings from INFOCOM '99. Eighteenth Annual Joint Conference of the IEEE Computer and Communications Societies*, vol. 9 (pp. 442-446). New York, NY: IEEE Press.

Yahya, B., & Ben-Othman, J. (2008). Achieving host mobility using DNS dynamic updating protocol . In *IEEE Local Computer Networks* (pp. 634–638). Montreal, Canada: IEEE Press. doi:10.1109/LCN.2008.4664258

Yang, S., Zhong, Z., & Xu, R. (2009). A cooperative vertical handover algorithm in integrated wireless and mobile network. *Proceedings of International Conference on Wireless Communication, Networking and Mobile Computing* (WiCom'09), vol. 1, (pp. 1-4).

Yang, X., Venkatachalam, M., & Mohanty, S. (2007). Exploiting the MAC layer flexibility of WiMAX to systematically enhance TCP performance. In *Proceeding IEEE Mobile WiMAX Symposium*. Orlando, FL: IEEE Press.

Yang, Y., & Kravets, R. (2004). Throughput guarantees for multi-priority traffic in ad hoc networks. *Proceeding of IEEE International Conference on Mobile Ad-Hoc and Sensor Systems*, (pp. 379–388). Ft. Lauderdale, FL: IEEE Press.

Yang-Hua, C., Rao, S. G., Seshan, S., & Hui, Z. (2002). A case for end system multicast. *IEEE Journal on Selected Areas in Communications*, *20*(8), 1456–1471. doi:10.1109/JSAC.2002.803066

Yang, Y., & Kravets, R. (2005). Contention-aware admission control for ad hoc networks. *IEEE Transactions on Mobile Computing*, *4*(4), 363–377. doi:10.1109/TMC.2005.52

Yang, Y., & Kravets, R. (2007). Throughput guarantees for multi-priority traffic in ad hoc networks. *Ad Hoc Networks*, *5*(2), 228–253. doi:10.1016/j.adhoc.2005.11.002

Yavuz, M., Diaz, S., Kapoor, R., Grob, M., Black, P., Tokgoz, Y., & Lott, C. (2006). VoIP over CDMA2000 1xEV-DO revision a. *IEEE Communications Magazine*, *44*(2), 50–57. doi:10.1109/MCOM.2006.1593550

Yee, Y. C., Choong, K. N., Low Andy, L. Y., & Tan, S. W. (2008). SIP-based proactive and adaptive mobility management framework for heterogeneous networks. *Journal of Network and Computer Applications*, *31*(4), 771–792. doi:10.1016/j.jnca.2007.09.002

Yilmaz, S., & Matta, I. (2001). On class-based isolation of UDP: Short-lived and long-lived TCP flows. *Proceedings of the 9th International Symposium on Modeling, Analysis, and Simulation of Computer and Telecommunication Systems*. Boston, MA: ACM Press.

Ying, W., Jun, Y., Yun, Z., Gen, L., & Ping, Z. (2008). Vertical handover decision in an enhanced media independent handover framework. *Proceedings of IEEE Wireless Communications and Networking Conference* (WCNC'08), vol. 1, (pp. 2693-2698).

Yi, Wang, Li, L., Zhang, P., & Liu1, Z. (2009). DFT-based channel estimation with symmetric extension for OFDMA Systems. *EURASIP Journal on Wireless Communications and Networking*, 2009.

Yoon, K., & Hwang, C. (1995). *Multiple attribute decision making: An introduction*. Sage Publications.

Young-June, C., & Saewoong, B. (2004). Scheduling for VoIP service in CDMA2000 1x EV-DO. *IEEE International Conference on Communications*, vol. 3, (pp. 1495-1499). IEEE Press.

Yuan, Y., Bahl, P., Chandra, R., Chou, P., Ferrell, J., & Moscibroda, T. ... Wu, Y. (2007). KNOWS: Kognitiv networking over white spaces. *IEEE DySPAN*, (pp. 416–427). IEEE Press.

Zadeh, L. A. (1996). Fuzzy logic = computing with words. *IEEE Transactions on Fuzzy Systems*, *4*(2), 103–111. doi:10.1109/91.493904

Zahran, A., Liang, B., & Saleh, A. (2006). Signal threshold adaptation for vertical handoff in heterogeneous wireless networks. [MONET]. *ACM Mobile Networks and Applications*, *11*(4), 625–640. doi:10.1007/s11036-006-7326-7

Zeinalipour, D., Aristeidou, S., & Kazeli, S. (2009). *IP quality of services* (in Greek). Burns College of Engineering (Data File). Retrieved from http://www.cs.ucr.edu/ csyiazti/downloads/papers/ipqos/ papers/ip-qos.pdf

Zhang, J., Stevens-Navarro, E., Wong, V., Chan, H., & Leung, V. (2008). Protocols and decision processes for vertical handovers. In Y. Zhang, L. Yang, J. Ma (Eds.), *Unlicensed Mobile Access Technology: Protocols, Architectures, Security, Standards and Applications* (123-145). Boca Raton, FL: Auercach Publications, Taylor & Francis Group.

Zhang, L. X., & Li, T. (2010). *Internet research task force routing research group* (RRG). IRTF RRG website. Retrieved July, 2010, from http://trac.tools.ietf.org/group/irtf/trac/wiki/RoutingResearchGroup

Zhang, P., Durresi, A., & Barolli, L. (2009). A survey of Internet mobility. In the *Proceeding of International Conference on Network-Based Information Systems* (147-154). Indianapolis, IN: IEEE Press.

Zhang, Q., Smit, G. J. M., Smit, L. T., Kokkeler, A., Hoeksema, F. W., & Heskamp, M. (2005). A reconfigurable platform for cognitive radio. In *Proceedings of 2nd International Conference on Mobile Technology, Applications, and Systems*, (pp. 1-5). Guangzhou, China: IEEE Press.

Zhang, W. (2004). Handover decision using fuzzy MADM in heterogeneous networks. *Proceedings of IEEE Wireless Communications and Networking Conference* (WCNC'04), vol. 2, (pp. 653-658).

Zhang, Y., Guo, H., Wang, X., & Zhang, X. (2010). *Coexistence analysis of LTE and WiMAX in uplink.* Paper presented at the 6th International Conference on Wireless Communications Networking and Mobile Computing (WiCOM).

Zhang, B., & Mouftah, H. T. (2005). QoS routing for wireless ad hoc networks: Problems, algorithms, and protocols. *IEEE Communications Magazine*, *43*(10), 110–115. doi:10.1109/MCOM.2005.1522133

Zhang, H. K., & Su, W. (2007). Fundamental research on the architecture of new network—Universal network and pervasive services. *Acta Electronica Sinica*, *35*(4), 593–598.

Zhang, Q., & Zhang, Y. (2008). Cross-layer design for QoS support in multihop wireless networks. *Proceedings of the IEEE*, *96*(1), 64–76. doi:10.1109/JPROC.2007.909930

Zhang, S., Wu, T., & Lau, V. (2009). A low-overhead energy detection based cooperative sensing protocol for cognitive radio systems. *IEEE Transactions on Wireless Communications*, *8*(11). doi:10.1109/TWC.2009.081106

Zhang, Y., & Fang, Y. (2007). A secure authentication and billing architecture for wireless mesh networks. *Wireless Networks*, *13*, 663–678. doi:10.1007/s11276-006-8148-z

Zhao, X., Lv, Y., Yeap, T. H., & Hou, B. (2009). A novel authentication and key agreement scheme for wireless mesh networks. *International Conference on Networked Computing and Advanced Information Management*, (pp. 471-474).

Zhao, L., Yao, G., & Mark, J. W. (2006). Mobile positioning based on relaying capability of mobile stations in hybrid wireless networks. *IEEE Proceedings-Communications*, *153*(5), 762–770. doi:10.1049/ip-com:20050142

Zheng, S., Liang, Y., Tham, C., & Kam, P. (2009). *Design of MAC with cooperative spectrum sensing in ad hoc cognitive radio networks.* Paper presented at the IEEE 20th International Symposium on Personal, Indoor, and Mobile Radio Communications. IEEE Press.

Zhou, H., Huang, C., Cheng, Y., & Wang, G. (2009). A new multi-metric QoS routing protocol in wireless mesh networks. In *2009 International Conference on Networks Security, Wireless Communications, and Trusted Computing,* volume 1, (pp. 459–467). Wuhan, China: IEEE Press.

Zhou, W., & Lam, W. H. (2009). A fast LMMSE channel estimation method for OFDM systems. *EURASIP Journal on Wireless Communications and Networking*, *2009*, 13. doi:10.1155/2009/752895

Zhu, Z., Wakikawa, R., & Zhang, L. (2010). *A survey of mobility support in the Internet. draft-zhu-mobility-survey-03.* IETF draft (work in progress).

Zimmermann, H. J. (1991). *Fuzzy sets theory – And its applications*. Boston, MA: Kluwer Academic Publishers.

Zinner, T., Tutschku, K., Nakao, A., & Tran-Gia, P. (2010). Using concurrent multipath transmission for transport virtualization: Analyzing path selection. In *Proceedings of the 22nd International Teletraffic Congress (ITC)*, Amsterdam, The Netherlands.

About the Contributors

Gabriel-Miro Muntean is a Lecturer with the School of Electronic Engineering, Dublin City University (DCU), Ireland, where he obtained his Ph.D. degree in 2003 for research on quality-oriented adaptive multimedia streaming over wired networks. He was awarded the B.Eng. and M.Sc. degrees in Software Engineering from the Computer Science Department, "Politehnica" University of Timisoara, Romania in 1996 and 1997 respectively. Dr. Muntean is co-director of the DCU Performance Engineering Laboratory research group and Director of the Network Innovations Centre, part of the RINCE Research Institute Ireland. His research interests include quality-oriented and performance-related issues of adaptive multimedia delivery, performance of wired and wireless communications, energy-aware networking and personalised e-learning. Dr. Muntean has published over 100 papers in prestigious international journals and conferences, has authored a book and six book chapters and has edited two other books. Dr. Muntean is an Associate Editor of the IEEE Transactions on Broadcasting and reviewer for other important international journals, conferences, and funding agencies. He is a member of IEEE, and IEEE Broadcast Technology Society.

Ramona Trestian is a Ph.D. researcher with the Performance Engineering Laboratory, School of Electronic Engineering, Dublin City University, Ireland. She was awarded the B.Eng. degree in Telecommunications from the Electronics, Telecommunications, and the Technology of Information Department, Technical University of Cluj-Napoca, Romania in 2007. Her research interests include wireless mobile networks, multimedia streaming over wireless access networks as well as handover and network selection strategies. She is student member of IEEE and member of the Network Innovations Centre, part of the RINCE Research Institute Ireland.

* * *

Iftekhar Ahmad is a lecturer at the School of Engineering, Edith Cowan University, Australia. Dr Ahmad received his PhD degree from Monash University, Australia in 2007. He worked as an assistant lecturer in the Faculty of Information Technology, Monash University, Australia. His research interests include 4G technologies, QoS in communication networks, WLANs and computational intelligence.

Anwer Al-Dulaimi received his BSc and MSc honours degrees in Telecommunication Engineering from University of Technology, Baghdad, in 1997 and 2003, respectively. He worked as laboratory assistant and then as assistant lecturer, in different universities. He is currently a PhD student in the Wireless Networks and Communication Centre (WCNC) at Brunel University, UK. His current research interests lie in the area of cognitive radio networks with special focus on network topology and optimisation, green communications, radio over fibre networks and Quality of Services solutions. He has published over 20 journal and conference papers in the area of telecommunications. Mr. Al-Dulaimi is a member in the IET and was recognized as a CEng in April 2010.

Yahya Al-Hazmi received his M.Sc. in Electrical Engineering with specific emphasis on Telecommunication Systems from Technical University of Berlin in 2009. Since then he has been Scientific Assistant in preparing his PhD at the Computer Networks and Communications Lab at the University of Passau. His main research interests include future Internet networking, network virtualization, wireless mesh networks, and IT energy efficiency. He is currently involved in European funded FP7 projects. He is member of the European Commission's Network of Excellence EuroNF.

Saba Al-Rubaye received her MSc (Hons) degrees in Telecommunications from University of Technology, Baghdad, in 2005. She was a Laboratory Instructor and then Assistant Lecturer in the department of Telecommunication Engineering in different universities. She taught different courses in communications and electronics engineering. She is currently a PhD candidate in the Wireless Network Communication Centre (WNCC) at Brunel University, UK. Her research interests lie in the area of wireless broadband access for next generation networks with special focus on cognitive Femtocell, future Internet, fiber-to-the-home and Quality of Services for Wireless Networks. She has published over ten journal and conference papers in the area of telecommunications.

Antonios Alexiou obtained his diploma from the Department of Electrical and Computer Engineering of the Aristotle University of Thessaloniki,Greece. Furthermore, he obtained his Master Degree and his PhD from the Computer Engineering and Informatics Department of Patras University, Greece. His research interests include data networks, mobile telecommunications networks, multicast routing and group management, radio resource management, radio network planning and optimization, and wireless networks. He has published over 40 papers in journals, international conferences, and book chapters.

Muhammad Ali is currently working as a research assistant, and a prospective Ph.D. candidate, at the Computer Networks and Communications Lab at the University of Passau. He has his basic degree in Computer Systems Engineering. Additionally, he also holds MBA (Management Information Systems) from IBA Karachi, Pakistan, and M.Sc. (Software Systems Engineering) from RWTH University, Germany. Before joining the University of Passau in November 2009, he worked as a professional for more than five years in the software development and payment systems industry in Pakistan. His current research interests include IT security and privacy issues in distributed systems, virtualized environments, and other modern networking paradigms.

Albert Banchs received his Telecommunications Engineering degree from the Polytechnical University of Catalonia in 1997, and the PhD degree from the same university in 2002. His Ph.D. received the national award for best thesis on broadband networks. He was a visitor researcher at ICSI, Berkeley, in 1997, worked for Telefonica I+D, in 1998, and for NEC Europe Ltd., Germany, from 1998 to 2003. Since 2003, he is with the University Carlos III of Madrid, and since 2009 he is also Deputy Director of IMDEA Networks. He has authored over 50 publications in peer-reviewed journals and conferences and four patents (two of them granted). He is senior editor for IEEE Communications Letters and area editor for Elsevier Computer Communications, has been guest editor for IEEE Wireless Communications and Computer Networks. He has served on the TPC of a number of conferences and workshops including IEEE INFOCOM, IEEE ICC, and IEEE GLOBECOM, and has been TPC chair for European Wireless 2010.

Oguz Bayat received the B.S. degree from Istanbul Technical University, Istanbul, Turkey, in 2000, and M.S degree from University of Hartford, USA, in 2002, and Ph.D. degree from Northeastern University, USA, in 2006, all in Electrical Engineering. He completed the Executive Certificate Program in Technical Management and Leadership at Massachusetts Institute of Technology, USA in 2009. Before attending the Communication and Digital Signal Processing Research Center at Northeastern University in 2002, he served as an adjunct faculty in the Department of Electrical and Computer Engineering at University of Hartford. From 2005 to 2010, he has worked as a Technical Leader/Manager on Nortel/ Ericsson project at Airvana Inc., USA, and he has been involved in the research and development of macro cell CDMA 1xEV-DO Rev 0, Rev A and Rev B radio node and radio node controller software products, and Femtocell micro base station products. From 2008 to 2009, he served as an Assistant Professor in the Department of Electronics and Communication Engineering of Beykent University, Turkey. He was recently appointed as an Assistant Professor in the Department of Electronics and Communication Engineering of Yeni Yuzyil University, Turkey. His current research interests are in the areas of wireless communication, wireless networks, digital signal processing, channel coding, channel estimation, channel equalization, interference cancellation, traffic modeling and scheduling in next generation wireless cellular networks.

Subir Biswas is an Associate Professor and the director of the Networked Embedded and Wireless Systems laboratory at the Michigan State University. Subir received his Ph.D. from University of Cambridge and held various research positions in NEC Research Institute, Princeton, AT&T Laboratories, Cambridge, and Tellium Optical Systems, NJ. He has published 120 peer-reviewed articles, books, and book chapters in the area of network protocols, and is a co-inventor of 6 USA patents, awarded and pending. His current research interests include the broad area reconfigurable embedded systems, wireless data networking, low-power network protocols, and application-specific sensor networks. He is a senior member of IEEE and a fellow of Cambridge Philosophical Society.

Christos Bouras obtained his Diploma and PhD from the Department of Computer Engineering and Informatics of Patras University, Greece. He is currently a Professor in the same department. He is also a scientific advisor to Research Unit 6 in the Computer Technology Institute & Press "Diophantus", Greece. His research interests include Analysis of Performance of Networking and Computer Systems, Computer Networks and Protocols, Telematics and New Services, QoS and Pricing for Networks and Services, e-Learning Networked Virtual Environments and WWW issues. He has extended professional experience in Design and Analysis of Networks, Protocols, Telematics, and New Services. He has published 250 papers in various well-known refereed conferences and journals. He is a co-author of 8 books in Greek. He has been a PC member and referee in various international journals and conferences. He has participated in R&D projects such as RACE, ESPRIT, TELEMATICS, EDUCATIONAL MULTIMEDIA, ISPO, EMPLOYMENT, ADAPT, STRIDE, EUROFORM, IST, GROWTH, and others. He is also a member of Experts in the Greek Research and Technology Network (GRNET), Advisory

Committee Member to the World Wide Web Consortium (W3C), IEEE - CS Technical Committee on Learning Technologies, IEEE ComSoc Radio Communications Committee, IASTED Technical Committee on Education WG6.4 Internet Applications Engineering of IFIP, ACM, IEEE, EDEN, AACE, New York Academy of Sciences, and Technical Chamber of Greece.

John Cosmas obtained a BEng honors degree in Electronic Engineering at Liverpool University in 1978 and a PhD in Image Processing and Pattern Recognition at Imperial College in 1987. He is a Professor of Multimedia Systems and became a Member (M) of IEEE in 1987 and a Member of IEE in 1977. His research interests are concerned with the design, delivery, and management of new fourth-generation TV and telecommunications services and networks, multimedia content and databases, and video/image processing. He has contributed towards eight EEC research projects and has published over 80 papers in refereed conference proceedings and journals. He leads the Wireless Networks and Communications Centre within the School of Engineering and Design at Brunel University.

Nuno Coutinho concluded in 2008 the 5 years Integrated M.Sc. in Electronics and Telecommunications Engineering at University of Aveiro. His master's dissertation, titled "Intelligence in Mobility Decisions," was about the selection of the best access network according to context information about user and network. Since October 2008, he is a PhD student at University of Aveiro and joined the Celfinet Innovation Department from October 2008 to February 2009. Currently, he is a researcher associated with the Institute of Telecommunications, being involved in national projects (MuMoMgt, GEN-CAN, UbiquiMesh) and European projects (C-CAST). His main research interests are related to Heterogeneous Networks, Multicast, Overlay Networks, context-awareness and Quality-of-Experience.

Debabrata Das received his Ph.D. degree from the Indian Institute of Technology, Kharagpur (IIT Kgp). At present he is serving as Hewlett Packard Chair and Associate Professor at IIIT-Bangalore. Before joining IIIT-B, he had served at G S Sanyal School of Telecommunication at IIT Kharagpur and later at Kirana Networks in New Jersey, USA. At present he is PI of Intel and Government India sponsored Research Projects in the area of Wireless Broadband Networks. His main areas of research interest are Wireless Access Network's MAC, QoS, and Green Mobile. Dr. Das has more than 50 peer reviewed papers in different journals and international conferences. He and his research team have contributed to IEEE 802.16m standard. He is also a board member of IT-Service Management Forum India, OCAC of Orissa Govt.; and a member of e-Governance committee under Govt. of Karnataka. He was General Chair of IEEE IMSAA-09. He is also TPC member of a number of international conferences and one of the reviewers of IEEE Transaction on Mobile Computing and IEEE COMSOC Magazine.

John Fitzpatrick received a Diploma in Applied Electronics from Dublin Institute of technology in 2003, a Degree in Telecommunications Engineering from Dublin City University and a PhD from University College Dublin in 2008. From 2008 to 2010, he was a postdoctoral researcher in the department of Computer Science and Informatics at University College Dublin where he worked on the EU FP7 project CARMEN. John is currently an IRCSET INSPIRE Marie Curie research fellow at NEC Laboratories Europe in Heidelberg, Germany, working in the area of VoIP, LTE and Femtocell networks.

Dominique Gaiti received the Ph.D. and the "Habilitation à Diriger des Recherches" degrees in Computer Science from the University of Paris VI and Paris IX on 1991 and 1995, respectively. She is currently a professor at the University of Technology in Troyes, France, and a member of the Institute Charles Delaunay (ICD). She is the leader of the team "Autonomic Networking" in this institute. Earlier, she was a research scientist at the University of Columbia, USA, from 1992-1994, and a researcher at the University of Paris 6, member of the LIP6 laboratory, France, from 1996-1997. She is the chairperson of the IFIP WG 6.7 on "Smart Networks." Her research interests include the smart networks, the intelligence in networks, and the control and management (through intelligent agents) in all types of networks. She is the author of one book and has edited several proceedings.

Gheorghita Ghinea received the B.Sc. and B.Sc. (Hons) degrees in Computer Science and Mathematics, in 1993 and 1994, respectively, and the M.Sc. degree in Computer Science, in 1996, from the University of the Witwatersrand, South Africa; he then received the Ph.D. degree in Computer Science from the University of Reading, UK, in 2000. He is a reader in the School of Information Systems, Computing and Mathematics, at Brunel University, UK. Dr. Ghinea has over 100 refereed publications and currently leads a team of 8 research students in his fields of interest, which span perceptual and networked multimedia, semantic media management, human computer interaction, and network security.

Jianfeng Guan received his B.S. degree from Northeastern University of China in July 2004, and received the Ph.D. degrees in Communications and Information Systems from the Beijing Jiaotong University in Jan. 2010. He is a Lecturer in the Institute of Network Technology at Beijing University of Posts and Telecommunications (BUPT), China. His main research interests focus around mobile IP, mobile multicast, IP security and next generation Internet.

Daryoush Habibi is an engineer with over 20 years of professional experience in industry and academia, much of it in leadership roles. He holds a Bachelor of Engineering with First Class Honours and a PhD from the University of Tasmania. He is currently the Professor of Engineering, Head of School of Engineering, and the Leader of the Centre for Communications Engineering Research at Edith Cowan University. Habibi has a keen interest in multidisciplinary research, but a significant portion of his research contributions have been around quality of service in communication networks. He has more than 100 refereed publications in conferences and journals. Habibi is a Fellow of Engineers Australia, a Senior Member of IEEE, the Editor-in-Chief of the Australian Journal of Electrical and Electronic Engineering, member of the ITEE College Board of Engineers, Australia, and member of the Electrical College Board of Engineers, Australia.

Andrea Hess is a Research Assistant at the Research Group Entertainment Computing (formerly Distributed Systems Group) at the University of Vienna, Austria. She is currently working toward a Ph.D. degree in Computer Science and received her Master's degree in Computer Science from the Vienna University of Technology in July 2008. Her current research interests lie in the area of mobile wireless networking; she is particularly interested in realistic mobility modelling and prediction for opportunistic networks.

Karin Anna Hummel received her Ph.D in 2005 from the Vienna University of Technology and is currently an Assistant Professor at the University of Vienna, Austria. Before joining academia, Hummel was a software engineer and project leader with Siemens Austria in the area of network management. Her research interests focus on mobility modelling and mobility-assisted, self-organized networking, context-aware distributed systems, and energy-efficient mobile networks. Hummel is a member of Euro-NF (EU FP7 Network of Excellence on the Network of the Future), and co-leader of working-group 2 and the focus group on energy-efficient wireless networks in the COST-action IC0804 on "Energy Efficiency in Large Scale Distributed Systems." She currently takes a leading role in two one-year Euro-NF specific research projects (MOPP-NET, EEWMI). Hummel is co-author of several journal and conference publications and co-editor of two proceedings (IWSOS'08/'09).

Balakrishnan K received his Master's degree from IIIT-Bangalore in 2010. He is currently working on his Ph.D degree at the Wireless Networks Laboratory at IIIT-Bangalore. His current research work includes modeling of heterogeneous network traffic, TCP optimization in wireless networks, and video streaming analysis in wireless networks.

Ritesh Kumar Kalle received the Masters' degree in Information Technology from IIIT-Bangalore, India, in 2007. He has been working towards the Ph.D degree at the Wireless Networks Laboratory at IIIT-Bangalore since 2007. His current research includes the modeling and analysis of QoS and power saving in heterogeneous wireless networks including WiMAX and Wi-Fi. His research work was supported by Microsoft Research India under the Microsoft Research India PhD Fellowship Award from 2008-2010. Currently he is employed as Senior Scientific Officer at IIIT-Bangalore on a Department of Information Technology, Govt. of India sponsored project on power saving in 4G Wireless technologies.

Lyes Khoukhi is an associate professor at the University of Technology of Troyes (France), since 2009. In 2008, he was researcher at the Computer Sciences department of the University of Montreal, Canada. He received his Ph.D degree in Electrical and Computer Engineering from the University of Sherbrooke, Canada, in 2007, and his M.Sc degree in Computer Engineering from the University of Versailles, France, in 2002. From 2003-2007, he worked in INTERLAB Communications Research Laboratory, Sherbrooke University. His research interests include wireless communications, multimedia networking, quality of service, and intelligent systems.

Vasileios Kokkinos obtained his diploma from the Physics Department of the University of Patras in October 2003. Next, he was accepted in the postgraduate program "Electronics and Information Processing" in the same department, and on March 2006 he obtained his Master Degree. In 2010, he received his PhD on Power Control in Mobile Telecommunication Networks from the Computer Engineering and Informatics Department. He has worked in the Research Unit 6 of the Computer Technology Institute & Press "Diophantus" since September 2006. His research interests include data networks, next generation mobile telecommunications networks, multicast routing and group management, and radio resource management. He has published more than 20 research papers in various well-known refereed conferences, books, and scientific journals.

Mathias Kretschmer received a Dipl-Ing. degree in Communications Engineering from the University of Applied Sciences Cologne in 1998. In 1998 he joined AT&T Labs Research where he was involved in a number of multimedia distribution/streaming projects focusing on point-to-multipoint communication protocols as well as loss-resilient coding and transport. He then joined PreCache Inc. to work on content based routing and messaging. Since 2005, he has been with Fraunhofer FOKUS as a Senior Research Staff Member working on next generation wireless networks. There, he also serves as Deputy Department Head. He holds three patents in the field of Communication and Network Protocols. Since 2008, he is also a part-time PhD. student at Brunel University.

Johannes Lessmann works for the Mobile and Wireless Networks Group at NEC Labs Europe in Germany in the domains of wireless mobile backhauls, self-organization in mobile networks, resilience, QoS, and mobility. Before joining NEC, he studied Computer Science at the University of Paderborn (UPB), Germany, and Carleton University, Canada. He received both his M.Sc. and PhD with distinction from the UPB. His PhD research was done on cross-layer support for voice and best-effort data in wireless multi-hop networks in cooperation with and funded by Siemens. He is a voting member of the IEEE 802.21 WG where he co-initiated and chaired the study group on management of heterogeneous wireless networks.

Elsa María Macías López is an associate professor of Telecommunications at Las Palmas de Gran Canaria University, Department of Telematics Engineering, Spain. She received her Ph.D. in Telecommunications (2001) from Las Palmas of Gran Canaria University for her work on parallel computing on a LAN-WLAN cluster controlling at Runtime the variation of the number of processes. She received her M.S. in Telecommunications (1997) from the same university for her work on parallelization of diffuse IR radiation system simulation for indoor applications. Her research interests are in parallel and distributed computing and infrastructure wireless networks for collaborative computing. Her current research efforts have focused on the management of wireless channel disconnections to prevent abrupt endings of applications. She has published 8 papers in refereed journals, 40 papers in refereed conferences, 1 paper in a Spanish magazine, 1 educational book, and co-edited 1 book. She is a member of the Program and Organizing Committees and Chair sessions for several international and Spanish conferences. She has collaborated in several research projects. Professor Macías teaches telecommunications at the beginning, advanced, and graduate levels, and advises graduate theses in the area of wireless communications and parallel and distributed computing.

Paulo Ferrer Loureiro graduated from University of Aveiro, Portugal, in 2004, with a diploma in Computer Engineering and Telematics. He did his thesis at Instituto de Telecomunicacoes in Aveiro on the design and implementation of "Support of Service Layer (SIP and QoS) in Heterogeneous Networks." During his studies, he completed an internship at NEC Network Labs developing the protocol for fast mobility of the IST Daidalos project. Afterwards he worked for the European project EUQoS at Portugal Telecom Inovacao developing modules for adaptation of Medical Applications to the service layer signaling protocol. In September 2005 he joined NEC Network Laboratories in Heidelberg, Germany, where he is currently a Research Scientist. His main focus is on network based mobility protocols and extensions with active contributions to standards (IEFT and 3GPP), European projects and internal Business Unit projects. Prototype implementation of the designed protocols is also one of his main activities.

Domingo Marrero is a Telecommunication Engineer and Electronic System Technical Engineer at the University of Las Palmas of Gran Canaria, Spain. He has participated in multiples tasks of academic management in many sections of the university. He has presented several articles in International Congress and National Congress, Spain, about wireless networks and computing. Recently his work has been in quality of service in wireless local area networks and multimedia services. He has presented articles at the ICWN'05 (Las Vegas, USA) and the ICWN'07 (London, UK), and the IAENG International Journal of Computer Science.

Jose D. Martinez-Morales received his M.Sc. degree in Electrical Engineering from Universidad Autonoma de San Luis Potosi (UASLP), Mexico, in 2009. Currently, he is working toward his Ph.D. in Electrical Engineering at Universidad Autonoma de San Luis Potosi (UASLP). His current research interests include artificial intelligence and soft computing.

Ali El Masri is a PhD student in the University of Technology of Troyes, France. He received his M.Sc. degree in Computer Networks from the University Pierre et Marie Curie – Paris 6, in 2009 and his Enginnering Degree in Telecommunications from the Lebanese University in the same year. His current reaserch interests include wireless mesh and ad hoc networks, multihop cognitive radio networks, medium access control, routing, and directional antennas.

Ricardo Matos is a PhD student at the University of Aveiro. He concluded his Integrated M.Sc. in Electronics and Telecommunications Engineering from the Department of Electronics, Telecommunications and Informatics of University of Aveiro in July 2008. His Master's Thesis was titled, "Mobility Support in WiMAX Networks." In September 2008, he joined the Institute of Telecommunications, University of Aveiro, as a researcher and PhD student. His current research interests are in future Internet architectures, virtualization techniques, wireless mesh networking, self-management, and context-awareness.

Hermann de Meer is Full Professor at the University of Passau. He is head of the Computer Networks and Computer Communications Lab and member of the executive board of the Institute of IT Security and Security Law (ISL). Before joining the University of Passau, he was Assistant Professor at Hamburg University, Germany, Visiting Professor at Columbia University in New York, USA, and Reader at University College London, UK. His main research interests include network virtualization, IT energy efficiency, IT security, self-organization, and performance modelling. De Meer currently leads several national and European projects. In addition to various highly cited journal and conference publications, De Meer co-authored a textbook, *Queueing Networks and Markov Chains - Modeling and Performance Evaluation with Computer Science Applications*, published by John Wiley in 1998 and 2006. He is co-editor of several proceedings and special journal issues, such as IWQoS 2005, IWSOS 2006, ComNet 2009, e-Energy 2010. He is on the editorial board of Elsevier Computer Network Journal.

Christian Niephaus received his B.Sc. and M.Sc. in Computer Science from the University of Applied Sciences Bonn-Rhein-Sieg in 2006 and 2008, respectively. Since 2007, he has been with the Fraunhofer Institute for Open Communication Systems (FOKUS) where he is involved in a couple of different research projects in the area of heterogeneous wireless mesh networks with a special focus on QoS and routing. He also participates in the IEEE 802 standardization effort as a voting member and active contributor to the IEEE 802.21 Working Group.

Andreas Papazois obtained his diploma, MSc, and PhD from the Computer Engineering and Informatics Dept., University of Patras, Greece. He is currently an R&D Engineer at Research Unit 6: Networks Telematics and New Services, Computer Technology Institute & Press "Diophantus". He has also worked as Telecommunication Systems Engineer in Intracom Telecom S.A. His research interests include Web services, mobile telecommunication networks, error control techniques, quality of service, and multicast transmission. He has published several research papers in various well-known refereed conferences, books, and scientific journals. He has also been a reviewer for various international journals and conferences.

Ulises Pineda-Rico received the B.Sc. degree in Electronic Engineering with Major in Communications from Universidad Autonoma de San Luis Potosí (UASLP), Mexico, in 2000, the M.Sc. degree in Electronic Engineering with Major in Telecommunications from Instituto Tecnologico y de Estudios Superiores de Monterrey (ITESM), Campus Monterrey, Mexico, in 2002, and the Ph.D. degree in Electrical and Electronic Engineering from the University of Manchester, UK, in 2009. He is currently an Assistant Professor with the College of Science of UASLP. His research interests include MIMO systems and space-time coding. Dr. Pineda-Rico has been a member of the Mexican National Research System since 2010.

Anthony Plummer, Jr. received his Ph.D. in Electrical Engineering from Michigan State University in 2011. His interests and research are in the field of wireless networking; more specifically dynamic spectrum access and multiple channels, multiple interface networks. During the summers of 2003 to 2006, he interned at Hewlett Packard working in test and projector divisions. In the summer of 2007 he interned at Intel working on automating data collection systems. Plummer, Jr. received many honours, including the ECE Graduate Office Fellowship Award, GEM Master's and Ph.D. fellowships through Hewlett Packard and Intel, respectively, University Enrichment Fellowship at Michigan State University, and Best Paper Award at the Globecom 2009 conference. He received a Masters from Michigan State University in 2007 and Bachelors from Morgan State University in 2005, both in Electrical Engineering.

Rajeshkumar Ramraj received his Bachelor of Engineering degree in Electronics and Instrumentation Engineering from Madurai Kamaraj University, India, and Master of Engineering degree in Applied Electronics from Anna University, India, and Master of Science degree in Mobile Computer Systems from Staffordshire University, UK, in 2003, 2005, and 2008, respectively. He worked in the RVS College of Engineering and Technology, Tamilnadu, India, as a Lecturer / Senior Lecturer in the department of Electronics and Communication for 5 years. He is currently working towards his PhD in high speed vehicular communications at Centre for Communications Engineering Research (CCER), School of Engineering, at Edith Cowan University, Australia. He is a life member of ISTE, India. His research interests include quality of service in 4G technologies, WiMAX, call admission control, MANET's, cross layer design, digital electronics, and electronic circuits.

Sebastian Robitzsch was born in Freital, Germany in 1982. He received his degree in Media and Communication Technology from the University of Applied Sciences Merseburg, Germany, in 2008. Afterwards, he worked for Fraunhofer-Gesellschaft FOKUS.NET, Germany, as a research engineer for one year before joining the Performance Engineering Laboratory research group at University College Dublin, Ireland, as a Ph.D. candidate. Since March 2009, Robitzsch's research area is related to interference problems and self-configuration mechanisms in 802.11a wireless mesh networks.

Susana Sargento received her PhD in 2003 in Electrical Engineering. She joined the Department of Computer Science of the University of Porto in September 2002, and the University of Aveiro and the Institute of Telecommunications in February 2004. She has also beena Guest Faculty Member of the Department of Electrical and Computer Engineering, Carnegie Mellon University, USA, since August 2008, where she is currently performing a Faculty Exchange. She has been involved in several national and European projects, taking leaderships of several activities in the projects, such as the QoS and ad-hoc networks integration activity in the FP6 IST-Daidalos Project. She has been recently involved in several FP7 projects (4WARD, Euro-NF, C-Cast, WIP, Daidalos, C-Mobile), national projects, and CMU|Portugal projects (DRIVE-IN with the Carnegie Melon University). She has been TPC-Chair at several conferences, such as MONAMI'11, NGI'09, IEEE ISCC'07. She has also been on the Technical Program Committee of several international conferences and workshops (more than 20 in 2009), such as ACM MobiCom 2009 Workshop CHANTS and IEEE Globecom (2010). She has also been a reviewer of numerous international conferences and journals, such as IEEE Wireless Communications, IEEE Networks, IEEE Communications, Telecommunications Systems Journal, IEEE Globecom, IEEE ISCC, and IEEE VTC. Her main research interests are in the areas of next generation and future networks, more specifically QoS, mobility, self- and cognitive networks. She regularly acts as an Expert for European Research Programmes.

Pablo Serrano got his Telecommunication Engineering degree and his PhD from the University Carlos III of Madrid (UC3M) in 2002 and 2006, respectively. He has been with the Telematics Department of UC3M since 2002, where he currently holds the position of Associate Professor. In 2007 he was a Visiting Researcher at the Computer Network Research Group at the University of Massachusetts Amherst partially supported by the Spanish Ministry of Education under a Jose Castillejo grant. His current work focuses on performance evaluation of wireless networks. He has over 30 scientific papers in peer-reviewed international journals and conferences. He also serves as TPC member of several international conferences, including IEEE Globecom and IEEE INFOCOM.

Volkan Sevindik received the B.S. degree in Electronics and Telecommunication Engineering with distinction (Summa Cum Laude) from Istanbul Technical University, Turkey, in 2004, and the M.S. degree in Electronics and Electrical Engineering from Bogazici University, Turkey, in 2006. He received his Ph.D. in Electrical and Computer Engineering from the University of Massachusetts, Lowell, USA, in 2010.He has worked as a system engineer in the System Capacity and Performance group of Airvana Inc., USA for 3 years. He was actively involved in the design and implementation of next generation quality of service architectures for CDMA 1xEVDO Rev 0, Rev A, and Rev B standard. He is currently working as an application support engineer in the Engineering Development Group at The MathWorks Inc., USA. His current research interests are in the areas of communication networks, wireless networks, traffic control and resource allocation, resource management, quality of service, quality of experience, scheduling, policy and utility based pricing models, and next generation wireless technologies such as Femtocell. He is an active member of IEEE, International Multimedia Telecommunication Consortium (IMTC), International Organization for Standardization (ISO), and American Society for Quality (ASQ). He has 2 USA patents in the areas of Femtocell technology and Long Term Evolution (LTE) standard.

George D. Stamoulis received the Diploma in Electrical Engineering (1987, with highest honours) from the National Technical University of Athens, Greece, and the MS (1988) and PhD (1991) degrees in Electrical Engineering from the Massachusetts Institute of Technology, USA. From 1995 to 2000, he was an Assistant Professor at the Computer Science Department of the University of Crete, Greece, and a member of the Telecommunications and Networks Division of the Institute of Computer Science, Foundation for Research and Technology Hellas (ICS-FORTH). Since 2001, he has been a faculty member, currently Professor, in the Department of Informatics of Athens University of Economics and Business (AUEB). His research interests are in network economics and use of economic mechanisms for traffic and resource management and control, auction mechanisms for bandwidth, charging for telecommunications services, reputation mechanisms for electronic environments, and telecommunications regulation. He has published 55 articles in scientific journals such as *IEEE/ACM Transactions on Networks*, *Telecommunications Systems*, *Computer Communications*, *IEEE Transactions on Communications*, *Journal of the ACM*, and *IEEE Transactions on Information Theory*, as well as in conferences such as INFOCOM, ITC, ACM SIGMETRICS, GLOBECOM, ICQT and GP2PC. He has also collaborated with the Greek Regulatory Authorities for Communications and Power on auction design as well as with Greek telecom companies on service pricing.

Enrique Stevens-Navarro received the B.Sc. degree in Electronic Engineering with Major in Communications from Universidad Autonoma de San Luis Potosí (UASLP), Mexico, in 2000, his M.Sc. degree in Electronic Engineering with Major in Telecommunications from Instituto Tecnologico y de Estudios Superiores de Monterrey (ITESM), Campus Monterrey, Mexico, in 2002, and his Ph.D. in Electrical and Computer Engineering from the University of British Columbia (UBC), Canada, in 2008. He is currently an Assistant Professor with the College of Science of UASLP. His research interests include the design and evaluation of vertical handover algorithms for heterogeneous wireless networks as well as resource management and protocol design for communication networks. Dr. Stevens-Navarro has been a member of the IEEE Communications Society since 1999, a member of the IEICE Communications Society since 2010, and a member of the Mexican National Research System since 2010.

Alvaro Suarez Sarmiento is Full Professor of Telecommunications, University of Las Palmas de Gran, Canaria, Spain. He has been a member of the Experts Commission of research of the University of Las Palmas de Gran Canaria from 1999-2002. In 1990, he started working in systolic computing in the Technical University of Catalonia. Then he turned his attention to network computing and heterogeneous computing in 1994 when he returned to the University of Las Palmas de Gran Canaria where he founded the Concurrency and Architecture Group (GAC). His research interests are in parallel and heterogeneous distributed computing, infrastructure wireless networks for collaborative computing and collaborative frameworks. His current research efforts have focused on the management of wireless channel disconnections to prevent abrupt endings of applications. Professor Suarez teaches telecommunications at the beginning, advanced, and graduate levels, and advises graduate theses in the area of wireless communications and parallel and heterogeneous distributed computing. He also gives lectures at European universities. He lectured at the Multimedia and Ubiquitous Wireless Access Networks and Services Conference at the University Ca' Foscari di Venezia (May 2005).

Mahmoud Taghizadeh is a Ph.D. student in the Electrical and Computer Engineering department at Michigan State University. He received both his M.S. and B.S. degrees in Computer Engineering from Sharif University of Technology in Tehran, Iran. His current research interests are in the field of wireless networking, more specifically cooperative caching algorithms, data dissemination and vehicular networks. In recognition of his work, he was awarded the best paper award at COMSNET 2011. Before joining Michigan State University, he had worked as a software manager in Procom technology in Iran. He is also co-founder of Imen Afzar Sahand, which is a leading company in the area of network security.

Stavros Toumpis received his degree in Electrical and Computer Engineering from the National Technical University of Athens, Greece, in 1997, his M.S. degrees in Electrical Engineering and Mathematics from Stanford University, USA, in 1999 and 2002, respectively, and his Ph.D. in Electrical Engineering, also from Stanford, in 2003. From 1998 to 1999 he worked as a Research Assistant for the Mars Global Surveyor Radio Science Team, providing operational support. From 2000 to 2003 he was a Member of the Wireless Systems Laboratory at Stanford University. From 2003 to 2005 he was a Senior Researcher with the Telecommunications Research Center Vienna, Austria. From 2005 to 2009 he was a Lecturer at the Electrical and Computer Engineering Department of the University of Cyprus. Since 2009, he has been an Assistant Professor in the Computer Science Department of the Athens University of Economics and Business. His research is on wireless ad hoc networks, with emphasis on their capacity, the effects of mobility on their performance, medium access control, and theoretical information issues.

Ioannis Tselekounis received the B.Sc. degree in mathematics from the University of Athens, Greece, in 2008, and his M.S. degree in Computer Science from the Informatics Department of the Athens University of Economics and Business, in 2010. He is currently enrolled in the University of Athens working towards a M.Sc. degree in Logic, Algorithms, and Computation. His research interests include computer security, cryptography, and network optimization.

George Tsichritzis was born in 1987 in Kalamata, Greece. He graduated from the Computer Engineering and Informatics Department, University of Patras, Greece. His research interests include networking technologies, data mining, neuroinformatics, and cognitive science. He has published many papers in international conferences.

Changqiao Xu is a Lecturer in the Institute of Network Technology and Vice Director of the Next Generation Internet Research Center at Beijing University of Posts and Telecommunications (BUPT), China. He received his Ph.D. in Computer Applied Technology from the Institute of Software Chinese Academy of Sciences (ISCAS), China, in January 2009. He joined ISCAS in 2002, where he held roles as a research staff and project manager in the R&D area of communication networks. From 2007 to 2009, he worked as a researcher in the Software Research Institute, Athlone Institute of Technology, Ireland. He joined BUPT from ISCAS in December 2009. His research interests include computer networks, wireless communications, and next generation Internet technology.

Hongke Zhang received his M.S. and Ph.D. degrees in Electrical and Communication Systems from the University of Electronic Science and Technology of China (formerly known as Chengdu Institute of Radio Engineering) in 1988 and 1992, respectively. From Sep. 1992 to June 1994, he was a post-doctoral research associate at Beijing Jiaotong University (formerly known as Northern Jiaotong University). In July 1994, he jointed Beijing Jiaotong University, where he is a professor. He has published more than 100 research papers in the areas of communications, computer networks, and information theory. He is now the Chief Scientist of a National Basic Research Program ("973" program).

Huachun Zhou received his B.S. degree from the People's Police Officer University of China in 1986, and his M.S. and Ph.D. degrees from Beijing Jiaotong University of China in 1989 and 2008, respectively. He is currently a professor at the Institute of Electronic Information Engineering, Beijing Jiaotong University, China. His main research interests include mobility management, mobile and secure computing, routing protocols, network management technologies, and database applications.

Index

D

E

F

G

H

I

L

M

N

O

P

Q

R

S

T

U

V

W

CPSIA information can be obtained at www.ICGtesting.com
Printed in the USA
BVOW050049071211

277460BV00009B/4/P